Electromagnetics, Microwave Circuit and Antenna Design for Communications Engineering

For a complete listing of the *Artech House Antennas and Propagation Library*, turn to the back of this book.

Electromagnetics, Microwave Circuit and Antenna Design for Communications Engineering

Peter Russer

Artech House
Boston • London
www.artechhouse.com

Library of Congress Cataloging-in-Publication Data
Russer, P. (Peter), 1943–
Electromagnetics, microwave circuit and antenna design for communications engineering / Peter Russer.
p. cm.—(Artech House antennas and propagation library)
Includes bibliographical references and index.
ISBN 1-58053-532-1 (alk. paper)
1. Electromagnetic fields. 2. Radio waves. I. Title. II. Series.
TK454.4.E5R87 2003
621.382—dc22 2003060066

British Library Cataloguing in Publication Data
Russer, P. (Peter), 1943–
Electromagnetics, microwave circuit and antenna design for communications engineering. — (Artech House antennas and propagation library)
1. Electromagnetism 2. Telecommunication systems—Design and construction
I. Title
621.3'82

ISBN 1-58053-532-1

Cover design by Gary Ragaglia

Whatever is born,
animate or inanimate,
know them to be born
from the union of the field
and the field knower

Bhagadvadgita 13.26

685 Canton Street
Norwood, MA 02062

International Standard Book Number: 1-58053-532-1
Library of Congress Catalog Card Number: 2003060066

10 9 8 7 6 5 4 3 2 1

Contents

Preface **xi**

Chapter 1 Introduction **1**
References 6

Chapter 2 Basic Electromagnetics **9**
2.1 The Electromagnetic Field Concept 9
2.2 Field Intensities 12
2.3 Current and Flux Densities 16
2.4 The Charge Density 21
2.5 The Maxwell Puzzle 22
2.6 The Integral Form of Maxwell's Equations 25
2.7 The Electromagnetic Wave 28
2.7.1 The Wave Equation 33
2.7.2 The Polarization of Electromagnetic Waves 35
2.8 The Kirchhoff Laws 37
2.9 Maxwell's Equations in Local Form 40
2.10 Time-Harmonic Electromagnetic Fields 41
2.11 Maxwell's Equations in the Frequency Domain 42
2.12 Curvilinear Coordinates 44
2.13 Boundary Conditions 45
References 55

Chapter 3 Potentials and Waves **57**
3.1 The Electromagnetic Potentials 57
3.2 The Helmholtz Equation 61
3.3 TM and TE Waves 65

	3.4	Spherical Waves	67
	References		69
Chapter 4	**Concepts, Methods and Theorems**		**71**
	4.1	Energy and Power	71
	4.2	Field Theoretic Formulation of Tellegen's Theorem	78
	4.3	Sources of the Electromagnetic Field	80
	4.4	The Uniqueness Theorem	81
	4.5	The Equivalence Principle	83
	4.6	Reciprocity	85
		4.6.1 The Lorentz Reciprocity Theorem	85
		4.6.2 The Reciprocity Theorem for Impressed Sources	86
	References		88
Chapter 5	**Static and Quasistatic Fields**		**89**
	5.1	Conditions for Static and Quasistatic Fields	89
	5.2	Static and Quasistatic Electric Fields	91
		5.2.1 The Green's Function for the Static Electric Field	91
		5.2.2 Capacitance	93
	5.3	Static and Quasistatic Magnetic Fields	99
		5.3.1 The Green's Function for the Static Magnetic Field	99
		5.3.2 Inductance	102
	References		108
Chapter 6	**Waves at the Surface of Conducting Media**		**109**
	6.1	Transverse Magnetic Surface Waves	110
	6.2	Surface Currents	118
	6.3	Surface Current Losses	123
	6.4	Induced Surface Currents	126
	References		129
Chapter 7	**Transmission Lines**		**131**
	7.1	The Principles of Transmission Lines	131
	7.2	Phase and Group Velocity	134
	7.3	The Field Components	135
	7.4	Waveguides for Transverse Electromagnetic Waves	136
	7.5	Rectangular Waveguides	150
		7.5.1 Transverse Electric Waves	151
		7.5.2 Transverse Magnetic Waves	156
		7.5.3 Power Flow in the Waveguide	158
		7.5.4 Orthogonality of the Waveguide Modes	159

7.5.5 Generalized Currents and Voltages in Waveguides 160
7.5.6 Attenuation Due to Conductor Losses 163
7.6 Circular Cylindric Waveguides 166
7.6.1 The Circular Waveguide Modes 166
7.6.2 Power Flow and Attenuation in the TE_{01} Mode 172
7.7 Dielectric Waveguides and Optical Fibers 174
7.7.1 Homogeneous Planar Dielectric Waveguides 174
7.7.2 Dielectric Slab with Single-Sided Metallization 179
7.7.3 Circular Dielectric Waveguides with Step Index Profile 180
7.8 Planar Transmission Lines 187
7.8.1 The Microstrip Line 187
7.8.2 Quasistatic Approximation for the Microstrip Line 190
7.8.3 Coplanar Line and Coplanar Strip Line 193
References 194

Chapter 8 The Transmission Line Equations 195
8.1 The Transmission Line Concept 195
8.2 Generalized Voltages and Currents 197
8.3 Solution of the Transmission Line Equations 202
8.4 Wave Amplitudes 205
8.5 Reflection Coefficient and Smith Chart 206
References 212

Chapter 9 Resonant Circuits and Resonators 213
9.1 The Linear Passive One-Port 213
9.2 The Reactance Theorem 215
9.3 Resonant Circuits 217
9.4 The Transmission Line Resonator 221
9.5 Cavity Resonators 224
9.5.1 The Rectangular Cavity Resonator 224
9.5.2 The Circular Cylindrical Cavity Resonator 228
9.5.3 The Quality of the Circular Cylindrical Cavity Resonators 230
9.6 Coupling of Resonant Circuits and Resonators 231
9.6.1 The Loaded Quality Factor 231
9.6.2 Termination of a Transmission Line with a Resonant Circuit 232
9.6.3 Coupling of Cavity Resonators 234
References 236

Chapter 10 Microwave Circuits **237**
10.1 Linear Multiports 237
10.2 Source-Free Linear Multiports 238
10.3 The Chain Matrix 240
10.4 The Scattering Matrix 244
10.5 The Transmission Matrix 247
10.6 Tellegen's Theorem 249
10.7 The Power Properties 252
10.8 Reciprocal Multiports 253
10.9 The Symmetry Properties of Waveguide Junctions 255
10.9.1 Symmetric Three-Port Waveguide Junctions 256
10.9.2 Symmetric Four-Port Waveguide Junctions 259
References 263

Chapter 11 Antennas **265**
11.1 Introduction 265
11.2 The Green's Function 267
11.3 The Hertzian Dipole 271
11.4 Aperiodic Spherical Waves 276
11.5 Linear Antennas 280
11.6 The Loop Antenna 286
11.7 Receiving Antennas 290
11.7.1 The Hertzian Dipole as Receiving Antenna 290
11.7.2 The Loop Antenna as Receiving Antenna 291
11.7.3 The Linear Dipole Antenna as Receiving Antenna 292
11.8 Gain and Effective Antenna Aperture 296
11.9 Antenna Arrays 301
11.10 Aperture Antennas 304
11.10.1 Radiating Apertures 304
11.10.2 Horn Antennas 308
11.10.3 Gain and Effective Area of Aperture Antennas 311
11.10.4 Mirror and Lens Antennas 313
11.10.5 Slot Antennas 315
11.11 Microstrip Antennas 317
11.11.1 Planar Rectangular Patch Antenna 319
11.12 Broadband Antennas 321
References 323

Chapter 12 Numerical Electromagnetics **325**
12.1 Introduction 325
12.2 The Sturm-Liouville Equation 327
12.3 Spectral Representation of Green's Functions 330
12.4 The Method of Moments 332
12.5 The Integral Equation Method 337
12.6 The Integral Equation for the Linear Antenna 340
12.7 The Transmission Line Matrix Method 342
References 348

Appendix A Vectors and Differential Forms **351**
A.1 Vectors 351
A.2 Differential Forms 355
A.2.1 Products of Exterior Differential Forms 356
A.2.2 The Exterior Derivative 358
A.2.3 The Laplace Operator 359
A.3 The Stokes' theorem 360
A.4 Curvilinear Coordinates 364
A.4.1 Circular Cylindrical Coordinates 370
A.4.2 Spherical Coordinates 373
A.4.3 Twisted Forms 376
A.4.4 Integration of Differential Forms by Pullback 377
A.5 Double Differential Forms 377
A.6 Relations between Exterior Calculus and Conventional Vector Notation 378
A.6.1 Differential Operators 378
A.6.2 Maxwell's Equations 380
References 380

Appendix B Special Functions **381**
B.1 The Ordinary Bessel Functions 381
B.2 The Modified Bessel Functions 384
B.3 Spherical Bessel Functions 386
B.4 Legendre Polynomials 388
B.5 Spherical Harmonics 391
References 393

Appendix C Linear Algebra **395**
References 400

Appendix D Fourier Series and Fourier Transform **401**
D.1 The Fourier Series 401
D.2 The Fourier Integral 403
D.3 The Delta Distribution 405
References 408

List of Symbols **409**

About the Author **415**

Index **417**

Preface

The aim of this book is to deliver a comprehensive and concise introduction to the principles and basic theoretic concepts of electromagnetics. It is intended as a resource for understanding electromagnetics required in current, emerging and future broadband communications systems and high-speed analog and digital electronic circuits and systems. The reader is expected to attain an understanding of the principles and skills in electromagnetics that will enable him or her to apply modern electromagnetic design tools. It attempts to provide the background for the understanding of electromagnetic wave propagation, electromagnetic interference, radio-frequency, microwave and optical circuits and systems, antennas and high-speed digital circuits. Making use of geometric concepts in the presentation of electromagnetic theory the required mathematical framework is introduced. The visual element will be crucial to the method of presentation.

The book is intended to provide the framework for engineers and students to attain the necessary background in electromagnetics for solving design problems in microwave circuits and antennas. Especially, if electromagnetic CAD tools are used, the engineer needs a solid grounding in electromagnetics in order to apply the tools in a proper and efficient way. The book is addressed to readers with a background in analysis and electrical engineering corresponding to first-year undergraduate courses. No previous introductory course in electromagnetic theory is required. The book addresses students and engineers in communications engineering who want to acquire a working knowledge in electromagnetics for designing microwave circuits, antennas and systems. Modern curriculae in communications engineering and information technology show fewer courses in analysis and electrodynamics compared with former curriculae in electrical engineering. In the past the study of electromagnetics has been motivated primarily by the requirements of defense systems. The field of microwave engineering now is rapidly shifting toward commercial and consumer applications. Electromagnetic wave phenomena that in the past were in the domain of the microwave engineer are now becoming a limiting factor in digital circuit operation. As wireless communications is penetrating into the millimeter-wave frequencies, communications engineers need an improved background in microwaves. The study of electromagnetics is fundamental to the advancement of communications engineering and information technology to push the frontiers of the ultra-fast and the high bandwidth regime.

The book develops all the required concepts of electrodynamics, starting from the elementary phenomena. Throughout the book exterior differential forms are used to describe fields. Differential forms are an extension of the vector concept. Today in the mathematics community the exterior differential calculus of Hermann Grassmann and Elie Cartan is considered to be the most suitable framework for geometrical analysis and field theory. Exterior differential calculus has simple and concise rules for computation. Furthermore, the objects of differential calculus have a clear geometrical significance and the geometrical laws of electromagnetics assume a simple and elegant form. The use of differential forms does not mean giving up the vector concept and its physical interpretations. On the contrary the differential form representation supplies additional physical insight in addition to the conventional vector picture. Moving between the representations of exterior differential calculus and conventional vector analysis is straightforward.

I am pleased to acknowledge the assistance of many individuals in my work on this book. I thank Fabio Coccetti, Wolfgang Dressel, Masafumi Fujii, Petr Lorenz, Larissa Vietzorreck, and José Vagner Vital for a careful proofreading of the manuscript. I am also indebted to Leopold Felsen and Ke Wu for many helpful comments. Also the reviewer at Artech House has been of great assistence. Of course none of the above individuals is responsible for whatever inaccuracies remain. The text for this book was typed by Christiane Wangerek. She also proposed editorial changes when my grammar went astray. I would like to thank Igor Iline and Dzianis Lukashevich for making a number of drawings for field visualization, Do-Hong Tuan and Bruno Biscontini for assistance in preparing the figures, and Michael Zedler for advice and assistance in solving LaTeX typesetting problems. I thank Julie Lancashire, Tiina Ruonamaa and Judi Stone of Artech House for their work in planning and production.

Peter Russer

Chapter 1

Introduction

A compact definition of radio-frequency engineering states that this discipline deals with methods and techniques for the generation, processing, transmission, and technical applications of electromagnetic waves. Some decades ago, radio-frequency engineering covered all aspects of circuits and systems for wireless communications and other radio frequency applications. Today, methods of radio-frequency engineering are necessary in a wide area of electrical and electronic engineering. Radio-frequency engineering has developed from a product-oriented discipline towards a method-oriented field, providing the fundamentals for dealing with high frequencies in all branches of electrical and electronic engineering.

Radio-frequency engineering is the discipline dealing with the technical applications of electromagnetic fields. Radio-frequency engineering is based on Maxwell's theory of electromagnetic fields. Table 1.1 illustrates the history of the field theory from Huygens to Maxwell. In 1690, Christian Huygens published his *Traité de la Lumière* and presented there basic concepts of a wave theory of light [1]. Huygens formulated the principle that each point of an intermediate plane inserted in a propagating wave may be considered as the origin of a spherical wave. The secondary waves are determined by the envelopes of all these spherical waves.

At the end of this development, James C. Maxwell compiled all knowledge on electric and magnetic phenomena available at his time [2, 3]. Introducing the electric displacement current, he went beyond the physical experience. Maxwell's theory yielded predictions on new phenomena that could be verified in the following years and decades. In 1865 Maxwell concluded that electromagnetic waves propagating in free space must be possible. The experimental verification of this prediction was given by Heinrich Hertz in 1888 [4]. An excellent historical survey over the development of the electromagnetic theory is presented in [5].

A *frequency band* is a continuous range of frequencies extending between two limiting frequencies. The band designations as decided upon by the Atlantic City Ra-

Table 1.1: The Development of the Field Concept

Scientist	*Invention*
Christian Huygens (1629–1695)	Wave theory of light Huygens' principle
Isaac Newton (1643–1727)	Law of gravitation Action over distance
Daniel Bernoulli (1700–1727)	Hydrodynamics
Leonhard Euler (1707–1783)	Flowing liquids are described by fields of velocity and acceleration
Charles Coulomb (1736–1806)	Electrostatics and magnetostatics
André Ampère (1775–1836)	Force between conductors under current
Siméon Poisson (1781–1840)	Potential theory Electrostatics and magnetostatics
Hans Ch. Oerstedt (1777–1851)	Magnetic field induced by moving electric charge
Michael Faraday (1791–1879)	Electromagnetic induction
James C. Maxwell (1831–1879)	Theory of the electromagnetic field

dio Convention of 1947 and later modified by the Comité Consultatif International des Radiocommunications (CCIR) Recommendation No. 142 in 1953 are listed in Table 1.2. Today the Telecommunication Standardization Sector (ITU-T) of the International Telecommunication Union (ITU) combines the standards-setting activities of the predecessor organization CCIR. The frequency range N extends from $0.3 \cdot 10^N$ Hz to $3 \cdot 10^N$ Hz. Below the 1 millimeter wavelength we have the submillimeter-waves, succeeded by the far-infrared band, the infrared band and the visible optical region. Due to the development of semiconductor lasers and optical fibers the infrared region has gained importance for communications and sensor applications.

In this book we are dealing with methods based on the electromagnetic theory for the modelling of microwave circuits and antennas. The necessity to use field theoretic methods for circuit modelling depends on the frequency as well as on the size of the considered structures. The modelling of antennas and wave propagation requires field-theoretic methods already at low frequencies.

- At low frequencies, circuits may be described on the basis of currents and voltages. The properties of a circuit only depend on its topological structure. The theoretical framework for describing circuits at low frequencies is established by the *network concept*. The network concept is based on the Kirchhoff laws.
- At high frequencies, the geometric structure of the circuits plays an impor-

Table 1.2: Frequency bands as defined by CCIR

Region No.	*Frequency*	*Wavelength*	*Name*	
4	3...30 kHz	100...10 km	very low frequencies	VLF
5	30...300 kHz	10...1 km	low frequencies	LF
6	300...3000 kHz	1000...100 m	medium frequencies	MF
7	3...30 MHz	100...10 m	high frequencies	HF
8	30...300 MHz	10... 1 m	very high frequencies	VHF
9	300... 3000 MHz	100...10 cm	ultra high frequencies	UHF
10	3...30 GHz	10...1 cm	super high frequencies	SHF
11	30...300 GHz	10...1 mm	extremely high frequencies	EHF

tant role. An understanding of the circuit operation requires the knowledge of the interaction of electric and magnetic fields in the circuit. The theoretical framework for describing circuits at high frequencies is established by the *field concept*. The field concept is based on Maxwell's theory.

The electromagnetic field is determined by the geometric and material properties of a structure as well as by external sources and boundary conditions. In many cases, sources exist in a given structure, and only the amplitude of the source is subject to variation. In this case we obtain an infinite number of partial solutions for the electromagnetic field. The amplitudes of these partial solutions are integral field quantities and may be considered as generalized voltages and currents.

The relation between the field concept and network concept may be illustrated by Figure 1.1. Let us first consider the resonant circuit in Figure 1.1(a). The resonant circuit is formed by an inductor and a capacitor. Both elements are storing magnetic energy and electric energy, respectively. However, the magnetic energy is concentrated within the inductor, whereas the electric energy is concentrated within the capacitor. Within the network concept the capacitor and inductor are connected via two network nodes. The circuit may be described in terms of node voltages and node currents. A field interaction between the circuit elements is not considered within the network concept. If the inductor has an inductance L and the capacitor exhibits a capacitance C, the resonant frequency f and the corresponding angular frequency ω are given by

$$\omega = 2\pi f = \frac{1}{\sqrt{L\,C}}\,. \tag{1.1}$$

Increasing the resonant frequency can be done by reducing the inductance L or the capacitance C. Reducing the inductance may be done by reducing the number of coils of the spiral inductor. In Figure 1.1(b) the inductor has been reduced to one coil. A further reduction of the inductance may be achieved by circuiting inductors in par-

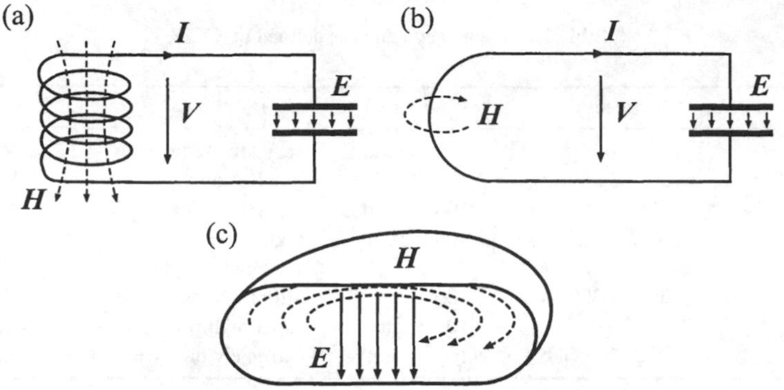

Figure 1.1: From the lumped circuit to the distributed circuit.

allel. Proceeding in that way we can circularly surround the capacitor with inductor coils. Connecting all these coils will create a closed conducting surface surrounding the capacitor. Now we also reduce the capacitance by increasing the distance between the plates of the capacitor. At the end of this morphing process we obtain the pillbox-shaped *resonator* shown in Figure 1.1(c). Electric and magnetic fields are interacting inside the resonator. We cannot easily describe this resonator in terms of voltages and currents. Defining, for example, an electric voltage via the path integral over the electric field from the bottom to the top of the resonator, this voltage will exhibit a maximum in the center of the resonator and decrease with the radial distance from the center. The current flowing on the top and bottom surfaces of the resonator in the radial direction increases with the radial distance from the center and vanishes at the center. The electric field is generated by the surrounding alternating magnetic field due to Faraday's law. The magnetic field is generated by the vertically flowing displacement current formed by the alternating electric field. A vertical line in the center of the cavity is surrounded by maximum magnetic flux, and therefore the electric field exhibits a maximum value in the center of the cavity. The magnetic field lines of a higher radius enclose a larger displacement current, and therefore the magnetic field increases with radius.

We can distinguish between the *low-frequency* case where electric and magnetic fields may be considered to be spatially separated and stored in *capacitors* and *inductors* respectively and the *high-frequency* case where the direct interaction of electric and magnetic fields has to be considered. Figure 1.1(a) illustrates the low-frequency case whereas Figure 1.1(c) is an example of the high-frequency case.

From this simple example we see that our conventional network concept is not applicable to describe the resonator. Circuits that are based on the direct interaction of electric and magnetic fields are called *distributed circuits*. The treatment of distributed circuits requires electromagnetic field modelling. We will, however, see later that it is possible to introduce *integral field quantities* also in the case of distributed circuits. These integral field quantities will be *generalized voltages* and *generalized currents*. Based on these generalized voltages and currents a network description is also possible for distributed circuits.

Excellent textbooks on electromagnetics at an intermediate undergraduate level include Stratton [6], Schelkunoff [7], Ramo, Whinnery and Van Duzer [8], Harrington [9], Collin [10–12], Kong [13], Balanis [14], Pozar [15] and Ishimaru [16].

Adding yet another book on electromagnetics to the existing literature should not be done without good reasons. The development of communications engineering in the past decade involved a tremendous increase in the body of knowledge to be acquired by communications engineers. Over decades electromagnetics occupied a large part of the education of electrical and electronic engineers. Today for electromagnetics, less space can be allocated in electrical and electronics engineering curriculae. However, the engineer developing circuits and systems for wireless communications needs a basic knowledge of electromagnetics. The goal in engineering education must be the accumulation of understanding, the stimulation of creativity and the promotion of the ability to solve problems. To solve problems in a systematic and efficient way the engineer must study the required theoretical framework. However, the engineer also needs intuition and creativity. These elements are strongly supported by imagery thinking [17].

The representation of electromagnetic theory can be simplified and the clarity can be improved by using geometrical methods. In 1844 Hermann Günter Grassmann published his book *Die lineale Ausdehnungslehre, ein neuer Zweig der Mathematik* [18], in which he developed the idea of an algebra in which the symbols representing geometric entities such as points, lines and planes are manipulated using certain rules. Grassmann introduced what is now called exterior algebra, based upon the *exterior product*

$$a \wedge b = -b \wedge a \,. \tag{1.2}$$

The work of Grassmann already contains most of the algebraic structures of modern exterior calculus. Élie Cartan applied Grassmann algebra to the theory of exterior differential forms in his book *Leçons sur les invariants intégraux* [19]. Exterior calculus allows for the solution of field theoretical problems easily and directly. Furthermore it establishes a direct connection to geometrical images and supplies additional physical insight. Let us demonstrate this for the example of the electric field. The usual physical interpretation of the electric field is the force applied to a small test charge. This leads in a natural way to the vector representation of the electric field and to a picture we may call the *force picture*. Considering the change of energy a test charge experiences as it is moved through the field yields a picture we may call the *energy picture*. The field lines in the force picture give the graphic representation of the field, and it is given by equipotential surfaces in the energy picture. The force picture is related to the vector representation whereas the energy picture is related to the differential form representation. However, in the differential form representation the vector representation is also visible. Therefore the differential form representation provides additional physical insight. We can represent one-forms graphically as surfaces in space. For electrostatic fields these surfaces are equipotential surfaces. Looking at the electric and magnetic flux densities and the electric current density, the differential form representation clearly points to the physical difference between field intensities and flux densities since flux densities are represented by two-forms. The graphical representation of a two-form is a bundle of tubes guiding the flux. This yields a much clearer physical interpretation than conventional vector notation. Finally, volume densities such as the charge density and the energy densities are represented by three-forms. Volume cells with cell size inversely proportional to the charge density give the geometric representation of a three-form form.

The exterior differential form calculus and its application to field theory is treated in a number of textbooks [20–29]. Textbooks on electromagnetics based on exterior calculus include [30] and [31]. An excellent approach for the use of differential forms as a tool for teaching electromagnetics has been given in [32].

In applying exterior calculus to an introductory level textbook our goal is to develop a working knowledge of the subject with as much of a theoretical framework as necessary and as much clarity as possible.

REFERENCES

[1] C. Huygens, *Traité de la Lumière*. Leyden, 1690.

[2] J. C. Maxwell, *A Treatise on Electricity and Magnetism*, vol. 1. New York: Oxford University Press, 1998.

[3] J. C. Maxwell, *A Treatise on Electricity and Magnetism*, vol. 2. New York: Oxford University Press, 1998.

[4] H. Hertz, *Gesammelte Werke, Untersuchungen über die Ausbreitung der elektrischen Kraft*, vol. 2. Leipzig: Johann Ambrosius Barth, 1894.

[5] R. S. Elliott, *Electromagnetics - History, Theory, and Applications*. New York: IEEE Press, 1991.

[6] J. A. Stratton, *Electromagnetic Theory*. New York: McGraw-Hill, 1941.

[7] S. Schelkunoff, *Electromagnetic Waves*. Princeton: Van Nostrand, 1943.

[8] S. Ramo, J. R. Whinnery, and T. van Duzer, *Fields and Waves in Communication Electronics*. New York: John Wiley & Sons, 1965.

[9] R. F. Harrington, *Time Harmonic Electromagnetic Fields*. New York: McGraw-Hill, 1961.

[10] R. E. Collin, *Field Theory of Guided Waves*. New York: McGraw-Hill, 1960.

[11] R. E. Collin, *Foundations of Microwave Engineering*. New York: McGraw-Hill, 1992.

[12] R. E. Collin, *Field Theory of Guided Waves*. New York: IEEE Press, 1991.

[13] J. A. Kong, *Electromagnetic Wave Theory*. New York: John Wiley & Sons, 1986.

[14] C. A. Balanis, *Advanced Engineering Electromagnetics*. New York: John Wiley & Sons, 1989.

[15] D. M. Pozar, *Microwave Engineering*. Reading: Addison-Wesley, 1990.

[16] A. Ishimaru, *Electromagnetic Wave Propagation, Radiation, and Scattering*. Englewood Cliffs, NJ: Prentice Hall, 1991.

[17] A. I. Miller, *Imagery in Scientific Thought*. Boston: Birkhäuser, 1984.

[18] L. K. H. Grassmann, *A New Branch of Mathematics: The "Ausdehnungslehre" of 1844 and Other Works*. Chicago: Open Court Publishing, 1995.

[19] E. Cartan, *Les systèmes différentielles extérieurs*. Paris: Hermann, 1945.

[20] H. Flanders, *Differential Forms*. New York: Academic Press, 1963.

[21] H. Cartan, *Formes différentielles*. Paris: Hermann, 1967.

[22] E. Heil, *Differentialformen*. Mannheim: Bibliographisches Institut, 1970.

[23] Y. Choquet-Bruhat, C. de Witt-Morette, and M. Dillard-Bleck, *Analysis, Manifolds and Physics*. Amsterdam: North-Holland, 1977.

[24] B. F. Schutz, *Geometrical Methods of Mathematical Physics*. Cambridge: Cambridge University Press, 1980.

[25] R. Abraham, J. E. Marsden, and T. Ratiu, *Manifolds, Tensor Analysis and Applications*. London: Addison-Wesley, 1983.

[26] W. L. Burke, *Applied Differential Geometry*. Cambridge: Cambridge University Press, 1985.

[27] P. Bamberg and S. Sternberg, *A Course in Mathematics for Students in Physics 2*. Cambridge: Cambridge University Press, 1990.

[28] T. Frankel, *The Geometry of Physics*. Cambridge: Cambridge University Press, 1997.

[29] S. Weintraub, *Differential Forms - A Complement to Vector Calculus*. New York: Academic Press, 1997.

[30] K. Meetz and W. Engl, *Elektromagnetische Felder*. Berlin: Springer, 1979.

[31] W. Thirring, *Lehrbuch der Mathematischen Physik*, vol. 2. Wien: Springer, 1978.

[32] K. F. Warnick, R. Selfridge, and D. Arnold, "Teaching electromagnetic field theory using differential forms," *IEEE Trans. Education*, vol. 40, pp. 53–68, Feb. 1997.

Chapter 2

Basic Electromagnetics

2.1 THE ELECTROMAGNETIC FIELD CONCEPT

In this chapter the framework of Maxwell's theory in differential form representation is introduced. The field quantities are discussed from a phenomenological point of view and related to the corresponding network quantities. Prior knowledge in field theory is not required but may be helpful [1–5]. The transfer of action between electrically charged matter via the *electromagnetic field* can be used to transfer energy as well as information. Electrically charged particles in rest interact via the *Coulomb force*. In the case of moving charged particles magnetic interaction also occurs. Electric and magnetic action propagates with a finite velocity, the speed of light, which is $2.998 \cdot 10^8\,\mathrm{ms}^{-1} \cong 3 \cdot 10^8\,\mathrm{ms}^{-1}$. To account for the finite velocity of the propagation of action the concept of the *field* has proven to be very powerful. The *field concept* means the assignment of physical quantities to the continuous space. A dynamic field is time-variable. The electromagnetic field can propagate in free space as well as in media. It does not require a medium of transmission. Electrodynamics is

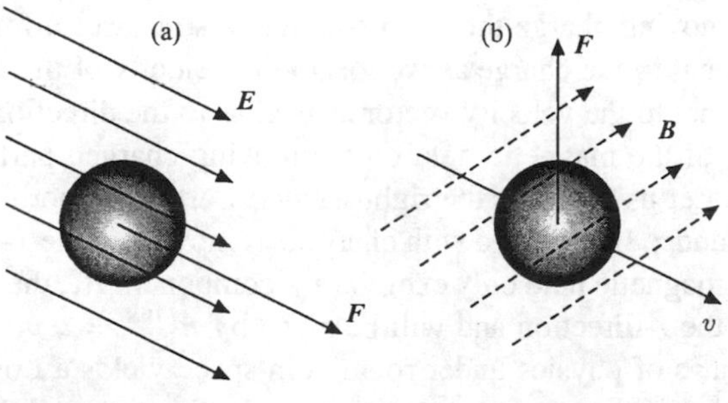

Figure 2.1: Electric and magnetic force on a charged particle.

concerned with interrelated electric and magnetic fields. This interrelation always occurs if the fields are time-varying. The electromagnetic field is generated by moving charged matter, and it acts on charged matter. Maxwell's theory summarizes the laws describing the electromagnetic field.

The complete description of the electromagnetic field is specified by two vector fields, i.e., directed fields. These fields are the electric field and the magnetic field. The electric field specifies the force on a point-like charged particle. Let us consider a particle with charge q at a point with the coordinates x, y and z at time t. If the electric field intensity exhibits in the x-, y- and z-direction at the point x, y and z at time t the components $E_x(x, y, z, t)$, $E_y(x, y, z, t)$ and $E_z(x, y, z, t)$ the three Coulomb force components $F_x^{el}(x, y, z, t)$, $F_y^{el}(x, y, z, t)$ and $F_z^{el}(x, y, z, t)$ are given by

$$\begin{aligned} F_x^{el}(x, y, z, t) &= q\, E_x(x, y, z, t)\,, \\ F_y^{el}(x, y, z, t) &= q\, E_y(x, y, z, t)\,, \\ F_z^{el}(x, y, z, t) &= q\, E_z(x, y, z, t)\,. \end{aligned} \tag{2.1}$$

The *electric field intensity* $\boldsymbol{E}$ has the dimension [Vm^{-1}]. The action of the electric field on the charged particle is illustrated in Figure 2.1(a). We summarize the components in vectors using the notation $\boldsymbol{x} = [x, y, z]^T$ where the superscript ^{T} denotes the transpose, i.e., the column vector corresponding to the written line vector

$$\boldsymbol{E}(\boldsymbol{x}, t) = [E_x(x, y, z, t), E_y(x, y, z, t), E_z(x, y, z, t)]^T\,. \tag{2.2}$$

This yields the more compact vector notation for (2.1):

$$\boldsymbol{F}^{el}(\boldsymbol{x}, t) = q\, \boldsymbol{E}(\boldsymbol{x}, t)\,. \tag{2.3}$$

In the case of a moving charge the magnetic field also causes a force. This *Lorentz force* is proportional to the charge as well as to the velocity of the particle and has a direction orthogonal to the velocity vector as well as to the direction of the magnetic field. The action of the magnetic field on the moving charged particle is illustrated in Figure 2.1(b). Let us consider the right-handed Cartesian coordinate system with coordinates x, y and z. If a particle with charge q is moving in the x-direction with velocity v_x and the magnetic field only exhibits a y-component B_y, the resulting Lorentz force will act in the z-direction and will be given by $F_z^{mag} = q\, v_x\, B_y$. Assuming invariance of the laws of physics under rotation in space yields a Lorentz force in the negative z-direction if the particle velocity has a y-direction and the magnetic field has an x-direction. In this case we obtain $F_z^{mag} = -q\, v_y\, B_x$. By superimposing all possible velocity components and magnetic field components we obtain

$$\begin{aligned} F_x^{mag} &= qv_y B_z - qv_z B_y \,, \\ F_y^{mag} &= qv_z B_x - qv_x B_z \,, \\ F_z^{mag} &= qv_x B_y - qv_y B_x \,. \end{aligned} \tag{2.4}$$

We can again summarize these three equations in vector notation by

$$\boldsymbol{F}^{mag}(\boldsymbol{x}, t) = q\boldsymbol{v} \times \boldsymbol{B}(\boldsymbol{x}, t) \,, \tag{2.5}$$

where the symbol $\times$ denotes the vector product

$$\begin{bmatrix} U_x \\ U_y \\ U_z \end{bmatrix} \times \begin{bmatrix} V_x \\ V_y \\ V_z \end{bmatrix} = \begin{bmatrix} U_y V_z - U_z V_y \\ U_z V_x - U_x V_z \\ U_x V_y - U_y V_x \end{bmatrix} . \tag{2.6}$$

The magnetic flux density $\boldsymbol{B}$ has the dimension [Vsm^{-2}]. It has to be pointed out that the fields $\boldsymbol{E}$ and $\boldsymbol{B}$ are of a different nature. $\boldsymbol{E}$ is a field intensity, whereas $\boldsymbol{B}$ is a flux density. The difference between these types of fields will be discussed later.

Let us now put together the electric force $\boldsymbol{F}^{el}$ and the magnetic force $\boldsymbol{F}^{mag}$. In this case a test particle with charge q and velocity $\boldsymbol{v}$ at point $\boldsymbol{x}$ and time t is influenced by the force

$$\boldsymbol{F}(\boldsymbol{x}, t) = q(\boldsymbol{E}(\boldsymbol{x}, t) + \boldsymbol{v} \times \boldsymbol{B}(\boldsymbol{x}, t)) \,. \tag{2.7}$$

For low frequencies the state of an electric circuit can be described by the currents flowing through the conductors and the voltages between the conductors. The *network concept* is following this description of circuits by voltages and currents. The properties of a circuit are only dependent on its topological structure, i.e. the connection structure of the network. The geometric arrangement of the network elements and the interconnections plays no role in a network. In the case of high frequencies, however, we make the observation that the geometric structure of the circuit may have a strong influence on the properties of a circuit or may even constitute its properties. The reason is that the abstract network describing the electrical properties of the circuit does not necessarily give a one-to-one mapping of the topology of physical high-frequency circuit. A one-to-one correspondence between the physical network and the abstract network only will hold if all circuit elements may be described by their relations between port currents and their port voltages and these current voltage relations provide a complete description of the circuit elements.

Network theory is based on the *Kirchhoff current law* and the *Kirchhoff voltage law* . The Kirchhoff laws allow an axiomatic foundation for the network theory. However, the Kirchhoff laws are not first principles, but may be derived from Maxwell's equations. For physical networks the Kirchhoff current law only is valid if the

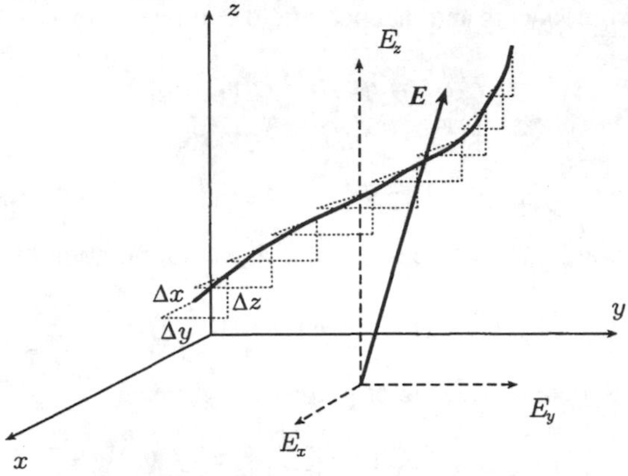

Figure 2.2: Line integral of the electric field over a curve C.

time derivative of the electric flux between the nodes can be neglected and if the time derivative of the magnetic flux through network loops can be neglected. The abstract network theory considers these conditions to be fulfilled exactly and takes the Kirchhoff laws as the fundamentals of the network theory.

Voltages applied to electric conductors cause an electric field between the conductors. The electric currents flowing through the conductors cause a magnetic field. At higher frequencies it is necessary also to take into consideration the electric flux between physical nodes and the magnetic flux through network loops. If frequencies are becoming very high, it will also be necessary to take into consideration the direct interaction of electric and magnetic fields. It is possible to describe the state of a circuit by specifying the electric and magnetic fields instead of voltages and currents. The field description is more general than the description by voltages and currents. However the field description is more complicated, since the electromagnetic field is a three-dimensional continuous quantity, and for the complete description of the electromagnetic field state it is necessary to specify the three electric field components and three magnetic field components throughout the three-dimensional continuum. Therefore, wherever electromagnetic effects may be described using the network concept, this description is much easier than a description based on the field concept.

2.2 FIELD INTENSITIES

The *electric field intensity* has the dimension [Vm^{-1}] and is described by a vector $\boldsymbol{E}(\boldsymbol{x}, t) = [E_x(\boldsymbol{x}, t), E_y(\boldsymbol{x}, t), E_z(\boldsymbol{x}, t)]^T$. As we have seen the meaning of the electric vector field is to assign some property to the space, namely to apply a force

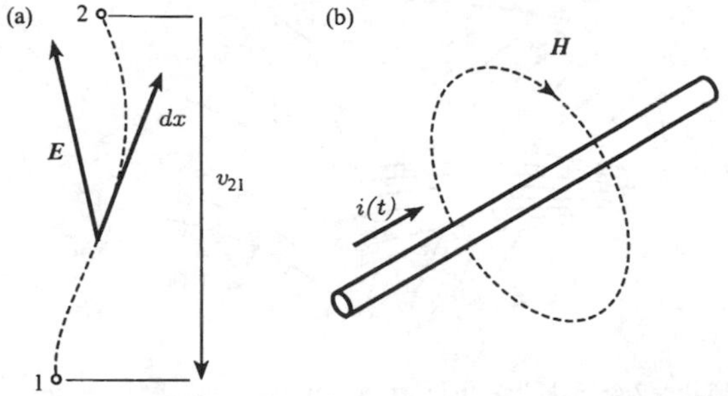

Figure 2.3: Path of integration (a) for the definition of the voltage and (b) for the definition of the current.

on charged matter. Let us consider in the following an electric field slowly varying with time. Moving a charged particle in the direction opposite to the direction of the field means that we have to supply *energy* equally to the product of force and distance of movement. If, for example, the electric field has an x-component E_x, the movement of a particle with charge q from position x to $x + \Delta x$ requires an energy $\Delta W = -qE_x\Delta x$. Moving the particle in three-dimensional space along any curve from point $\boldsymbol{x}_1 = (x_1, y_1, z_1)$ to $\boldsymbol{x}_2 = (x_2, y_2, z_2)$ means that we have to sum up the infinitesimal contributions $-qE_x\Delta x$, $-qE_y\Delta y$ and $-qE_z\Delta z$. This is illustrated in Figure 2.2. The path C is approximated by infinitesimal line elements in the x-, y- and z-direction. We are summing all contributions in the x-, y- and z-direction and with $\Delta x, \Delta y, \Delta z \to 0$ we obtain the energy W_{21} required for moving the charge from $\boldsymbol{x}_1$ to $\boldsymbol{x}_2$:

$$W_{21}(t) = -q \int_{\boldsymbol{x}_1}^{\boldsymbol{x}_2} E_x(x, y, z, t)\, \mathrm{d}x + E_y(x, y, z, t)\, \mathrm{d}y + E_z(x, y, z, t)\, \mathrm{d}z\,. \quad (2.8)$$

The *electric potential difference* or *voltage* v_{21} between $\boldsymbol{x}_2$ and $\boldsymbol{x}_1$ is given by

$$v_{21}(t) = -\int_{\boldsymbol{x}_1}^{\boldsymbol{x}_2} E_x(x, y, z, t)\, \mathrm{d}x + E_y(x, y, z, t)\, \mathrm{d}y + E_z(x, y, z, t)\, \mathrm{d}z\,. \quad (2.9)$$

For moving a charge q from $\boldsymbol{x}_1$ to $\boldsymbol{x}_2$ the energy $W_{21}(t)$ is related to the potential difference $v_{21}(t)$ via

$$W_{21}(t) = q\, v_{21}(t)\,. \quad (2.10)$$

Introducing the so-called *electric field differential form*

$$\mathcal{E} = E_x(x, y, z, t)\, \mathrm{d}x + E_y(x, y, z, t)\, \mathrm{d}y + E_z(x, y, z, t)\, \mathrm{d}z \quad (2.11)$$

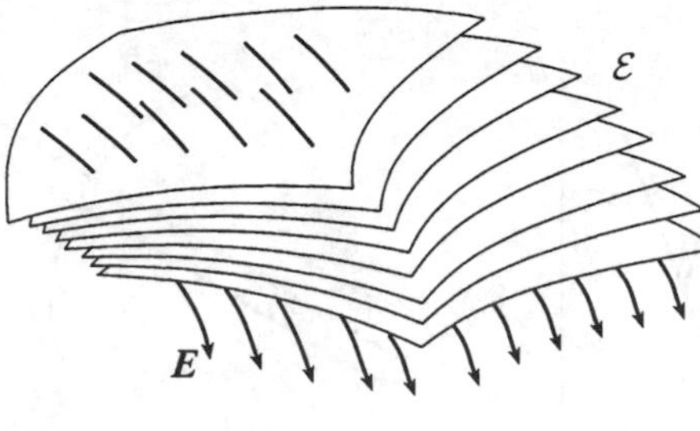

Figure 2.4: Field line and surface representations of a one-form.

we also may write

$$v_{21}(t) = -\int_{x_1}^{x_2} \mathcal{E}\,. \tag{2.12}$$

With a differential form we mean the complete expression under an integral sign, including also the differentials, e.g., $\mathrm{d}x$, $\mathrm{d}y$ and $\mathrm{d}z$, respectively [6–9]. In order to distinguish the differential form (2.11) from others to be introduced later, we call this differential form a *one-form.*

Figure 2.3(a) shows the path of integration for the definition of the voltage v_{21} from node 2 to node 1. The line integral sums up the projection of the field vector on the vectorial path element. The contribution of the integrand is proportional to the product of the magnitudes of field vector with the infinitesimal path element and the cosine of the angle enclosed between them.

The common physical interpretation of the electric field is related to the force on a point-like unit charge. This *force picture* yields in a natural way to the vector representation and to the visualization of the electric field via *field lines*. The field lines are curves having the property that the field vector is tangential at all points of the curve. Another viewpoint is to consider the energy of a charge moved through the field. We can visualize the field via the change of the energy of a test charge moved through the field. Figure 2.4 shows the representation of the field via field lines and via the surfaces of constant test charge energy or constant electric potential, respectively. The *energy picture* is more related to differential forms. For an electrostatic field the surfaces associated with the one-form $\mathcal{E}$ are equipotentials. Since the dimension of the differential form $\mathcal{E}$ is [V] the differential form $\mathcal{E}$ expresses the change of electric potential over an infinitesimal path element. The field lines are orthogonal to the potential surfaces. Depending on the properties of the field the potential surfaces also may end or join. Figure 2.5 shows the surface representation of the three *fundamental one-forms* $\mathrm{d}x$, $\mathrm{d}y$ and $\mathrm{d}z$. Figure 2.6 shows a situation we will encounter in time variable fields. In the center of the structure the field intensity is higher than at its edges. In this case the integral (2.12) will depend on the path from x_1 to x_2 and

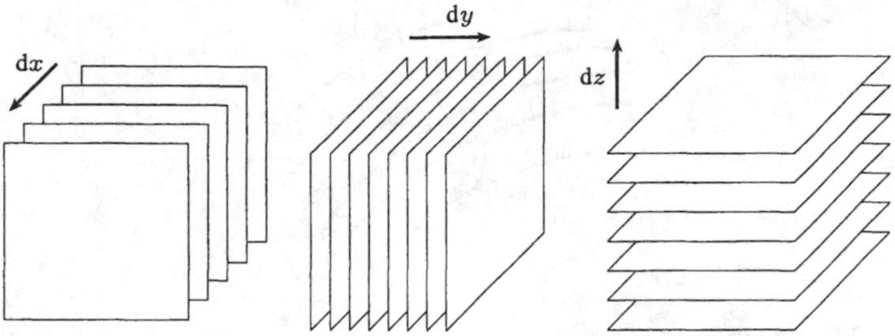

Figure 2.5: The fundamental one-forms in cartesian coordinates.

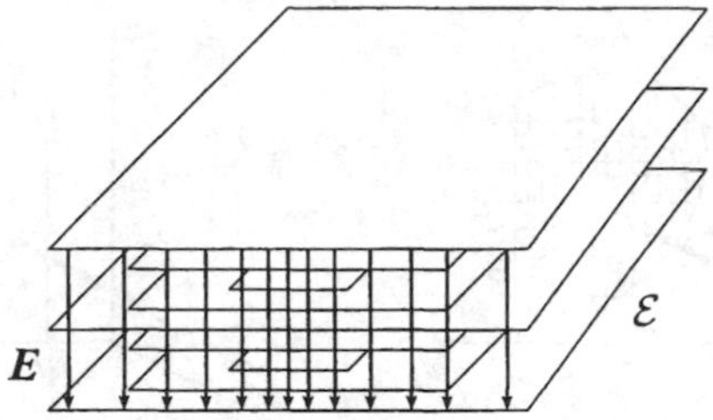

Figure 2.6: One-form with ending surfaces.

we cannot assign a scalar potential to the field.

The vector $\boldsymbol{H}(\boldsymbol{x},t) = [H_x(\boldsymbol{x},t),\, H_y(\boldsymbol{x},t),\, H_z(\boldsymbol{x},t)]^T$ describes the *magnetic field intensity* and has the dimension [Am^{-1}]. An *electric current* i, which is slowly varying with time, and the magnetic field generated by this current are related via

$$i(t) = \oint_{\partial A} \mathcal{H} \tag{2.13}$$

with the *magnetic field differential form*

$$\mathcal{H} = H_x(x,y,z,t)\,\mathrm{d}x + H_y(x,y,z,t)\,\mathrm{d}y + H_z(x,y,z,t)\,\mathrm{d}z\,. \tag{2.14}$$

The circle on the integral symbol denotes the integration over a closed boundary. Figure 2.3(b) shows the path of integration for the definition of the current i. The relation between the direction of reference for the current and the orientation of the path of integration is shown in Figure 2.3(b). The current is counted positive if its direction coincides with the direction of reference.

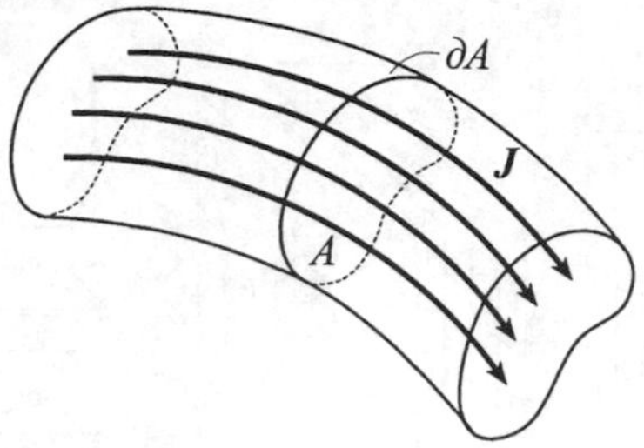

Figure 2.7: Current flow.

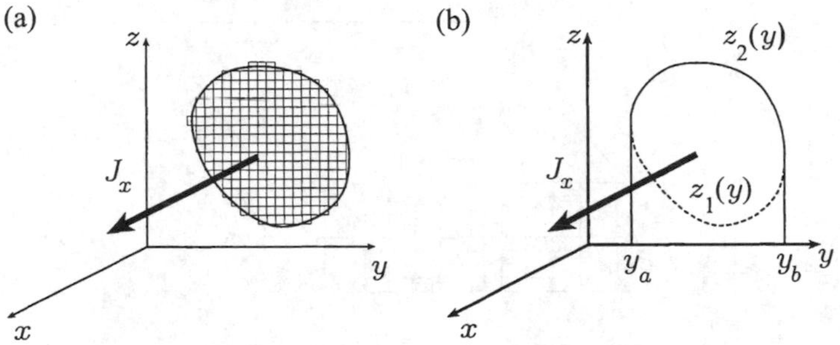

Figure 2.8: The integration over an area.

2.3 CURRENT AND FLUX DENSITIES

The current flowing in a conductor varies through the cross-section. We describe the flow of the current by a *current density* vector field $\boldsymbol{J}(\boldsymbol{x}) = [J_x(\boldsymbol{x}), J_y(\boldsymbol{x}), J_z(\boldsymbol{x})]^T$. The current I is flowing through a tube formed by the current density field lines going through the boundary ∂A of the area A as shown in Figure 2.7. Let us at first assume a current flowing in the x-direction only as shown in Figure 2.8(a). In this case to compute the total current we have to integrate over the surface A in the yz-plane. The integration may be performed by subdividing the area A in small elements as depicted in Figure 2.8(a), multiplying the current density with the area of the area elements and summing all these contributions.

$$i = \int_A J_x \, \mathrm{d}y \, \mathrm{d}z \,. \tag{2.15}$$

If, for example the boundary ∂A can be represented by two functions $z_1(y)$ and $z_2(y)$,

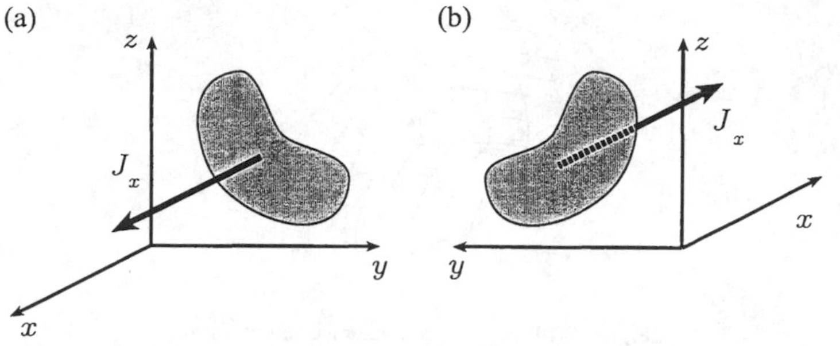

Figure 2.9: The orientation of an area.

as shown in Figure 2.8(b), we can bring the integral (2.15) into the form

$$i = \int_{y_a}^{y_b} \left[\int_{z_1(y)}^{z_2(y)} J_x \, \mathrm{d}z \right] \mathrm{d}y \,. \tag{2.16}$$

If we integrate a current density over an area not perpendicular to a coordinate axis we have to consider the orientation of the area. If in Figure 2.8(a) the current density J_x is positive, the current i also will be positive. Inverting the direction of J_x will yield a negative current. This inversion may be performed by mirroring the coordinates with respect to the yz-plane. How do we know whether a *surface integral* is positive or negative? The answer is: We have to define a positive orientation. A positive-oriented or right-handed Cartesian coordinate system is specified as follows: If we are looking in the z-direction on the xy-plane the x-axis may be rotated clockwise by 90° into the y-axis. In Figure 2.9 the vector component J_x is pointing in a positive orientation. On the right side of Figure 2.9 the coordinate system as well as the vector field were rotated by 180° around the z-axis. Physically nothing has changed. However in the left figure, the vector pointing towards the observer is positive, whereas in the right figure the vector pointing away from the observer is positive.

We now introduce a notation that takes into account the orientation of a coordinate system. The so-called *exterior differential form* $\mathrm{d}y \wedge \mathrm{d}z$ has the property

$$\mathrm{d}y \wedge \mathrm{d}z = -\,\mathrm{d}z \wedge \mathrm{d}y \,. \tag{2.17}$$

The product denoted by the symbol $\wedge$ is called an *exterior product* or *wedge product*. Exterior differential forms consisting of wedge products of two differentials or sums of such products are called *two-forms*. We may decide either $\mathrm{d}y \wedge \mathrm{d}z = \mathrm{d}y\,\mathrm{d}z$ or $\mathrm{d}y \wedge \mathrm{d}z = -\,\mathrm{d}y\,\mathrm{d}z$. Deciding

$$\mathrm{d}y \wedge \mathrm{d}z = \mathrm{d}y\,\mathrm{d}z \tag{2.18}$$

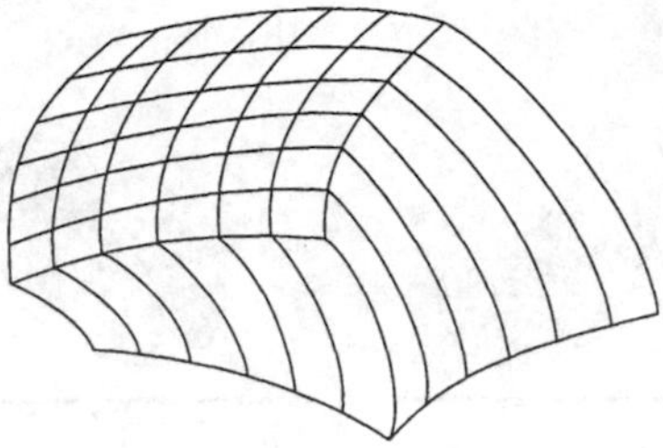

Figure 2.10: Tube representation of a two-form.

assigns to $dy \wedge dz$ the positive orientation and to $dz \wedge dy$ the negative orientation. The integral (2.15) can now be written in the orientation-independent form

$$i = \int_A J_x \, dy \wedge dz \,. \tag{2.19}$$

Figure 2.10 shows the *tube representation* of a two-form. The two-form is visualized by a bundle of tubes carrying the current. The current density is inversely proportional to the cross-sectional area of the tubes. Figure 2.11 shows the tube representations of the *fundamental two-forms* $dy \wedge dz$, $dz \wedge dx$, $dx \wedge dy$.

If the surface A is an arbitrarily oriented curved surface in three-dimensional space and the current density vector has the x-, y- and z-components J_x, J_y and J_z, we have to perform the integration over

$$i = \int_A J_x \, dy \wedge dz + J_y \, dz \wedge dx + J_z \, dx \wedge dy \,. \tag{2.20}$$

The first term of the integrand concerns the integration of the x-component of the current density over the projection of the surface A on the yz-plane and so forth.

Let us introduce the *current density form* $\mathcal{J}$ by the exterior differential form

$$\mathcal{J} = J_x \, dy \wedge dz + J_y \, dz \wedge dx + J_z \, dx \wedge dy. \tag{2.21}$$

The current i may be expressed in a compact notation as the integral of the differential form $\mathcal{J}$

$$i = \int_A \mathcal{J} \,. \tag{2.22}$$

In the case of electric and magnetic field quantities we distinguish between *field intensities* and *flux densities*. A field intensity usually occurs in a path integral whereas a flux density occurs in surface integrals. Field intensities are related to the flux densities via the *constitutive relations*. The constitutive equations depend on the metric

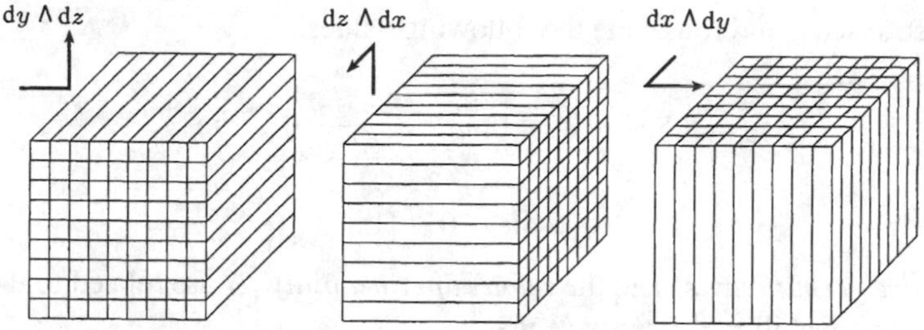

Figure 2.11: The fundamental two-forms in cartesian coordinates.

properties of the space or the chosen coordinate system and on the *macroscopic material parameters*.

On the microscopic scale of atomic dimensions the electric and magnetic fields exhibit strong spatial variations. However, the knowledge of these variations is not relevant. We are dealing with the averages of fields and sources over volumes large compared with the volume occupied by a single atom or field. Such averaged quantities are called *macroscopic fields* [1,2]. In a dielectric by an applied primary electric field the negative electronic charge is shifted spatially relative to the positive background charge. This yields an *electric polarization* of the medium contributing to the *electric flux density*. In the magnetic field the *magnetic polarization* determines the relation between magnetic field and magnetic flux density.

The *electric flux density* in the literature usually is called *electric displacement*, and the magnetic flux density $\boldsymbol{B}(\boldsymbol{x}) = [B_x(\boldsymbol{x}), B_y(\boldsymbol{x}), B_z(\boldsymbol{x})]^T$ usually is called *magnetic induction*. The denomination "flux density" stresses the geometric properties of these quantities and therefore is preferred in the following. The electric flux density $\boldsymbol{D}(\boldsymbol{x}) = [D_x(\boldsymbol{x}), D_y(\boldsymbol{x}), D_z(\boldsymbol{x})]^T$ and the magnetic flux density $\boldsymbol{B}(\boldsymbol{x}) = [B_x(\boldsymbol{x}), B_y(\boldsymbol{x}), B_z(\boldsymbol{x})]^T$ are related to the field intensities $\boldsymbol{E}$ and $\boldsymbol{H}$ via the *material equations* or *constitutive equations*. The electric flux density $\boldsymbol{D}$ has the dimension [Asm^{-2}], and the magnetic flux density $\boldsymbol{B}$ has the dimension [Vsm^{-2}]. For homogeneous and isotropic media the constitutive equations are given by

$$\boldsymbol{D} = \varepsilon \boldsymbol{E} \,, \tag{2.23}$$

$$\boldsymbol{B} = \mu \boldsymbol{H} \,, \tag{2.24}$$

where ε is the *permittivity* and μ is the *permeability*.

ε	Permittivity	[AsV^{-1}m^{-1}], [Fm^{-1}]
μ	Permeability	[VsA^{-1}m^{-1}], [Hm^{-1}]

In the free space ε and μ assume the following values:

$$\varepsilon_0 = 8.854 \cdot 10^{-12}\,\mathrm{Fm}^{-1} \cong \frac{1}{36\pi} 10^{-9}\,\mathrm{Fm}^{-1}, \tag{2.25}$$

$$\mu_0 = 4\pi \cdot 10^{-7}\,\mathrm{Hm}^{-1}. \tag{2.26}$$

The *relative permittivity* ε_r and the *relative permeability* μ_r are related to the permittivity and permeability of free space via

$$\varepsilon_r = \frac{\varepsilon}{\varepsilon_0}, \tag{2.27a}$$

$$\mu_r = \frac{\mu}{\mu_0}. \tag{2.27b}$$

The surface integral of the electric flux density $\boldsymbol{D}$ over an area A yields the *electric flux* Ψ. Introducing the *electric flux form*

$$\mathcal{D} = D_x\,\mathrm{d}y \wedge \mathrm{d}z + D_y\,\mathrm{d}z \wedge \mathrm{d}x + D_z\,\mathrm{d}x \wedge \mathrm{d}y \tag{2.28}$$

we can write

$$\Psi = \int_A \mathcal{D}. \tag{2.29}$$

The surface integral of the magnetic flux density $\boldsymbol{B}$ over an area A yields the *magnetic flux* Φ. Introducing the *magnetic flux form*

$$\mathcal{B} = B_x\,\mathrm{d}y \wedge \mathrm{d}z + B_y\,\mathrm{d}z \wedge \mathrm{d}x + B_z\,\mathrm{d}x \wedge \mathrm{d}y \tag{2.30}$$

we can write

$$\Phi = \int_A \mathcal{B}. \tag{2.31}$$

We have seen that field intensities are described by one-forms whereas current densities and flux densities are described by two-forms. The field intensities $\boldsymbol{E}$ and $\boldsymbol{H}$ are related to the flux densities $\boldsymbol{D}$ and $\boldsymbol{B}$ via the material equations (2.23) and (2.24). We introduce the *star operator* $\star$ or *Hodge operator*, defined by

$$\begin{aligned} \star f &= f\,\mathrm{d}x \wedge \mathrm{d}y \wedge \mathrm{d}z\,, \\ f &= \star(f\,\mathrm{d}x \wedge \mathrm{d}y \wedge \mathrm{d}z)\,, \\ \star(A_x\,\mathrm{d}x + A_y\,\mathrm{d}y + A_z\,\mathrm{d}z) &= A_x\,\mathrm{d}y \wedge \mathrm{d}z + A_y\,\mathrm{d}z \wedge \mathrm{d}x + A_z\,\mathrm{d}x \wedge \mathrm{d}y\,, \\ A_x\,\mathrm{d}x + A_y\,\mathrm{d}y + A_z\,\mathrm{d}z &= \star(A_x\,\mathrm{d}y \wedge \mathrm{d}z + A_y\,\mathrm{d}z \wedge \mathrm{d}x + A_z\,\mathrm{d}x \wedge \mathrm{d}y)\,. \end{aligned} \tag{2.32}$$

The star operator has the property

$$\star\star = 1\,. \tag{2.33}$$

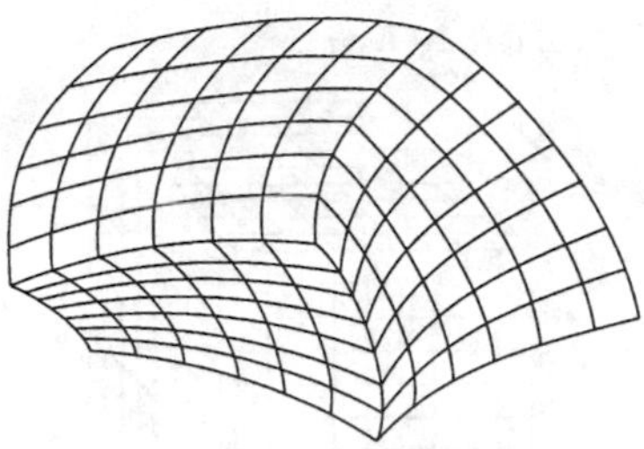

Figure 2.12: Volume element representation of a three-form.

The star operator allows to bring the material equations (2.23) and (2.24) into the form

$$\mathcal{D} = \varepsilon \star \mathcal{E}\,, \tag{2.34}$$

$$\mathcal{B} = \mu \star \mathcal{H}\,. \tag{2.35}$$

2.4 THE CHARGE DENSITY

The *electric charge* q is given by the volume integral over the *electric charge density* ρ. For the electric charge density we may introduce a *three-form*, the so-called *charge density form*

$$\mathcal{Q} = \rho\, \mathrm{d}x \wedge \mathrm{d}y \wedge \mathrm{d}z\,. \tag{2.36}$$

We obtain the charge q by performing the volume integral over the three-form $\mathcal{Q}$:

$$q = \int_V \mathcal{Q}\,. \tag{2.37}$$

We note that the exterior product $\mathrm{d}x \wedge \mathrm{d}y \wedge \mathrm{d}z$ changes its sign if two factors are interchanged. Figure 2.12 shows the graphic visualization of a three-form by subdividing the volume into cells. The cell volume is inversely proportional to the charge density. Figure 2.13 shows the *fundamental three-form* $\mathrm{d}x \wedge \mathrm{d}y \wedge \mathrm{d}z$.

The star operator applied to a three-form yields a *zero-form* and vice versa. A zero-form is a *true scalar* as, for example, the scalar potential. A true scalar is invariant under coordinate transformations, whereas a three-form may depend on the coordinate system.

The electric flux flowing through the boundary ∂V of a volume V is related to the charge stored in the volume V via

$$\oint_{\partial V} \mathcal{D} = \int_V \mathcal{Q}\,. \tag{2.38}$$

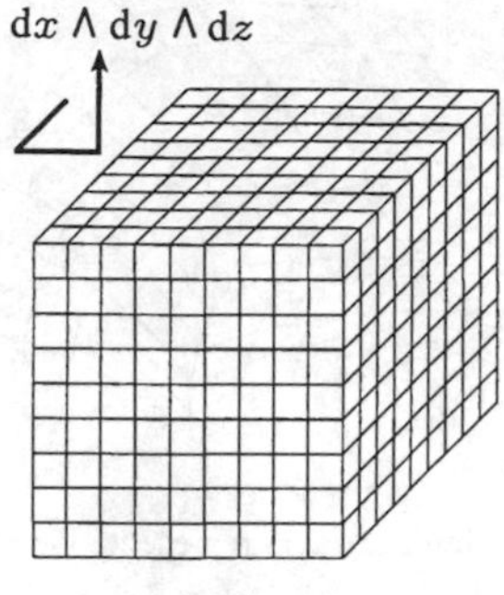

Figure 2.13: The fundamental three-form in cartesian coordinates.

To calculate the electric field of a point charge q we consider the point charge q in the center of a sphere with radius r. For symmetry reasons the flux density is homogeneous over the sphere and is directed radially. Therefore

$$\oint_{\partial V} \mathcal{D} = 4\pi r^2 D_r = q \,. \tag{2.39}$$

From this we obtain

$$E_r = \frac{q}{4\pi \varepsilon r^2} \,. \tag{2.40}$$

The electric field vector is given by

$$\boldsymbol{E} = \frac{q\boldsymbol{r}}{4\pi \varepsilon r^3} \,. \tag{2.41}$$

There exists no magnetic charge. Therefore over any closed boundary ∂V of a volume V we obtain

$$\oint_{\partial V} \mathcal{B} = 0 \,. \tag{2.42}$$

2.5 THE MAXWELL PUZZLE

Let us assume the electromagnetic field to be slowly varying. In this case we can consider electric and magnetic fields to be independent from each other. In lumped element circuits we know elementary circuit elements, which are based either on electric field concentration or magnetic field concentration. Capacitors store electric field energy and inductors store magnetic field energy. Capacitors as well as inductors may be considered as lumped circuit elements within the network concept.

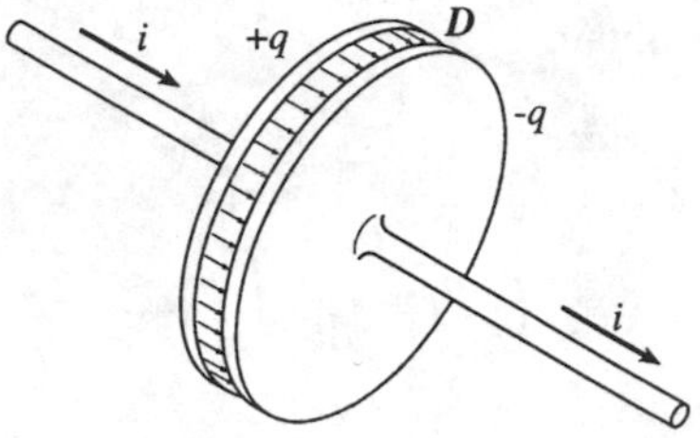

Figure 2.14: Plate capacitor.

Ampère's law relates the current flowing through a surface A to the magnetic field tangential to the boundary ∂A of the surface A:

$$\oint_{\partial A} \mathcal{H} = \int_{A} \mathcal{J} . \tag{2.43}$$

In the case of rapidly varying electromagnetic fields, however, we have to consider the direct mutual influence of electric and magnetic fields. To demonstrate this we consider the plate capacitor depicted in Figure 2.14. This capacitor is permeable for alternating current. The current flowing within a positive half-wave into the left capacitor plate builds up a positive charge. This positive charge influences a negative charge of equal magnitude in the right capacitor plate. The current flowing on the left side into the capacitor is equal to the current flowing out from the right side of the capacitor. With the electric field $\boldsymbol{E}$ between the capacitor plates there is a related electric flux density $\boldsymbol{D}$ given by (2.23) and (2.34), respectively. We assume the diameter of the plate capacitor to be large compared with the plate distance b. In this case the electric field is essentially concentrated between the capacitor plates and is homogeneous within this region. The electric flux Ψ is the product of the magnitude of electric flux density $\boldsymbol{D}$ and the capacitor area A

$$|\Psi| = \int_{A} \mathcal{D} = A\,|\boldsymbol{D}| . \tag{2.44}$$

The total electric flux Ψ is equal to the electric charge q stored in the capacitor

$$\Psi = \int_{V} \mathcal{Q} = q . \tag{2.45}$$

The rate of change of the electric charge q is equal to the current

$$i = \frac{\mathrm{d}q}{\mathrm{d}t} . \tag{2.46}$$

With (2.22) and (2.38) we obtain

$$\int_A \mathcal{J} = \frac{\mathrm{d}}{\mathrm{d}t}\int_A \mathcal{D}. \tag{2.47}$$

The time derivative of the electric flux $\mathrm{d}\Psi/\mathrm{d}t$ is interpreted as the *displacement current*. The displacement current was introduced by Maxwell, who considered for the first time the concept that variations in the position of bound charge were equivalent in their effect to a conduction current [1,10]. The *conduction current*, i.e., the current carried by moving charges and flowing through the conductor is continued by the displacement current between the capacitor plates. Since the displacement current is equal to the rate of change of the electric flux, the displacement current is proportional to the frequency. It has been the achievement of James Clerk Maxwell to recognize that the displacement current can give rise to a magnetic field in the same way as the conduction current. Therefore he added in Ampère's law (2.43) the displacement current term to the conduction current term and obtained

$$\oint_{\partial A} \mathcal{H} = \frac{\mathrm{d}}{\mathrm{d}t}\int_A \mathcal{D} + \int_A \mathcal{J}\,. \tag{2.48}$$

Especially in the case of high frequencies the displacement current can give a considerable contribution for the generation of the magnetic field.

Michael Faraday discovered that a time-varying magnetic field generates an electric field. This law is called *Faraday's law* or *law of induction*

$$\oint_{\partial A} \mathcal{E} = -\frac{\mathrm{d}}{\mathrm{d}t}\int_A \mathcal{B}\,. \tag{2.49}$$

As a consequence in a rapidly varying electromagnetic field the electric and magnetic fields are directly interacting in space.

Table 2.1: The Field Quantities

Name	*Field Quantity*	*Dimension*	*Differential Form*	*Dimension*
Electric field intensity	$\boldsymbol{E}$	[Vm^{-1}]	$\mathcal{E}$	[V]
Magnetic field intensity	$\boldsymbol{H}$	[Am^{-1}]	$\mathcal{H}$	[A]
Electric flux density	$\boldsymbol{D}$	[As m^{-2}]	$\mathcal{D}$	[As]
Magnetic flux density	$\boldsymbol{B}$	[Vs m^{-2}]	$\mathcal{B}$	[Vs]
Current density	$\boldsymbol{J}$	[Am^{-2}]	$\mathcal{J}$	[A]
Charge density	ρ	[As m^{-3}]	$\mathcal{Q}$	[As]

2.6 THE INTEGRAL FORM OF MAXWELL'S EQUATIONS

Let us now summarize Maxwell's equations. The integral form of Maxwell's equations is given by:

$$\oint_{\partial A} \mathcal{H} = \frac{\mathrm{d}}{\mathrm{d}t}\int_A \mathcal{D} + \int_A \mathcal{J}, \qquad \text{Ampère's law} \qquad (2.50)$$

$$\oint_{\partial A} \mathcal{E} = -\frac{\mathrm{d}}{\mathrm{d}t}\int_A \mathcal{B}, \qquad \text{Faraday's law} \qquad (2.51)$$

$$\oint_{\partial V} \mathcal{B} = 0, \qquad \text{Magnetic flux continuity} \qquad (2.52)$$

$$\oint_{\partial V} \mathcal{D} = \int_V \mathcal{Q}. \qquad \text{Gauss' law} \qquad (2.53)$$

Equations (2.50) and (2.51) are named as Ampère's law and Faraday's law; (2.52) describes the flux continuity and (2.53) is named Gauss' law. The field quantities and the corresponding differential forms occurring in these equations are summarized in Table 2.1.

In (2.50) and (2.51) line integrals over the boundary of the surface A are related to surface integrals over the area A. Figure 2.15(a) shows the relation between the orientation of the area A and the contour ∂A. The line integral over the closed contour ∂A is called *circulation*. In (2.52) and (2.53) the surface integrals are performed over the boundary ∂V of the volume V. Figure 2.15(b) shows the orientation of the boundary surface ∂V.

The conservation of charge is embodied in the *continuity equation* following directly from Maxwell's equations. Let us apply Ampère's law (2.50) on a surface A which is the boundary of a volume V, i.e., $A = \partial V$. Since a boundary has no

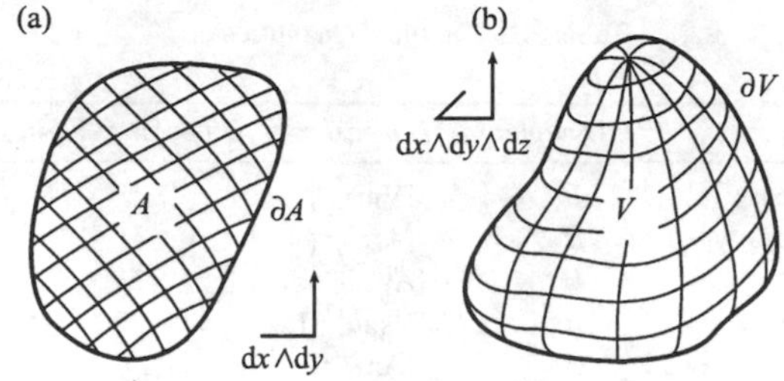

Figure 2.15: (a) Area A with boundary ∂A and (b) volume V with boundary ∂V.

boundary it follows $\partial A = \partial(\partial V) = 0$ and therefore we obtain from (2.50)

$$\frac{\mathrm{d}}{\mathrm{d}t}\oint_{\partial V}\mathcal{D} + \oint_{\partial V}\mathcal{J} = 0\,. \tag{2.54}$$

Inserting Gauss' law (2.53) yields the *integral form* of the *continuity equation*

$$\frac{\mathrm{d}}{\mathrm{d}t}\int_{V}\mathcal{Q} + \oint_{\partial V}\mathcal{J} = 0\,. \tag{2.55}$$

From this it follows that any current flow through the boundary ∂V of a volume V must be accompanied by an appropriate change of the charge in V. The total charge is conserved.

We shall demonstrate below that Maxwell's equations exhibit plane wave solutions with a free-space propagation velocity, i.e., the *speed of light* c_0, given by

$$c_0 = \frac{1}{\sqrt{\varepsilon_0\mu_0}} = 2.998\cdot 10^8\,\mathrm{ms}^{-1} \cong 3\cdot 10^8\,\mathrm{ms}^{-1}. \tag{2.56}$$

In electric conductive media the current density form $\mathcal{J}$ is related to the electric field form $\mathcal{E}$ via *Ohm's law*

$$\mathcal{J} = \sigma\star\mathcal{E} \tag{2.57}$$

where σ is the *conductivity* with dimension $[\mathrm{AV}^{-1}\mathrm{m}^{-1}]$. If there also exists an *impressed current density* with the current density differential form $\mathcal{J}_0$, the total current density is given by

$$\mathcal{J} = \mathcal{J}_0 + \sigma\star\mathcal{E}\,. \tag{2.58}$$

To every field quantity there exists a corresponding integral field quantity. The integral field quantities are obtained from the corresponding differential field quantities by integration over n one-dimensional or multidimensional spatial domain. Integral

Table 2.2: Field Quantities and Network Quantities

Field Quantity	*Network Quantity*	*Name*	*Dimension*
$\boldsymbol{E}$	v	Electric voltage	[V]
$\boldsymbol{H}$	v_m	Magnetic voltage	[A]
$\boldsymbol{D}$	Ψ	Electric flux	[As]
$\boldsymbol{B}$	Φ	Magnetic flux	[Vs]
$\boldsymbol{J}$	i	Electric current	[A]
ρ	q	Electric charge	[As]

field quantities may be interpreted as network quantities. Table 2.2 summarizes the differential field quantities and the corresponding integral field quantities or network quantities, respectively. The *voltage* v is defined by

$$v_{21} = -\int_1^2 \mathcal{E} \tag{2.59}$$

as the negative line integral over the electric field intensity. Figure 2.3(a) shows the orientation of the voltage v_{21} and the path of integration. In the same way we may define a *magnetic voltage*

$$v_{m21} = -\int_1^2 \mathcal{H}\,. \tag{2.60}$$

The electric flux Ψ, the magnetic flux Φ and the current i are given by

$$\Psi = \int_A \mathcal{D}\,, \tag{2.61}$$

$$\Phi = \int_A \mathcal{B}\,, \tag{2.62}$$

$$i = \int_A \mathcal{J}\,. \tag{2.63}$$

The electric charge q is given by

$$q = \int_V \mathcal{Q}\,. \tag{2.64}$$

2.7 THE ELECTROMAGNETIC WAVE

James C. Maxwell was the first to predict the existence of *electromagnetic waves*. In 1864 he proposed that light is an electromagnetic disturbance in the form of waves [10]. In an electromagnetic wave the magnetic field is built up by the displacement current due to the time-varying electric field and, vice versa, the electric field is built up by the time-varying electric field. In 1887 Heinrich Hertz provided experimental verification of electromagnetic waves [11].

Following [12] we demonstrate how a propagating *electromagnetic wave* develops by the mutual influence of electric and magnetic fields. For simplicity we consider the case of an electromagnetic plane wave, where the electric and magnetic field components are constant in planes transverse to the direction of propagation. If we are choosing in a cartesian coordinate system the z-direction as the direction of propagation, all field components will only depend on the coordinate z and time t. Within some finite cross-section an electromagnetic wave emitted from a far-distant source can be approximated by a plane wave.

The following considerations are not a derivation of an electromagnetic wave, since we are already making very detailed pre-assumptions. However, we want to visualize how the physical phenomena interact in order to establish wave propagation. First we assume that in a far distant transverse plane located in the negative z-direction, at a certain moment a magnetic field $\boldsymbol{H}$ is suddenly turned on. We assume that this magnetic field is directed in the y-direction and is homogeneous throughout the whole transverse plane. This may be realized when in a large conducting plane a homogeneous surface current directed in the x-direction is turned on at a moment. Now let us assume that a physical action can propagate with a maximum velocity, which we will name c. Let us assume that a plane wave front is propagating now in positive the z-direction and that the space in front of the wave front is free of field. The space behind the wave front is filled by a homogeneous magnetic field H_y in the y-direction.

We now, in a first step, are going to show that this propagating magnetic field due to Faraday's law will induce an electric field. Let us consider an area element of height h and length l corresponding to Figure 2.16. During the time of propagation of the wave front through this area element, the magnetic flux Φ flowing through the area element is increasing linearly with time. We obtain from (2.62)

$$\Phi = \mu\, h\, \Delta z\, H_y. \tag{2.65}$$

For Δz increasing with time the velocity c of the wave front is given by

$$c = \frac{\mathrm{d}\Delta z}{\mathrm{d}t}. \tag{2.66}$$

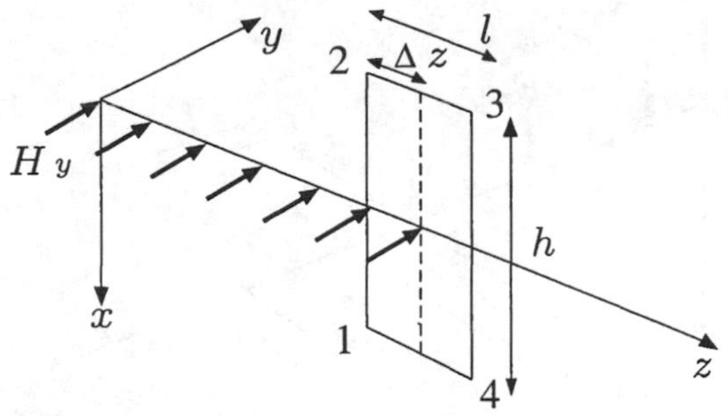

Figure 2.16: Generation of the electric field by the magnetic field.

Therewith we obtain from (2.65) for the time interval, in which the wave front is marching through the area element

$$\frac{d\Phi}{dt} = \mu\, h\, c\, H_y\,. \tag{2.67}$$

This change of the flux with time is due to Faraday's law; (2.51) has to be related with the circulation integral of the induced electrical field $\boldsymbol{E}$. Now we may assume that the induced electric field has no component in the z-direction, since such a field should have the same direction throughout the complete transverse plane for symmetry reasons. Such a longitudinal electromagnetic field would not be divergence-free. Therefore a component E_z cannot occur, since we did not assume any electric charge in front of the wave front. We also may assume that there is no electric field in front of the wave front. Therefore only a homogeneous electric field E_x between point 1 and point 2 of the path of integration gives a contribution to the integral. Therefore we obtain

$$\oint_{\partial A} \mathcal{E} = -h\, E_x\,. \tag{2.68}$$

With (2.51) and (2.67) we obtain

$$E_x = \mu\, c\, H_y\,. \tag{2.69}$$

We did assume that the magnetic field propagates in the positive z-direction without discussing the rules governing this propagation. We now want to show that the propagating electric field gives rise to a magnetic field. For this purpose we consider a horizontal area element of length l and width a according to Figure 2.17. In the spatial region inside the wave front exists a homogeneous electric field E_x. In the time interval in which the wave front propagates through the area element the electric flux through this area element is given by

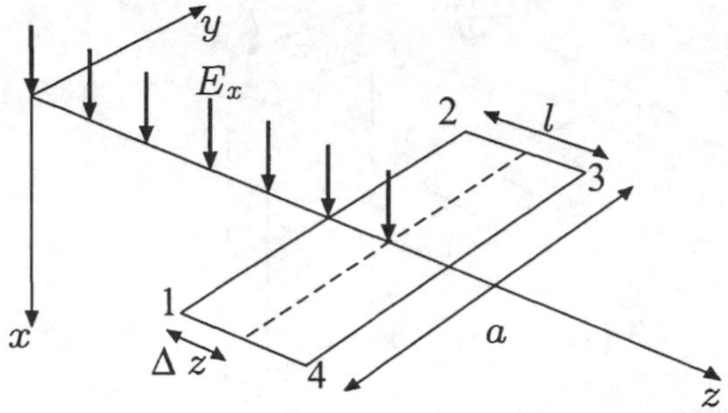

Figure 2.17: Generation of the magnetic field by the electric field.

$$\Psi = \varepsilon\, a\, \Delta z\, E_x\,. \tag{2.70}$$

With (2.66) we obtain

$$\frac{\mathrm{d}\Psi}{\mathrm{d}t} = \varepsilon\, a\, c\, E_x\,. \tag{2.71}$$

The displacement current $\mathrm{d}\Psi/\,\mathrm{d}t$ generates a magnetic field. The circulation integral of the magnetic field over the boundary of the area element is related to the displacement current $\mathrm{d}\Psi/\,\mathrm{d}t$ according to (2.50). Only the homogeneous magnetic field H_y between points 1 and 2 in Figure 2.17 gives a contribution to the integral. We therefore obtain

$$\oint_{\partial A} \mathcal{H} = a\, H_y\,. \tag{2.72}$$

With (2.50) and (2.71) we obtain

$$H_y = \varepsilon\, c\, E_x\,. \tag{2.73}$$

From (2.69) and (2.73) we obtain the propagation velocity c of the plane electromagnetic wave

$$c = \frac{1}{\sqrt{\varepsilon\mu}}\,. \tag{2.74}$$

This is the speed of light in the medium with permittivity ε and permeability μ. A dielectric material with $\varepsilon \neq \varepsilon_0$ and $\mu = \mu_0$ may be characterized by the *refractive index*

$$n = \sqrt{\varepsilon_r}\,. \tag{2.75}$$

The refractive index gives the ratio of the free-space velocity c_0 of the transverse electromagnetic wave to its velocity c in the medium:

$$n = \frac{c_0}{c}\,. \tag{2.76}$$

For free space the propagation velocity is the speed of light c_0 given in (2.56). Furthermore, from (2.73) and (2.74) we obtain the ratio of electric and magnetic field intensities, given by

$$\frac{E_x}{H_y} = \sqrt{\frac{\mu}{\varepsilon}}\,. \tag{2.77}$$

The ratio E_x/H_y has the dimension [VA^{-1}]. We define the *wave impedance* Z_F by

$$Z_F = \sqrt{\frac{\mu}{\varepsilon}}\,. \tag{2.78}$$

The *free-space wave impedance* is given by

$$Z_{F0} = \sqrt{\frac{\mu_0}{\varepsilon_0}} = 377\,\Omega \cong 120\pi\,\Omega\,. \tag{2.79}$$

The considerations we have presented here naturally are independent from the choice of the coordinates. Therefore we may assume in general that in the case of the *plane electromagnetic wave* the direction of the electric field, the direction of the magnetic field, and the direction of propagation form an orthogonal tripod. Up to now we only have considered a *step wave* defined by

$$E_x(z,t) = Z_{F0} H_y(z,t) \begin{cases} 0 & \text{for } ct - z < 0 \\ E_0 & \text{for } ct - z \geq 0 \end{cases}\,. \tag{2.80}$$

Due to the linearity of Ampère's and Faraday's law, however, we can use the principle of superposition and thereby construct plane waves of arbitrary shape.

By superposition of a step wave, Figure 2.18(a), and a time-delayed step-wave with opposite amplitude, Figure 2.18(b), we obtain a rectangular wave as shown in Figure 2.18(c). Putting together such rectangular waves, we can construct wave forms as depicted in Figure 2.18(d). If we reduce the width of the rectangular segments to zero, we may construct continuous waves of arbitrary shape.

We therefore may assume waves of arbitrary shape propagating at a velocity c with a stable wave-form. The spatial dependence and the time dependence of a plane electromagnetic wave propagating in the positive z-direction with the electric field directed in the x-direction is given by

$$E_x(z,t) = E_x(z - ct)\,, \tag{2.81}$$

$$H_y(z,t) = Z_F^{-1}\, E_x(z - ct)\,. \tag{2.82}$$

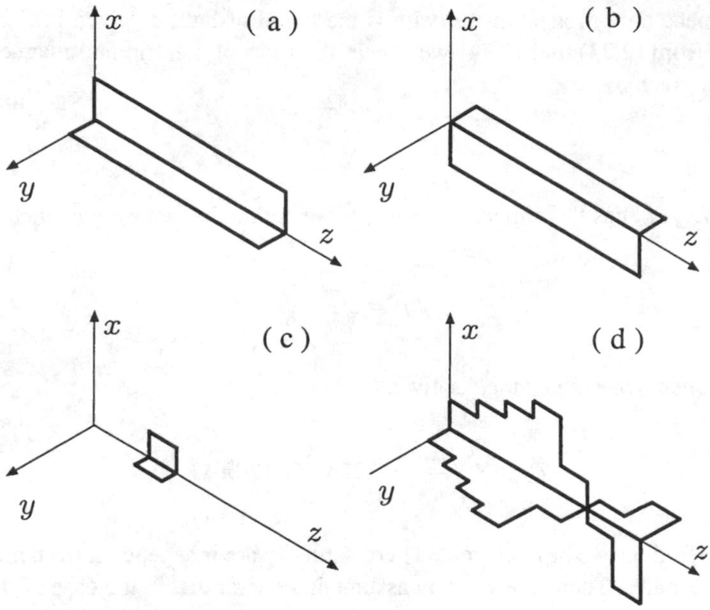

Figure 2.18: Superposition of step waves: (a) positive step wave, (b) negative step wave, (c) rectangular pulse wave and (d) step wave train.

If the wave is propagating in the negative z-direction we obtain

$$E_x(z,t) = E_x(z+ct)\,, \tag{2.83}$$

$$H_y(z,t) = -Z_F^{-1}\, E_x(z+ct)\,. \tag{2.84}$$

Assuming a sinusoidal time dependence of the electromagnetic field, we obtain the *time-harmonic electromagnetic wave*. With frequency f, angular frequency ω, wavelength λ and wave number k, given by

$$\omega = 2\pi f\,, \tag{2.85}$$

$$k = \frac{\omega}{c} = 2\pi/\lambda \tag{2.86}$$

we obtain the field components

$$E_x(z,t) = E_{0x}^{+}\cos(\omega t - kz + \varphi_0)\,, \tag{2.87}$$

$$H_y(z,t) = H_{0y}^{+}\cos(\omega t - kz + \varphi_0)\,. \tag{2.88}$$

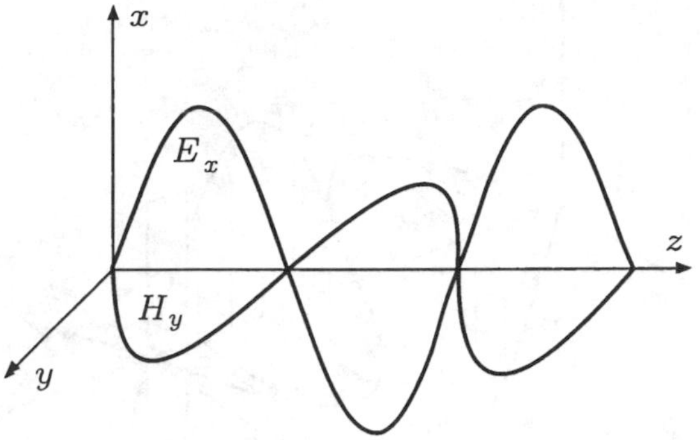

Figure 2.19: Linearly polarized time-harmonic electromagnetic wave.

E_{0x}^{+} and H_{0y}^{+} are the amplitudes of the time-harmonic electromagnetic wave propagating in the positive z-direction. The electric and magnetic field amplitudes are related via

$$E_{0x}^{+} = Z_F \, H_{0y}^{+} \,. \tag{2.89}$$

Figure 2.19 shows the field components E_{0x}^{+} and H_{0y}^{+} of the time-harmonic electromagnetic wave. If the electrical field is directed in one direction only, the wave exhibits *linear polarization*. The direction of the electric field vector is always called the direction of polarization.

The superposition of an electromagnetic wave propagating in the positive z-direction with an electromagnetic wave propagating at the same frequency in the negative z-direction is given by

$$E_x(z,t) = E_{0x}^{+} \cos(\omega t - kz + \varphi_0^{+}) + E_{0x}^{-} \cos(\omega t + kz + \varphi_0^{-}) \,, \tag{2.90}$$

$$H_y(z,t) = H_{0y}^{+} \cos(\omega t - kz + \varphi_0^{+}) + H_{0y}^{-} \cos(\omega t + kz + \varphi_0^{-}) \,. \tag{2.91}$$

The field amplitudes E_{0x}^{-} and H_{0y}^{-} of the wave propagating in the negative z-direction are related by

$$E_{0x}^{-} = -Z_F \, H_{0y}^{-} . \tag{2.92}$$

The negative sign in (2.92) is due to the circumstance that also for the wave propagating in the negative z-directions of $\boldsymbol{E}$ and $\boldsymbol{H}$ the direction of propagation form a right-handed orthogonal tripod.

2.7.1 The Wave Equation

To derive a wave equation for the plane electromagnetic wave we consider the continuous plane wave according to Figure 2.20. We assume that the area elements consid-

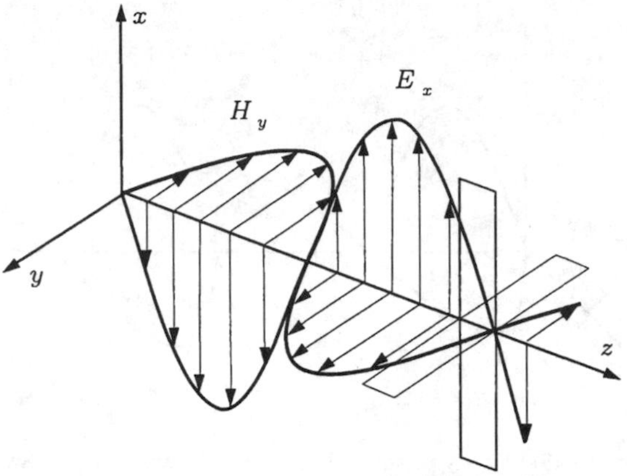

Figure 2.20: Linearly polarized time-harmonic electromagnetic wave.

ered in Figures 2.16 and 2.17 exhibit an infinitesimal length Δz. The magnetic flux varies continuously, and we obtain from (2.62)

$$\frac{\mathrm{d}\Phi}{\mathrm{d}t} = \mu\, h\, \Delta z\, \frac{\mathrm{d}H_y}{\mathrm{d}t}\,. \tag{2.93}$$

The circulation integral of the electric field over the contour of the vertical area element is given by

$$\oint_{\partial A} \mathcal{E} = -h\,(E_x(z) - E_x(z + \Delta z))\,. \tag{2.94}$$

From (2.51), (2.93) and (2.94) we obtain for $\Delta z \to 0$ the differential equation

$$\frac{\partial E_x}{\partial z} = -\mu\, \frac{\partial H_y}{\partial t}\,. \tag{2.95}$$

The operator ∂ denotes the *partial derivation*. The partial derivative $\partial E_x/\partial z$ is formed as if E_x is only dependent on z. For the change of the electric flux with time we obtain

$$\frac{\mathrm{d}\Psi}{\mathrm{d}t} = \varepsilon\, a\, \Delta z\, \frac{dE_x}{\mathrm{d}t}\,. \tag{2.96}$$

The circulation integral of the magnetic field around the area element is given by

$$\oint_{\partial A} \mathcal{H} = a\left(H_y(z) - H_y(z + \Delta z)\right)\,. \tag{2.97}$$

For $\Delta z \to 0$, we obtain

$$\frac{\partial H_y}{\partial z} = -\varepsilon \frac{\partial E_x}{\partial t} . \tag{2.98}$$

From (2.95) and (2.98), we obtain under consideration of (2.74) the wave equation

$$\frac{\partial^2}{\partial z^2} \begin{Bmatrix} E_x \\ H_y \end{Bmatrix} - \frac{1}{c^2} \frac{\partial^2}{\partial t^2} \begin{Bmatrix} E_x \\ H_y \end{Bmatrix} = 0 . \tag{2.99}$$

This wave equation is valid for E_x and H_y as well. It is easy to verify that all the above-presented solutions fulfill this equation.

2.7.2 The Polarization of Electromagnetic Waves

Let us again consider a plane electromagnetic wave propagating in the positive z-direction. We now assume that the electric field exhibits an x component as well as a y component given by

$$E_x^{(+)} \left(t - \frac{z}{c}\right) = E_{x0}^{(+)} \cos(\omega t - \beta z + \psi_x) , \tag{2.100}$$

$$E_y^{(+)} \left(t - \frac{z}{c}\right) = E_{y0}^{(+)} \cos(\omega t - \beta z + \psi_y) . \tag{2.101}$$

We allow different amplitudes $E_{x0}^{(+)}$ and $E_{y0}^{(+)}$ different phases, ψ_x and ψ_y for both electric field components. Since $E_x^{(+)}$ and $E_y^{(+)}$ may be chosen independently every superposition of (2.100) and (2.101) is a solution of the Maxwell's equations.

Let us consider the superposition of these two waves. We first consider the case where ψ_x is equal to ψ_y, i.e., both components $E_x^{(+)}$ and $E_y^{(+)}$ have the same phase. In this case

$$\frac{E_x^{(+)} \left(t - \frac{z}{c}\right)}{E_y^{(+)} \left(t - \frac{z}{c}\right)} = \frac{E_{x0}^{(+)}}{E_{y0}^{(+)}} \quad \text{for } \psi_x = \psi_y \tag{2.102}$$

is valid. The ratio between the x component and the y component of the electric field is independent from space and time. The angle

$$\theta = \arctan \frac{E_y^{(+)} \left(t - \frac{z}{c}\right)}{E_x^{(+)} \left(t - \frac{z}{c}\right)} \tag{2.103}$$

in Figure 2.21 is constant. Such a wave is called *linearly polarized.* In general the condition for linear polarization is given by $\psi_y = \psi_x + m\pi$ where m is an integer.

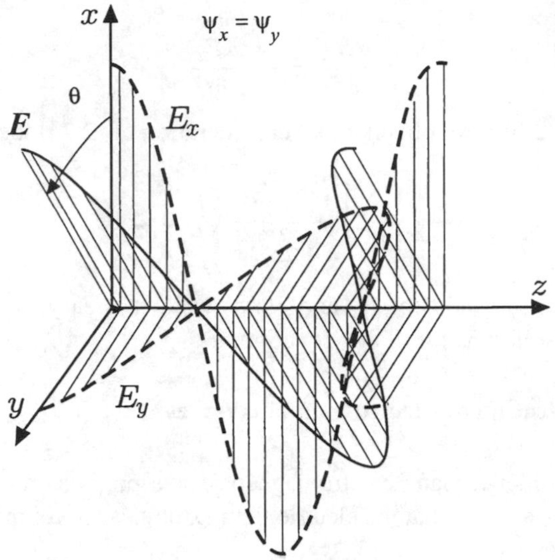

Figure 2.21: Linearly polarized plane electromagnetic wave.

We consider the special case given by $\psi_y = \psi_x \pm \frac{\pi}{2}$ and $E_{y0}^{(+)} = E_{x0}^{(+)} = E_0^{(+)}$. From (2.100), (2.101) and (2.102) we obtain

$$\left.\begin{aligned} |\boldsymbol{E}^{(+)}\left(t - \frac{z}{c}\right)| = |E_0^{(+)}| = \text{const.} \\ \theta = \pm(\omega t - \beta z + \psi_x) \end{aligned}\right\} \quad \begin{aligned} &\text{for } \psi_y = \psi_x \pm \frac{\pi}{2} \\ &\text{and } E_0^{(+)} = E_{y0}^{(+)} = E_{x0}^{(+)} . \end{aligned} \tag{2.104}$$

The magnitude $|\boldsymbol{E}^{(+)}|$ of the electrical field intensity is constant, whereas the direction of the electric field rotates around the z-axis. The rotation occurs with the angular frequency ω. The rotation in space has a period given by the wave number β. This wave exhibits *circular polarization*. Figure 2.22 illustrates this case. If we are looking into the direction of propagation of this wave, the electric field vector is rotating anticlockwise. The wave is left-handed circularly polarized. We have to consider that the sense of rotation for constant space is opposite to the sense of rotation for constant time. For $\psi_y = \psi_x - \frac{\pi}{2}$ the electric field vector $\boldsymbol{E}$ is rotating clockwise, if we are looking in the direction of propagation. The wave is left-handed circularly polarized. If the rotation of a circularly polarized wave with respect to time and the direction of propagation corresponds to a right-handed system, the wave is called right-handed circularly polarized. In the most common case, if $\psi_y - \psi_x$ and $E_{x0}^{(+)}$ and $E_{y0}^{(+)}$ are assuming arbitrary values, the electric field vector performs an elliptic motion. In this case the polarization is called elliptic.

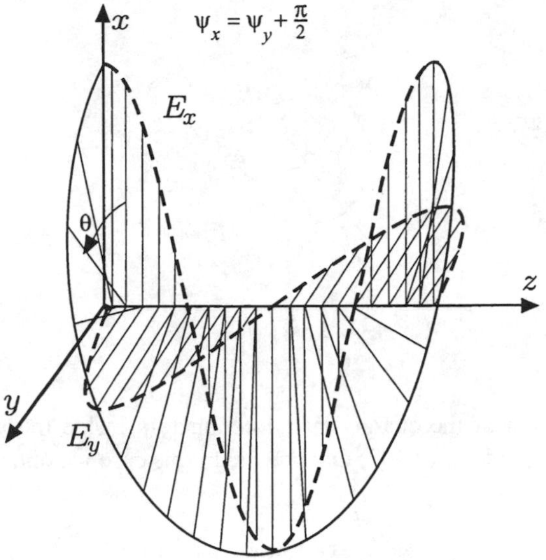

Figure 2.22: Left-hand circularly polarized plane electromagnetic wave.

2.8 THE KIRCHHOFF LAWS

The fundamental equations of network theory may be derived from the Maxwell's equations. Figure 2.23 shows a network node with n conductors. We enclose the network node in a volume V and apply the integral form (2.55) of the equation of continuity

$$\oint_{\partial V} \mathcal{J} = -\frac{\mathrm{d}}{\mathrm{d}t} \int_V \mathcal{Q} . \tag{2.105}$$

On the left side of (2.105) only the integration over the surface areas of the n conductors gives a contribution. The surface integral over ∂V may be subdivided into n integrals over the cross-section areas A_k of the conductors. Each of these integrals corresponds to a current i_k flowing from the node.

$$\oint_{\partial V} \mathcal{J} = \sum_{k=1}^{n} \int_{A_k} \mathcal{J} = \sum_{k=1}^{n} \mathrm{i}_k . \tag{2.106}$$

From (2.64), (2.105) and (2.106) we obtain

$$\sum_{k=1}^{n} i_k = -\frac{\mathrm{d}q}{\mathrm{d}t} . \tag{2.107}$$

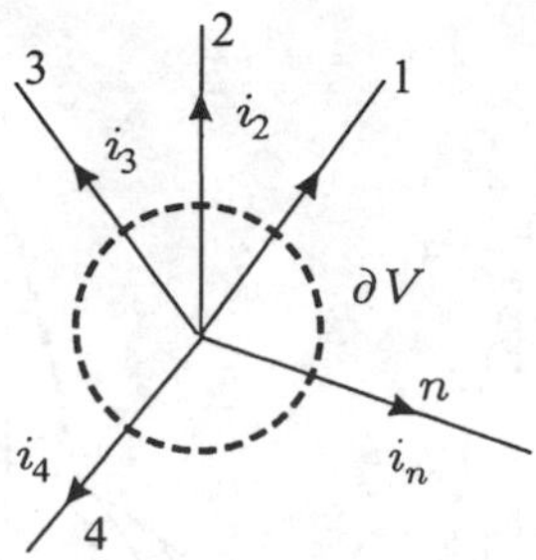

Figure 2.23: Network node.

If the time variation of the charge on the conductors inside the volume V can be neglected, the right side of (2.107) disappears. In this case we obtain

$$\sum_{k=1}^{n} i_k = 0 \quad \text{for} \quad \frac{\mathrm{d}q}{\mathrm{d}t} = 0 \,. \tag{2.108}$$

The sum of the node currents vanishes. This is the *Kirchhoff current law* . Since for a given charge q the magnitude of $\mathrm{d}q/\mathrm{d}t$ increases in proportion to the frequency, at higher frequencies the term $\mathrm{d}q/\mathrm{d}t$ may not be neglected any more. The displacement current flowing in a real network from the node, may be considered by insertion of one or several capacitors in an equivalent circuit. Via these capacitors the displacement current flowing between the conductors of the real network are modelled. Figure 2.24 shows a network in which three capacitors are inserted in order to consider the displacement current flowing between nodes a, b and c.

To derive the Kirchhoff voltage law we apply the integral form of Faraday's law (2.51) to the loop depicted in Figure 2.25. The area or virtual surface bounded by this loop is A. The line integral of the electrical field intensity $\boldsymbol{E}$ over the boundary ∂A of the loop area A is equal to the negative sum of the branch voltages:

$$\oint_{\partial A} \mathcal{E} = -v_{21} - v_{32} - v_{43} - v_{54} - v_{15} \,. \tag{2.109}$$

From (2.51), (2.63) and (2.109) we obtain

$$v_{21} + v_{32} + v_{43} + v_{54} + v_{15} = \frac{\mathrm{d}\Phi}{\mathrm{d}t} \,. \tag{2.110}$$

If the time variation of the magnetic flux through the loop can be neglected, the sum

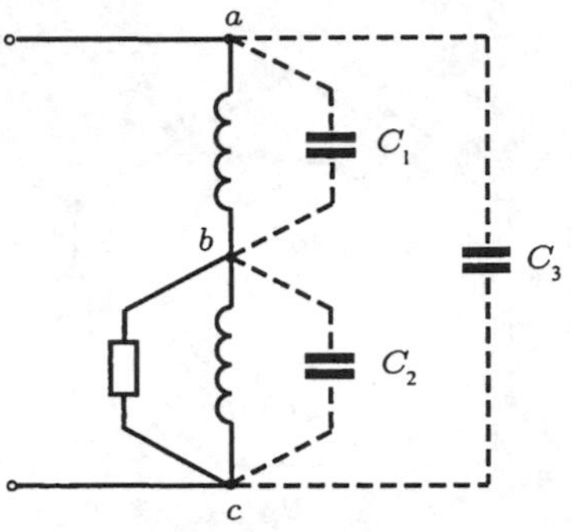

Figure 2.24: Equivalent circuit with parasitic capacitors.

of the loop voltages disappears.

$$\sum_{loop} v_{kl} = 0 \quad \text{for} \quad \frac{d\Phi}{dt} = 0\,. \tag{2.111}$$

This is the *Kirchhoff voltage law* . Since the quantity $d\Phi/\,dt$ is also increasing with frequency, at higher frequencies we can no longer neglect the magnetic flux through the loop. This magnetic flux may be considered by insertion of inductors into the loop branches. The voltage induced from one branch into another thereby is considered via a mutual inductance.

Inductors and capacitors, which do not correspond to real circuit elements, but are introduced via the geometry of the conductors, are called *parasitic capacitors* and *parasitic inductors*. At higher frequencies, these *parasitic network elements* cannot be neglected anymore. Usually such parasitic elements are disturbing. By careful geometric design of the circuit and the paths of the lines, such elements have to be kept under control. As long as only parasitic capacitors and parasitic inductors occur, these network elements may be considered within the framework of the network concept. The situation becomes more complicated as soon as the dimensions of the circuits reach the order of magnitude of the wavelength. In this case electric and magnetic fields may be directly linked with each other. In this case it is not possible to describe the influence of electric and magnetic fields independently via equivalent capacitors and inductors. However, also in these cases an equivalent circuit may be established after solving the electromagnetic field problem.

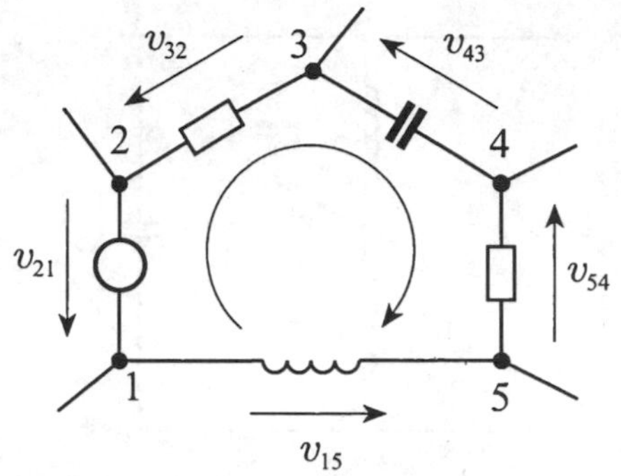

Figure 2.25: Network loop.

2.9 MAXWELL'S EQUATIONS IN LOCAL FORM

Any measurement of electromagnetic field quantities is related to an integration or averaging over a finite spatial domain. However the field concept is essentially a local concept. That means, in using the field concept we are considering the local relations between field quantities within an infinitesimally small neighborhood. In the following we will bring the Maxwell's equations into their local form.

We introduce the *exterior derivative* $d\mathcal{U}$ of an exterior differential form $\mathcal{U}$ by

$$\mathrm{d}\mathcal{U} = \sum_i \mathrm{d}x_i \wedge \frac{\partial \mathcal{U}}{\partial x_i}. \tag{2.112}$$

For the exterior differential we have to consider the following rules:

$$\mathrm{d}(\mathcal{U} + \mathcal{V}) = \mathrm{d}\mathcal{U} + \mathrm{d}\mathcal{V}, \tag{2.113}$$

$$\mathrm{d}(\mathcal{U} \wedge \mathcal{V}) = \mathrm{d}\mathcal{U} \wedge \mathcal{V} + (-1)^{(\deg \mathcal{U})}\mathcal{U} \wedge \mathrm{d}\mathcal{V}, \tag{2.114}$$

where the *degree* of the differential form $\mathcal{U}$ is $\deg \mathcal{U} = p$ if $\mathcal{U}$ is a p-form.

The *Stokes' theorem* relates the integration of a p-form $\mathcal{U}$ over the closed p-dimensional boundary ∂V of an $p+1$-dimensional volume V to the volume integral of $\mathcal{U}$ over V via

$$\oint_{\partial V} \mathcal{U} = \int_V \mathrm{d}\mathcal{U}. \tag{2.115}$$

Applying Stokes' theorem to the integral form of Maxwell's equations (2.50) to (2.53) we obtain

$$\int_A \mathrm{d}\mathcal{H} = \frac{\mathrm{d}}{\mathrm{d}t}\int_A \mathcal{D} + \int_A \mathcal{J}, \tag{2.116}$$

$$\int_A \mathrm{d}\mathcal{E} = -\frac{\mathrm{d}}{\mathrm{d}t}\int_A \mathcal{B}, \tag{2.117}$$

$$\int_V \mathrm{d}\mathcal{B} = 0, \tag{2.118}$$

$$\int_V \mathrm{d}\mathcal{D} = \int_V \mathcal{Q} \tag{2.119}$$

and from this the differential representation of Maxwell's equations:

$$\mathrm{d}\mathcal{H} = \frac{\mathrm{d}}{\mathrm{d}t}\mathcal{D} + \mathcal{J}, \quad \text{Ampère's law} \tag{2.120}$$

$$\mathrm{d}\mathcal{E} = -\frac{\mathrm{d}}{\mathrm{d}t}\mathcal{B}, \quad \text{Faraday's law} \tag{2.121}$$

$$\mathrm{d}\mathcal{B} = 0, \quad \text{Magnetic flux continuity} \tag{2.122}$$

$$d\mathcal{D} = \mathcal{Q}. \quad \text{Gauss' law} \tag{2.123}$$

Applying the Stokes' theorem (2.115) to the integral form of the continuity equation (2.55) yields the *differential form* of the *continuity equation*

$$\frac{\mathrm{d}}{\mathrm{d}t}\mathcal{Q} + \mathrm{d}\mathcal{J} = 0. \tag{2.124}$$

2.10 TIME-HARMONIC ELECTROMAGNETIC FIELDS

As far as the considered quantities exhibit a sinusoidal time dependence, a representation of these quantities by complex *phasors* is useful [13]. The quantity

$$\boldsymbol{E}(\boldsymbol{x}, t) = \boldsymbol{E}_0(\boldsymbol{x}) \cos(\omega t + \varphi(\boldsymbol{x})) \tag{2.125}$$

can be represented in the following way

$$\boldsymbol{E}(\boldsymbol{x}, t) = \Re\left\{\underline{\boldsymbol{E}}(\boldsymbol{x})\, \mathrm{e}^{\mathrm{j}\omega t}\right\}. \tag{2.126}$$

We mark phasors by underlining. The complex phasor $\underline{\boldsymbol{E}}(\boldsymbol{x})$ is given by

$$\underline{\boldsymbol{E}}(\boldsymbol{x}) = \boldsymbol{E}_0(\boldsymbol{x})\, \mathrm{e}^{\mathrm{j}\varphi(\boldsymbol{x})}. \tag{2.127}$$

Table 2.3: Electric Material Properties

Material	$\sigma/[\mathrm{Sm}^{-1}]$	ε_r'	$f_c/[\mathrm{Hz}]$
Copper	$5.8 \cdot 10^7$	1	–
Germanium (pure)	2.2	16	$2.5 \cdot 10^9$
Sea water	4	81	$8.9 \cdot 10^8$
Water	10^{-3}	81	$2.2 \cdot 10^5$
Wet earth	10^{-3}	10	$1.8 \cdot 10^6$
Dry earth	10^{-5}	5	$3.6 \cdot 10^4$

Due to

$$\frac{\partial}{\partial t}\boldsymbol{E}(\boldsymbol{x},t) = \Re\left\{\mathrm{j}\omega\underline{\boldsymbol{E}}(\boldsymbol{x})\,\mathrm{e}^{\mathrm{j}\omega t}\right\} \tag{2.128}$$

the partial derivation with respect to time may be replaced by a multiplication with the factor $\mathrm{j}\omega$.

We introduce the complex phasor $\underline{\mathcal{E}}$ of the differential form $\mathcal{E}$. For a time-harmonic field described by a differential form

$$\mathcal{E}(\boldsymbol{x},t) = \mathcal{E}_0\cos(\omega t + \varphi(\boldsymbol{x})) \tag{2.129}$$

we introduce

$$\underline{\mathcal{E}}(\boldsymbol{x},t) = \mathcal{E}_0\,\mathrm{e}^{\mathrm{j}\varphi(\boldsymbol{x})} \tag{2.130}$$

and obtain

$$\mathcal{E}(\boldsymbol{x},t) = \Re\left\{\underline{\mathcal{E}}(\boldsymbol{x})\,\mathrm{e}^{\mathrm{j}\omega t}\right\}, \tag{2.131}$$

$$\frac{\partial}{\partial t}\mathcal{E}(\boldsymbol{x},t) = \Re\left\{\mathrm{j}\omega\,\underline{\mathcal{E}}(\boldsymbol{x})\,\mathrm{e}^{\mathrm{j}\omega t}\right\}. \tag{2.132}$$

2.11 MAXWELL'S EQUATIONS IN THE FREQUENCY DOMAIN

Replacing in (2.116) – (2.119) the time derivative by multiplication with $\mathrm{j}\omega$ we obtain the complex *phasor representation* of the Maxwell's equations [14]

$$\mathrm{d}\underline{\mathcal{H}} = \mathrm{j}\omega\underline{\mathcal{D}} + \underline{\mathcal{J}}, \qquad \text{Ampère's Law} \tag{2.133}$$

$$\mathrm{d}\underline{\mathcal{E}} = -\mathrm{j}\omega\underline{\mathcal{B}}, \qquad \text{Faraday's law} \tag{2.134}$$

$$\mathrm{d}\underline{\mathcal{B}} = 0, \qquad \text{Magnetic flux continuity} \tag{2.135}$$

$$\mathrm{d}\underline{\mathcal{D}} = \underline{\mathcal{Q}}. \qquad \text{Gauss' law} \tag{2.136}$$

Table 2.4: Complex Permittivity of some Materials at 25° C

Material		10^2	*Frequency* 10^4	f/[Hz] 10^6	10^8	10^{10}
Teflon	ε_r'	2.1	2.1	2.1	2.1	2.0
	ε_r''	$1.1{\cdot}10^{-3}$	$7{\cdot}10^{-4}$	$4{\cdot}10^{-4}$	$3{\cdot}10^{-4}$	$8{\cdot}10^{-4}$
Polystyrole	ε_r'	2.56	2.56	2.56	2.56	2.54
	ε_r''	$1.3{\cdot}10^{-4}$	$1.3{\cdot}10^{-4}$	$1.8{\cdot}10^{-4}$	$3{\cdot}10^{-4}$	$1.1{\cdot}10^{-3}$
Quartz	ε_r'	3.78	3.78	3.78	3.78	3.78
	ε_r''	$3.2{\cdot}10^{-3}$	$2.3{\cdot}10^{-3}$	$7.5{\cdot}10^{-4}$	$4{\cdot}10^{-4}$	$4{\cdot}10^{-4}$

In a homogeneous isotropic medium with the permittivity ε, the permeability μ and the conductivity σ, we obtain with (3.34), (3.35) and (2.58)

$$\mathrm{d}\underline{\mathcal{H}} = (\mathrm{j}\omega\varepsilon + \sigma) \star \underline{\mathcal{E}} + \underline{\mathcal{J}}_0\,, \tag{2.137}$$

$$\mathrm{d}\underline{\mathcal{E}} = -\mathrm{j}\omega\mu \star \underline{\mathcal{H}}\,. \tag{2.138}$$

For insulating materials usually $\omega\varepsilon \gg \sigma$ is valid. We may consider the losses in an insulator via complex permittivity $\underline{\varepsilon}$.

$$\underline{\varepsilon} = \varepsilon' + \frac{\sigma}{\mathrm{j}\omega} = \varepsilon' - j\varepsilon''. \tag{2.139}$$

If the losses in a dielectric material are due to ohmic conduction, σ will be independent from ω, and ε'' will be proportional to $1/\omega$. If the losses have other origins, e.g., polarization losses in a dielectric material, ε'' is not proportional to $1/\omega$. Sometimes also magnitude $|\underline{\varepsilon}|$ and phase δ_e of the complex permittivity are specified.

$$\underline{\varepsilon} = |\underline{\varepsilon}|\,\mathrm{e}^{-\mathrm{j}\delta_e} = \varepsilon'(1 - \mathrm{j}\tan\delta_e). \tag{2.140}$$

The phase δ_e is named the *dielectric loss angle*, and $\tan\delta_e$ is the *dielectric loss factor*. For small losses we obtain $|\underline{\varepsilon}| \cong \varepsilon'$ and $\tan\delta_e \cong \delta_e$. Magnetic losses may be described by a *complex permeability* $\underline{\mu}$.

$$\underline{\mu} = \mu' - \mathrm{j}\mu'' = |\underline{\mu}|\,\mathrm{e}^{-\mathrm{j}\delta_m} = \mu'(1 - \mathrm{j}\tan\delta_m). \tag{2.141}$$

The phase δ_m is the *magnetic loss angle*, and $\tan\delta_m$ the *magnetic loss factor*.

The *relative permittivity* is the ratio of the permittivity to the free-space dielectric constant.

$$\underline{\varepsilon}_r = \varepsilon_r' - \mathrm{j}\varepsilon_r'' = \frac{\underline{\varepsilon}}{\varepsilon_0}. \tag{2.142}$$

In the same way, the *relative permeability* μ_r is given by

$$\underline{\mu}_r = \mu_r' - \mathrm{j}\mu_r'' = \frac{\underline{\mu}}{\mu_0}. \tag{2.143}$$

With the complex permittivity $\underline{\varepsilon}$ and the complex permeability $\underline{\mu}$, we obtain from (2.137) and (2.138)

$$\mathrm{d}\underline{\mathcal{H}} = \mathrm{j}\omega\underline{\varepsilon} \star \underline{\mathcal{E}} + \underline{\mathcal{J}}_0, \tag{2.144}$$

$$\mathrm{d}\underline{\mathcal{E}} = -\mathrm{j}\omega\underline{\mu} \star \underline{\mathcal{H}}. \tag{2.145}$$

For non-magnetic materials $\underline{\mu} = \mu_0$ is valid. A material for which, in the considered frequency domain, $\varepsilon' \gg \varepsilon''$ is valid is named *quasi-dielectric*, for $\varepsilon' \ll \varepsilon''$ the material is called a *quasi-conductor*. The same material may be a quasi-conductor at low frequencies, and a quasi-dielectric at higher frequencies. The *material cut-off frequency* f_c between these regions is given by

$$f_c = \frac{\sigma}{2\pi\varepsilon'}. \tag{2.146}$$

In Table 2.3, ε_r', σ and ω_c are given for some materials. For metals ω_c is far beyond the optical frequency region. Table 2.4 specifies ε_r' for some important dielectric materials independent from the frequency.

2.12 CURVILINEAR COORDINATES

It is one principal advantage of vector calculus and exterior calculus that the equations defining and describing the properties may be formulated without reference to a specific coordinate system. Depending on the problem the choice of a specific coordinate system may simplify the problem solution considerably. A detailed treatment of curvilinear coordinates is given in Appendix A.4.

We introduce an *orthogonal curvilinear coordinate system*

$$u = u(x, y, z), \quad v = v(x, y, z), \quad w = w(x, y, z). \tag{2.147}$$

The *coordinate curves* are obtained by setting two of the three coordinates u, v and w constant. *Coordinate surfaces* are defined by setting one of the three coordinates constant. In an orthogonal coordinate system in any point (except singular points), of the space the three coordinate curves are orthogonal. The same holds for the three coordinate surfaces going through any point. The differentials $\mathrm{d}x$, $\mathrm{d}y$, $\mathrm{d}z$ by the differentials $\mathrm{d}u$, $\mathrm{d}v$, $\mathrm{d}w$ are related to

$$\mathrm{d}x = \frac{\partial x}{\partial u}\mathrm{d}u + \frac{\partial x}{\partial v}\mathrm{d}v + \frac{\partial x}{\partial w}\mathrm{d}w\,, \tag{2.148a}$$

$$\mathrm{d}y = \frac{\partial y}{\partial u}\mathrm{d}u + \frac{\partial y}{\partial v}\mathrm{d}v + \frac{\partial y}{\partial w}\mathrm{d}w\,, \tag{2.148b}$$

$$\mathrm{d}z = \frac{\partial z}{\partial u}\mathrm{d}u + \frac{\partial z}{\partial v}\mathrm{d}v + \frac{\partial z}{\partial w}\mathrm{d}w\,. \tag{2.148c}$$

The rules for transformation of the Cartesian *basis two-forms* $\mathrm{d}x \wedge \mathrm{d}y$, $\mathrm{d}y \wedge \mathrm{d}z$, $\mathrm{d}z \wedge \mathrm{d}x$ and the Cartesian *basis three-form* $\mathrm{d}x \wedge \mathrm{d}y \wedge \mathrm{d}z$ follow directly from the above equations by applying the rules of the exterior product and are given explicitly in (A.95) and (A.96). Using the *metric coefficients* g_1, g_2 and g_3, defined in (A.95) *unit one-forms* (A.107)

$$s_1 = g_1\,\mathrm{d}u\,, \quad s_2 = g_2\,\mathrm{d}v\,, \quad s_3 = g_3\,dw \tag{2.149}$$

are introduced. The integral of $s_1 = g_1\,\mathrm{d}u$ along any path with v and w constant yields the length of the path.

In a *circular cylindric coordinate system*, defined by (A.131) the unit differential forms (A.134) are

$$s_1 = \mathrm{d}r\,, \quad s_2 = r\,\mathrm{d}\varphi\,, \quad s_3 = \mathrm{d}z\,. \tag{2.150}$$

In a *spherical coordinate system*, defined by (A.148) the unit differential forms (A.151) are

$$s_1 = \mathrm{d}r\,, \quad s_2 = r\,\mathrm{d}\vartheta\,, \quad s_3 = r\sin\vartheta\,\mathrm{d}\varphi\,. \tag{2.151}$$

For the curvilinear unit differentials the *Hodge operator* as defined in (2.32) is

$$\begin{aligned}
\star f &= f\,s_1 \wedge s_2 \wedge s_3\,,\\
\star(A_u s_1 + A_v s_2 + A_w s_3) &= A_u s_2 \wedge s_3 + A_v s_3 \wedge s_1 + A_w s_1 \wedge s_2\,,\\
\star(A_u s_2 \wedge s_3 + A_v s_3 \wedge s_1 + A_w s_1 \wedge s_2) &= A_u s_1 + A_v s_2 + A_w s_3\,,\\
\star(f\,s_1 \wedge s_2 \wedge s_3) &= f\,.
\end{aligned} \tag{2.152}$$

2.13 BOUNDARY CONDITIONS

Usually we are considering electromagnetic structures assembled from various materials with different material properties. At a boundary surface between two materials

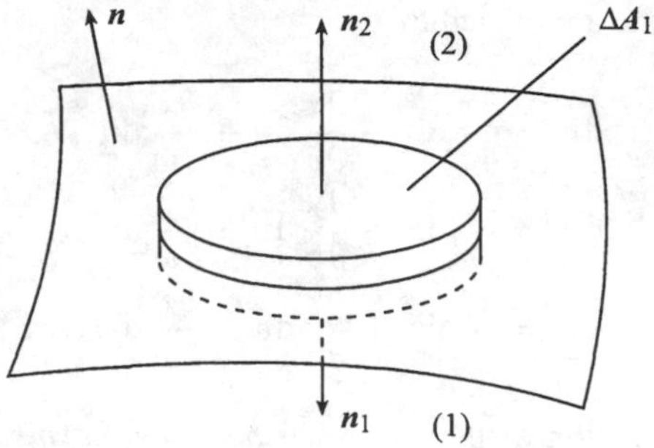

Figure 2.26: The normal boundary conditions.

the material parameters undergo a discontinuous change. At the boundary surfaces the field quantities have to fulfill *boundary conditions*. If the materials are homogeneous within same spatial domain, we can try to find solutions of the Maxwell's equations in the following way: We seek the solutions in each subdomain and fit these solutions along the boundaries. In the following we show that at boundary surfaces the tangential components of the field intensities and the normal components of the flux densities and the current density fulfill boundary conditions. Figure 2.26 shows a section of the boundary surface between spatial regions 1 and 2. We are now considering an area element ΔA_1 of this boundary surface. We introduce a local orthonormal right-handed curvilinear coordinate system u, v, n with the coordinates u and v tangential to the boundary surface and the coordinate n normal to the boundary surface. In this coordinate system the magnetic flux density differential form is given by

$$\mathcal{B} = B_u \mathsf{s}_2 \wedge \mathsf{n} + B_v \mathsf{n} \wedge \mathsf{s}_1 + B_n \mathsf{s}_1 \wedge \mathsf{s}_2 \,. \tag{2.153}$$

We construct a small volume ΔV that is formed as a small circular cylinder like a pillbox. The bottom surface of the cylinder is located in region 1, whereas the top surface of the cylinder is located in region 2. The pillbox encloses a volume ΔV. We now apply the magnetic flux continuity law (2.52) to the cylinder volume ΔV. In a limiting process $\Delta V \to 0$, the side surface of the pillbox will go to zero by a higher order than the bottom surface as well as the top surface and may be neglected. We therefore obtain

$$\oint_{\partial(\Delta V)} \mathcal{B} = \int_{\Delta A_1} \left(\mathcal{B}^{(2)} - \mathcal{B}^{(1)} \right) = \int_{\Delta A_1} \left(B_n^{(2)} - B_n^{(1)} \right) \mathsf{s}_1 \wedge \mathsf{s}_2 = 0 \,. \tag{2.154}$$

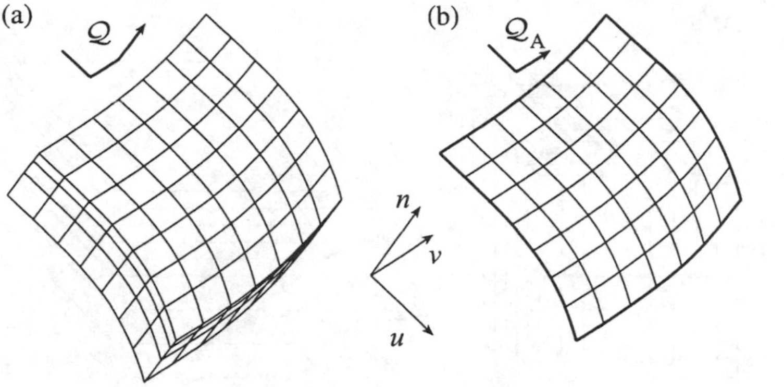

Figure 2.27: (a) Charge form $\mathcal{Q}$ and (b) surface charge form$\mathcal{Q}_A$.

$\mathcal{B}^{(1)}$ and $\mathcal{B}^{(2)}$ are the magnetic flux density differential forms in region 1 and region 2 respectively. The above condition is fulfilled if and only if

$$n \wedge \left(\mathcal{B}^{(2)} - \mathcal{B}^{(1)}\right) = 0 . \tag{2.155}$$

This is the boundary condition for the magnetic flux density.

Let us now compute the surface integral of the electric flux density over the boundary of the pillbox. Performing the same limiting process as above, we obtain

$$\oint_{\partial(\Delta V)} \mathcal{D} = \int_{\Delta A} \left(\mathcal{D}^{(2)} - \mathcal{D}^{(1)}\right) = \int_{\Delta A} \left(D_n^{(2)} - D_n^{(1)}\right) s_1 \wedge s_2 . \tag{2.156}$$

We introduce a *surface charge density* ρ_A (dimension [As/m^2]). This means, the infinitely thin boundary surface contains a finite charge. We can describe the surface charge density by the *surface charge differential form*

$$\mathcal{Q}_A = \rho_A \, s_1 \wedge s_2 . \tag{2.157}$$

The surface charge density is related to the volume charge density via

$$\mathcal{Q} = \delta(n) \, n \wedge \mathcal{Q}_A , \tag{2.158}$$

where n is the coordinate normal to the area and $\delta(n)$ is the *Dirac delta distribution* defined by

$$\int_{x_1}^{x_2} \delta(x) \, \mathrm{d}x = \begin{cases} 1 & \text{for } x \in [x_1, x_2] \\ 0 & \text{for } x \notin [x_1, x_2] \end{cases} . \tag{2.159}$$

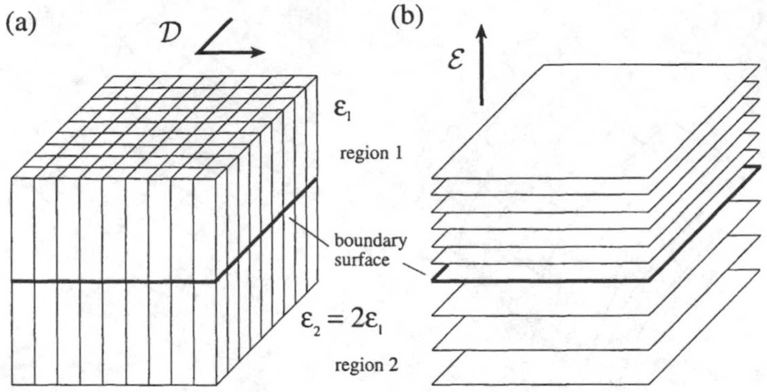

Figure 2.28: The normal boundary conditions: (a) dielectric flux density and (b) electric field.

Let A be an area on the boundary surface and V be a volume supported by A and exhibiting an extension normal to the surface from $n = -\Delta n/2$ to $n = +\Delta n/2$. In this case

$$\int_V \mathcal{Q} = \int_A \mathcal{Q}_A \tag{2.160}$$

is valid. Figure 2.27(a) shows the geometric picture of a charge form $\mathcal{Q}$ and Figure 2.27(b) depicts the corresponding surface charge form $\mathcal{Q}_A$. Inserting this in Gauss' law (2.53) yields

$$\oint_{\partial V} \mathcal{D} = \int_A \mathcal{Q}_A\,, \tag{2.161}$$

and

$$\int_{\Delta A} \left(D_n^{(2)} - D_n^{(1)}\right)\, s_1 \wedge s_2 = \int_{\Delta A} \rho_A\, s_1 \wedge s_2\,. \tag{2.162}$$

This yields the boundary condition for the electric flux density.

$$n \wedge \left(\mathcal{D}^{(2)} - \mathcal{D}^{(1)}\right) = n \wedge \mathcal{Q}_A\,. \tag{2.163}$$

Equations (2.155) and (2.163) are called the *normal boundary conditions*, since they give information about the normal components of the flux densities $\boldsymbol{B}$ and $\boldsymbol{D}$.

Figure 2.28(a) shows the dielectric flux through the boundary for a homogeneous electric field normal to the boundaries and no surface charge in the boundary. The permittivity ε_2 in region 2 is assumed to be twice the permittivity ε_1 of region 1. The cross-section of the flux tubes remains unchanged when the flux tubes are crossing the boundary surface. For same flux densities in both regions we obtain in region 1 twice the electric field intensity as in region 2. Therefore the potential planes in region 1 have twice the density as in region 2, see Figure 2.28(b).

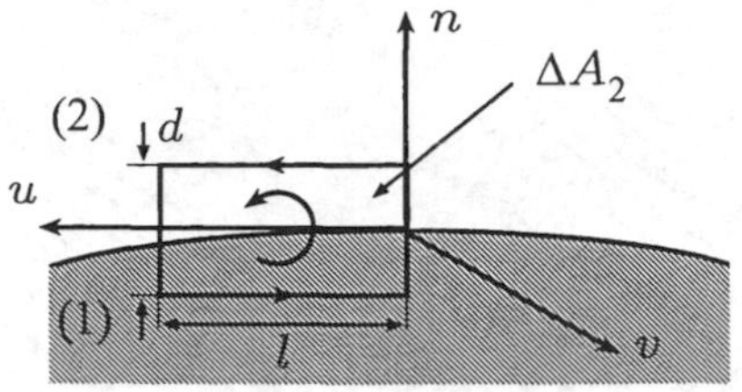

Figure 2.29: The tangential boundary conditions.

We now investigate the *tangential boundary conditions*. Figure 2.29 shows a normal cut through the boundary surface. In our local coordinate system the magnetic field differential form $\mathcal{H}$ is given by

$$\mathcal{H} = H_u s_1 + H_v s_2 + H_n n \,. \tag{2.164}$$

The rectangular area element ΔA_2 has the side lengths d and l, where the longer side l is in parallel to u. One of these longer sides is totally embedded in region 1, and the other one on region 2. Computing the circulation integral of the magnetic field intensity over the contour ΔA_2, we obtain after for $d \to 0$:

$$\oint_{\partial \Delta A_2} \mathcal{H} = \int_{\Delta l} \left(\mathcal{H}^{(2)} - \mathcal{H}^{(1)} \right) . \tag{2.165}$$

$\mathcal{H}^{(1)}$ and $\mathcal{H}^{(2)}$ are the magnetic field intensities in regions 1 and 2 respectively. Let us also assume a current flowing on the infinitely thin boundary layer. The current density exhibits only components tangential to the boundary layer. The current density differential form $\mathcal{J}$ therefore is given by

$$\mathcal{J} = J_u s_2 \wedge n + J_v n \wedge s_1 \,. \tag{2.166}$$

If the current is concentrated in the boundary layer the dependence in the normal direction n is given by the delta distribution $\delta(n)$ and we obtain

$$\mathcal{J} = -J_{Au}(u, v)\delta(n) n \wedge s_2 + J_{Av}(u, v)\delta(n) n \wedge s_1 \,. \tag{2.167}$$

We introduce the *surface current density*

$$\mathcal{J}_A(u, v) = J_{Av}(u, v) s_1 - J_{Au}(u, v) s_2 \,. \tag{2.168}$$

The surface current density form $\mathcal{J}_A$ and the current density form $\mathcal{J}$ are related via

$$\mathcal{J}(u, v, n) = \delta(n) n \wedge \mathcal{J}_A(u, v) \,. \tag{2.169}$$

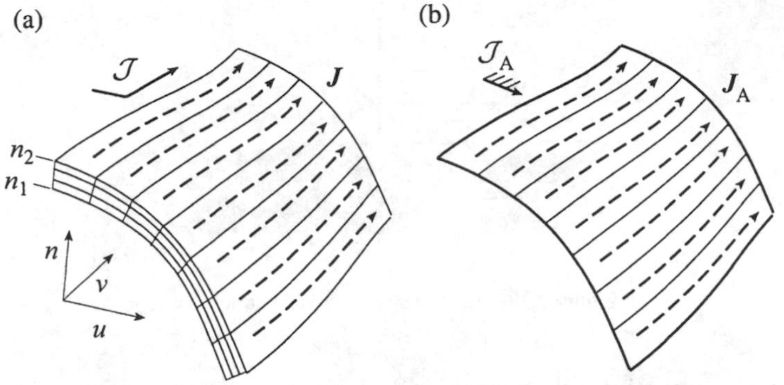

Figure 2.30: (a) Current form $\mathcal{J}$ and (b) surface current form $\mathcal{J}_A$.

The surface current density form $\mathcal{J}_A$ describes a finite current flowing in an infinitely thin surface. Figure 2.30 shows the geometric representation of the current density and the surface current density. The surface current density form $\mathcal{J}_A$ is a one-form. However, different from one-forms discussed up to now, the direction of integration is not same as the direction of the field but orthogonal to the corresponding field direction. We call such a one-form a *twisted one-form*. As a symbol for a twisted form we introduce a line with an arrow indicating the direction of integration with thin lines to the side of the arrow indicating the direction of the surface flux flow as depicted in Figure 2.30(b).

To transform an ordinary one-form into a twisted one-form and vice versa we introduce the *twist operator* by

$$\perp_n \mathcal{U} = \star (n \wedge \mathcal{U}) \ . \tag{2.170}$$

The index n of the twist operator $\perp_n$ denotes the axis of rotation. If u, v are the coordinates tangential to the surface and n is the coordinate normal to the surface, the application of the twist operator $\perp_n$ to a one-form tangential to the surface rotates the one-form around n by 90° in positive direction,

$$\perp_n (U_u s_1 + U_v s_2) = -U_v s_1 + U_u s_2 \, , \tag{2.171a}$$

$$\perp_n (U_v s_1 - U_u s_2) = \ \ U_u s_1 + U_v s_2 \, . \tag{2.171b}$$

If a current is flowing tangential to a surface, given by n = const., and if the current is flowing within a small interval (n_1, n_2), as depicted in Figure 2.30(a) we can approximate the current distribution by a surface current distribution shown in Figure 2.30(b). The surface current twisted one-form $\mathcal{J}_A$ is obtained by integrating the current density two-form over the normal coordinate n from n_1 to n_2. To integrate a differential form over a single coordinate we first bring the corresponding coordinate differential to the left. This follows from the circumstance that the integration is the inverse operation to the exterior derivative, and the exterior derivation attaches a coordinate differential to the left of a differential form.

$$\mathcal{J}_A(u, v) = \int_{n_1}^{n_2} \mathcal{J}(u, v, n)\,. \tag{2.172}$$

Inserting (2.165) and (2.169) in Ampère's law (2.50) gives

$$\int_{\partial A_2} \mathcal{H} = \int_{\Delta l} \left(\mathcal{H}^{(2)} - \mathcal{H}^{(1)}\right) = \int_{\Delta l} \mathcal{J}_A\,. \tag{2.173}$$

This yields the boundary condition for the magnetic field

$$n \wedge \left(\mathcal{H}^{(2)} - \mathcal{H}^{(1)}\right) = n \wedge \mathcal{J}_A\,. \tag{2.174}$$

$\mathcal{J}_A$ is the sum of the impressed surface current density and the surface current density induced by the field. A field-induced surface current density only may occur, if we assume infinite conductivity of the medium. In case the media in both subspaces exhibit finite conductivity and if no surface current density is impressed, we obtain

$$n \wedge \left(\mathcal{H}^{(2)} - \mathcal{H}^{(1)}\right) = 0 \quad \text{for} \quad \mathcal{J}_\mathrm{A} = 0\,. \tag{2.175}$$

For an *electric surface polarization* $\mathcal{M}_{eA}$ in the boundary surface related to $\mathcal{J}_A$ via

$$\mathcal{J}_A = \frac{\partial}{\partial t} \mathcal{M}_{eA} \tag{2.176}$$

we obtain from (2.174) the *tangential boundary condition for the magnetic field intensity*

$$n \wedge \left(\mathcal{H}^{(2)} - \mathcal{H}^{(1)}\right) = n \wedge \frac{\partial}{\partial t} \mathcal{M}_{eA}\,. \tag{2.177}$$

In the same way we obtain the *tangential boundary condition for the electric field intensity*:

$$n \wedge \left(\mathcal{E}^{(2)} - \mathcal{E}^{(1)}\right) = -n \wedge \frac{\partial}{\partial t} \mathcal{M}_{mA}\,. \tag{2.178}$$

In this equation $\mathcal{M}_{mA}$ is the *magnetic surface polarization* in the boundary surface. If there is no surface polarization in the boundary surface, we obtain

$$n \wedge \left(\mathcal{E}^{(2)} - \mathcal{E}^{(1)}\right) = 0\,. \tag{2.179}$$

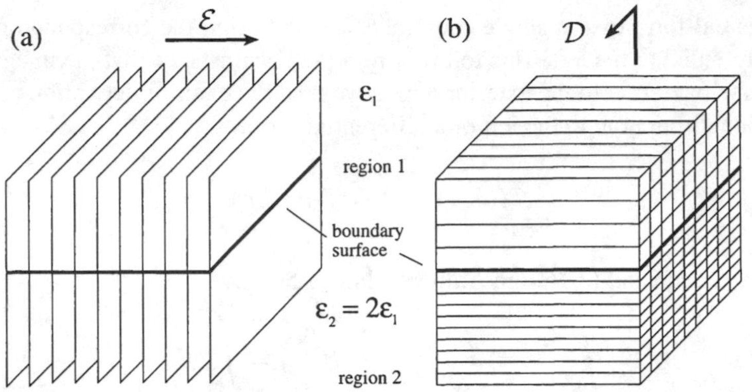

Figure 2.31: The tangential boundary conditions: (a) electric field and (b) dielectric flux density.

In Figure 2.31 we consider an electric field parallel to the boundary for no magnetic surface polarization in the boundary. The electric potential surfaces are normal to the boundary plane and are not changing their spacing when crossing the boundary. Assuming $\varepsilon_2 = 2\varepsilon_1$ yields in region 2 twice the density of dielectric flux tubes as in region 1.

For a field in an arbitrary direction with respect to the boundary the spacing of the potential planes remains unchanged in directions parallel to the boundary surface whereas the spacing normal to the boundary surface becomes smaller in the region with the lower permittivity, see Figure 2.32(a), indicating that the tangential component of the electric field remains unchanged, whereas the normal component of the electric field intensity is larger in the region with the lower permittivity. Figure 2.32(b) shows the change tilt in the flux tubes when crossing the boundary surface. The area of the cuts of the flux tubes with surfaces parallel to the boundary remains unchanged. This indicates that the flux component normal to the boundary is maintained. The area of the cuts of the flux tubes with planes normal to the boundary surface is proportional to the permittivity.

Introducing the *contraction* $\mathcal{A} \lrcorner \mathcal{B}$ of two differential forms $\mathcal{A}$ and $\mathcal{B}$ defined by

$$s_i \lrcorner s_j = \delta_{ij} \tag{2.180}$$

$$\mathcal{A} \lrcorner (\mathcal{B} \wedge \mathcal{C}) = (\mathcal{A} \lrcorner \mathcal{B}) \wedge \mathcal{C} + (-1)^{deg(\mathcal{A})} \mathcal{B} \wedge (\mathcal{A} \lrcorner \mathcal{C}) \tag{2.181}$$

we can bring the boundary conditions in an explicit form with respect to the sources impressed in the boundaries. The symbol $\lrcorner$ is named "angle" and the contraction also is called the *angle product*. The angle product was introduced by Burke [7]. We use the modified form given by Warnick [15]. It may be shown easily that the following

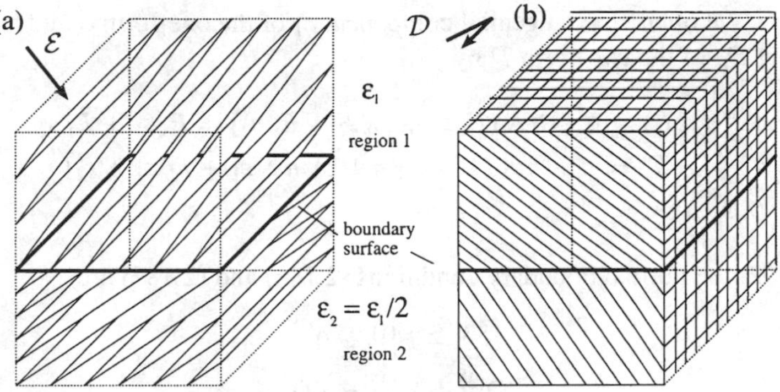

Figure 2.32: Boundary conditions for field in arbitrary direction, (a) electric field and (b) dielectric flux density.

relations hold

$$\mathcal{Q}_A = n \lrcorner (n \wedge \mathcal{Q}_A)\,, \tag{2.182}$$
$$\mathcal{J}_A = n \lrcorner (n \wedge \mathcal{J}_A)\,, \tag{2.183}$$
$$\mathcal{M}_{eA} = n \lrcorner (n \wedge \mathcal{M}_{eA})\,, \tag{2.184}$$
$$\mathcal{M}_{mA} = n \lrcorner (n \wedge \mathcal{M}_{mA})\,. \tag{2.185}$$

The normal boundary conditions (2.155) and (2.163) now can be written as

$$n \lrcorner \left[n \wedge \left(\mathcal{B}^{(2)} - \mathcal{B}^{(1)} \right) \right] = 0\,, \tag{2.186}$$
$$n \lrcorner \left[n \wedge \left(\mathcal{D}^{(2)} - \mathcal{D}^{(1)} \right) \right] = \mathcal{Q}_A\,, \tag{2.187}$$

and the tangential boundary conditions (2.174), (2.177) and (2.178) are

$$n \lrcorner \left[n \wedge \left(\mathcal{H}^{(2)} - \mathcal{H}^{(1)} \right) \right] = \mathcal{J}_A\,, \tag{2.188}$$
$$n \lrcorner \left[n \wedge \left(\mathcal{H}^{(2)} - \mathcal{H}^{(1)} \right) \right] = \frac{\partial}{\partial t} \mathcal{M}_{eA}\,, \tag{2.189}$$
$$n \lrcorner \left[n \wedge \left(\mathcal{E}^{(2)} - \mathcal{E}^{(1)} \right) \right] = -\frac{\partial}{\partial t} \mathcal{M}_{mA}\,. \tag{2.190}$$

In the above notation we have brought tangential and normal boundary conditions to a unified form. We note that the expression $n \lrcorner (n \wedge \mathcal{U})$ filters the tangential component of $\mathcal{U}$ if $\mathcal{U}$ is a one-form, and same operation filters the normal component of $\mathcal{U}$ if $\mathcal{U}$ is a two-form. We obtain the tangential component $\mathcal{E}_t$ of the one-form $\mathcal{E}$ and the normal component $\mathcal{D}_n$ of the two-form $\mathcal{D}$ by

$$\mathcal{E}_t = n \lrcorner (n \wedge \mathcal{E}) = n \lrcorner [n \wedge (E_u s_1 + E_v s_2 + E_n n)] = E_u s_1 + E_v s_2 , \tag{2.191}$$

$$\begin{aligned} \mathcal{D}_n &= n \lrcorner (n \wedge \mathcal{D}) = n \lrcorner [n \wedge (D_u s_2 \wedge n + D_v n \wedge s_1 + D_n s_1 \wedge s_2] \\ &= D_n s_1 \wedge s_2 . \end{aligned} \tag{2.192}$$

With (2.192) the normal boundary conditions (2.186) and (2.187) are

$$\mathcal{B}_n^{(2)} - \mathcal{B}_n^{(1)} = 0 , \tag{2.193}$$

$$\mathcal{D}_n^{(2)} - \mathcal{D}_n^{(1)} = \mathcal{Q}_A , \tag{2.194}$$

and with (2.192) the tangential boundary conditions (2.174), (2.177) and (2.178) are

$$\mathcal{H}_t^{(2)} - \mathcal{H}_t^{(1)} = \mathcal{J}_A , \tag{2.195}$$

$$\mathcal{H}_t^{(2)} - \mathcal{H}_t^{(1)} = \frac{\partial}{\partial t} \mathcal{M}_{eA} , \tag{2.196}$$

$$\mathcal{E}_t^{(2)} - \mathcal{E}_t^{(1)} = -\frac{\partial}{\partial t} \mathcal{M}_{mA} . \tag{2.197}$$

For complex phasors the normal boundary conditions corresponding to (2.186) and (2.187) are

$$n \lrcorner \left[n \wedge \left(\underline{\mathcal{B}}^{(2)} - \underline{\mathcal{B}}^{(1)} \right) \right] = 0 , \tag{2.198}$$

$$n \lrcorner \left[n \wedge \left(\underline{\mathcal{D}}^{(2)} - \underline{\mathcal{D}}^{(1)} \right) \right] = \underline{\mathcal{Q}}_A . \tag{2.199}$$

The tangential boundary conditions corresponding to (2.188), (2.189) and (2.190) are

$$n \lrcorner \left[n \wedge \left(\underline{\mathcal{H}}^{(2)} - \underline{\mathcal{H}}^{(1)} \right) \right] = \underline{\mathcal{J}}_A , \tag{2.200}$$

$$n \lrcorner \left[n \wedge \left(\underline{\mathcal{H}}^{(2)} - \underline{\mathcal{H}}^{(1)} \right) \right] = \mathrm{j}\omega \, \underline{\mathcal{M}}_{eA} , \tag{2.201}$$

$$n \lrcorner \left[n \wedge \left(\underline{\mathcal{E}}^{(2)} - \underline{\mathcal{E}}^{(1)} \right) \right] = -\mathrm{j}\omega \, \underline{\mathcal{M}}_{mA} . \tag{2.202}$$

Introducing the normal components $\underline{\mathcal{B}}_n$ and $\underline{\mathcal{D}}_n$ of the flux densities via (2.192) tangential components of the field intensities and the tangential components $\underline{\mathcal{H}}_t$ and $\underline{\mathcal{E}}_t$ of the field intensities via (2.191) yields for the normal boundary conditions

$$\underline{\mathcal{B}}_n^{(2)} - \underline{\mathcal{B}}_n^{(1)} = 0 , \tag{2.203}$$

$$\underline{\mathcal{D}}_n^{(2)} - \underline{\mathcal{D}}_n^{(1)} = \underline{\mathcal{Q}}_A \tag{2.204}$$

and for the tangential boundary conditions

$$\underline{\mathcal{H}}_t^{(2)} - \underline{\mathcal{H}}_t^{(1)} = \underline{\mathcal{J}}_A\,, \tag{2.205}$$

$$\underline{\mathcal{H}}_t^{(2)} - \underline{\mathcal{H}}_t^{(1)} = \mathrm{j}\omega\,\underline{\mathcal{M}}_{eA}\,, \tag{2.206}$$

$$\underline{\mathcal{E}}_t^{(2)} - \underline{\mathcal{E}}_t^{(1)} = -\mathrm{j}\omega\,\underline{\mathcal{M}}_{mA}\,. \tag{2.207}$$

REFERENCES

[1] R. S. Elliott, *Electromagnetics - History, Theory, and Applications*. New York: IEEE Press, 1991.

[2] J. D. Jackson, *Classical Electrodynamics*. New York: John Wiley & Sons, 1975.

[3] H. A. Haus and J. R. Melcher, *Electromagnetic Fields and Energy*. Englewood Cliffs, NJ: Prentice Hall, 1989.

[4] M. N. Sadiku, *Elements of Electromagnetics*. Orlando: Saunders, 1989.

[5] Z. Popović and B. D. Popović, *Introductory Electromagnetics*. Englewood Cliffs, NJ: Prentice Hall, 2000.

[6] H. Flanders, *Differential Forms*. New York: Academic Press, 1963.

[7] W. L. Burke, *Applied Differential Geometry*. Cambridge: Cambridge University Press, 1985.

[8] P. Bamberg and S. Sternberg, *A Course in Mathematics for Students in Physics 2*. Cambridge: Cambridge University Press, 1990.

[9] S. Weintraub, *Differential Forms - A Complement to Vector Calculus*. New York: Academic Press, 1997.

[10] J. C. Maxwell, *A Treatise on Electricity and Magnetism*, vol. 2. New York: Oxford University Press, 1998.

[11] H. Hertz, *Gesammelte Werke, Untersuchungen über die Ausbreitung der elektrischen Kraft*, vol. 2. Leipzig: Johann Ambrosius Barth, 1894.

[12] R. P. Feynman, *Lectures on Physics 2, Electromagnetics and Matter*. London: Addison-Wesley, 1977.

[13] L. Chua, C. Desoer, and E. Kuh, *Linear and Nonlinear Circuits*. New York: McGraw-Hill, 1987.

[14] R. F. Harrington, *Time Harmonic Electromagnetic Fields*. New York: McGraw-Hill, 1961.

[15] K. F. Warnick, R. Selfridge, and D. Arnold, "Electromagnetic boundary conditions and differential forms," *IEE Proc. Microwaves, Antennas, Propagat.*, vol. 142, pp. 326–332, Aug. 1995.

Chapter 3

Potentials and Waves

3.1 THE ELECTROMAGNETIC POTENTIALS

The Maxwell's equations (2.120) – (2.123) are a system of 12 coupled scalar partial differential equations. The introduction of *electromagnetic potentials* allows a systematic solution of the Maxwell's equations [1–3]. We are distinguishing between *scalar potentials* and *vector potentials*. After solution of the wave equation for a potential, all field quantities may be derived from this potential.

According to (2.122), i.e., $\mathrm{d}\mathcal{B} = 0$, the magnetic flux density is free of divergence. Therefore, due to *Poincaré's lemma* (A.59), $\mathcal{B}$ may be represented as the exterior derivative of a one-form $\mathcal{A}$:

$$\mathcal{B} = \mathrm{d}\mathcal{A} \,. \tag{3.1}$$

The corresponding vector field $\boldsymbol{A}$ is called the *magnetic vector potential* and $\mathcal{A}$ is called the *magnetic vector potential form.* Any two-form $\mathcal{B}$ with a vanishing exterior derivative can be expressed as the exterior derivative of a one-form. Such a two-form describes a so-called *solenoidal field.* Such a field has neither source nor sink of flux. The flux tubes of a solenoidal field are continuous, neither originating nor ending anywhere. The flux tubes of $\mathcal{B}$ entering any closed surface must also leave it. Inserting (3.1) into the second Maxwell's equation (2.121) yields

$$\mathrm{d}\left(\mathcal{E} + \frac{\partial}{\partial t}\mathcal{A}\right) = 0 \,. \tag{3.2}$$

Since the exterior derivative of the one-form inside the brackets vanishes, due to Poincaré's lemma (A.59) we may express this one-form as the exterior derivative of the *scalar potential* Φ and obtain

$$\mathcal{E} = -\,\mathrm{d}\Phi - \frac{\partial}{\partial t}\mathcal{A} \,. \tag{3.3}$$

The negative sign of Φ has been chosen due to the physical convention in definition of potentials. Whereas in electrostatics the electric field may be computed from a scalar potential Φ in the case of rapidly varying electromagnetic fields, we also need the vector potential $\boldsymbol{A}$.

A field that may be expressed as the exterior derivative of a scalar potential only, for example,

$$\mathcal{E} = -\,\mathrm{d}\Phi \tag{3.4}$$

is called an *irrotational field*. From Poincaré's Lemma (A.59) and Stokes' theorem (A.88) we obtain

$$\oint_{\partial A} \mathcal{E} = -\int_{A} \mathrm{d}\,\mathrm{d}\Phi = 0\,. \tag{3.5}$$

The circulation of an irrotational field $\mathcal{E}$ around any closed path ∂A is identically zero. This implies that the line integral of an irrotational field $\mathcal{E}$ is independent from the chosen path. Therefore an irrotational field is also called a *conservative field*. A vector field may be either irrotational or solenoidal or neither irrotational nor solenoidal.

The two potentials $\boldsymbol{A}$ and Φ are not defined in an unambiguous way. Due to Poincaré's lemma, adding the exterior derivative of a zero-form Ψ to the vector potential $\boldsymbol{A}$ does not influence the magnetic induction $\boldsymbol{B}$. The electric field $\boldsymbol{E}$ also remains unchanged, if $\boldsymbol{A}$ and Φ together are transformed in the following way:

$$\mathcal{A}_1 = \mathcal{A} + \mathrm{d}\Psi\,, \tag{3.6}$$

$$\Phi_1 = \Phi - \frac{\partial \Psi}{\partial t}\,. \tag{3.7}$$

This transformation is called a *gauge transformation*. The one-form $\mathcal{A}$ may be defined in an unambiguous way, if we are prescribing its exterior derivative.

Inserting (3.1) and (3.3) in consideration of (2.34), (2.35) and (2.58) into the first Maxwell's equation (2.120), we obtain

$$\star\,\mathrm{d}\star\,\mathrm{d}\mathcal{A} + \mu\varepsilon\frac{\partial^2}{\partial t^2}\mathcal{A} + \mu\sigma\frac{\partial}{\partial t}\mathcal{A} + \mu\,\mathrm{d}\left(\varepsilon\frac{\partial \Phi}{\partial t} + \sigma\Phi\right) = \star\,\mu\mathcal{J}_0\,. \tag{3.8}$$

Inserting (3.3) and (2.34) into (2.123), yields

$$\star\,\mathrm{d}\star\,\mathrm{d}\Phi + \star\,\mathrm{d}\star\frac{\partial}{\partial t}\mathcal{A} = -\frac{1}{\varepsilon}\star\mathcal{Q}\,. \tag{3.9}$$

Since we may choose the exterior derivative of $\star\mathcal{A}$ arbitrarily, we can make use of this option in order to decouple the differential equations for $\mathcal{A}$ and Φ. We impose the so-called *Lorentz condition* given by

$$\star\,\mathrm{d}\star\mathcal{A} + \mu\left(\varepsilon\frac{\partial}{\partial t}\Phi + \sigma\Phi\right) = 0\,. \tag{3.10}$$

In this way, we obtain from (3.8) and (3.9) the equations

$$(\mathrm{d}\star\,\mathrm{d}\star - \star\,\mathrm{d}\star\,\mathrm{d})\mathcal{A} - \mu\varepsilon\frac{\partial^2}{\partial t^2}\mathcal{A} - \mu\sigma\frac{\partial}{\partial t}\mathcal{A} = -\star\mu\mathcal{J}_0\,, \tag{3.11}$$

$$\star\,\mathrm{d}\star\,\mathrm{d}\Phi - \mu\varepsilon\frac{\partial^2}{\partial t^2}\Phi - \mu\sigma\,\frac{\partial}{\partial t}\Phi = -\frac{1}{\varepsilon}\star\mathcal{Q}\,. \tag{3.12}$$

We define the *covariant derivative*, expressed by the operator $\tilde{\mathrm{d}}$, via

$$\tilde{\mathrm{d}}\mathcal{U} = (-1)^{\deg\mathcal{U}+1}\star\mathrm{d}\star\mathcal{U}\,. \tag{3.13}$$

Using the covariant derivative we introduce the *Laplace operator* Δ defined by

$$\Delta = \tilde{\mathrm{d}}\,\mathrm{d} + \mathrm{d}\,\tilde{\mathrm{d}}\,. \tag{3.14}$$

Applying the Laplace operator to a zero-form Φ and a one-form $\mathcal{A}$ respectively yields

$$\Delta\,\Phi = \star\,\mathrm{d}\star\,\mathrm{d}\Phi\,, \tag{3.15}$$

$$\Delta\,\mathcal{A} = (\mathrm{d}\star\,\mathrm{d}\star - \star\,\mathrm{d}\star\,\mathrm{d})\,\mathcal{A}\,. \tag{3.16}$$

With the Laplace operator Δ we can write (3.11) and (3.12) in the form

$$\Delta\,\mathcal{A} - \mu\varepsilon\frac{\partial^2}{\partial t^2}\mathcal{A} - \mu\sigma\frac{\partial}{\partial t}\mathcal{A} = -\star\,\mu\mathcal{J}_0\,, \tag{3.17}$$

$$\Delta\Phi - \mu\varepsilon\frac{\partial^2}{\partial t^2}\Phi - \mu\sigma\,\frac{\partial}{\partial t}\Phi = -\frac{1}{\varepsilon}\star\mathcal{Q}\,. \tag{3.18}$$

The field intensities $\boldsymbol{E}$ and $\boldsymbol{H}$ derived from $\boldsymbol{A}$ and Φ satisfy the four Maxwell's equations (2.120) – (2.123). Equations (3.17) and (3.18) are called *wave equations*, since their solutions describe propagating waves. Equation (3.17) is a *vector wave equation*, whereas (3.18) is a *scalar wave equation*.

It is possible to derive both potentials $\boldsymbol{A}(\boldsymbol{x}, t)$ and $\Phi(\boldsymbol{x}, t)$ from one vector, the so-called *electric Hertz vector* $\boldsymbol{\Pi}_e(\boldsymbol{x}, t)$. We introduce the *electric Hertz differential form*

$$\mathit{\Pi}_e = \Pi_{ex}\,\mathrm{d}x + \Pi_{ey}\,\mathrm{d}y + \Pi_{ez}\,\mathrm{d}z\,. \tag{3.19}$$

The Lorentz condition (3.10) is fulfilled, if $\mathcal{A}$ and Φ are derived from the electric Hertz differential form $\mathit{\Pi}_e$ via

$$\mathcal{A} = \mu\varepsilon\frac{\partial}{\partial t}\,\mathit{\Pi}_e + \mu\sigma\,\mathit{\Pi}_e\,, \tag{3.20}$$

$$\Phi = -\tilde{\mathrm{d}}\,\mathit{\Pi}_e\,. \tag{3.21}$$

Inserting (3.20) into (3.17), we obtain

$$\mu\left(\varepsilon\frac{\partial}{\partial t}+\sigma\right)\left(\Delta\Pi_e-\mu\varepsilon\frac{\partial^2}{\partial t^2}\Pi_e-\mu\sigma\frac{\partial}{\partial t}\Pi_e\right)=-\mu\star\mathcal{J}_0\,. \tag{3.22}$$

For $\mathcal{J}_0=0$, i.e., without impressed current sources we obtain the *homogeneous wave equation*

$$\Delta\Pi_e-\mu\varepsilon\frac{\partial^2}{\partial t^2}\Pi_e-\mu\sigma\frac{\partial}{\partial t}\Pi_e=0\,. \tag{3.23}$$

The field intensity differential forms $\mathcal{E}$ and $\mathcal{H}$ follow from (2.35), (3.1), (3.3), (3.20) and (3.21):

$$\mathcal{E}=\mathrm{d}\tilde{\mathrm{d}}\,\Pi_e-\mu\varepsilon\frac{\partial^2}{\partial t^2}\Pi_e-\mu\sigma\frac{\partial}{\partial t}\Pi_e\,, \tag{3.24}$$

$$\mathcal{H}=\star\,\mathrm{d}\left(\varepsilon\frac{\partial}{\partial t}\Pi_e+\sigma\,\Pi_e\right). \tag{3.25}$$

Subtracting from (3.24) the wave equation (3.23) we obtain

$$\mathcal{E}=-\tilde{\mathrm{d}}\,\mathrm{d}\,\Pi_e \qquad \text{for} \quad \mathcal{J}_0=0\,. \tag{3.26}$$

Let us now consider the lossless case with impressed current sources. In this case it is advantageous to use the *impressed electric polarization* $\boldsymbol{M}_{e0}(\boldsymbol{x},t)$ instead of the impressed current density $\boldsymbol{J}_0(\boldsymbol{x},t)$. The corresponding differential form is

$$\mathcal{M}_{e0}=M_{ex}\,\mathrm{d}y\wedge\mathrm{d}z+M_{ey}\,\mathrm{d}z\wedge\mathrm{d}x+M_{ez}\,\mathrm{d}x\wedge\mathrm{d}y\,. \tag{3.27}$$

The *impressed electric polarization form* $\mathcal{M}_{e0}$ is related to an impressed electric current $\mathcal{J}_0$ via

$$\frac{\partial}{\partial t}\mathcal{M}_{e0}=\mathcal{J}_0\,. \tag{3.28}$$

In this way it follows from (3.22)

$$\frac{\partial}{\partial t}\left(\Delta\Pi_e-\mu\varepsilon\frac{\partial^2}{\partial t^2}\Pi_e\right)=-\frac{1}{\varepsilon}\star\frac{\partial}{\partial t}\mathcal{M}_{e0} \quad \text{for } \sigma=0\,. \tag{3.29}$$

By integration over t we obtain

$$\Delta\Pi_e-\mu\varepsilon\frac{\partial^2}{\partial t^2}\Pi_e=-\frac{1}{\varepsilon}\star\mathcal{M}_{e0} \quad \text{for } \sigma=0\,. \tag{3.30}$$

Since the source of the Hertz vector field is an impressed electric polarization, the Hertz vector also is called the *electric polarization potential*. From the solution of (3.30) we obtain $\mathcal{E}$ and $\mathcal{H}$ via (3.24) and (3.25). From (3.24) and (3.30) we obtain

$$\mathcal{E} = -\tilde{\mathrm{d}}\,\mathrm{d}\,\Pi_e - \frac{1}{\varepsilon}\star\mathcal{M}_{e0}\,. \tag{3.31}$$

In the general case $\mathcal{J}_0 \neq 0$ and $\sigma \neq 0$, we will obtain an equation containing time derivatives up to third order. This difficulty can be avoided by using the frequency domain representation treated in the next section.

3.2 THE HELMHOLTZ EQUATION

For time-harmonic fields we can replace the operator $\partial/\partial t$ by multiplication with the factor $\mathrm{j}\omega$. This transforms the wave equation into the *Helmholtz equation*. Using Maxwell's equations in phasor representation as introduced in Section 2.11 allows us to consider complex frequency-dependent permittivity and permeability. From (2.144) and (2.145) we obtain

$$\mathrm{d}\underline{\mathcal{H}} = \mathrm{j}\omega\left(\underline{\varepsilon}\star\underline{\mathcal{E}} + \underline{\mathcal{M}}_{e0}\right), \tag{3.32}$$

$$\mathrm{d}\underline{\mathcal{E}} = -\mathrm{j}\omega\underline{\mu}\star\underline{\mathcal{H}}, \tag{3.33}$$

where $\underline{\mathcal{M}}_{e0}$ is the *impressed electric polarization phasor*. Assuming also a complex permeability in order to consider the magnetic losses according to (3.22) and (3.28) after transforming into the frequency domain and using (2.142) and generalizing for complex permittivity, we obtain

$$\Delta\underline{\Pi}_e + \omega^2\underline{\mu\varepsilon}\,\underline{\Pi}_e = -\frac{1}{\underline{\varepsilon}}\star\underline{\mathcal{M}}_{e0}\,, \tag{3.34}$$

where $\underline{\mathcal{M}}_{e0}$ is the *electric Hertz differential form phasor*. This type of equation that we have obtained from the wave equation via $\partial/\partial t \to \mathrm{j}\omega$ is called *Helmholtz equation*. In a similar way to (3.24), (3.25) and (3.31), we obtain the complex electric and magnetic field forms $\underline{\mathcal{E}}$ and $\underline{\mathcal{H}}$

$$\underline{\mathcal{E}} = \mathrm{d}\tilde{\mathrm{d}}\,\underline{\Pi}_e + \omega^2\underline{\mu\varepsilon}\,\underline{\Pi}_e = -\tilde{\mathrm{d}}\,\mathrm{d}\,\underline{\Pi}_e - \frac{1}{\underline{\varepsilon}}\star\underline{\mathcal{M}}_{e0}\,, \tag{3.35}$$

$$\underline{\mathcal{H}} = \mathrm{j}\omega\underline{\varepsilon}\star\mathrm{d}\,\underline{\Pi}_e\,. \tag{3.36}$$

If the electromagnetic field is generated by an impressed magnetic polarization $\underline{\mathcal{M}}_{m0}$, we obtain – instead of (3.32), (3.34) – the equations

$$\mathrm{d}\underline{\mathcal{H}} = \mathrm{j}\omega\underline{\varepsilon}\star\underline{\mathcal{E}}, \tag{3.37}$$

$$\mathrm{d}\underline{\mathcal{E}} = -\mathrm{j}\omega\left(\underline{\mu}\star\underline{\mathcal{H}} + \underline{\mathcal{M}}_{m0}\right). \tag{3.38}$$

There exists a *duality* relationship [4, 5] between (3.32) and (3.33) on one hand, and (3.37) and (3.38) on the other hand. We may transform one pair of equations into the dual one by performing the substitution

$$\underline{\mathcal{E}}_2 = -Z_F\,\underline{\mathcal{H}}_1, \tag{3.39}$$

$$\underline{\mathcal{H}}_2 = \frac{1}{Z_F}\,\underline{\mathcal{E}}_1, \tag{3.40}$$

$$\underline{\mathcal{M}}_{m02} = Z_F\,\underline{\mathcal{M}}_{e01}. \tag{3.41}$$

The field quantities of the pair of (3.32), (3.34) are marked by the index 1, whereas the index 2 is assigned to field quantities of the pair of (3.37), (3.38). The *wave impedance* Z_F is given by

$$Z_F = \sqrt{\frac{\underline{\mu}}{\underline{\varepsilon}}}. \tag{3.42}$$

We can make use of the duality principle to derive from one set of field solutions a dual set of field solutions with interchanged $\underline{\boldsymbol{E}}$ and $\underline{\boldsymbol{H}}$. We also may introduce a vector potential dual to the Hertz vector. This vector potential is generated by an impressed magnetic polarization. We call this vector potential the *magnetic Hertz vector* or magnetic polarization potential. This dual magnetic Hertz vector $\boldsymbol{\Pi}_m$ is related to the electric Hertz vector $\boldsymbol{\Pi}_e$ by

$$\underline{\boldsymbol{\Pi}}_m = \frac{1}{Z_F}\,\underline{\boldsymbol{\Pi}}_e. \tag{3.43}$$

The corresponding differential forms are related via

$$\underline{\mathit{\Pi}}_m = \frac{1}{Z_F}\,\underline{\mathit{\Pi}}_e. \tag{3.44}$$

From (3.34) and (3.39) – (3.41), we obtain the inhomogeneous Helmholtz equation for the *magnetic Hertz form* $\underline{\mathcal{M}}_m$:

$$\Delta\underline{\mathit{\Pi}}_m + \omega^2\underline{\mu\varepsilon}\,\underline{\mathit{\Pi}}_m = -\frac{1}{\underline{\mu}}\star\underline{\mathcal{M}}_{m0}. \tag{3.45}$$

After inserting (3.39) into (3.42) and (3.44) into (3.35) and (3.36), we obtain

$$\underline{\mathcal{E}} = -\mathrm{j}\omega\underline{\mu} \star \mathrm{d}\,\underline{\Pi}_m\,, \tag{3.46}$$

$$\underline{\mathcal{H}} = \mathrm{d}\tilde{\mathrm{d}}\,\underline{\Pi}_m + \omega^2\underline{\mu\varepsilon}\,\underline{\Pi}_m = -\tilde{\mathrm{d}}\,\mathrm{d}\,\underline{\Pi}_m - \frac{1}{\underline{\mu}} \star \underline{\mathcal{M}}_{m0}\,. \tag{3.47}$$

For the source-free region with $\underline{\mathcal{M}}_{e0} = 0$ and $\underline{\mathcal{M}}_{m0} = 0$ the Helmholtz equation (3.34) for the Hertz form $\underline{\Pi} = \underline{\Pi}_e,\ \underline{\Pi}_m$ is given by

$$\Delta\underline{\Pi} + \omega^2\underline{\mu\varepsilon}\,\underline{\Pi} = 0\,. \tag{3.48}$$

Using (3.14) and (3.16) yields

$$(\mathrm{d}\tilde{\mathrm{d}} + \tilde{\mathrm{d}}\,\mathrm{d})\,\underline{\Pi} + \omega^2\underline{\mu\varepsilon}\,\underline{\Pi} = (\mathrm{d}\star\mathrm{d}\star - \star\,\mathrm{d}\star\mathrm{d})\,\underline{\Pi} + \omega^2\underline{\mu\varepsilon}\,\underline{\Pi} = 0\,. \tag{3.49}$$

For Cartesian coordinates the Laplace operator for one-forms is given by

$$\Delta\,\underline{\Pi} = \frac{\partial^2\underline{\Pi}}{\partial x^2} + \frac{\partial^2\underline{\Pi}}{\partial y^2} + \frac{\partial^2\underline{\Pi}}{\partial z^2}\,, \tag{3.50}$$

and the Helmholtz equation can be put into the form

$$\frac{\partial^2\underline{\Pi}}{\partial x^2} + \frac{\partial^2\underline{\Pi}}{\partial y^2} + \frac{\partial^2\underline{\Pi}}{\partial z^2} + \omega^2\underline{\mu\varepsilon}\,\underline{\Pi} = 0\,. \tag{3.51}$$

The complete solution of this equation is given by

$$\underline{\Pi}(\boldsymbol{x}) = \underline{\Pi}\mathrm{e}^{\pm j\boldsymbol{k}\cdot\boldsymbol{x}} \tag{3.52}$$

with

$$\boldsymbol{k}^2 = k_x^2 + k_y^2 + k_z^2 = \omega^2\underline{\mu\varepsilon}\,. \tag{3.53}$$

The *wave vector* $\boldsymbol{k} = [k_x, k_y, k_z]^T$ with dimension $[m^{-1}]$ is complex for complex $\underline{\varepsilon}$ and complex $\underline{\mu}$. In a lossy medium a complex wave vector $\boldsymbol{k}$ describes an attenuated plane wave. The real time-dependent Hertz form is given by

$$\Pi(\boldsymbol{x}, t) = \Re\left\{\underline{\Pi}_0\,\mathrm{e}^{\mathrm{j}(\omega t \pm \boldsymbol{k}\cdot\boldsymbol{x})}\right\}. \tag{3.54}$$

The quantity

$$k = \sqrt{k_x^2 + k_y^2 + k_z^2} \tag{3.55}$$

is called the *wave number*. The wave vector $\boldsymbol{k}$ is the product of the wave number and the unit vector in direction of propagation. Let us now consider a plane wave propagating in the z-direction.

$$\boldsymbol{k} = k\,\boldsymbol{e}_z\,. \tag{3.56}$$

Due to (3.55), the sign of $\boldsymbol{k}$ is not determined. Since, due to (3.52), to each sign of $\boldsymbol{k}$ a corresponding solution of the wave equation exists, we may specify

$$\Re\{k\} > 0 \tag{3.57}$$

without a loss of generality. According to (3.54) and (3.56) the Hertz form

$$\Pi^{(+)}(z,t) = \Re\left\{\underline{\Pi}_0^{(+)}\,\mathrm{e}^{\mathrm{j}(\omega t - kz)}\right\} \tag{3.58}$$

describes a plane wave propagating in the positive z-direction, whereas the Hertz form

$$\Pi^{(-)}(z,t) = \Re\left\{\underline{\Pi}_0^{(-)}\,\mathrm{e}^{\mathrm{j}(\omega t + kz)}\right\} \tag{3.59}$$

describes a plane wave propagating in the negative z-direction. The imaginary parts of $\underline{\varepsilon}$ and $\underline{\mu}$ are negative for passive media, $\varepsilon'' \geq 0$ and $\mu'' \geq 0$. Hence, for a plane wave under the condition given by (3.57) it follows that $\Im\{k\} \leq 0$. This means that the Hertz vector and therefore also the field quantities, are decaying exponentially in the direction of propagation. Figure 3.1(a) shows the amplitude of the Hertz vector of an attenuated wave propagating in positive z-direction, and Figure 3.1(b) shows the wave amplitude of the wave propagating in negative z-direction. Instead of the *wave number* k the complex *propagation coefficient* γ is also used. The propagation coefficient γ is defined by

$$\gamma = \mathrm{j}\,k\,. \tag{3.60}$$

The convention

$$\Im\{\gamma\} > 0 \tag{3.61}$$

corresponds to the convention specified in (3.57). The real part α and the imaginary part β of the propagation coefficient γ are given by

$$\gamma = \alpha + \mathrm{j}\beta. \tag{3.62}$$

The real part α is called the *attenuation coefficient*, and the imaginary part β is called the *phase coefficient*. The dimension of γ is $[m^{-1}]$. The attenuation coefficient α can also be specified in decibels (dB) or nepers. For a wave propagating over a length l we obtain

$$\text{Attenuation in nepers:} \qquad \alpha l\,, \tag{3.63}$$

$$\text{Attenuation in decibels:} \qquad 20\log \mathrm{e}^{\alpha l} \cong 8.69\alpha l\,. \tag{3.64}$$

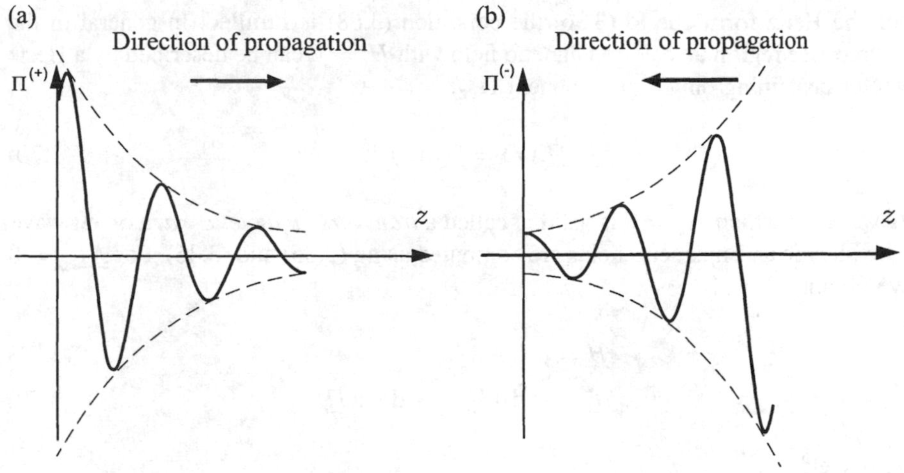

Figure 3.1: Magnitude of the Hertz vector for an attenuated plane wave in (a) positive and (b) negative z-direction.

3.3 TM AND TE WAVES

Let us seek a solution for which the z-component of the magnetic field vanishes, i.e., $H_z = 0$. From (2.135) it follows

$$\mathrm{d}\underline{\mathcal{B}} = \underline{\mu}\,\mathrm{d}\star\underline{\mathcal{H}} = 0\,. \tag{3.65}$$

From this we obtain

$$\frac{\partial \underline{H}_x}{\partial x} + \frac{\partial \underline{H}_y}{\partial y} + \frac{\partial \underline{H}_z}{\partial z} = 0 \tag{3.66}$$

and

$$\frac{\partial \underline{H}_x}{\partial x} = -\frac{\partial \underline{H}_y}{\partial y} \quad \text{for} \quad \underline{H}_z = 0\,. \tag{3.67}$$

This is the condition that the system of differential equations

$$\frac{\partial f}{\partial y} = \underline{H}_x\,,$$
$$\frac{\partial f}{\partial x} = -\underline{H}_y\,. \tag{3.68}$$

can be integrated. Due to (3.67) a function $f(\boldsymbol{x})$ must exist, from which $\underline{H}_x$ and $\underline{H}_y$ can be derived using (3.68). Making the ansatz

$$\underline{\Pi}_e(\boldsymbol{x}) = \frac{f(\boldsymbol{x})}{\mathrm{j}\omega\varepsilon}\,\mathrm{d}z \tag{3.69}$$

for the Hertz form due to (3.36) the condition (3.68) is fulfilled. In general in the source-free region an electromagnetic field with $H_z = 0$ can be described by a Hertz vector containing only a z component Π_{ez}:

$$\underline{\Pi}_e(\boldsymbol{x}) = \underline{\Pi}_{ez}(\boldsymbol{x})\,\mathrm{d}z\,. \tag{3.70}$$

A wave for which $H_z = 0$ is valid is called a *transverse magnetic wave* or TM wave. The electric and magnetic fields are computed using (3.35) and (3.36). For $\underline{\mathcal{M}}_{e0} = 0$ we obtain

$$\underline{\mathcal{H}} = \mathrm{j}\omega\underline{\varepsilon} \star \mathrm{d}\,\underline{\Pi}_e\,, \tag{3.71}$$

$$\underline{\mathcal{E}} = -\tilde{\mathrm{d}}\,\mathrm{d}\,\underline{\Pi}_e = \star\,\mathrm{d}\star\mathrm{d}\,\underline{\Pi}_e\,. \tag{3.72}$$

This yields

$$\underline{\mathcal{H}} = \mathrm{j}\omega\underline{\varepsilon}\left(\frac{\partial\underline{\Pi}_{ez}(\boldsymbol{x})}{\partial y}\,\mathrm{d}x - \frac{\partial\underline{\Pi}_{ez}(\boldsymbol{x})}{\partial x}\,\mathrm{d}y\right), \tag{3.73}$$

$$\underline{\mathcal{E}} = \frac{\partial^2\underline{\Pi}_{ez}}{\partial x\partial z}\,\mathrm{d}x + \frac{\partial^2\underline{\Pi}_{ez}}{\partial y\partial z}\,\mathrm{d}y - \left(\frac{\partial^2\underline{\Pi}_{ez}}{\partial x^2} + \frac{\partial^2\underline{\Pi}_{ez}}{\partial y^2}\right)\mathrm{d}z\,. \tag{3.74}$$

In the same way electromagnetic fields in a source-free region, fulfilling the condition $E_z = 0$, may be derived from a magnetic Hertz vector that exhibits only a z component Π_{mz}:

$$\underline{\Pi}_m(\boldsymbol{x}) = \underline{\Pi}_{mz}(\boldsymbol{x})\,\mathrm{d}z\,. \tag{3.75}$$

A wave for which $E_z = 0$ is valid is called a *transverse electric wave* or TE wave. The electric and magnetic field are computed using (3.46) and (3.47). For $\underline{\mathcal{M}}_{m0} = 0$ we obtain

$$\underline{\mathcal{E}} = -\mathrm{j}\omega\underline{\mu} \star \mathrm{d}\,\underline{\Pi}_m\,, \tag{3.76}$$

$$\underline{\mathcal{H}} = -\tilde{\mathrm{d}}\,\mathrm{d}\,\underline{\Pi}_m\,. \tag{3.77}$$

This yields

$$\underline{\mathcal{E}} = -\mathrm{j}\omega\underline{\mu}\left(\frac{\partial\underline{\Pi}_{mz}(\boldsymbol{x})}{\partial y}\,\mathrm{d}x - \frac{\partial\underline{\Pi}_{mz}(\boldsymbol{x})}{\partial x}\,\mathrm{d}y\right), \tag{3.78}$$

$$\underline{\mathcal{H}} = \frac{\partial^2\underline{\Pi}_{mz}}{\partial x\partial z}\,\mathrm{d}x + \frac{\partial^2\underline{\Pi}_{mz}}{\partial y\partial z}\,\mathrm{d}y - \left(\frac{\partial^2\underline{\Pi}_{mz}}{\partial x^2} + \frac{\partial^2\underline{\Pi}_{mz}}{\partial y^2}\right)\mathrm{d}z\,. \tag{3.79}$$

3.4 SPHERICAL WAVES

Waves emitted from point-like sources are called *spherical waves* since surfaces of constant phases are spheres. Spherical waves occur in radiation problems [1–3, 6]. The Helmholtz equation in spherical coordinates has a *complete set of spherical solutions*. Waves emitted from a closed volume of limited spatial extension usually are represented by a superposition of spherical waves.

The solution of the Helmholtz equation (3.34) in spherical coordinates is a difficult task since in spherical coordinates all three coordinates of the Hertz form are mixed in the Laplace operator (A.162). However, also in the case of spherical solutions the TE- and TM-wave solutions may be derived from the solution of the *scalar Helmholtz equation* in spherical coordinates. The homogeneous scalar Helmholtz equation in spherical coordinates is given by

$$\frac{1}{r^2}\frac{\partial}{\partial r}\left(r^2\frac{\partial\Psi}{\partial r}\right)+\frac{1}{r^2\sin\vartheta}\frac{\partial}{\partial\vartheta}\left(\sin\vartheta\frac{\partial\Psi}{\partial\vartheta}\right)+\frac{1}{r^2\sin^2\vartheta}\frac{\partial^2\Psi}{\partial\varphi^2}+k^2\Psi=0\,. \quad (3.80)$$

Introducing

$$\Psi(r,\vartheta,\varphi)=z_n(kr)Y_n^m(\vartheta,\varphi) \quad (3.81)$$

the homogeneous scalar Helmholtz equation in spherical coordinates (3.80) is separated into the two differential equations (B.82) and (B.38)

$$r^2\frac{\mathrm{d}^2 z_n(kr)}{\mathrm{d}r^2}+2r\frac{\mathrm{d}z_n(kr)}{\mathrm{d}r}+(k^2r^2-n^2)z_n(kr)=0\,, \quad (3.82a)$$

$$\frac{1}{\sin\vartheta}\frac{\mathrm{d}}{\mathrm{d}\vartheta}\left(\sin\vartheta\,\frac{\mathrm{d}Y_n^m(\vartheta,\varphi)}{\mathrm{d}\vartheta}\right)+\frac{1}{\sin^2\vartheta}\frac{\mathrm{d}^2Y_n^m(\vartheta,\varphi)}{\mathrm{d}\varphi^2}+n(n+1)Y_n^m(\vartheta,\varphi)=0\,, \quad (3.82b)$$

where $z_n(kr)$ is a *spherical Bessel function* $j_n(kr)$, $y_n(kr)$, $h_n^{(1)}(kr)$ or $h_n^{(2)}(kr)$ as defined in (B.41) to (B.44), and $Y_n^m(\vartheta,\varphi)$ is the *Spherical harmonic*, defined in (B.86). For real k the solutions $j_n(kr)$ and $y_n(kr)$ represent standing waves with respect to radial direction, $h_n^{(2)}(kr)$ represents outward travelling waves and $h_n^{(1)}(kr)$ represents inward travelling waves.

If both $\Psi_1(r,\vartheta,\varphi)$ and $\Psi_2(r,\vartheta,\varphi)$ are two independent solutions of the scalar Helmholtz equation (3.80), we can use (A.167) to construct solutions of the vector Helmholtz equation by

$$\underline{\mathcal{E}}_1=\star(\mathrm{d}r\wedge\mathrm{d}\Psi_1)\,, \quad (3.83a)$$

$$\underline{\mathcal{H}}_1=-\frac{1}{\mathrm{j}\omega\mu}\star\mathrm{d}\underline{\mathcal{E}}_1\,. \quad (3.83b)$$

$$\underline{\mathcal{H}}_2 = \star\,(\mathrm{d}r \wedge \mathrm{d}\Psi_2)\,, \tag{3.84a}$$

$$\underline{\mathcal{E}}_2 = \frac{1}{\mathrm{j}\omega\varepsilon} \star \mathrm{d}\underline{\mathcal{H}}_2\,. \tag{3.84b}$$

From this we obtain

$$\underline{E}_r = -\frac{n(n+1)}{\mathrm{j}\omega\varepsilon\, r}\Psi_2\,, \tag{3.85a}$$

$$\underline{E}_\vartheta = -\frac{1}{\sin\vartheta}\frac{\partial\Psi_1}{\partial\varphi} - \frac{1}{\mathrm{j}\omega\varepsilon\, r}\frac{\partial}{\partial r}\left(r\frac{\partial\Psi_2}{\partial\vartheta}\right), \tag{3.85b}$$

$$\underline{E}_\varphi = \frac{\partial\Psi_1}{\partial\vartheta} - \frac{1}{\mathrm{j}\omega\varepsilon\, r\sin\vartheta}\frac{\partial}{\partial r}\left(r\frac{\partial\Psi_2}{\partial\varphi}\right), \tag{3.85c}$$

$$\underline{H}_r = \frac{n(n+1)}{\mathrm{j}\omega\mu\, r}\Psi_1\,, \tag{3.85d}$$

$$\underline{H}_\vartheta = \frac{1}{\mathrm{j}\omega\mu\, r}\frac{\partial}{\partial r}\left(r\frac{\partial\Psi_1}{\partial\vartheta}\right) - \frac{1}{\sin\vartheta}\frac{\partial\Psi_2}{\partial\varphi}\,, \tag{3.85e}$$

$$\underline{H}_\varphi = \frac{1}{\mathrm{j}\omega\mu\, r\sin\vartheta}\frac{\partial}{\partial r}\left(r\frac{\partial\Psi_1}{\partial\varphi}\right) + \frac{\partial\Psi_2}{\partial\vartheta}\,. \tag{3.85f}$$

Solutions containing only Ψ_1 represent the TE waves and solutions containing only Ψ_2 represent the TM waves.

If all sources of the electromagnetic field are located within a sphere of radius R outside this sphere only outward travelling waves exist. The outward travelling waves may be described by a superposition of waves with the radial dependence given by $h_n^{(2)}(kr)$. The operator $\partial/\partial r$ when applied to $\mathrm{e}^{-\mathrm{j}kr}$ gives $-\mathrm{j}k$. Application of the $\partial/\partial r$ to all other terms gives rise to terms of the order $(1/r)^2$ or higher. In the *far-field approximation* $kr \ll 1$ we keep only the terms of the order of $1/r$. Therefore we can replace the operator $\partial/\partial r$ by $\mathrm{j}k$ and obtain the far-field approximation

$$\underline{E}_r = 0\,, \tag{3.86a}$$

$$\underline{E}_\vartheta = -\frac{1}{\sin\vartheta}\frac{\partial\Psi_1}{\partial\varphi} + Z_F\frac{\partial\Psi_2}{\partial\vartheta}\,, \tag{3.86b}$$

$$\underline{E}_\varphi = \frac{\partial\Psi_1}{\partial\vartheta} + \frac{Z_F}{\sin\vartheta}\frac{\partial\Psi_2}{\partial\varphi}\,, \tag{3.86c}$$

$$\underline{H}_r = 0\,, \tag{3.86d}$$

$$\underline{H}_\vartheta = -\frac{1}{Z_F}\frac{\partial\Psi_1}{\partial\vartheta} - \frac{1}{\sin\vartheta}\frac{\partial\Psi_2}{\partial\varphi}\,, \tag{3.86e}$$

$$\underline{H}_\varphi = -\frac{1}{Z_F\sin\vartheta}\frac{\partial\Psi_1}{\partial\varphi} + \frac{\partial\Psi_2}{\partial\vartheta}\,. \tag{3.86f}$$

From these equations we obtain in the far-field

$$\underline{E}_\vartheta \cong Z_F \underline{H}_\varphi \,, \tag{3.87a}$$

$$\underline{E}_\varphi \cong -Z_F \underline{H}_\vartheta \,. \tag{3.87b}$$

We can summarize (3.87a) and (3.87b) in the *Sommerfeld radiation condition*

$$\lim_{r\to\infty} (\mathcal{E} + Z_F \star \mathrm{d}r \wedge \mathcal{H}) = 0 \,. \tag{3.88a}$$

The Sommerfeld radiation condition may also be written in the form

$$\lim_{r\to\infty} \left(\mathcal{H} - Z_F^{-1} \star \mathrm{d}r \wedge \mathcal{E}\right) = 0 \,. \tag{3.88b}$$

It can be shown that the Sommerfeld radiation condition holds for an electromagnetic field generated by any source distribution confined in a volume V of finite extension around $r = 0$. The proof may be given by expanding the field outside V into spherical waves. A general proof of the Sommerfeld radiation condition is given for example in [7]. The radiation condition requires that the electric and magnetic fields bear the relation to each other found in wave propagation in regions remote from the sources.

REFERENCES

[1] J. A. Stratton, *Electromagnetic Theory*. New York: McGraw-Hill, 1941.

[2] R. F. Harrington, *Time Harmonic Electromagnetic Fields*. New York: McGraw-Hill, 1961.

[3] J. A. Kong, *Electromagnetic Wave Theory*. New York: John Wiley & Sons, 1986.

[4] H. Carlin and A. Giordano, *Network Theory*. Englewood Cliffs, NJ: Prentice Hall, 1964.

[5] C. A. Balanis, *Advanced Engineering Electromagnetics*. New York: John Wiley & Sons, 1989.

[6] L. Felsen and N. Marcuvitz, *Radiation and Scattering of Waves*. Englewood Cliffs, NJ: Prentice Hall, 1972.

[7] R. S. Elliott, *Electromagnetics - History, Theory, and Applications*. New York: IEEE Press, 1991.

Chapter 4

Concepts, Methods and Theorems

4.1 ENERGY AND POWER

The field concept is based upon the hypothesis that electromagnetic energy is distributed over the space. The electric and magnetic fields carry energy and changing electric and magnetic energy densities are related to power flow in space [1–5]. We introduce the *electric energy density* $w_e(\boldsymbol{x}, t)$ and the *magnetic energy density* $w_m(\boldsymbol{x}, t)$ with the corresponding three-forms

$$\mathcal{W}_e = w_e(\boldsymbol{x}, t)\, \mathrm{d}x \wedge \mathrm{d}y \wedge \mathrm{d}z\,, \tag{4.1}$$

$$\mathcal{W}_m = w_m(\boldsymbol{x}, t)\, \mathrm{d}x \wedge \mathrm{d}y \wedge \mathrm{d}z\,. \tag{4.2}$$

The energy densities are given by

$$\mathcal{W}_e = \frac{1}{2}\, \mathcal{E} \wedge \mathcal{D} = \frac{1}{2}\, \left(E_x D_x + E_y D_y + E_z D_z\right) \mathrm{d}x \wedge \mathrm{d}y \wedge \mathrm{d}z\,, \tag{4.3}$$

$$\mathcal{W}_m = \frac{1}{2}\, \mathcal{H} \wedge \mathcal{B} = \frac{1}{2}\, \left(H_x B_x + H_y B_y + H_z B_z\right) \mathrm{d}x \wedge \mathrm{d}y \wedge \mathrm{d}z\,. \tag{4.4}$$

Figure 4.1 visualizes the exterior product of the field one-form $\mathcal{E}$ and the flux density two-form $\mathcal{D}$. The resulting three-form is visualized by the subdivision of the space into cells. The number of cells per unit of volume is proportional to the electric energy density.

In order to investigate energy storage and power flow in the electromagnetic field, we start again with the Maxwell's equations (2.120) and (2.121). Exterior multiplication of Ampère's law from the left with $-\mathcal{E}$ and of Faraday's law from the right

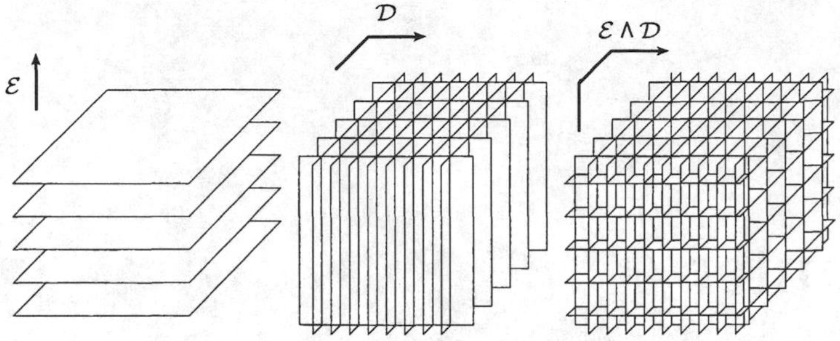

Figure 4.1: The exterior product of the field form $\mathcal{E}$ and the flux density form $\mathcal{D}$.

with $\mathcal{H}$ yields

$$-\mathcal{E}\wedge \Big| \qquad d\mathcal{H} = \dot{\mathcal{D}} + \mathcal{J}\,, \tag{4.5}$$

$$d\mathcal{E} = -\dot{\mathcal{B}} \qquad \Big| \wedge \mathcal{H}\,, \tag{4.6}$$

where the dot means the partial derivative with respect to t, i.e., $\dot{\mathcal{D}} = (\partial/\partial t)\mathcal{D}$. This yields

$$\mathrm{d}\,(\mathcal{E} \wedge \mathcal{H}) = -\mathcal{E} \wedge \dot{\mathcal{D}} - \mathcal{H} \wedge \dot{\mathcal{B}} - \mathcal{E} \wedge \mathcal{J}\,. \tag{4.7}$$

This equation can be brought into the form

$$\mathrm{d}\,(\mathcal{E} \wedge \mathcal{H}) = -\frac{\partial}{\partial t}\left(\frac{1}{2}\,\mathcal{E} \wedge \mathcal{D} + \frac{1}{2}\,\mathcal{H} \wedge \mathcal{B}\right) - \mathcal{E} \wedge \mathcal{J}\,. \tag{4.8}$$

The *power loss density* $p_L(\boldsymbol{x}, t)$ with the corresponding differential form

$$\mathcal{P}_L(\boldsymbol{x}, t) = p_L(\boldsymbol{x}, t)\,\mathrm{d}x \wedge \mathrm{d}y \wedge \mathrm{d}z\,. \tag{4.9}$$

is given by

$$\mathcal{P}_L = \mathcal{E} \wedge \sigma \star \mathcal{E}\,. \tag{4.10}$$

Due to the impressed current density $\boldsymbol{J}_0$, a power per unit of volume $p_0(\boldsymbol{x}, t)$ is added to the electromagnetic field. With the differential form

$$\mathcal{P}_0(\boldsymbol{x}, t) = p_0(\boldsymbol{x}, t)\,\mathrm{d}x \wedge \mathrm{d}y \wedge \mathrm{d}z \tag{4.11}$$

the power added to the field by the impressed current $\mathcal{J}_0$ is given by

$$\mathcal{P}_0 = -\mathcal{E} \wedge \mathcal{J}_0\,. \tag{4.12}$$

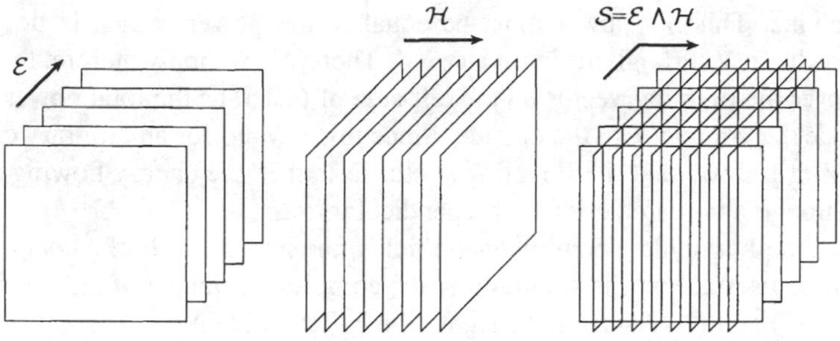

Figure 4.2: The Poynting form $\mathcal{S}$ as the product of the field forms $\mathcal{E}$ and $\mathcal{H}$.

We introduce the *Poynting vector* $\boldsymbol{S}(\boldsymbol{x}, t)$ with the corresponding *Poynting differential form*

$$\mathcal{S}(\boldsymbol{x}, t) = S_x \, \mathrm{d}y \wedge \mathrm{d}z + S_y \, \mathrm{d}z \wedge \mathrm{d}x + S_z \, \mathrm{d}x \wedge \mathrm{d}y \,, \tag{4.13}$$

given by

$$\mathcal{S} = \mathcal{E} \wedge \mathcal{H} \,. \tag{4.14}$$

Figure 4.2 visualizes the Poynting two-form as the exterior product of the electric and magnetic field one-forms $\mathcal{E}$ and $\mathcal{H}$. The potential planes of the electric and magnetic fields together form the tubes of the Poynting form. The distance of the electric and magnetic potential planes exhibit the dimensions [V] and [A], respectively. The cross-sectional areas of the flux tubes have the dimension [VA]. This yields the assumption that the Poynting form describes a power flow through these flux tubes. The proof for this assumption will be given in the following.

Inserting (4.3), (4.4), (4.10) and (4.12) into (4.8) yields the local form of *Poynting's theorem*:

$$\mathrm{d}\mathcal{S} = -\frac{\partial}{\partial t}\mathcal{W}_e - \frac{\partial}{\partial t}\mathcal{W}_m - \mathcal{P}_L + \mathcal{P}_0 \,. \tag{4.15}$$

Integrating (4.15) over a volume V and transforming the integral over S into a surface integral over the boundary ∂V, we obtain the integral form of *Poynting's Theorem*:

$$\oint_{\partial V} \mathcal{S} = \int_V \mathcal{P}_0 - \frac{\mathrm{d}}{\mathrm{d}t}\int_V \mathcal{W}_e - \frac{\mathrm{d}}{\mathrm{d}t}\int_V \mathcal{W}_m - \int_V \mathcal{P}_L . \tag{4.16}$$

The first term on the right side of (4.16) describes the power added into the volume V via impressed currents. The second and the third term describe the time variation of the electric and magnetic energy stored in the volume. The last term describes the conductive losses occurring inside the volume V. The right side of the equation comprises the total electromagnetic power generated within the volume V minus the power losses in the volume minus the increase of electric and magnetic energy stored in the volume. This net power must be equal to the power, which is flowing out from the volume V through the boundary ∂V. Therefore we may interpret the surface integral over the pointing vector on the left side of (4.16) as the total power flowing from inside the volume V to the outside. Since this is valid for an arbitrary choice of volume V, it follows that the Poynting vector describes the energy flowing by units of time through a unit area oriented perpendicular to $\boldsymbol{S}$.

For time-harmonic electromagnetic fields, the introduction of a *complex Poynting vector* is useful. For this we multiply the complex conjugate of (2.133) from the left with $(-\underline{\mathcal{E}})$ and (2.134) from the right with $\underline{\mathcal{H}}^*$ and obtain

$$-\underline{\mathcal{E}}\wedge \Big| \qquad d\underline{\mathcal{H}}^* = -\mathrm{j}\omega\underline{\varepsilon}^* \star \underline{\mathcal{E}}^* + \underline{\mathcal{J}}_0^* \,, \tag{4.17}$$

$$d\underline{\mathcal{E}} = -\mathrm{j}\omega\underline{\mu} \star \underline{\mathcal{H}} \qquad \Big| \wedge \underline{\mathcal{H}}^* \,. \tag{4.18}$$

Forming the sum of both equations, we obtain

$$\mathrm{d}\left(\frac{1}{2}\underline{\mathcal{E}} \wedge \underline{\mathcal{H}}^*\right) = -2\mathrm{j}\omega\left(\frac{\underline{\mu}}{4}\underline{\mathcal{H}}^* \wedge \star\, \underline{\mathcal{H}} - \frac{\underline{\varepsilon}^*}{4}\underline{\mathcal{E}} \wedge \star\, \underline{\mathcal{E}}^*\right) - \frac{1}{2}\underline{\mathcal{E}} \wedge \underline{\mathcal{J}}_0^* \,. \tag{4.19}$$

We now introduce the *complex Poynting vector* $\boldsymbol{T}$ with the corresponding differential form

$$\mathcal{T} = T_x \,\mathrm{d}y \wedge \mathrm{d}z + T_y \,\mathrm{d}z \wedge \mathrm{d}x + T_z \,\mathrm{d}x \wedge \mathrm{d}y \,, \tag{4.20}$$

given by

$$\mathcal{T} = \frac{1}{2}\underline{\mathcal{E}} \wedge \underline{\mathcal{H}}^* \,. \tag{4.21}$$

We have to note that $\mathcal{T}$ is not the phasor corresponding to $\mathcal{S}$. Therefore we have used a different character to distinguish between the complex Poynting vector and the real Poynting vector. In order to give an interpretation of the complex Poynting vector $\boldsymbol{T}$, we compute first the time-dependent Poynting vector form $\mathcal{S}$ for a harmonic electromagnetic field

$$\mathcal{E}(\boldsymbol{x}, t) = \Re\left\{\underline{\mathcal{E}}(\boldsymbol{x})\,\mathrm{e}^{\mathrm{j}\omega t}\right\} = \frac{1}{2}\left(\underline{\mathcal{E}}(\boldsymbol{x})\,\mathrm{e}^{\mathrm{j}\omega t} + \underline{\mathcal{E}}^\star(\boldsymbol{x})\,\mathrm{e}^{-\mathrm{j}\omega t}\right) , \tag{4.22}$$

$$\mathcal{H}(\boldsymbol{x}, t) = \Re\left\{\underline{\mathcal{H}}(\boldsymbol{x})\,\mathrm{e}^{\mathrm{j}\omega t}\right\} = \frac{1}{2}\left(\underline{\mathcal{H}}(\boldsymbol{x})\,\mathrm{e}^{\mathrm{j}\omega t} + \underline{\mathcal{H}}^\star(\underline{\boldsymbol{x}})\,\mathrm{e}^{-\mathrm{j}\omega t}\right) . \tag{4.23}$$

Inserting into (4.14) we obtain

$$\mathcal{S}(\boldsymbol{x}, t) = \frac{1}{2}\Re\left\{\underline{\mathcal{E}}(\boldsymbol{x}) \wedge \underline{\mathcal{H}}^\star(\boldsymbol{x})\right\} + \frac{1}{2}\Re\left\{\underline{\mathcal{E}}(\boldsymbol{x}) \wedge \underline{\mathcal{H}}(\boldsymbol{x})\,\mathrm{e}^{2\mathrm{j}\omega t}\right\} . \tag{4.24}$$

The first term on the right side of (4.24) is equal to the real part of the complex Poynting form $\mathcal{T}$ according to (4.21). This term is independent of time. The second term on the right side of (4.24) oscillates with the double frequency of the alternating electromagnetic field. The time average of this part vanishes. Therefore the real part of the complex Poynting vector form $\mathcal{T}$ is the time average of the Poynting form $\mathcal{S}$.

$$\overline{\mathcal{S}(\boldsymbol{x},t)} = \Re\{\mathcal{T}(\boldsymbol{x})\}. \tag{4.25}$$

The real part of the complex Poynting vector $\boldsymbol{T}$ denotes the power flowing through a unit area surface element oriented perpendicular to $\boldsymbol{T}$.

For time-harmonic fields the time averages of the electric and magnetic energy densities $\overline{w}_e$ and $\overline{w}_m$ and the corresponding differential forms are related via

$$\overline{\mathcal{W}}_e = \overline{w}_e \, \mathrm{d}x \wedge \mathrm{d}y \wedge \mathrm{d}z\,, \tag{4.26}$$

$$\overline{\mathcal{W}}_m = \overline{w}_m \, \mathrm{d}x \wedge \mathrm{d}y \wedge \mathrm{d}z\,. \tag{4.27}$$

The *time-average electric* and *magnetic energy density forms* are given by

$$\overline{\mathcal{W}}_e = \frac{\varepsilon'}{4}\underline{\mathcal{E}} \wedge \star\, \underline{\mathcal{E}}^* = \frac{\varepsilon'}{4}\left(|\underline{E}_x|^2 + |\underline{E}_y|^2 + |\underline{E}_z|^2\right) \mathrm{d}x \wedge \mathrm{d}y \wedge \mathrm{d}z\,, \tag{4.28}$$

$$\overline{\mathcal{W}}_m = \frac{\mu'}{4}\underline{\mathcal{H}} \wedge \star\, \underline{\mathcal{H}}^* = \frac{\mu'}{4}\left(|\underline{H}_x|^2 + |\underline{H}_y|^2 + |\underline{H}_z|^2\right) \mathrm{d}x \wedge \mathrm{d}y \wedge \mathrm{d}z\,. \tag{4.29}$$

We have to consider that the quantities ε' and μ' in the complex representation correspond to the quantities ε and μ in the time-dependent formulation. The *time-average electric power dissipation density* $\overline{p}_{Le}$ with the differential form

$$\overline{\mathcal{P}}_{Le} = \overline{p}_{Le} \, \mathrm{d}x \wedge \mathrm{d}y \wedge \mathrm{d}z \tag{4.30}$$

is given by the differential form

$$\overline{\mathcal{P}}_{Le} = \frac{\sigma}{2}\, \underline{\mathcal{E}} \wedge \star\, \underline{\mathcal{E}}^* = \frac{\omega\varepsilon''}{2}\, \underline{\mathcal{E}} \wedge \star\, \underline{\mathcal{E}}^*. \tag{4.31}$$

The introduction of the complex permittivity $\underline{\mu}$ allows also to consider the magnetic losses with the average magnetic power dissipation density $\overline{p}_{Lm}$ with the differential form

$$\overline{\mathcal{P}}_{Lm} = \overline{p}_{Lm} \, \mathrm{d}x \wedge \mathrm{d}y \wedge \mathrm{d}z \tag{4.32}$$

given by

$$\overline{\mathcal{P}}_{Lm} = \frac{\omega\mu''}{2}\, \underline{\mathcal{H}} \wedge \star\, \underline{\mathcal{H}}^*. \tag{4.33}$$

The total average power dissipation density is described by the differential form

$$\overline{\mathcal{P}}_{L} = \frac{\omega\varepsilon''}{2}\, \underline{\mathcal{E}} \wedge \star\, \underline{\mathcal{E}}^* + \frac{\omega\mu''}{2}\, \underline{\mathcal{H}} \wedge \star\, \underline{\mathcal{H}}^*. \tag{4.34}$$

The complex power added to the field per unit volume, $P_{c0}(\boldsymbol{x})$ due to the impressed current density $\boldsymbol{J}_0$ is described by the differential form

$$\mathcal{P}_{c0} = P_{c0}\,\mathrm{d}x \wedge \mathrm{d}y \wedge \mathrm{d}z\,. \tag{4.35}$$

The differential form describing the complex power added via the impressed current $\underline{\mathcal{J}}_0$ is

$$\mathcal{P}_{c0} = \frac{1}{2}\underline{\mathcal{E}} \wedge \underline{\mathcal{J}}_0^*\,. \tag{4.36}$$

The real part of P_{c0} equals the time average $\overline{P}_0$ according to equation (4.25).

$$\overline{P}_0 = \Re\{P_{c0}\}\,. \tag{4.37}$$

The proof is similar to the one of (4.25).

After inserting of (4.21), (4.28), (4.29), (4.34) and (4.36) into (4.19), we obtain the local form of the *complex Poynting's theorem*:

$$\mathrm{d}\mathcal{T} = -2\mathrm{j}\omega(\overline{\mathcal{W}}_m - \overline{\mathcal{W}}_e) - \overline{\mathcal{P}}_L + \mathcal{P}_{c0}\,. \tag{4.38}$$

By integration over a volume V, we obtain the *integral form of the complex Poynting's theorem*

$$\oint_{\partial V} \mathcal{T} = \int_V \mathcal{P}_{c0} - 2\mathrm{j}\omega \int_V (\overline{\mathcal{W}}_m - \overline{\mathcal{W}}_e) - \int_V \overline{\mathcal{P}}_L\,. \tag{4.39}$$

We consider first the real part of (4.39):

$$\Re\left\{\oint_{\partial V} \mathcal{T}\right\} = \Re\left\{\int_V \mathcal{P}_{c0}\right\} - \int_V \overline{\mathcal{P}}_L\,. \tag{4.40}$$

The left side of (4.40) equals the active power radiated from inside the volume V through the boundary ∂V. On the right side of this equation, the first term denotes the power added via the impressed current density $\boldsymbol{J}_0$; the second term describes the conductive losses, the dielectric losses and the magnetic losses inside the volume V. The imaginary part of (4.40) is

$$\Im\left\{\oint_{\partial V} \mathcal{T}\right\} = \Im\left\{\int_V \mathcal{P}_{c0}\right\} - 2\omega \int_V (\overline{\mathcal{W}}_m - \overline{\mathcal{W}}_e)\,. \tag{4.41}$$

The first term on the right side gives the reactive power added into the volume V via the impressed current density $\boldsymbol{J}_0$. Let us first consider the case where the second term on the right side is vanishing. In this case we see that the left side of (4.41) denotes the power radiated from volume V. Since the volume V can be chosen arbitrarily,

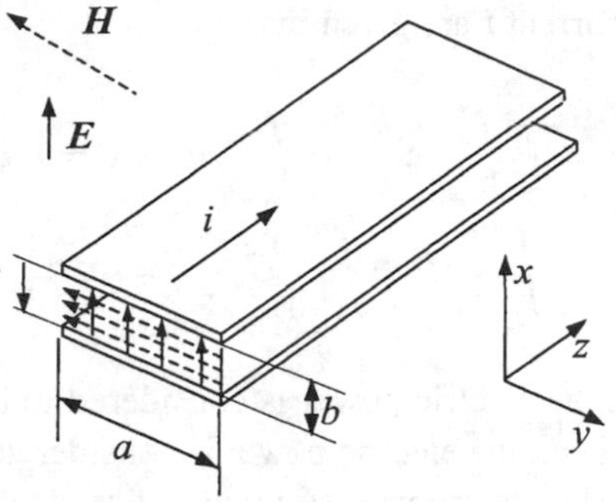

Figure 4.3: Power flow in a strip line.

it follows that the imaginary part of the complex Poynting vector $\boldsymbol{T}$ describes the reactive power radiated through a unit area normally oriented to the vector $\boldsymbol{T}$. The second term on the right side of (4.41) contains the product of the double angle of frequency with the difference of the average stored magnetic and electric energies. This term yields no contribution, if the magnetic energy stored in the volume V equals the average electric energy stored in V. The magnetic energy as well as electric energy oscillate with an angular frequency 2ω.

The field energy is permanently converted between electric energy and magnetic energy. If the averages $\overline{w}_e$ and $\overline{w}_m$ are equal, electric and magnetic energies may be mutually converted completely. In this case the energy oscillates between electric and magnetic fields inside the volume V. If the average electric and magnetic energies are not equal, energy as well oscillates between volume V and the space outside V. In this case there is a power flow between V and the outer region. For $\overline{w}_m > \overline{w}_e$ the reactive power flowing into volume V is positive, whereas for $\overline{w}_m < \overline{w}_e$ the reactive power flowing into V is negative.

To give an example for the relation between the description by voltages and currents and the electromagnetic field description let us consider the *strip line* depicted in Figure 2.13. We assume a transverse electromagnetic wave to propagate in positive z-direction. In the case of a transverse electromagnetic wave neither the electric field nor the magnetic field exhibits a z component. Let us furthermore assume the distance b between both conductor strips to be small compared with the width a of the strips. The electric power P transported via the line in the positive z-direction is given by

$$P = v\,i\,. \tag{4.42}$$

The voltage v and the current i are given by

$$v = -\int_{x=0}^{b} \mathcal{E} = -\int_{x=0}^{b} E_x \,\mathrm{d}x = b\,E_x\,, \tag{4.43}$$

$$i = \int_{y=0}^{a} \mathcal{H} = \int_{y=0}^{a} H_y \,\mathrm{d}y = a\,H_y\,. \tag{4.44}$$

In the *network concept* the electric power is considered to be carried by voltage and current. In the *field concept*, the electric power is considered to be transported via the electromagnetic field. The electromagnetic power flux density is represented by the *Poynting form* $\mathcal{S}$ given by

$$\mathcal{S} = E_x\,H_y\,\mathrm{d}x \wedge \mathrm{d}y. \tag{4.45}$$

If $\boldsymbol{E}$ and $\boldsymbol{H}$ are orthogonal, the vectors $\boldsymbol{E}$, $\boldsymbol{H}$ and $\boldsymbol{S}$ form a positive-oriented orthogonal tripod. In Figure 4.3 the Poynting vector $\boldsymbol{S}(\boldsymbol{x}, t)$ is directed in the positive z-direction, with z-component S_z.

$$P = \int_A \mathcal{S} = \int_A E_x\,H_y\,\mathrm{d}x \wedge \mathrm{d}y. \tag{4.46}$$

We obtain the power P flowing through the strip line by multiplying the power density S with the cross-sectional area $a\,b$ of the strip line and obtain

$$P = a\,b\,E_x\,H_y\,. \tag{4.47}$$

Inserting (4.43) and (4.44) into this equation yields (4.42). The network concept and the field concept give the same result.

4.2 FIELD THEORETIC FORMULATION OF TELLEGEN'S THEOREM

Complex electromagnetic structures may be subdivided into several spatial subdomains. Comparing a distributed circuit represented by an electromagnetic structure with a lumped element circuit represented by a network, the spatial subdomains may be considered as the circuit elements whereas the complete set of boundary surfaces separating the subdomains corresponds to the connection circuit [6]. Figure 4.4 shows the segmentation of an electromagnetic structure into different regions $\mathcal{R}_l$ separated by boundaries B_{lk}. The dashed curves denote the boundaries. The regions $\mathcal{R}_l$ may contain any electromagnetic substructure. In our network analogy the two-dimensional manifold of all boundary surfaces B_{lk} represents the connection circuit whereas the subdomains V_l are representing the circuit elements. We can establish a

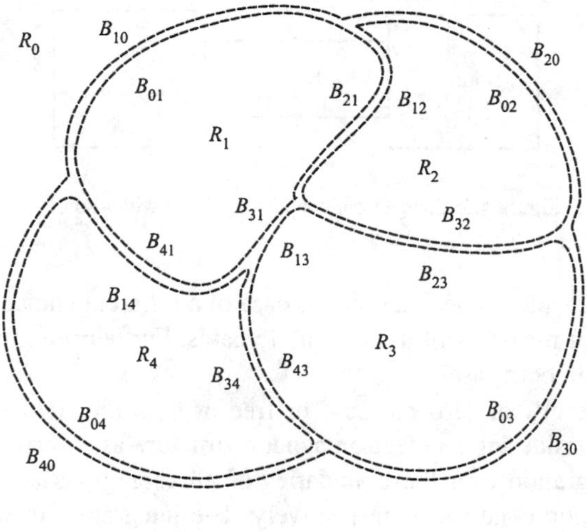

Figure 4.4: Segmentation of an electromagnetic structure.

field representation of the *Tellegen's theorem* relating the tangental electric and magnetic fields on the two-dimensional manifolds of boundaries B_{lk} [7].

Tellegen's theorem states fundamental relations between voltages and currents in a network and is of considerable versatility and generality in network thoory [7–9]. A noticeable property of this theorem is that it is only based on Kirchhoff's current and voltage laws, i.e., on topological relationships and that it is independent from the constitutive laws of the network. The same reasoning that yields from Kirchhoff's laws to Tellegen's theorem allows us to directly derive a field form of Tellegen's theorem from Maxwell's equations [7].

In order to derive Tellegen's theorem for partitioned electromagnetic structures let us consider two electromagnetic structures based on the same partition by equal boundary surfaces. The subdomains of either electromagnetic structure, however, may be filled with different materials. The connection network is established via the relations of the tangential field components on both sides of the boundaries. Since the connection network exhibits zero volume no field energy is stored therein and no power loss occurs therein.

Starting directly from Maxwell's equations we may derive for a closed volume V_n with boundary surface ∂V_n and relative normal vector $\boldsymbol{n}$ the following relation:

$$\oint_{\partial V_n} \mathcal{E}'(\boldsymbol{x}, t') \wedge \mathcal{H}''(\boldsymbol{x}, t'') = -\oint_{V_n} \mathcal{E}'(\boldsymbol{x}, t') \wedge \mathcal{J}''(\boldsymbol{x}, t'')$$
$$-\int_{V_n} \mathcal{E}'(\boldsymbol{x}, t') \wedge \frac{\partial \mathcal{D}''(\boldsymbol{x}, t'')}{\partial t''} - \int_{V_n} \mathcal{H}'(\boldsymbol{x}, t') \wedge \frac{\partial \mathcal{B}''(\boldsymbol{x}, t'')}{\partial t''}. \quad (4.48)$$

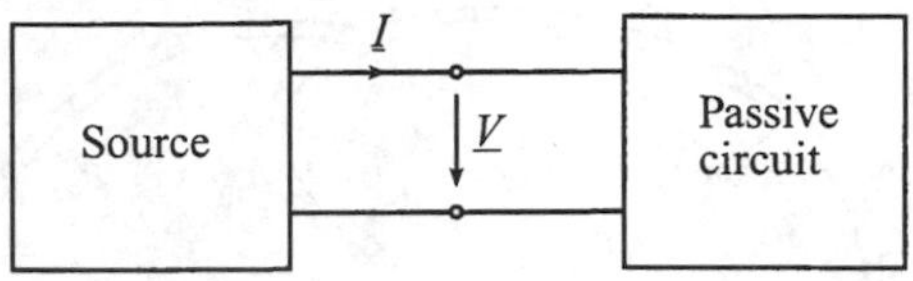

Figure 4.5: Connection of a passive circuit with a source.

The prime $'$ and double prime $''$ denote the case of a different choice of sources and a different choice of materials filling the subdomains. Furthermore the time argument may be different in both cases.

For volumes V_n of zero measure or free of field the right side of this equation vanishes. Considering an electromagnetic structure as shown in Figure 4.4, we perform the integration over the boundaries of all subregions not filled with ideal electric or magnetic conductors, respectively. The integration over both sides of a boundary yields zero contribution to the integrals on the right side of (4.48). Also the integration over finite volumes filled with ideal electric or magnetic conductors gives no contribution to these integrals. We obtain the *field form of Tellegen's theorem*:

$$\oint_{\partial V_n} \mathcal{E}'(\boldsymbol{x}, t') \wedge \mathcal{H}''(\boldsymbol{x}, t'') = 0\,. \tag{4.49}$$

Tellegen's theorem is a very powerful theorem yielding important applications in the segmentation of electromagnetic structures and in circuit theory. In Section 10.6 Tellegen's theorem will be treated from the network point of view.

4.3 SOURCES OF THE ELECTROMAGNETIC FIELD

In the network concept, electric sources are modelled using ideal current sources or ideal voltage sources, respectively. Ideal current sources or voltage sources impress a current or a voltage, respectively, into the network. Sources may be modelled by parallel circuiting an ideal current source and an admittance or series circuiting an ideal voltage source and an impedance. In this fashion, we obtain a simple phenomenological description of sources without having to consider the complicated structure of real sources in detail. In the same way we can establish a simple phenomenological description of electromagnetic field sources [2].

Figure 4.5 shows the connection of a network source with a passive network. The network source and the passive network are connected via one pair of nodes. From the network source a current $\underline{I}$ is flowing into the passive network. At the pair of nodes a voltage $\underline{V}$ is occurring. The node current $\underline{I}$ and node voltage $\underline{V}$ are related

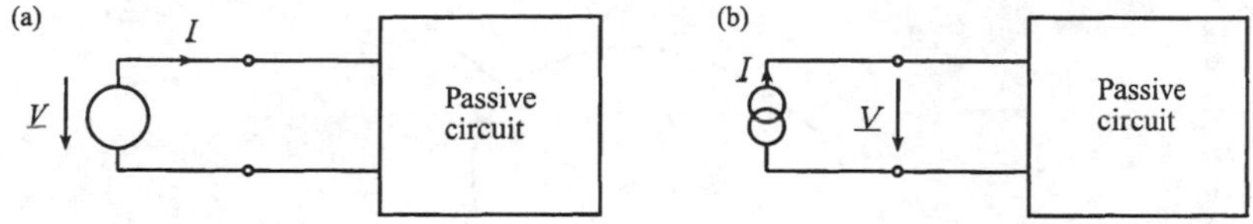

Figure 4.6: Introduction of (a) an equivalent voltage source $\underline{V}$ and (b) an equivalent current source $\underline{I}$.

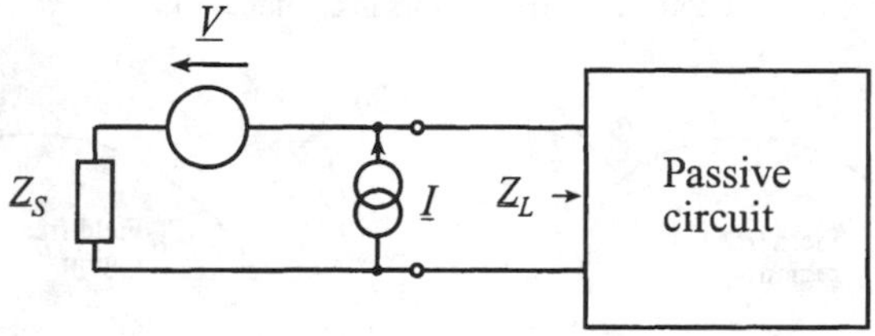

Figure 4.7: Introduction of an equivalent voltage source $\underline{V}$ and an equivalent current source $\underline{I}$.

via

$$\underline{V} = Z_L \underline{I}, \tag{4.50}$$

where Z_L is the impedance of the passive network. If the node current $\underline{I}$ and the node voltage $\underline{V}$ are known, it is possible to replace the source with a voltage source $\underline{V}$ or a current source $\underline{I}$. Figure 4.6(a) shows the introduction of an equivalent voltage source $\underline{V}$, and Figure 4.6(b) shows the introduction of an equivalent current source $\underline{I}$. It is also possible to introduce an equivalent current source $\underline{I}$ and an equivalent voltage source $\underline{V}$ together as shown in Figure 4.7. Whereas in Figure 4.6(a) the impedance of the source is replaced with a short circuit, and in Figure 4.6(b) the impedance of the source is replaced by an open circuit, we can insert a source impedance Z_S in Figure 4.7. Due to (4.50) in Figure 4.7, no current is flowing through Z_S and no voltage is applied to Z_S. This means that we can replace Z_S by an arbitrary impedance including also a short circuit and an open circuit. Replacing Z_S with a short circuit, we can omit the current source in Figure 4.7 and we have reduced the source to an impressed voltage source as in Figure 4.6(a). If, however, Z_S is replaced by an open circuit, we can omit the impressed voltage source $\underline{V}$, and we obtain an equivalent circuit as shown in Figure 4.6(b).

4.4 THE UNIQUENESS THEOREM

Figure 4.8 shows a number of field sources within a volume V_1. The volume V_1 is bounded by the virtual boundary ∂V_1. In our network model of the source we

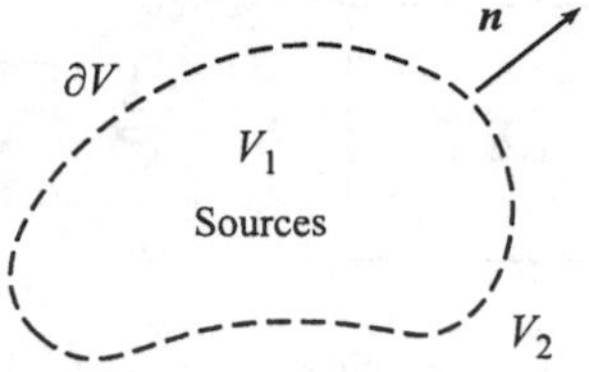

Figure 4.8: Field sources in a volume V_1.

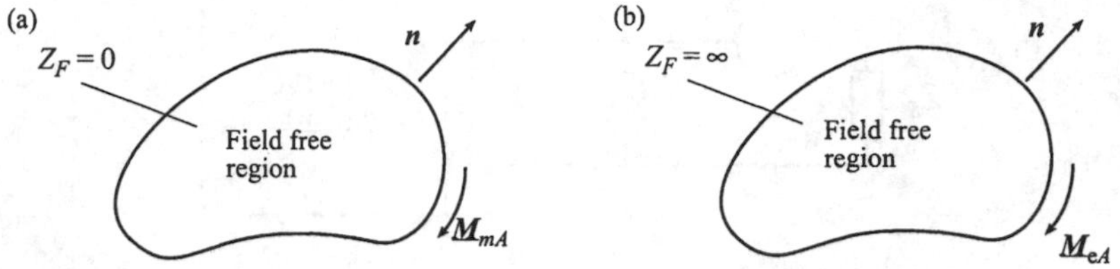

Figure 4.9: (a) Magnetic area polarization $\underline{M}_{mA}$ impressed on an ideal electric conductor and (a) electric area polarization $\underline{M}_{eA}$ impressed on an ideal magnetic conductor.

only were interested in investigating the passive circuit and decided not to investigate the source in detail. Therefore we have chosen a simple model for the source. We have an analogous situation in our field model, where we are only interested in the field outside V_1, which is generated by the sources located within the volume V_1. We will try to replace the field sources inside volume V_1 by impressed equivalent polarizations at the boundary surface ∂V_1. To do this, we first prove the so-called *uniqueness theorem* [2,4], which states:

> *The electromagnetic field in the source-free outside region V_2 is determined in a unique way, if the tangential component of either the electric field intensity or the magnetic field intensity is known on the boundary surface ∂V_1.*

To prove the uniqueness theorem, we will show that the opposite assumption will yield a wrong result. Let us assume that in the outer region V_2 there exist two different field solutions $\underline{E}_a$ and $\underline{H}_a$ on the one hand, and $\underline{E}_b$ and $\underline{H}_b$ on the other hand, that exhibit the same tangential electric and magnetic field components on ∂V. The difference between field solutions $\delta\underline{E} = \underline{E}_a - \underline{E}_b$, $\delta\underline{H} = \underline{H}_a - \underline{H}_b$ due to the linearity of the field equations must also be a field solution. In this case, however, either $\delta\underline{E}$ or $\delta\underline{H}$ has no tangential field component on ∂V. Consequently the complex Poynting differential form $\mathcal{T}$ due to (4.21) has no component normal to ∂V, and the surface integral of $\mathcal{T}$ over ∂V vanishes. We now apply the real part of (4.39) to the outer

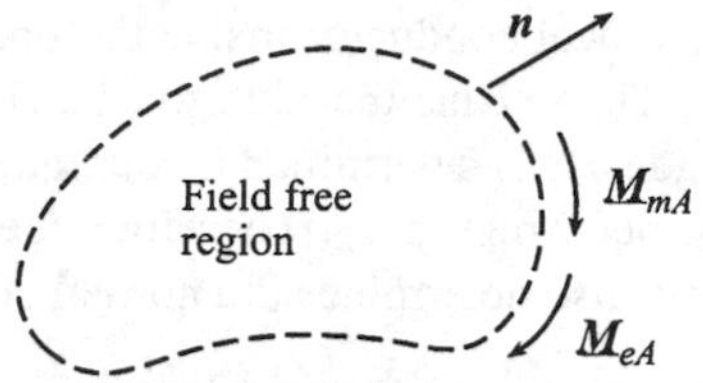

Figure 4.10: Electric area polarization $\underline{M}_{eA}$ and magnetic area polarization $\underline{M}_{mA}$ impressed in a virtual boundary surface.

region V_2. Since the left side of (4.40) vanishes, and since V_2 is free of sources, also the first term on the right side vanishes, and the integral of the average power loss density p_v over V_2 also must vanish. Due to (4.34) the integrand is positive definite and can in the case of arbitrarily small electric and magnetic losses only vanish, if in the complete outer region V_2, $\delta \underline{E} = \mathbf{0}$ as well as $\delta \underline{H} = \mathbf{0}$ are fulfilled. This means that in V_2 the identities $\underline{E}_a = \underline{E}_b$ as well as $\underline{H}_a = \underline{H}_b$ are fulfilled and consequently the definition of the tangential components of either $\underline{E}_a$ or $\underline{H}_a$ on ∂V determines the electromagnetic field in the outer space in an unique way. This proves the uniqueness theorem.

4.5 THE EQUIVALENCE PRINCIPLE

Many source distributions outside a certain region may yield the same field distribution inside that region. Two source distributions outside a region producing the same field distribution inside this region are said to be *equivalent* within that region. This is called the *equivalence principle* [2]. The equivalence principle follows directly from the uniqueness theorem. Due to the uniqueness theorem different source distributions inside a volume V that produce the same tangential electric or magnetic field distribution on the boundary ∂V will produce the same electromagnetic field outside the volume V.

The uniqueness theorem makes it possible to replace the sources in the volume V_1 by area polarizations impressed into the boundary surface ∂V. In the same way as we have replaced the source in Figure 4.5 by a current source in series with a short circuit in Figure 4.6(b), now we may replace the virtual boundary surface ∂V in Figure 4.8 by a real boundary surface ∂V formed by an ideal conductor as shown in Figure 4.9(a). We impress a magnetic area polarization $\underline{\mathcal{M}}_{mA}$ given by

$$\underline{\mathcal{M}}_{mA}(x') = -\frac{1}{j\omega}\, n \lrcorner \left(n \wedge \underline{\mathcal{E}}\right) = -\frac{1}{j\omega}\underline{\mathcal{E}}_t\,, \tag{4.51}$$

on this conductor where $\underline{E}$ is the electric field intensity on ∂V as specified in Figure 4.8. Since we assumed an ideal conductor inside the boundary, now the electric field within V_1 is vanishing. This means that due to (2.207), the tangential component of the electric field outside ∂V is determined in a unique way. We have assumed that there is no source in the outer region V_2. Therefore due to (4.21) the tangential component of the electric field also determines the normal component.

The replacement of the source by an equivalent current source in the network model in Figure 4.6(b) corresponds to the replacement of the boundary surface ∂V by an ideal magnetic conductor ($Z_F = \infty$) and impressing an electric area polarization $\underline{\mathcal{M}}_{eA}$ on ∂V. The electric area polarization is given by

$$\underline{\mathcal{M}}_{eA}(x') = \frac{1}{\mathrm{j}\omega} n \lrcorner \left(n \wedge \underline{\mathcal{H}}\right) = \frac{1}{\mathrm{j}\omega} \underline{\mathcal{H}}_t . \tag{4.52}$$

Since the magnetic field inside the boundary ∂V vanishes, due to (2.206) for an electric area polarization $\underline{\mathcal{M}}_{eA}$ impressed on the boundary ∂V the tangential magnetic field on the outer surface of ∂V is determined in a unique way. Figure 4.9(b) illustrates the impression of electric area polarization sources on an ideal magnetic conductor. The ideal conductor in support of the impressed magnetic polarization as shown in Figure 4.9(a) corresponds to the short-circuit in series to the ideal voltage source in Figure 4.6(a) and the ideal magnetic conductor in support of the impressed electric polarization according to Figure 4.9(b) corresponds to the open-circuit in series to the ideal voltage source in Figure 4.6(b).

The combination of impressed voltage sources and impressed current sources due to Figure 4.7 yields an analogous field model. We remove in Figure 4.8 the field sources in the inner region V_1 and impress in ∂V magnetic as well as electric area polarizations $\underline{\boldsymbol{M}}_{mA}$ and $\underline{\boldsymbol{M}}_{eA}$, respectively, such that in the inner region V_1 the electromagnetic field vanishes completely, and in the outer region V_2 the electromagnetic field corresponds to the electromagnetic field as generated before by the sources in V_1. Figure 4.10 illustrates this arrangement. The replacement of the inner field sources by the impressed area polarizations $\underline{\boldsymbol{M}}_{mA}$ and $\underline{\boldsymbol{M}}_{eA}$ is accomplished by choosing the impressed area polarizations according to (4.51) and (4.52). Due to (2.206) and (2.207) this choice of the equivalent polarizations is compatible with a vanishing field in the inner region V_1 and the field produced by the original field sources in the outer region V_2. Since in Figure 4.10 both regions V_1 as well as V_2 contain no sources due to the uniqueness theorem this solution also is the only existing solution. Since V_2 is free of field, any medium without sources introduced into the region V_1 does not influence the field in V_2. This case corresponds to the introduction of a voltage source and a current source in the network model according to Figure 4.7.

4.6 RECIPROCITY

4.6.1 The Lorentz Reciprocity Theorem

We consider the electromagnetic field excited by various field sources. We combine these field sources into sets of field sources. Each of these sets of field sources may contain an arbitrary distribution of impressed electric polarization sources $\underline{\boldsymbol{M}}_{e0i}(\boldsymbol{x})$ and an arbitrary distribution of impressed magnetic polarizations $\underline{\boldsymbol{M}}_{m0i}(\boldsymbol{x})$ with $i = 1, 2, \ldots n$. Let $\underline{\boldsymbol{E}}_i(\boldsymbol{x})$ and $\underline{\boldsymbol{H}}_i(\boldsymbol{x})$ be the electromagnetic field excited by the polarization distributions $\underline{\boldsymbol{M}}_{e0i}$ and $\underline{\boldsymbol{M}}_{m0i}$. From (3.32) and (3.38) it follows

$$\mathrm{d}\underline{\mathcal{H}}_i = \mathrm{j}\omega\left(\underline{\varepsilon} \star \underline{\mathcal{E}}_i + \underline{\mathcal{M}}_{e0i}\right), \tag{4.53}$$

$$\mathrm{d}\underline{\mathcal{E}}_i = -\mathrm{j}\omega\left(\underline{\mu} \star \underline{\mathcal{H}}_i + \underline{\mathcal{M}}_{m0i}\right). \tag{4.54}$$

Let us now compute the expression

$$\begin{aligned} -\mathrm{d}(\underline{\mathcal{E}}_i \wedge \underline{\mathcal{H}}_j) &= \mathrm{d}\underline{\mathcal{E}}_i \wedge \underline{\mathcal{H}}_j - \underline{\mathcal{E}}_i \wedge \mathrm{d}\underline{\mathcal{H}}_j = \\ &= \mathrm{j}\omega\left[\underline{\varepsilon}\underline{\mathcal{E}}_i \wedge (\star\, \underline{\mathcal{E}}_j) + \underline{\mu}(\star\, \underline{\mathcal{H}}_i) \wedge \underline{\mathcal{H}}_j + \underline{\mathcal{M}}_{m0i} \wedge \underline{\mathcal{H}}_j + \underline{\mathcal{E}}_i \wedge \underline{\mathcal{M}}_{e0j}\right]. \end{aligned} \tag{4.55}$$

After interchanging i and j and forming the difference of both equations, we obtain

$$\begin{aligned} -\,\mathrm{d}(\underline{\mathcal{E}}_i \wedge \underline{\mathcal{H}}_j - \underline{\mathcal{E}}_j \wedge \underline{\mathcal{H}}_i) &= \\ = \mathrm{j}\omega\left[\underline{\mathcal{E}}_i \wedge \underline{\mathcal{M}}_{e0j} - \underline{\mathcal{H}}_i \wedge \underline{\mathcal{M}}_{m0j} - \underline{\mathcal{E}}_j \wedge \underline{\mathcal{M}}_{e0i} + \underline{\mathcal{H}}_j \wedge \underline{\mathcal{M}}_{m0i}\right]&. \end{aligned} \tag{4.56}$$

In a source-free subregion of the space the right side of (4.56) disappears and we obtain the so-called *Lorentz reciprocity theorem*, which states the following: In source-free regions and for isotropic materials the electromagnetic fields $\underline{\boldsymbol{E}}_i$, $\underline{\boldsymbol{H}}_i$ and $\underline{\boldsymbol{E}}_j$, $\underline{\boldsymbol{H}}_j$, respectively, excited from different sets of sources $\underline{\boldsymbol{M}}_{e0i}$, $\underline{\boldsymbol{M}}_{m0i}$ and $\underline{\boldsymbol{M}}_{e0j}$, $\underline{\boldsymbol{M}}_{m0j}$, respectively, satisfy the equation

$$\mathrm{d}(\underline{\mathcal{E}}_i \wedge \underline{\mathcal{H}}_j - \underline{\mathcal{E}}_j \wedge \underline{\mathcal{H}}_i) = 0. \tag{4.57}$$

Integrating this equation over a source-free subdomain V of the space and converting the volume integral into a surface integral by using the Stokes' law we obtain the integral form of the *Lorentz reciprocity theorem*

$$\oint_{\partial V} (\underline{\mathcal{E}}_i \wedge \underline{\mathcal{H}}_j - \underline{\mathcal{E}}_j \wedge \underline{\mathcal{H}}_i) = 0. \tag{4.58}$$

4.6.2 The Reciprocity Theorem for Impressed Sources

We now derive from (4.56) another useful form of the theorem of reciprocity. For this purpose we integrate (4.56) over a volume V, where we are transforming the left-hand side into a surface integral over the boundary ∂V.

$$
\begin{aligned}
-\oint_{\partial V} (\underline{\mathcal{E}}_i \wedge \underline{\mathcal{H}}_j - \underline{\mathcal{E}}_j \wedge \underline{\mathcal{H}}_i) = \\
= \mathrm{j}\omega \int_V \left[\underline{\mathcal{E}}_i \wedge \underline{\mathcal{M}}_{e0j} - \underline{\mathcal{H}}_i \wedge \underline{\mathcal{M}}_{m0j} - \underline{\mathcal{E}}_j \wedge \underline{\mathcal{M}}_{e0i} + \underline{\mathcal{H}}_j \wedge \underline{\mathcal{M}}_{m0i} \right] . \qquad (4.59)
\end{aligned}
$$

$\underline{\boldsymbol{M}}_{e0i}$, $\underline{\boldsymbol{M}}_{m0i}$, and $\underline{\boldsymbol{M}}_{e0j}$, $\underline{\boldsymbol{M}}_{m0j}$, respectively, are mutually independent distributions of impressed field sources of the source-sets i and j. $\underline{\boldsymbol{E}}_i$, $\underline{\boldsymbol{H}}_i$, and $\underline{\boldsymbol{E}}_j$, $\underline{\boldsymbol{H}}_j$, respectively, are the fields excited by the sets of field source sets i and j. We now are choosing a volume of integration V such that all field sources of the ith group as well as the field sources of the jth group are located in the volume V and choose the boundary ∂V sufficiently far in the farfield so that all the field sources of the ith and the jth set form together a point-like source, if they are observed from any point on the boundary ∂V. Since we may choose the volume V arbitrarily this condition may be fulfilled with an arbitrary accuracy.

We now embed our electromagnetic structure in a spherical volume V with radius r_V and let $r_V \to \infty$. From the Sommerfeld radiation condition (3.88b) we obtain on the boundary ∂V of the volume V

$$
\underline{\mathcal{E}}_i \wedge \underline{\mathcal{H}}_j = \underline{\mathcal{E}}_j \wedge \underline{\mathcal{H}}_i = \frac{1}{Z_{F0}} \left(\underline{E}_{\vartheta i} \underline{E}_{\vartheta j} + \underline{E}_{\varphi i} \underline{E}_{\varphi j} \right) r^2 \sin\vartheta \, \mathrm{d}\vartheta \wedge \mathrm{d}\varphi . \qquad (4.60)
$$

Since the field components are of order $(1/r)$ this expression remains finite for $r \to \infty$. Therefore the left-hand side of (4.59) vanishes, if we are expanding the volume V into infinity. Integrating over the complete space, the following relation is fulfilled exactly

$$
\mathrm{j}\omega \int_V \left[\underline{\mathcal{E}}_i \wedge \underline{\mathcal{M}}_{e0j} - \underline{\mathcal{H}}_i \wedge \underline{\mathcal{M}}_{m0j} \right] = \mathrm{j}\omega \int_V \left[\underline{\mathcal{E}}_j \wedge \underline{\mathcal{M}}_{e0i} - \underline{\mathcal{H}}_j \wedge \underline{\mathcal{M}}_{m0i} \right] . \qquad (4.61)
$$

The integral on the left side describes the *reaction* [2, 10] of the field $\underline{\boldsymbol{E}}_i$, $\underline{\boldsymbol{H}}_i$ on the sources $\underline{\boldsymbol{M}}_{e0j}$, $\underline{\boldsymbol{M}}_{m0j}$. This is a further useful formulation of the reciprocity theorem. We define the so-called *reaction* R_{ij} *of a field* $\underline{\boldsymbol{E}}_i$, $\underline{\boldsymbol{H}}_i$ *on the sources* $\underline{\boldsymbol{M}}_{e0j}$, $\underline{\boldsymbol{M}}_{m0j}$ by

$$
\mathsf{R}_{ij} = \mathrm{j}\omega \int_V \left[\underline{\mathcal{E}}_i \wedge \underline{\mathcal{M}}_{e0j} - \underline{\mathcal{H}}_i \wedge \underline{\mathcal{M}}_{m0j} \right] . \qquad (4.62)
$$

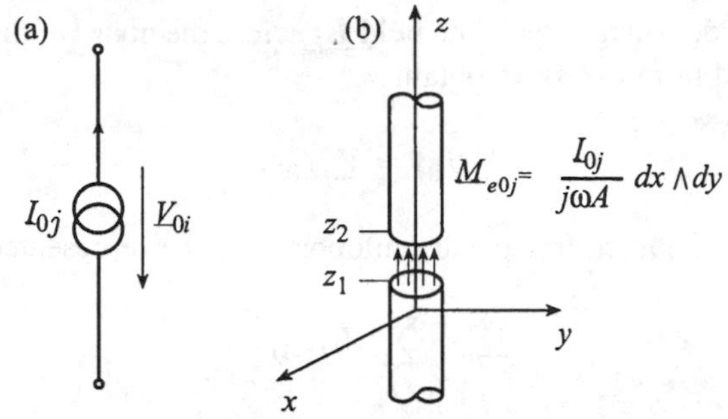

Figure 4.11: Impressed current source (a) in the network concept and (b) in the field concept.

The *reciprocity theorem* according to (4.61) now can be written in the form

$$\mathsf{R}_{ij} = \mathsf{R}_{ji} . \tag{4.63}$$

The reaction of a field $\underline{\boldsymbol{E}}_i, \underline{\boldsymbol{H}}_i$ on the sources $\underline{\boldsymbol{M}}_{e0j}, \underline{\boldsymbol{M}}_{m0j}$ is equal to the reaction of a field $\underline{\boldsymbol{E}}_j, \underline{\boldsymbol{H}}_j$ on the sources $\underline{\boldsymbol{M}}_{e0i}, \underline{\boldsymbol{M}}_{m0i}$. As an example we compute the reaction of a voltage $\underline{V}_i$ on a current source $\underline{I}_{0j}$ according to Figure 4.11(a). In a field an impressed current source $\underline{I}_{0j}$ can be represented, e.g., by a polarization density impressed into a gap of a conductor. Let us impress a polarization

$$\underline{\mathcal{M}}_{e0j} = \frac{\underline{I}_{0j}}{\mathrm{j}\omega A} \, \mathrm{d}x \wedge \mathrm{d}y \tag{4.64}$$

into the gap between the two conductors in Figure 4.11. Inserting (11.27) and integrating over the cross-sectional area A verifies this relation. According to (4.62) the reaction of the field $\underline{\boldsymbol{E}}_i$ on the network-source $\underline{I}_{0j}$ is given by

$$\mathsf{R}_{ij} = \frac{\underline{I}_{0j}}{A} \int_{V_{0j}} \underline{E}_{iz} \, dx \wedge dy \wedge dz \,. \tag{4.65}$$

Performing the integration over the gap region where the impressed polarization $\underline{\boldsymbol{M}}_{e0j}$ exhibits a non-zero value, yields

$$\int_{z_1}^{z_2} \underline{E}_{iz} dz = -\underline{V}_i \,. \tag{4.66}$$

From this we obtain

$$\mathsf{R}_{ij} = -\underline{V}_i \underline{I}_{0j} \,, \tag{4.67}$$

where $\underline{V}_i$ is the node voltage due to the field $\underline{\boldsymbol{E}}_i$ across the nodes of the current source $\underline{I}_{0j}$. From this and from (4.63) we obtain

$$\underline{V}_i \underline{I}_{0j} = \underline{V}_j \underline{I}_{0i} \,. \tag{4.68}$$

The port voltages of source-free linear multiports may be represented by the system of equations

$$\underline{V}_i = \sum_j Z_{ij} \underline{I}_{0j} \,, \tag{4.69}$$

if the $\underline{I}_{0j}$ are the impressed port currents. With (4.63) we obtain from this the condition

$$Z_{ij} = Z_{ji} \,. \tag{4.70}$$

This is the *network form of the reciprocity theorem.*

REFERENCES

[1] J. A. Stratton, *Electromagnetic Theory*. New York: McGraw-Hill, 1941.

[2] R. F. Harrington, *Time Harmonic Electromagnetic Fields*. New York: McGraw-Hill, 1961.

[3] J. D. Jackson, *Classical Electrodynamics*. New York: John Wiley & Sons, 1975.

[4] J. A. Kong, *Electromagnetic Wave Theory*. New York: John Wiley & Sons, 1986.

[5] H. A. Haus and J. R. Melcher, *Electromagnetic Fields and Energy*. Englewood Cliffs, NJ: Prentice Hall, 1989.

[6] P. Russer, M. Mongiardo, and L. Felsen, "Electromagnetic field representations and computations in complex structures III: Network representations of the connection and subdomain circuits," *Int. J. Numer. Modeling*, vol. 15, pp. 127–145, 2002.

[7] P. Penfield, R. Spence, and S. Duinker, *Tellegen's theorem and electrical networks*. Cambridge, Massachusetts: MIT Press, 1970.

[8] B. Tellegen, "A general network theorem with applications," *Philips Research Reports*, vol. 7, pp. 259–269, 1952.

[9] B. Tellegen, "A general network theorem with applications," *Proc. Inst. Radio Engineers*, vol. 14, pp. 265–270, 1953.

[10] V. Rumsey, "The reaction concept in electromagnetic theory," *Phys. Rev., ser. 2*, vol. 94, pp. 1483–1491, June 1954.

Chapter 5

Static and Quasistatic Fields

5.1 CONDITIONS FOR STATIC AND QUASISTATIC FIELDS

A field invariant with time is called a *static field*. For static fields the time derivatives in Maxwell's equations (2.120) to (2.123) vanish and we obtain

$$\mathrm{d}\mathcal{H} = \mathcal{J}\,, \qquad (5.1a) \qquad\qquad \mathrm{d}\mathcal{B} = 0\,, \qquad (5.1b)$$

$$\mathrm{d}\mathcal{E} = 0\,, \qquad (5.2a) \qquad\qquad \mathrm{d}\mathcal{D} = \mathcal{Q}\,. \qquad (5.2b)$$

In the static case electric and magnetic fields are not coupled with each other. The source of the electric field is the electric charge and the source of the magnetic field is the electric current. Since in the static case electric and magnetic fields are not coupled with each other the *electrostatic field* and the *magnetostatic field* may be treated independently. Electrostatic phenomena involve time-independent distributions of electric charge and electric field. On the other hand, there are no free magnetic charges. Therefore magnetic phenomena are quite different from electric phenomena. Magnetostatic phenomena involve time-independent distribution of electric current and magnetic field. A detailed treatment of electrostatic and magnetostatic fields is given in [1–5].

An example for an electrostatic problem is an arrangement of two or more conductors insulated from each other at different time-constant potential levels and with no impressed currents. In this case an electric field, but no magnetic field exists. Figure 5.1(a) shows an arrangement of two narrowly spaced conducting plates. The electric field and the stored electric energy are mainly concentrated in the space between the conducting plates. A structure optimized for storing electric energy is called a *capacitor*. An example for a magnetostatic problem is a conductor coil or solenoid with an impressed time-constant current. The current flowing through the conductor gives rise to a static magnetic field only. Figure 5.1(b) shows a solenoid. A current impressed into the solenoid gives rise to a magnetic field mainly concentrated

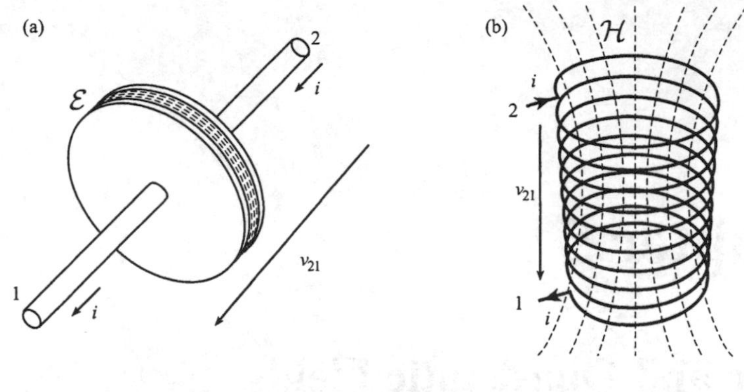

Figure 5.1: (a) Capacitor and (b) inductor.

inside the solenoid. A structure designed for storing magnetic energy when a current is impressed is called an *inductor*.

Maxwell's equations describe the most complex electromagnetic wave phenomena occurring at short time scales or at high frequencies. For problems dealing with relatively long time scales and low frequencies truncated versions of Maxwell's equations may be applied. To establish a measure for the vague characterization "slow" we consider the time an electromagnetic wave needs to propagate through a typical dimension of the system of interest. If this time is small compared with the time scale of field evolution in the system, we may consider the field as a so-called *quasistatic field.*

Slowly varying fields in many cases may be treated as *quasistatic fields* [3, 6]. The quasistatic laws are obtained by neglecting either the time derivative of the magnetic induction or the electric displacement current. The first approximation is called the *electroquasistatic approximation*, the second one the *magnetoquasistatic approximation*. From Maxwell's equations (2.120) – (2.123) we obtain

Electroquasistatic approximation

$$\mathrm{d}\mathcal{H} = \frac{\partial}{\partial t}\mathcal{D} + \mathcal{J}\,, \tag{5.3a}$$

$$\mathrm{d}\mathcal{E} \cong 0\,, \tag{5.4a}$$

$$\mathrm{d}\mathcal{B} = 0\,, \tag{5.5a}$$

$$\mathrm{d}\mathcal{D} = \mathcal{Q}\,, \tag{5.6a}$$

Magnetoquasistatic approximation

$$\mathrm{d}\mathcal{H} \cong \mathcal{J}\,, \tag{5.3b}$$

$$\mathrm{d}\mathcal{E} = -\frac{\partial}{\partial t}\mathcal{B}\,. \tag{5.4b}$$

$$\mathrm{d}\mathcal{B} = 0\,. \tag{5.5b}$$

$$\mathrm{d}\mathcal{D} = \mathcal{Q}\,. \tag{5.6b}$$

Due to (5.4a) the electroquasistatic field is essentially irrotational whereas (5.5b) requires the magnetoquasistatic field to be solenoidal.

To give an example for an electroquasistatic field we consider again the capacitor depicted in Figure 5.1(a). Impressing a slowly time-varying voltage gives rise to a time-varying electric flux between the capacitor plates and by that way to a displacement current flowing between the plates. To describe the capacitor in the case of an applied time-varying voltage we have to consider the time variation of the electric flux density $\mathcal{D}$. However, in the low-frequency case we usually can neglect the time variation of the magnetic induction $\mathcal{B}$ due to the current flowing through the capacitor. Capacitors may be treated within the electroquasistatic approximation specified by (5.3a) and (5.4a).

The inductor in Figure 5.1(b) may be considered within the magnetoquasistatic approximation specified by (5.3b) and (5.4b). The time variation of the current creates a time variation of the magnetic induction $\mathcal{B}$ in the solenoid and this time-varying flux induces a voltage in the solenoid. In the case of the inductor at low frequencies the displacement current may be neglected.

5.2 STATIC AND QUASISTATIC ELECTRIC FIELDS

5.2.1 The Green's Function for the Static Electric Field

In 1828 George Green published the *Essay on the Application of Mathematical Analysis to the Theories of Electricity and Magnetism* [7]. In this work Green developed a technique to solve Poisson's equation of potential theory. Green's function gives the potential of a point or line source of unit strength. Since arbitrary source distributions may be considered as superpositions of point or line sources Green's function is a powerful tool for solving field problems [8,9].

Due to (5.2a) and (5.4a) in the electrostatic or quasielectrostatic case the electric field is irrotational and may be derived from the scalar potential Φ as discussed in Section 3.1 from

$$\mathcal{E} = -\,\mathrm{d}\Phi\,. \tag{5.7}$$

We obtain this equation also from (3.3) for the static case. From (5.6a), (5.7) and the constitutive relation (2.23) we obtain the *Poisson equation*

$$\Delta\Phi = -\frac{1}{\varepsilon}\star\mathcal{Q} = -\frac{1}{\varepsilon}q\,. \tag{5.8}$$

We note that the Poisson equation also follows from (3.18) for the static case.

We calculate the scalar potential field due to a point-like unit charge located at the point $\boldsymbol{x}'$. A point-like source may be described by a *three-dimensional Dirac delta distribution*

$$\delta(\boldsymbol{x}-\boldsymbol{x}') = \delta(x-x')\,\delta(y-y')\,\delta(z-z')\,. \tag{5.9}$$

From (D.45) we obtain

$$\int_V \delta(\boldsymbol{x}-\boldsymbol{x}')\,\mathrm{d}x\wedge\mathrm{d}y\wedge\mathrm{d}z = \begin{cases} 1 & \text{for} \quad x' \in V \\ 0 & \text{for} \quad x' \notin V \end{cases} . \tag{5.10}$$

For an arbitrary smooth scalar function $f(\boldsymbol{x})$ we obtain as the three-dimensional generalization of (D.47) the relation

$$\int f(\boldsymbol{x})\delta(\boldsymbol{x}-\boldsymbol{x}')\,\mathrm{d}x\wedge\mathrm{d}y\wedge\mathrm{d}z = \begin{cases} f(\boldsymbol{x}') & \text{for} \quad x' \in V \\ 0 & \text{for} \quad x' \notin V \end{cases} . \tag{5.11}$$

The electric charge form $\mathcal{Q}(\boldsymbol{x})$ describing a point-like unit charge located at the point $\boldsymbol{x}'$ is given by

$$\mathcal{Q}(\boldsymbol{x}) = \delta(\boldsymbol{x}-\boldsymbol{x}')\,\mathrm{d}x\wedge\mathrm{d}y\wedge\mathrm{d}z\,. \tag{5.12}$$

Inserting this into (5.8) yields

$$\Delta\, G_0(\boldsymbol{x},\boldsymbol{x}') = -\frac{1}{\varepsilon}\delta(\boldsymbol{x}-\boldsymbol{x}')\,. \tag{5.13}$$

We have renamed the unknown function $\Phi(\boldsymbol{x})$ by $G_0(\boldsymbol{x},\boldsymbol{x}')$, where $\boldsymbol{x}$ is the variable, and $\boldsymbol{x}'$ is the constant coordinate vector denoting the location of the unit source. The function $G_0(\boldsymbol{x},\boldsymbol{x}'))$ is called *Green's function.* It relates the *source space* $\boldsymbol{x}'$ to the *observation space* $\boldsymbol{x}$ and gives the potential at point $\boldsymbol{x}$ created by a unit source at point $\boldsymbol{x}'$. Multiplying this equation with $\mathcal{Q}(\boldsymbol{x}')$ and integrating over the primed coordinates yields

$$\Delta \int_V^{\prime} G_0(\boldsymbol{x},\boldsymbol{x}')\mathcal{Q}(\boldsymbol{x}') = -\frac{1}{\varepsilon}q(\boldsymbol{x})\,. \tag{5.14}$$

Comparing this with (5.8) yields

$$\Phi(x) = \int_V^{\prime} G_0(\boldsymbol{x},\boldsymbol{x}')\mathcal{Q}(\boldsymbol{x}')\,. \tag{5.15}$$

We apply the Hodge operator on both sides of (5.13) and integrate over a spherical volume of radius r and center point $\boldsymbol{x}'$. On the left side we apply Stokes' theorem (A.88) to transform the volume integral into a surface integral.

$$\oint_{\partial V} \star\, \mathrm{d}G_0(\boldsymbol{x},\boldsymbol{x}') = -\frac{1}{\varepsilon}\int_V \delta(\boldsymbol{x}-\boldsymbol{x}')\,dx\wedge dy\wedge dz = -\frac{1}{\varepsilon}\,. \tag{5.16}$$

We now introduce spherical coordinates (r, ϑ, φ) around the center point $\boldsymbol{x}'$. Due to the spherical symmetry of the problem, $G(\boldsymbol{x},\boldsymbol{x}')$ only depends on $r = |\boldsymbol{x}-\boldsymbol{x}'|$ and

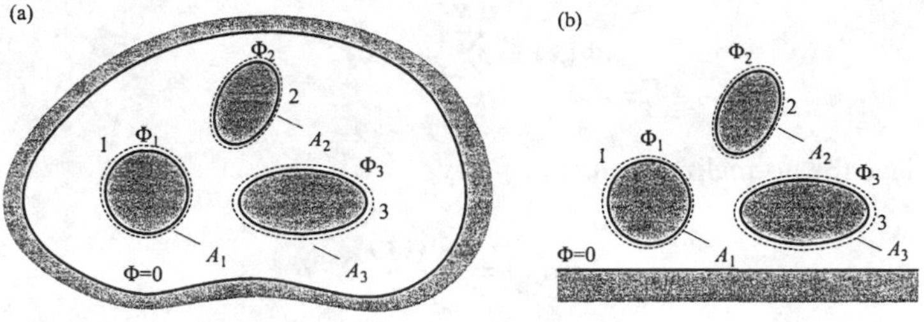

Figure 5.2: (a) Closed multiconductor structure and (b) open multiconductor structure.

we can write $G(r)$ instead of $G(\boldsymbol{x}, \boldsymbol{x}')$. Since G_0 is constant over the surface of the sphere, performing the integration on the left side of (5.16) yields

$$\oint_{\partial V} \star \, \mathrm{d}G_0(\boldsymbol{x}, \boldsymbol{x}') = \oint_{\partial V} \frac{\partial G_0}{\partial r} r^2 \sin\vartheta \, d\vartheta \wedge d\varphi = 4\pi r^2 \frac{\partial G_0}{\partial r} \,. \tag{5.17}$$

From (5.16) and (5.17) we obtain

$$\frac{\partial G_0}{\partial r} = -\frac{1}{4\pi \varepsilon r^2} \tag{5.18}$$

and after integration the *scalar Green's function* for the electrostatic problem

$$G_0(r) = \frac{1}{4\pi \varepsilon r} \,. \tag{5.19}$$

The Green's function relates the observation space $\boldsymbol{x}$ to the source space $\boldsymbol{x}'$.

5.2.2 Capacitance

Applying an electric voltage between two conductors yields an electric field and an associated electric flux density between these conductors. The flux density induces positive and negative charges of same magnitude in both conductors. The ratio of the induced charge to the applied voltage is called *capacitance*.

Figure 5.2 shows closed and open multiconductor structures. We consider a multiconductor structure with $N + 1$ conductors. The conductor 0 is assumed to exhibit the potential $\Phi_0 = 0$. Let the voltage between conductor k and conductor 0 be v_k. The potential field for the case when the kth conductor exhibits voltage v_k and all other conductors exhibit voltage $v_l = 0$ for $l \neq k$ is named $\Phi_k(\boldsymbol{x})$. Applying the superposition principle yields the total potential

$$\Phi(\boldsymbol{x}) = \sum_{k=1}^{N} \Phi_k(\boldsymbol{x})\,. \tag{5.20}$$

We introduce the normalized potential ϕ_k by

$$\phi_k(\boldsymbol{x}) = \frac{\Phi_k(\boldsymbol{x})}{v_k}\,. \tag{5.21}$$

With this we obtain from (5.20)

$$\Phi(\boldsymbol{x}) = \sum_{k=1}^{N} v_k \phi_k(\boldsymbol{x})\,. \tag{5.22}$$

The charge of conductor l is given by the integral of the electric flux density over the surface A_l enclosing conductor l.

$$q_l = \oint_{A_l} \mathcal{D} = -\varepsilon \oint_{A_l} \mathrm{d}\Phi = -\sum_{k=1}^{N} v_k \varepsilon \oint_{A_l} \mathrm{d}\phi_k\,. \tag{5.23}$$

The charges induced on conductor l due to the voltages v_k on conductors k is given by

$$q_l = \sum_{k=1}^{N} C_{lk}\, v_k\,. \tag{5.24}$$

where the *partial capacitance* C_{lk} is given by

$$C_{lk} = -\varepsilon \oint_{A_l} \mathrm{d}\phi_k\,. \tag{5.25}$$

The current i_l flowing into the lth conductor is given by

$$i_l = \frac{\mathrm{d}q_l}{\mathrm{d}t} = \sum_{k=1}^{N} C_{lk} \frac{\mathrm{d}v_k}{\mathrm{d}t}\,. \tag{5.26}$$

Let us now explicitly calculate the capacitance of some geometrically simple conductor structures. We compute the electric field of an electrically charged spherical conductor of radius a depicted in Figure 5.3(b). We assume spherically symmetric charge distribution over the surface of the spherical conductor. Due to the spherical symmetry of the problem the electric field as well as the electric flux density will only

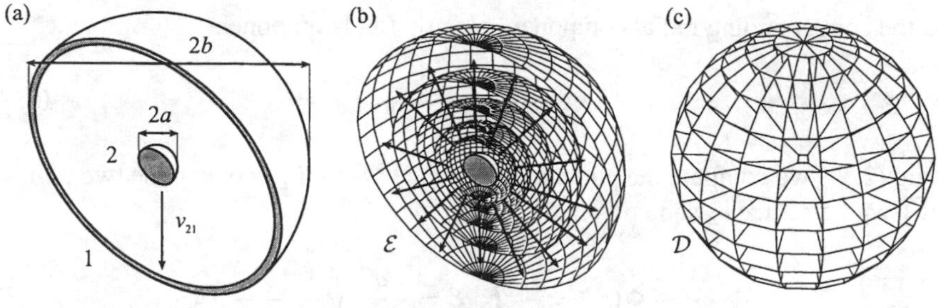

Figure 5.3: (a) Spherical capacitor, (b) electric field and (c) electric flux density.

exhibit radial components $E_r(r)$ and $D_r(r)$, where r is the distance from the center of the sphere. In spherical coordinates the electric flux form is given by

$$\mathcal{D}(r) = D_r(r)\, r^2 \sin\vartheta \, \mathrm{d}\vartheta \wedge \mathrm{d}\varphi \,. \tag{5.27}$$

Applying Gauss' law (2.53) and integrating the flux density over a sphere with radius $r \geq a$ yields the charge q

$$q = \int_V \mathcal{Q} = \oint_{\partial V} \mathcal{D} = r^2 D_r(r) \oint_{\partial V} \mathrm{d}\vartheta \wedge \mathrm{d}\varphi = 4\pi r^2 D_r(r)\,. \tag{5.28}$$

From this we obtain the radial component of the flux density

$$D_r(r) = \frac{q}{4\pi r^2}\,. \tag{5.29}$$

and the flux differential form

$$\mathcal{D}(r) = \frac{q}{4\pi} \sin\vartheta \, \mathrm{d}\vartheta \wedge \mathrm{d}\varphi \,. \tag{5.30}$$

Figure 5.3(c) visualizes the flux differential form $\mathcal{D}$ by cones originating at the center of the sphere. The cross-section of the cones are proportional to $1/r^2$. We consider that it is impossible to give a spherically symmetric tessellation of the surface of a sphere. Therefore the picture in Figure 5.3(c) is not spherically symmetric. However, the shape of the flux tubes is of no account. If all flux tubes have the same cross-section at a given distance r from the center of the sphere, the dielectric flux density exhibits spherical symmetry. With (2.34) we obtain

$$\mathcal{E}(r) = \frac{1}{\varepsilon} \star \mathcal{D}(r) = \frac{q}{4\pi\varepsilon r^2} \,\mathrm{d}r\,. \tag{5.31}$$

and the corresponding radial component electric field component

$$E_r(r) = \frac{q}{4\pi\varepsilon r^2}\,. \tag{5.32}$$

Using (2.59) we compute the potential difference $\Phi_2 - \Phi_1$ between the two points 2 and 1 with distance r_2 and r_1 from the origin

$$\Phi(r_2) - \Phi(r_1) = -\int_1^2 \mathcal{E} = \frac{q}{4\pi\varepsilon}\left(\frac{1}{r_2} - \frac{1}{r_1}\right)\,. \tag{5.33}$$

Choosing the potential of reference at $\Phi(\infty) = 0$, the potential $\Phi(r)$ is given by

$$\Phi(r) = \frac{q}{4\pi\varepsilon r}\,. \tag{5.34}$$

Figure 5.3(b) shows the equipotential spheres. For an equal potential difference between neighboring spheres the distance between the spheres must be proportional to $1/r$. Drawing the electric field lines normal to the equipotential spheres, the number of field lines piercing an equipotential sphere per unit of area is proportional to $1/r^2$.

The potential of the surface of a metallic sphere with radius a and charge q is $\Phi(a)$. The *capacitance* is defined as the ratio of the charge and the potential,

$$C = \frac{q}{\Phi}\,. \tag{5.35}$$

A sphere with radius a has a capacitance

$$C = 4\pi\varepsilon a\,. \tag{5.36}$$

A *spherical capacitor* is formed by two concentric conducting spheres as depicted in Figure 5.3(a). The inner sphere 2 has a radius a and the inner radius of the outer spherical conductor 1 is b. For a charge q on sphere 2 we obtain from (5.33) the voltage

$$v_{21} = \frac{q}{4\pi\varepsilon}\left(\frac{1}{a} - \frac{1}{b}\right)\,. \tag{5.37}$$

and from this with (5.35) the capacitance of the spherical capacitor

$$C = 4\pi\varepsilon\frac{a\,b}{b-a}\,. \tag{5.38}$$

The charge q on the inner sphere 2 induces a charge $-q$ on the outer sphere 1.

Figure 5.4 shows an electric *dipole* consisting of two spherical conductors of radius a in a distance h. We assume $h \gg a$. This means that we can neglect the

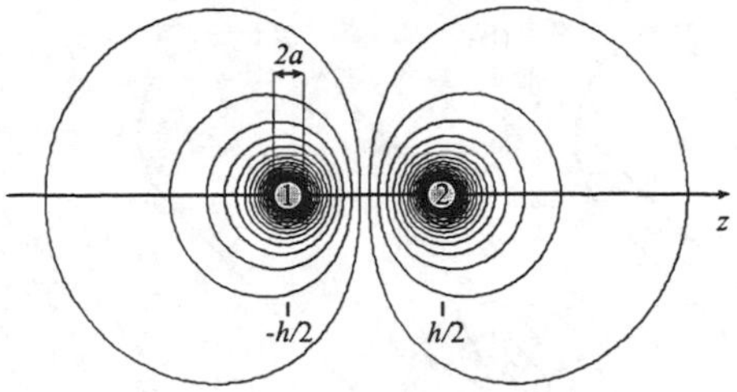

Figure 5.4: Electrostatic field of a dipole.

perturbation of the field of sphere 1 when bringing the uncharged sphere 2 into the field created by sphere 1. In this case the potentials Φ_1 and Φ_2 on spheres 1 and 2 are approximately given by

$$\Phi_1 = \frac{1}{4\pi\varepsilon}\left(\frac{q_1}{a} + \frac{q_2}{h}\right), \qquad \Phi_2 = \frac{1}{4\pi\varepsilon}\left(\frac{q_2}{a} + \frac{q_1}{h}\right), \tag{5.39}$$

where q_1 and q_2 are the charges of sphere 1 and sphere 2. If both spheres contain charge of equal magnitude and opposite sign, $q = q_1 = -q_2$, the potential difference between the spheres is

$$v_{21} = \Phi_2 - \Phi_1 = \frac{q}{2\pi\varepsilon}\left(\frac{1}{a} - \frac{1}{h}\right). \tag{5.40}$$

With (5.35) we obtain the capacitance

$$C = 2\pi\varepsilon\frac{a}{1 - a/h}. \tag{5.41}$$

For $h \gg a$ the capacitance C becomes independent of h and is half the capacitance of a single sphere. This means the dipole structure of Figure 5.4 in this case behaves like two spheres connected in series.

The coaxial cylindric structure shown in Figure 5.5(a) is formed by an inner conductor with outer radius a and an outer conductor with inner radius b. We consider a segment of this structure of length l. Let q be the charge on a segment of length l of the inner conductor. Due to the cylindric symmetry of the structure, the electric field exhibits only an r component in cylindric coordinate system. The electric flux form $\mathcal{D}$ is given by

$$\mathcal{D}(r) = D_r(r)\, r\, \mathrm{d}\varphi \wedge \mathrm{d}z\,. \tag{5.42}$$

Using (2.53) and integrating the flux density over a cylinder of length l and radius r $a \le r \le b$ yields the charge q on the inner conductor segment of length l,

$$q = \oint_{\partial V} \mathcal{D} = r D_r(r) \oint_{\partial V} \mathrm{d}\varphi \wedge \mathrm{d}z = 2\pi r l D_r(r)\,. \tag{5.43}$$

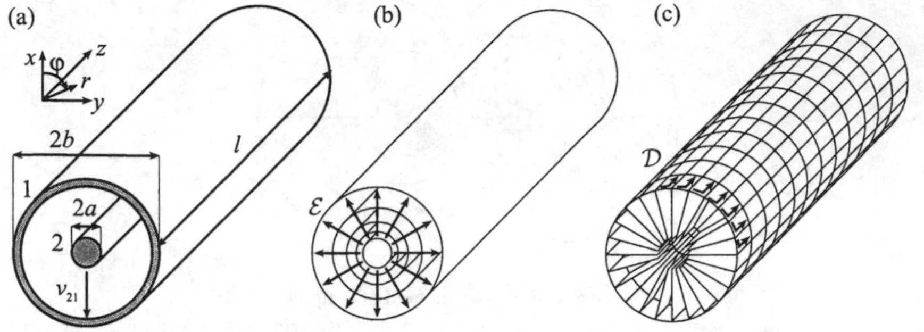

Figure 5.5: (a) Coaxial line segment, (b) electric field and (c) electric flux density.

The charge per unit of length q' is given by

$$q' = \frac{q}{l} = \pi r D_r(r)\,. \tag{5.44}$$

From this we obtain the radial component of the flux density

$$D_r(r) = \frac{q'}{2\pi r}\,. \tag{5.45}$$

and the flux differential form

$$\mathcal{D}(r) = \frac{q'}{2\pi}\,\mathrm{d}\varphi \wedge \mathrm{d}z\,. \tag{5.46}$$

With (2.34) we obtain

$$\mathcal{E}(r) = \frac{1}{\varepsilon} \star \mathcal{D}(r) = \frac{q'}{2\pi\varepsilon r}\,\mathrm{d}r\,. \tag{5.47}$$

and the corresponding radial component E_r of the electric field,

$$E_r(r) = \frac{q'}{2\pi\varepsilon r}\,. \tag{5.48}$$

The voltage v_{21} between inner conductor 2 and outer conductor 1 is given by

$$v_{21} = \Phi_2 - \Phi_1 = -\int_a^b \mathcal{E} = \frac{q'}{2\pi\varepsilon}\ln\frac{b}{a}\,. \tag{5.49}$$

From this we obtain the *capacitance per unit of length*

$$C' = \frac{2\pi\varepsilon}{\ln\frac{b}{a}}\,. \tag{5.50}$$

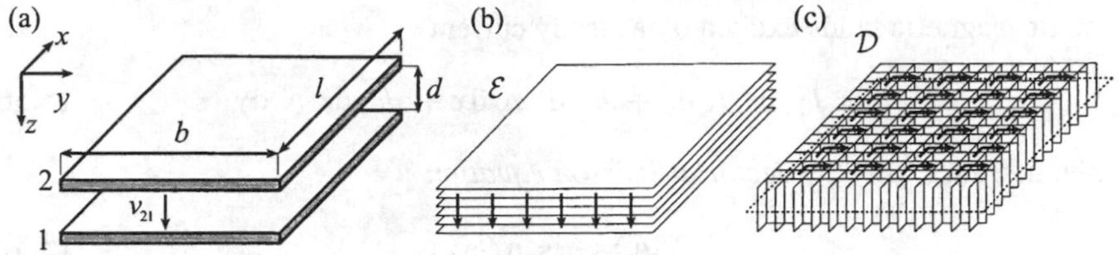

Figure 5.6: (a) Plate capacitor, (b) electric field (c) electric flux density.

Figure 5.6 shows two parallel plates with distance d. In case of infinite extension of these plates we obtain a homogeneous electric field between the two plates if a voltage v_{21} is applied between both plates. The field solution for infinite extension will also give a good approximation for plates of finite transverse extension if $b, l \gg d$. For symmetry reasons the electric field is homogeneous and only exhibits a component normal to the plates. The plates 1 and 2 exhibit uniform surface charge densities $\mathcal{Q}_{A1}$ and $\mathcal{Q}_{A2}$. With (2.194) we obtain

$$\mathcal{D} = -\mathcal{Q}_{A1} = \mathcal{Q}_{A2} = q_A \, \mathrm{d}x \wedge \mathrm{d}y \,. \tag{5.51}$$

From (2.34) we obtain

$$\mathcal{E} = \frac{1}{\varepsilon} \star \mathcal{D} = \frac{q_A}{\varepsilon} \, \mathrm{d}z \,. \tag{5.52}$$

With (2.59) we compute the potential difference v_{21} between the top plate 2 and the bottom plate 1.

$$v_{21} = -\int_1^2 \mathcal{E} = -\frac{q_A}{\varepsilon} \int_1^2 \mathrm{d}z = \frac{q_A d}{\varepsilon} \,. \tag{5.53}$$

From this we obtain the *capacitance per unit of area*

$$C'' = \frac{q_A}{v_{21}} = \frac{\varepsilon}{d} \,. \tag{5.54}$$

5.3 STATIC AND QUASISTATIC MAGNETIC FIELDS

5.3.1 The Green's Function for the Static Magnetic Field

Following (3.1) we obtain the magnetic induction form $\mathcal{B}$ from the magnetic vector potential form $\mathcal{A}$ via

$$\mathcal{B} = \mathrm{d}\mathcal{A} \,. \tag{5.55}$$

For static magnetic fields excited by a steady current

$$\mathcal{J}_0 = J_{0x}\,\mathrm{d}y \wedge \mathrm{d}z + J_{0y}\,\mathrm{d}z \wedge \mathrm{d}x + J_{0z}\,\mathrm{d}x \wedge \mathrm{d}y \tag{5.56}$$

we obtain from (3.17) the *vectorial Poisson equation*

$$\Delta\,\mathcal{A} = -\star\,\mu\mathcal{J}_0\,. \tag{5.57}$$

The solution of the vector field problem with unit point-like vectorial source is given by a *dyadic Green's function* with the components G_{ij} relating the ith component of the field vector to the jth component of the source vector [8, 9]. In differential form calculus dyadics may be represented by *double forms.* To compute the field at a point $\boldsymbol{x}$ excited by a point-like source located at $\boldsymbol{x}'$ we use the *Green's double one-form* [10, 11]. A *double one-form* $\mathcal{G}$ is defined by

$$\begin{aligned}\mathcal{G}(\boldsymbol{x},\boldsymbol{x}') = {} & G_{11}\,\mathrm{d}x\,\mathrm{d}x' + G_{12}\,\mathrm{d}x\,\mathrm{d}y' + G_{13}\,\mathrm{d}x\,\mathrm{d}z' \\ & + G_{21}\,\mathrm{d}y\,\mathrm{d}x' + G_{22}\,\mathrm{d}y\,\mathrm{d}y' + G_{23}\,\mathrm{d}y\,\mathrm{d}z' \\ & + G_{31}\,\mathrm{d}z\,\mathrm{d}x' + G_{32}\,\mathrm{d}z\,\mathrm{d}y' + G_{33}\,\mathrm{d}z\,\mathrm{d}z'\,.\end{aligned} \tag{5.58}$$

The Green's double form relates the observation space $\boldsymbol{x}$ to the source space $\boldsymbol{x}'$. Primed and unprimed differentials $\mathrm{d}x_i'$ and $\mathrm{d}x_j$ commute, i.e., in products they may be interchanged without changing the sign. The rules are

$$\mathrm{d}x_i\,\mathrm{d}x_j' = \mathrm{d}x_i'\,\mathrm{d}x_j \quad \text{with } \mathrm{d}x_i = \mathrm{d}x,\ \mathrm{d}y,\ \mathrm{d}z\,. \tag{5.59}$$

We now can introduce the so-called *identity kernel*

$$\mathcal{I}(\boldsymbol{x},\boldsymbol{x}') = \delta(\boldsymbol{x}-\boldsymbol{x}')\,(\,\mathrm{d}x\,\mathrm{d}x' + \mathrm{d}y\,\mathrm{d}y' + \mathrm{d}z\,\mathrm{d}z')\,. \tag{5.60}$$

With the identity kernel we can map any one-form $\mathcal{U}$ and any two-form $\mathcal{V}$ from the source space to the observation space, i.e., the respective form is mapped in itself and the primed differentials are replaced by unprimed differentials. We obtain

$$\int' \mathcal{I}(\boldsymbol{x},\boldsymbol{x}') \wedge \star\,\mathcal{U}(\boldsymbol{x}') = \mathcal{U}(\boldsymbol{x})\,, \tag{5.61a}$$

$$\star \int' \mathcal{I}(\boldsymbol{x},\boldsymbol{x}') \wedge \mathcal{V}(\boldsymbol{x}') = \mathcal{V}(\boldsymbol{x})\,. \tag{5.61b}$$

The primed integration symbol denotes that the integration is performed over the primed variables. For the integration the unprimed differentials are treated as constants. Using double forms we can express the vectorial Poisson equation (5.57) for a point-like unit source at $\boldsymbol{x}'$ by

$$\Delta\mathcal{G}(\boldsymbol{x},\boldsymbol{x}') = -\mu\,\mathcal{I}(\boldsymbol{x},\boldsymbol{x}')\,, \tag{5.62}$$

In (5.62) the Laplace operator acts on the unprimed coordinate variables whereas the primed coordinate variables are treated as constants. Forming the exterior product with $\mathcal{J}_0(\boldsymbol{x}')$, integrating over the primed variables and using (5.61b) yields

$$\star \int' \Delta\, \mathcal{G}(\boldsymbol{x}, \boldsymbol{x}') \wedge \mathcal{J}_0(\boldsymbol{x}') = -\mu \mathcal{J}_0(\boldsymbol{x}) , \tag{5.63}$$

Since the sequence of partial derivation with respect to the unprimed coordinates and integration over the primed coordinates may be interchanged under very general conditions and considering (A.62) we obtain

$$\Delta \int' \mathcal{G}(\boldsymbol{x}, \boldsymbol{x}') \wedge \mathcal{J}_0(\boldsymbol{x}') = - \star\, \mu \mathcal{J}_0(\boldsymbol{x}) . \tag{5.64}$$

Comparing this equation with (5.57) yields

$$\mathcal{A}(\boldsymbol{x}) = \int' \mathcal{G}(\boldsymbol{x}, \boldsymbol{x}') \wedge \mathcal{J}_0(\boldsymbol{x}') . \tag{5.65}$$

In cartesian coordinates the Laplace operator for one-forms (A.71) is symmetric in its three components and leaves the three components uncoupled. Therefore (5.64) may be solved by

$$\mathcal{G}(\boldsymbol{x}, \boldsymbol{x}') = G(\boldsymbol{x}, \boldsymbol{x}')(\,\mathrm{d}x\,\mathrm{d}x' + \mathrm{d}y\,\mathrm{d}y' + \mathrm{d}z\,\mathrm{d}z') , \tag{5.66}$$

where $G(\boldsymbol{x}, \boldsymbol{x}')$ is a *scalar Green's function*, obtained by solving the scalar Poisson equation

$$\Delta G(\boldsymbol{x}, \boldsymbol{x}') = -\mu\, \delta(\boldsymbol{x} - \boldsymbol{x}') . \tag{5.67}$$

This problem has already been treated in Section 5.2.1. In analogy to the solution (5.19) of (5.8) we obtain

$$G(r) = \frac{\mu}{4\pi r} . \tag{5.68}$$

Inserting (5.56), (5.67) and (5.68) in (5.65) yields

$$\mathcal{A}(\boldsymbol{x}) = \frac{\mu}{4\pi} \int' \frac{J_{0x}(\boldsymbol{x}')\,\mathrm{d}x + J_{0y}(\boldsymbol{x}')\,\mathrm{d}y + J_{0z}(\boldsymbol{x}')\,\mathrm{d}z}{|\boldsymbol{x} - \boldsymbol{x}'|}\, \mathrm{d}x' \wedge \mathrm{d}y' \wedge \mathrm{d}z' . \tag{5.69}$$

To compute the vector potential $\mathcal{A}$ excited by a thin wire carrying a current I_0 we have to perform the volume integration in (5.65) over the volume of the wire. The integral may be performed first over the cross-sectional area of the wire, which is transverse to the curve followed by the wire and than in the direction tangential to the wire. After the integration over the cross-sectional area we obtain

$$\mathcal{A}(\boldsymbol{x}) = I_0 \int'_{C_1} \mathcal{G}(\boldsymbol{x}, \boldsymbol{x}') . \tag{5.70}$$

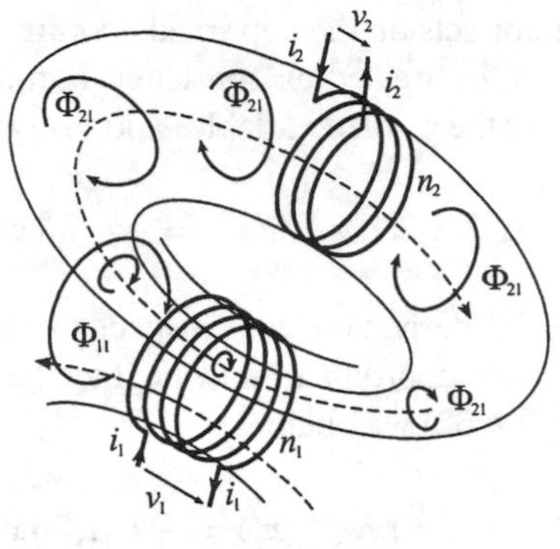

Figure 5.7: Transformer.

5.3.2 Inductance

A current flowing through a conductor creates a magnetic field. The magnetic induction or flux density associated with this field yields a magnetic flux through a coil formed by the conductor. The ratio of this magnetic flux to the current creating the flux is called the *inductance* of the conductor. If the magnetic field that is created by the current in one conductor induces a magnetic flux through a coil formed by another conductor. The ratio of the flux through this coil formed by the other conductor to the current through the first conductor is called the *mutual inductance* between both conductors.

Figure 5.7 illustrates the inductive coupling of two coils 1 and 2. The current i_1 flowing through coil 1 creates a magnetic field and a total flux Φ_{11}. The part of the magnetic flux flowing through the coil 2 is denominated with Φ_{21}. If current i_1 changes with time, the corresponding time variations of Φ_{11} and Φ_{21} respectively, induce voltages u_1 and u_2 across the poles of coil 2. The arrangement of two inductive coupled coils is called a *transformer*.

We consider two wire coils 1 and 2 following the closed curves ∂A_1 and ∂A_2, respectively. These closed curves may be interpreted as the boundaries of the surfaces A_1 and A_2. The vector potential due to a current i_{01} flowing in coil 1 is given by

$$\mathcal{A}(x) = i_1 \int'_{\partial A_1} \mathcal{G}(x, x') \,. \tag{5.71}$$

The magnetic flux Φ_{21} through coil 2 due to the current i_1 flowing in coil 1 is

$$\Phi_{21} = \int_{A_2} \mathcal{B}(x) = \int_{\partial A_2} \mathcal{A}(x) \,. \tag{5.72}$$

Inserting (5.71) into (5.72) yields

$$\Phi_{21} = i_1 \int_{\partial A_2} \int'_{\partial A_1} \mathcal{G}(\boldsymbol{x}, \boldsymbol{x}') \,. \tag{5.73}$$

The ratio M_{21} of flux Φ_{21} through coil 2 to the exciting current i_1 through coil 1 is called the *mutual inductance* between coils i and 2.

$$M_{21} = \frac{\Phi_{21}}{i_1} = \int_{\partial A_2} \int'_{\partial A_1} \mathcal{G}(\boldsymbol{x}, \boldsymbol{x}') \,. \tag{5.74}$$

We note that the expression for the mutual inductance is symmetric with respect to coils 1 and 2. From this it follows

$$M_{21} = M_{12} \,. \tag{5.75}$$

If both coils exhibit the same orientation with respect to the flux flow, the mutual inductance is positive; otherwise it is negative. Interchanging the poles of one coil changes the sign of the mutual inductance. The ratio of flux and current through the same coil is called *self-inductance* or simply *inductance*. The inductance of coil 1 is given by

$$L_1 = \int_{\partial A_1} \int'_{\partial A_1} \mathcal{G}(\boldsymbol{x}, \boldsymbol{x}') \,. \tag{5.76}$$

If two coils k and l are coupled magnetically, the magnetic flux Φ_l through coil l, excited by a current i_k through coil k is given by

$$\Phi_l = M_{lk} \, i_k \,. \tag{5.77}$$

The voltage v_k induced in coil k is given by

$$\frac{\mathrm{d}\Phi_l}{\mathrm{d}t} = M_{lk} \frac{\mathrm{d}i_k}{\mathrm{d}t} \,. \tag{5.78}$$

In the following we will discuss the inductance of some simple conductor structures. Figure 5.8 shows a solenoid of length l and diameter $2a$ and n turns. It is composed of a thin, perfectly conducting wire, wound in a helix. We assume $l \gg 2a$. From Ampère's law (5.3b) we obtain for magnetoquasistatic fields

$$\oint_{\partial A} \mathcal{H} = \int_A \mathcal{J} \,. \tag{5.79}$$

The area A enclosed by the closed path of integration in Figure 5.8(b) is pierced n times by the conductor of the solenoid. Therefore the path of integration encloses

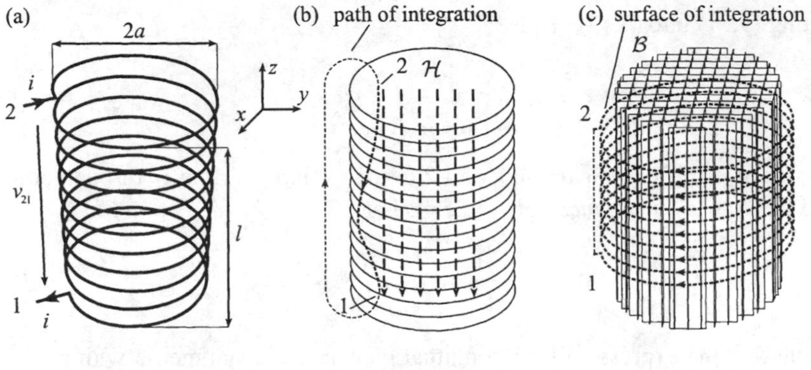

Figure 5.8: (a) Solenoid, (b) magnetic field and (c) magnetic flux density.

the total current ni. Due to the assumption $l \gg 2a$ the magnetic field created by the current is mainly concentrated inside the solenoid. For this reason we only need to integrate $\mathcal{H}$ over the part of the path inside the solenoid from the top 2 to the bottom 1. We obtain

$$\int_2^1 \mathcal{H} = ni\,. \tag{5.80}$$

and since the field is homogeneous inside the solenoid

$$\mathcal{H} = -\frac{ni}{l}\,\mathrm{d}z\,, \tag{5.81}$$

With (2.34) we obtain

$$\mathcal{B} = \mu \star \mathcal{H} = B_z\,\mathrm{d}x \wedge \mathrm{d}y = -\frac{n\mu i}{l}\,\mathrm{d}x \wedge \mathrm{d}y\,. \tag{5.82}$$

The coil and therewith the path of integration is linked n times with the flux. Therefore the area of integration A in the above integral is n times the cross-sectional area of the solenoid.The flux Φ through the solenoid is given by

$$\Phi = \int_A \mathcal{B} = a^2\pi\,B_z = \mu\,\frac{na^2\pi}{l}\,i\,. \tag{5.83}$$

The surface A_L bounded by the solenoid exhibits n layers as depicted in Figure 5.8(c). Therefore the magnetic flux flows n times through cross-sectional area A and we obtain the *flux linkage*

$$\Phi_L = \int_{A_L} \mathcal{B} = n \int_A \mathcal{B} = na^2\pi\,B_z = \mu\,\frac{n^2a^2\pi}{l}\,i\,. \tag{5.84}$$

by summing the flux contributions of all the turns of the coil. The flux linked by the coil is due to the current i itself. Since the flux linkage Φ_L is proportional to the current i the ratio of the flux linkage to the current characterizes the solenoid. The *inductance* L is defined as

$$L = \frac{\Phi_L}{i}\,. \tag{5.85}$$

The inductance of the solenoid is

$$L = \mu\,\frac{n^2 a^2 \pi}{l}\,. \tag{5.86}$$

We apply Faraday's law (2.51)

$$\oint_{\partial A_L} \mathcal{E} = -\frac{\mathrm{d}}{\mathrm{d}t}\int_{A_L} \mathcal{B}\,. \tag{5.87}$$

The boundary ∂A_L follows the conductor solenoid and is closed via the path marked by the thin dashed line outside the conductor. Since the electric field component tangential to the solenoid conductor vanishes, only the part of the path outside the conductor region from node 1 to node 2 contributes to the path integral over $\mathcal{E}$ and gives the voltage between nodes 2 and 1:

$$\oint_{\partial A_L} \mathcal{E} = \int_1^2 \mathcal{E} = -v_{21}\,. \tag{5.88}$$

The voltage v_{21} is independent of the path of integration as long as this path is not wound around the solenoid.

The inductance L is the ratio between the voltage v and the time derivative $\mathrm{d}i/\,\mathrm{d}t$ of the current:

$$v_{21} = L\,\frac{\mathrm{d}i}{\mathrm{d}t}\,. \tag{5.89}$$

A coaxial line structure also exhibits an *inductance per unit of length.* We assume a current i_2 flowing in the z-direction in the inner conductor 2 of the coaxial line depicted in Figure 5.9(a). The magnetic field must exhibit cylindrical symmetry and therefore depend only on r. Due to (5.5b) the r- and z-component of the flux density and the magnetic field vanish and we obtain

$$\mathcal{H} = H_\varphi(r)\, r\, \mathrm{d}\varphi\,. \tag{5.90}$$

From Ampère's law (5.1a) we obtain by integrating over the path C in Figure 5.9(b)

$$\oint_C \mathcal{H} = 2\pi r\, H_\varphi(r) = i_2 \tag{5.91}$$

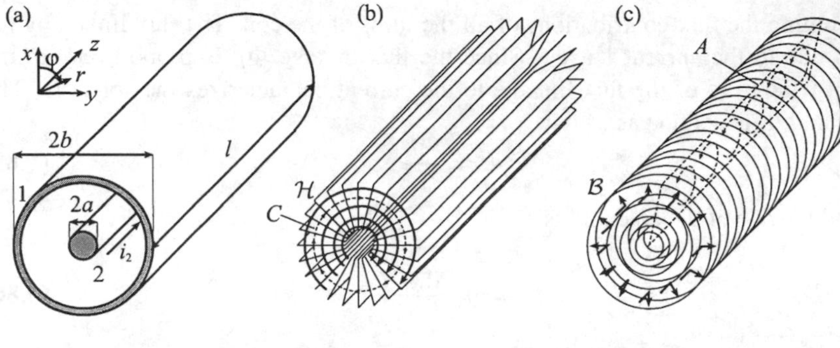

Figure 5.9: (a) Coaxial line segment, (b) magnetic field and (c) magnetic flux density.

and from this

$$\mathcal{H} = \frac{i_2}{2\pi r} r \, \mathrm{d}\varphi \, . \tag{5.92}$$

With (2.34) we obtain the magnetic flux density form

$$\mathcal{B} = \mu \star \mathcal{H} = \frac{\mu i_2}{2\pi r} \mathrm{d}z \wedge \mathrm{d}r \, . \tag{5.93}$$

The flux Φ between the inner conductor 2 and the outer conductor 1 over a length l of the coaxial line is given by

$$\Phi = \int_A \mathcal{B} = \frac{\mu l i_2}{2\pi} \int_2^1 \frac{\mathrm{d}r}{r} = \frac{\mu l i_2}{2\pi} \ln \frac{b}{a} \, . \tag{5.94}$$

The flux per unit of length Φ' is given by

$$\Phi' = \Phi / l = \frac{\mu i_2}{2\pi} \ln \frac{b}{a} \tag{5.95}$$

and the *inductance per unit of length* L' of the coaxial line is

$$L' = \Phi' / i_2 = \Phi' / l = \frac{\mu}{2\pi} \ln \frac{b}{a} \, . \tag{5.96}$$

We now consider the parallel plate structure shown in Figure 5.10. On the inner surface of both plates surface currents with surface current densities $\boldsymbol{J}_{A1}$ and $\boldsymbol{J}_{A2}$ are impressed. We assume the surface current to flow in the positive or negative x-direction. In this case the surface current differential forms are given by

$$\mathcal{J}_{A1} = -J_{A1x} \, \mathrm{d}y \, , \qquad \mathcal{J}_{A2} = J_{A2x} \, \mathrm{d}y \, . \tag{5.97}$$

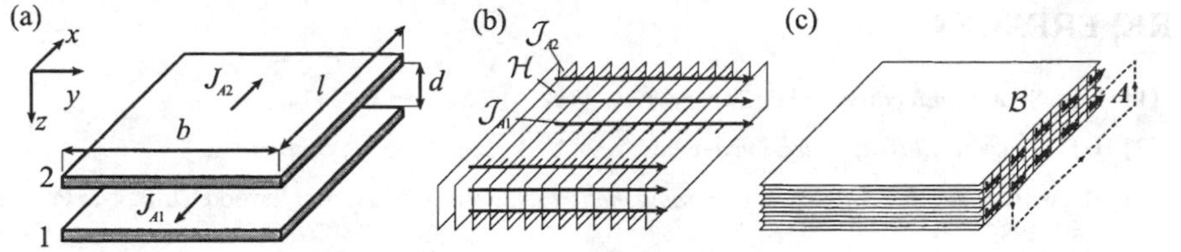

Figure 5.10: (a) Parallel plate structure, (b) magnetic field and (c) magnetic flux density.

We make the point that the direction of the surface current flow, the orientation of the twisted one-form, and the orientation of the surface normal direction form a positive oriented orthogonal tripod. This explains the different signs in the equations above. Due to the symmetry of the structure and also due to (5.5b) between the parallel plates only a tangential magnetic field exists and from (2.188) it follows

$$\mathcal{H} = \mathcal{J}_{A1} = \mathcal{J}_{A2} \tag{5.98}$$

and from this $J_{A1x} = -J_{A2x}$. With (2.34) we obtain the magnetic flux density form

$$\mathcal{B} = \mu \star \mathcal{H} = \mu\, J_{A2x}\, \mathrm{d}z \wedge \mathrm{d}x\,. \tag{5.99}$$

The magnetic flux Φ through the area element A of length l and height d (Figure 5.10(c)) is given by

$$\Phi = \int_A \mathcal{B} = \mu\, d\, l\, J_{A2x}. \tag{5.100}$$

The current i_2 flowing on conductor 2 in the positive x-direction within a strip of width b is

$$i_2 = \int_1^2 \mathcal{J}_{A2} = b\, J_{A2x}. \tag{5.101}$$

From this we obtain the inductance $L_{l/b}$ of a parallel plate segment of width b and length l

$$L_{l/b} = \Phi / i_2 = \mu\, \frac{d\, l}{b}\,. \tag{5.102}$$

Introducing the inductance $L_{\square}$ of a quadratic parallel plate element

$$L_{\square} = \mu\, d \tag{5.103}$$

we obtain

$$L_{l/b} = \frac{\Phi}{i_2} = \frac{l}{b} L_{\square}\,. \tag{5.104}$$

REFERENCES

[1] J. A. Stratton, *Electromagnetic Theory*. New York: McGraw-Hill, 1941.

[2] J. D. Jackson, *Classical Electrodynamics*. New York: John Wiley & Sons, 1975.

[3] H. A. Haus and J. R. Melcher, *Electromagnetic Fields and Energy*. Englewood Cliffs, NJ: Prentice Hall, 1989.

[4] M. N. Sadiku, *Elements of Electromagnetics*. Orlando: Saunders, 1989.

[5] Z. Popović and B. D. Popović, *Introductory Electromagnetics*. Englewood Cliffs, NJ: Prentice Hall, 2000.

[6] H. A. Haus and J. R. Melcher, "Fields are always dynamic," *IEEE Trans. Education*, pp. 35–46, Feb. 1990.

[7] D. Cannel, *George Green Miller and Mathematician 1793 - 1841*. Nottingham: City of Nottingham Arts Department, 1988.

[8] R. S. Elliott, *Electromagnetics - History, Theory, and Applications*. New York: IEEE Press, 1991.

[9] R. E. Collin, *Field Theory of Guided Waves*. New York: IEEE Press, 1991.

[10] G. de Rham, *Differentiable Manifolds*. New York: Springer, 1984.

[11] K. F. Warnick and D. Arnold, "Electromagnetic Green functions using differential forms," *J. Electromagn. Waves and Appl.*, vol. 10, no. 3, pp. 427–438, 1996.

Chapter 6

Waves at the Surface of Conducting Media

We consider the propagation of an electromagnetic wave along a plane to be an infinite conducting surface. According to Figure 6.1 the space is subdivided into two half spaces 1 and 2 by a plane surface at $x = 0$. The half spaces 1 and 2 each are filled with homogeneous and isotropic media. This problem is encountered when the propagation of electromagnetic waves along the Earth's surface or the propagation of electromagnetic waves along metallic surfaces is considered. Assuming region 1 to be filled with an ideal conductor a TEM wave with electric field perpendicular to the plane $z = 0$ and magnetic field parallel to this plane fulfills the boundary conditions. In the conductor surface the tangential magnetic field induces a surface current, shielding region 1 from the magnetic field. The tangential magnetic field is directed perpendicular to the direction of wave propagation, whereas the surface current is flowing in the direction of wave propagation.

If, however the region 1 is filled with a conductor of finite conductivity the electromagnetic field and the shielding current are penetrating into the conductor. Due to the finite conductivity the current flowing in the direction of propagation gives rise to a longitudinal electric field component. The field in both regions 1 and 2 is transverse-magnetic. Due to the conductor losses the electromagnetic wave is attenuated in the direction of propagation. In transverse direction the field is strongly decaying in the conductor region 1 and weakly decaying in region 2 if region 2 is free space or filled by a dielectric. We call such a wave a *surface wave*. Zenneck was the first to give a solution of the Maxwell's equations describing a surface wave guided by a plane interface separating any two media [1]. Treatment of surface waves is also presented in [2–5].

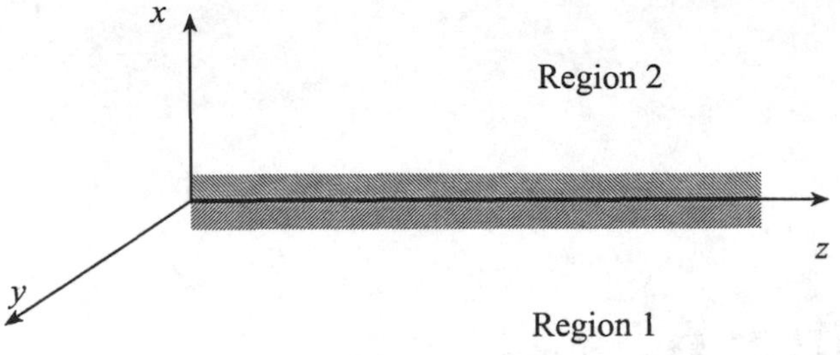

Figure 6.1: The plane surface.

6.1 TRANSVERSE MAGNETIC SURFACE WAVES

We seek solutions of the field equations that describe a plane wave propagating parallel to the surface. Without any restriction in generality we assume the z-direction to be the direction of propagation of the electromagnetic wave. According to (3.70) a transverse magnetic wave can be derived from a Hertz form that has a z component only. Therefore we make the ansatz

$$\underline{\Pi}_{ei}(\boldsymbol{x}) = \underline{\Pi}_{ezi}(\boldsymbol{x})\,dz \tag{6.1}$$

for the Hertz form in either half spaces $i = 1, 2$. We are only considering a wave propagating in positive z-direction and can therefore describe the z dependence of the Hertz vector by a factor $\mathrm{e}^{-\gamma z}$. This assumption corresponds with an electromagnetic field generated from a source at $z = -\infty$. Thereby we obtain the following ansatz for the z component of the Hertz vector for both half spaces 1 and 2:

$$\underline{\Pi}_{ei}(x, z) = \Pi_{ez0i}(x)\,\mathrm{e}^{-\gamma z} dz\,. \tag{6.2}$$

The propagation coefficient in the z-direction must be equal in either spaces 1 and 2, since only in this way the solution in both half spaces can be matched along the boundary at $x = 0$. From (3.51) and (6.2) we obtain the Helmholtz equation for the z component of the Hertz vector for both half spaces 1 and 2:

$$\frac{\partial^2 \Pi_{ezi}}{\partial x^2} + \frac{\partial^2 \Pi_{ezi}}{\partial z^2} - \gamma_{0i}^2 \Pi_{ezi} = 0\,. \tag{6.3}$$

In this equation γ_{01} and γ_{02} are the complex propagation coefficients of the plane wave in medium 1 and 2, respectively. According to (3.53) and (3.60) we obtain

$$\gamma_{0i}^2 = -\omega^2 \mu_i \varepsilon_i\,. \tag{6.4}$$

The sign of the γ_{0i} is determined by (3.61). We now solve (6.3) for either half spaces 1 and 2. Afterwards we match these partial solutions along the boundary surface $x = 0$. Inserting (6.2) into (6.3) yields

$$\frac{\mathrm{d}^2\Pi_{e0i}}{\mathrm{d}x^2} + (\gamma^2 - \gamma_{0i}^2)\Pi_{e0i} = 0\,. \tag{6.5}$$

Introducing the propagation coefficient in the x-direction χ_i defined by

$$\chi_i^2 = \gamma_{0i}^2 - \gamma^2 \tag{6.6}$$

we obtain

$$\frac{\mathrm{d}^2\Pi_{e0i}}{\mathrm{d}x^2} - \chi_i^2\Pi_{e0i} = 0. \tag{6.7}$$

The solution of this equation is given by

$$\Pi_{e0i} = A_i\,\mathrm{e}^{\chi_i x} + B_i\,\mathrm{e}^{-\chi_i x}\,, \tag{6.8}$$

where A_i and B_i are complex amplitudes not yet determined. The sign of χ_i is determined via the condition

$$\Re\{\chi_i\} \geq 0 \tag{6.9}$$

The solutions exponentially decaying for $|x| \to \infty$ are given by

$$\Pi_{ez1} = A_1 \exp(\chi_1 x - \gamma z)\,, \tag{6.10}$$

$$\Pi_{ez2} = B_2 \exp(-\chi_2 x - \gamma z)\,. \tag{6.11}$$

With (3.71) and (3.72) we obtain the following relations for the field intensities $\boldsymbol{E}$ and $\boldsymbol{H}$ in both subspaces:

$$\underline{E}_{xi} = \frac{\partial^2\Pi_{ezi}}{\partial x\,\partial z}\,, \tag{6.12a}$$

$$\underline{E}_{yi} = 0\,, \tag{6.12b}$$

$$\underline{E}_{zi} = -\frac{\partial^2\Pi_{ezi}}{\partial x^2}\,, \tag{6.12c}$$

$$\underline{H}_{xi} = 0\,, \tag{6.12d}$$

$$\underline{H}_{yi} = -\mathrm{j}\omega\varepsilon_i\frac{\partial\Pi_{ezi}}{\partial x}\,, \tag{6.12e}$$

$$\underline{H}_{zi} = 0\,. \tag{6.12f}$$

Inserting (6.10) and (6.11), we obtain the field quantities

$$\underline{E}_{x1} = -\chi_1 \gamma A_1 \exp(\chi_1 x - \gamma z)\,, \tag{6.13a}$$
$$\underline{E}_{z1} = -\chi_1^2 A_1 \exp(\chi_1 x - \gamma z)\,, \tag{6.13b}$$
$$\underline{H}_{y1} = -\mathrm{j}\omega\varepsilon_1 \chi_1 A_1 \exp(\chi_1 x - \gamma z)\,, \tag{6.13c}$$
$$\underline{E}_{x2} = \chi_2 \gamma B_2 \exp(-\chi_2 x - \gamma z)\,, \tag{6.13d}$$
$$\underline{E}_{z2} = -\chi_2^2 B_2 \exp(-\chi_2 x - \gamma z)\,, \tag{6.13e}$$
$$\underline{H}_{y2} = \mathrm{j}\omega\varepsilon_2 \chi_2 B_2 \exp(-\chi_2 x - \gamma z) \tag{6.13f}$$

in both subspaces. Due to the continuity to the tangential components of the electric field intensity at the boundary surface,

$$\underline{E}_{z1}\Big|_{x=0} = \underline{E}_{z2}\Big|_{x=0}\,, \tag{6.14}$$

we obtain

$$\chi_1^2 A_1 = \chi_2^2 B_2\,. \tag{6.15}$$

In the same way from the continuity of the tangential components of the magnetic field intensities at the boundary surface,

$$\underline{H}_{y1}\Big|_{x=0} = \underline{H}_{y2}\Big|_{x=0}\,, \tag{6.16}$$

we obtain

$$\varepsilon_1 \chi_1 A_1 = -\varepsilon_2 \chi_2 B_2\,. \tag{6.17}$$

From (6.4), (6.6), (6.15) and (6.17) we obtain the equation for the propagation coefficient γ of the wave propagating along the boundary surface:

$$\gamma^2 = \gamma_{02}^2 \frac{1 - \frac{\gamma_{02}^2}{\gamma_{01}^2}\frac{\mu_1^2}{\mu_2^2}}{1 - \frac{\gamma_{02}^4}{\gamma_{01}^4}\frac{\mu_1^2}{\mu_2^2}}\,. \tag{6.18}$$

From γ and from (6.6), we also may determine the transverse propagation coefficients χ_1 and χ_2. In the important special case, where both half spaces exhibit the same permeability, (6.18) reduces to

$$\gamma^2 = \frac{\gamma_{01}^2 \gamma_{02}^2}{\gamma_{01}^2 + \gamma_{02}^2}\,, \qquad \text{for} \quad \mu_1 = \mu_2\,. \tag{6.19}$$

Before discussing the solution we will make further simplifying assumptions. We consider the case where half space 1 is filled by a non-magnetic conductor and half space 2 is free space. For this case we obtain

$$\omega\varepsilon_1' \ll \sigma_1\,, \quad \mu_1 = \mu_0\,, \quad \varepsilon_2 = \varepsilon_0\,, \quad \mu_2 = \mu_0\,. \tag{6.20}$$

In half space 1 we can neglect $\omega\varepsilon_1'$ compared with σ_1, and with

$$\sqrt{\mathrm{j}} = \mathrm{e}^{\mathrm{j}\frac{\pi}{4}} = \frac{1}{\sqrt{2}}(1+\mathrm{j}) \tag{6.21}$$

we obtain

$$\gamma_{01} = \sqrt{\mathrm{j}\omega\mu_0\sigma_1} = \sqrt{\frac{\omega\mu_0\sigma_1}{2}}(1+\mathrm{j})\,. \tag{6.22}$$

For the free-space region 2 we obtain

$$\gamma_{02} = \mathrm{j}\omega\sqrt{\mu_0\varepsilon_0} = \mathrm{j}\beta_0 = \mathrm{j}\frac{\omega}{c_0}\,, \tag{6.23}$$

where β_0 is the phase coefficient of the free space and c_0 is the propagation velocity of the plane electromagnetic wave in free space. Inserting (6.22) and (6.23) into (6.19) yields

$$\gamma = \mathrm{j}\beta_0\frac{1}{\sqrt{1+\mathrm{j}\frac{\omega\varepsilon_0}{\sigma_1}}}\,. \tag{6.24}$$

Under the assumption $\omega\varepsilon_0 \ll \sigma_1$, we obtain the following approximation

$$\gamma = \alpha + \mathrm{j}\beta = \beta_0\left(\frac{\omega\varepsilon_0}{2\sigma_1} + \mathrm{j}\right)\,. \tag{6.25}$$

The transverse propagation coefficient in the quasi-conductor χ_1 can be computed from (6.6). Due to $\omega\varepsilon_0 \ll \sigma_1$ we can neglect γ^2 compared with γ_{01}^2 in (6.6) and with $\chi_1^2 = \gamma_{01}^2$ we obtain from (6.22):

$$\chi_1 = \sqrt{\frac{\omega\mu_0\sigma_1}{2}}(1+\mathrm{j})\,. \tag{6.26}$$

To compute the transverse propagation coefficient in the free space, χ_2, we eliminate from (6.4), (6.15) and (6.17) the ε_i as well as the B_i and A_i and obtain

$$\chi_2 = -\chi_1\frac{\gamma_{02}^2}{\gamma_{01}^2}\,. \tag{6.27}$$

After inserting of (6.22) and (6.23) we obtain

$$\chi_2 = \beta_0 \sqrt{\frac{\omega\varepsilon_0}{2\sigma_1}}(1 - \mathrm{j}). \tag{6.28}$$

The field is decaying with increasing distance from the boundary surface in the free space as well as in the conductor. Inside the conductor the field decays with

$$d_0 = \frac{1}{\Re\{\chi_1\}} \tag{6.29}$$

by a factor 1/e. This effect is called the *skin effect*, and d_0 is called the *skin penetration depth*. Within the chosen approximation we obtain from (6.26) and (6.29):

$$d_0 = \sqrt{\frac{2}{\omega\mu_0\sigma_1}}. \tag{6.30}$$

Also in the free space the field intensity decays with increasing distance from the boundary surface. Within the *transverse extending height* h_0, given by

$$h_0 = \frac{1}{\Re\{\chi_2\}}, \tag{6.31}$$

the field intensity decays by a factor 1/e. From (6.28) and (6.31) we obtain within our approximation

$$h_0 = \frac{1}{\beta_0}\sqrt{\frac{2\sigma_1}{\omega\varepsilon_0}}. \tag{6.32}$$

In Figure 6.2 the frequency dependence of the skin penetration depth d_0 and the transverse extending height h_0 are shown for copper and wet earth. For the frequency dependence of the skin penetration depth in copper, we obtain

$$d_0 = \frac{66}{\sqrt{f/\mathrm{Hz}}}, \quad \text{for copper}. \tag{6.33}$$

We now compute the field intensities $\boldsymbol{E}$ and $\boldsymbol{H}$ by inserting (6.25), (6.26) and (6.28) into (6.13a) to (6.13f). Furthermore we use (6.16) and express the complex amplitudes A_1 and B_2 by the complex magnetic field amplitude $\underline{H}_{y0}$ at $x = 0, z = 0$, given by

$$\underline{H}_{y0} = \underline{H}_{y1}(x = 0, z = 0) = \underline{H}_{y2}(x = 0, z = 0). \tag{6.34}$$

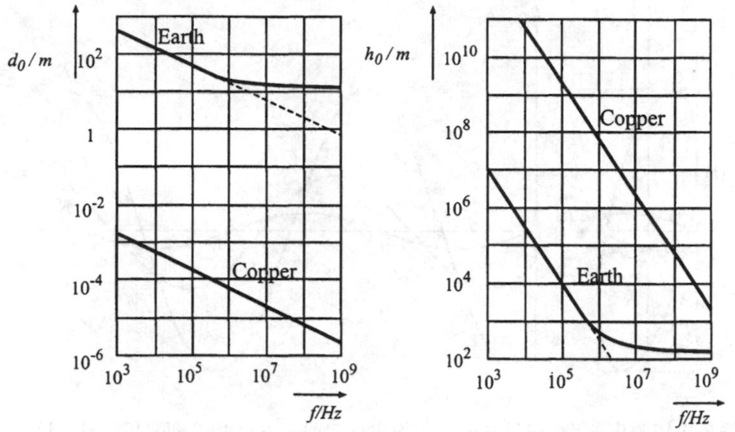

Figure 6.2: Skin depth d_0 and transverse extending height h_0 dependence on frequency f for copper ($\sigma = 5.8 \cdot 10^7 \mathrm{Sm}^{-1}$) and wet earth ($\sigma = 10^{-3}\mathrm{Sm}^{-1}$, $\varepsilon_r = 10$).

We obtain

$$\underline{E}_{x1} = \frac{Z_{F1}^2}{Z_{F0}} \underline{H}_{y0} \exp(\chi_1 x - \gamma z) , \tag{6.35a}$$

$$\underline{E}_{z1} = Z_{F1}\underline{H}_{y0} \exp(\chi_1 x - \gamma z) , \tag{6.35b}$$

$$\underline{H}_{y1} = \underline{H}_{y0} \exp(\chi_1 x - \gamma z) , \tag{6.35c}$$

$$\underline{E}_{x2} = Z_{F0}\underline{H}_{y0} \exp(-\chi_2 x - \gamma z) , \tag{6.35d}$$

$$\underline{E}_{z2} = Z_{F1}\underline{H}_{y0} \exp(-\chi_2 x - \gamma z) , \tag{6.35e}$$

$$\underline{H}_{y2} = \underline{H}_{y0} \exp(-\chi_2 x - \gamma z) . \tag{6.35f}$$

Within our approximation, $\omega\varepsilon_1 \ll \sigma_1$, the wave impedance Z_{F1} of the quasi-conductor is given by

$$Z_{F1} = \sqrt{\frac{\omega\mu_0}{2\sigma_1}}(1 + \mathrm{j}) = \sqrt{\frac{\omega\mu_0}{\sigma_1}}\, \mathrm{e}^{\mathrm{j}\frac{\pi}{4}} . \tag{6.36}$$

The wave impedance of the free space, Z_{F0}, is given by

$$Z_{F0} = \sqrt{\frac{\mu_0}{\varepsilon_0}} . \tag{6.37}$$

Figure 6.3 shows the time dependence of the electric field intensity at the boundary in the free space as well as in the conductor. The longitudinal component $\underline{E}_z$ of the electric field intensity is continuous at the boundary plane, whereas the normal component of the electric field $\underline{E}_x$ is discontinuous. Usually the magnitude of the wave

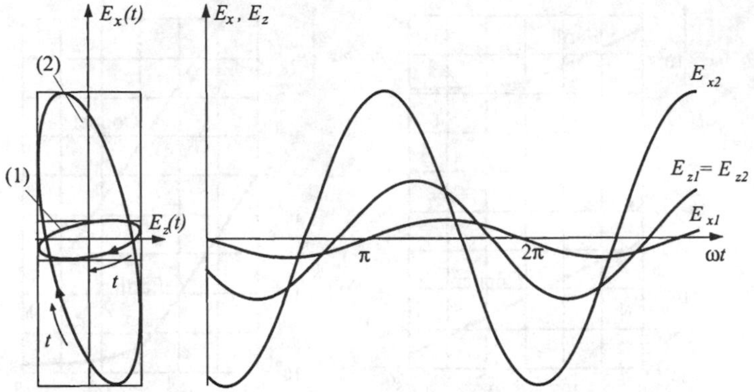

Figure 6.3: Time dependence of electric field intensity in the quasi-conductor and in free space.

impedance Z_{F1} of the quasi-conductor is by order of magnitudes smaller than the wave impedance Z_{F0} of free space. From (6.35a), (6.35b), (6.35d) and (6.35e) it follows that the magnitude $|\underline{E}_{x2}|$ of the transverse component of the electric field in free space is considerably larger than the magnitude $|\underline{E}_{z2}|$ of the longitudinal component, whereas in the quasi-conductor the magnitude of the transverse component of the electric field $|\underline{E}_{x1}|$ is by orders of magnitude smaller than the magnitude $|\underline{E}_{z1}|$ of the longitudinal component of the electric field. Furthermore longitudinal and transverse components of the electric fields are – neither in free space nor in the quasi-conductor – in the phase with transverse component of the electric field. In free space the transverse component of the electric field is delayed by 45° with respect to the longitudinal component, whereas in the quasi-conductor the transverse component advances the longitudinal component of the electric field by 45° with respect to the longitudinal component. The electric field vectors move on elliptic traces.

We now consider the surfaces of constant phase and constant amplitude of the surface wave. From (6.35a) – (6.35c), we obtain

$$\exp(\chi_1 x - \gamma z) = \exp\left(\sqrt{\frac{\omega\mu_0\sigma_1}{2}}x - \frac{\omega^2\varepsilon_0}{2\sigma_1 c_0}z\right)\exp\left[\mathrm{j}\left(\sqrt{\frac{\omega\mu_0\sigma_1}{2}}x - \frac{\omega}{c_0}z\right)\right]. \tag{6.38}$$

The first exponential term at the right side of (6.38) gives the space dependence of the amplitude, whereas the second exponential term describes the space dependence of the phase. From (6.35d) to (6.35f) we obtain for region 2

$$\exp(-\chi_2 x - \gamma z) = \exp\left[-\frac{\omega}{c_0}\sqrt{\frac{\omega\varepsilon_0}{2\sigma_1}}\left(x + \sqrt{\frac{\omega\varepsilon_0}{2\sigma_1}}z\right)\right]\exp\left[\mathrm{j}\frac{\omega}{c_0}\left(\sqrt{\frac{\omega\varepsilon_0}{2\sigma_1}}x - z\right)\right]. \tag{6.39}$$

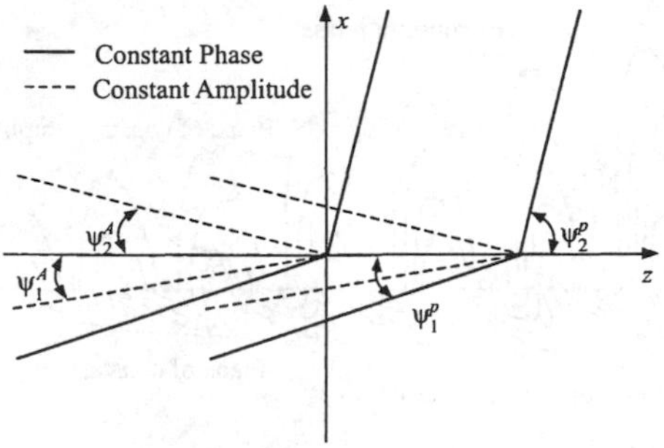

Figure 6.4: Planes of constant amplitude and constant phase.

We obtain the areas of constant amplitude in both regions by setting the first exponential factor on the right side of (6.38) or (6.39), respectively, constant. The surfaces of constant phase we obtain by setting the second exponential factor constant in either equation. The surfaces of constant amplitude as well as constant phase are all planes parallel to the y-axis. Figure 6.4 depicts the planes of constant amplitude and constant phase for either half spaces. The planes of constant amplitude enclose angles Ψ_1^A and Ψ_2^A, respectively, with the plane $x = 0$. The planes of constant phase enclose the angels Ψ_1^P and Ψ_2^P with the plane $x = 0$. From (6.38) and (6.39) we obtain by setting constant either constant exponential terms:

$$\tan \Psi_1^A = \sqrt{\frac{1}{2}\left(\frac{\omega\varepsilon_0}{\sigma_1}\right)^3}, \qquad (6.40a) \qquad \tan \Psi_1^P = \sqrt{\frac{2\varepsilon_0\omega}{\sigma_1}}, \qquad (6.40b)$$

$$\tan \Psi_2^A = \sqrt{\frac{\omega\varepsilon_0}{2\sigma_1}}, \qquad (6.41a) \qquad \tan \Psi_2^P = \sqrt{\frac{2\sigma_1}{\omega\varepsilon_0}}. \qquad (6.41b)$$

For earth with $\sigma_1 = 10^{-2}$ S/m we obtain at a frequency $f = 1$ MHz the angles $\Psi_2^A = 3°$, $\Psi_2^P = 87°$, $\Psi_1^A = 1'$, $\Psi_1^P = 6°$, where the superscript prime denotes the arc minute. In case of a good conductor with $\sigma_1 \gg \omega\varepsilon_1'$, the planes of constant phase in free space are nearly perpendicular to the boundary surface, whereas the planes of constant phase in the conductor are nearly parallel to the boundary plane. The planes of constant amplitude are nearly parallel to the boundary plane in either subspaces.

The electric field lines are depicted in Figure 6.5. In metallic conductors the depth of penetration is very small. The transverse extending height exceeds the dimensions of circuit elements or systems considerably and may be considered to be infinite. In free space the electromagnetic field distribution is not considerably influenced by a finite conductivity of the conductors. Therefore in the electromagnetic modelling of transmission lines and distributed circuits the metallic may be considered to exhibit infinite conductivity. In a further step of the analysis the power losses

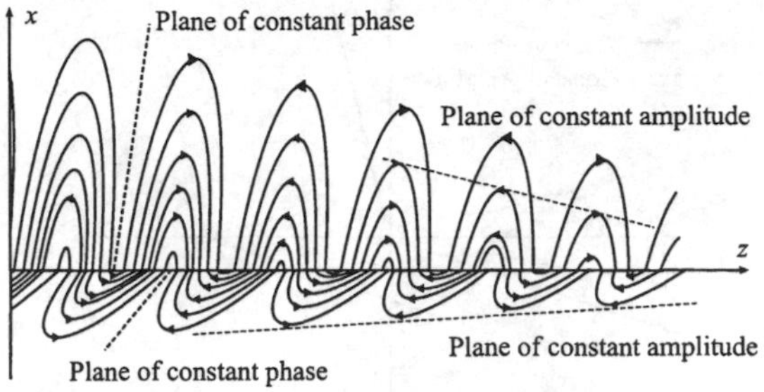

Figure 6.5: Electric field lines of the surface wave.

may be computed from the surface current distribution and the finite surface conductivity due to the skin effect. The surface current distribution for the power loss computation is directly obtained from the tangential magnetic field distribution at the metallic surfaces.

The behavior of electromagnetic surface waves also is important for the understanding of wave propagation phenomena along the Earth's surface. The surface wave may be considerably attenuated by losses in the Earth's surface. In this case the transverse extending height cannot be neglected. Surface waves can only be used up to the medium frequency ranges for radio transmission. Below about a 10 m wavelength the surface wave will be attenuated within a short distance. In the short-wave range long-distance propagation is achieved by making use of the ionospheric reflection of the wave radiated into space, whereas in the wavelength region down from meter-wave range only line-of-sight propagation is possible.

6.2 SURFACE CURRENTS

We now consider the case of metallic conductors in which σ exceeds $\omega\varepsilon'$ by orders of magnitude. In this case the electric field in free space is nearly normal to the conductor surface. Using (6.25), (6.26), (6.30) and (6.35b) we obtain the longitudinal component of the electric field in the conductor,

$$\begin{aligned} \underline{E}_{z1}(x, z, t) &= \Re\left\{\underline{E}_{z1}(x, z)\,\mathrm{e}^{\mathrm{j}\omega t}\right\} \\ &= |Z_{F1}\underline{H}_{y0}| \exp\left(-\alpha z + \frac{x}{d_0}\right) \cos\left(\omega t + \varphi_0 - \beta z + \frac{x}{d_0}\right), \end{aligned} \tag{6.42}$$

where φ_0 is the phase of $Z_{F1}\underline{H}_{y0}$. The x dependence of $\underline{E}_{z1}$ is shown in Figure 6.6(a). Since the field as well as the current penetrates into the metal only to a depth of the

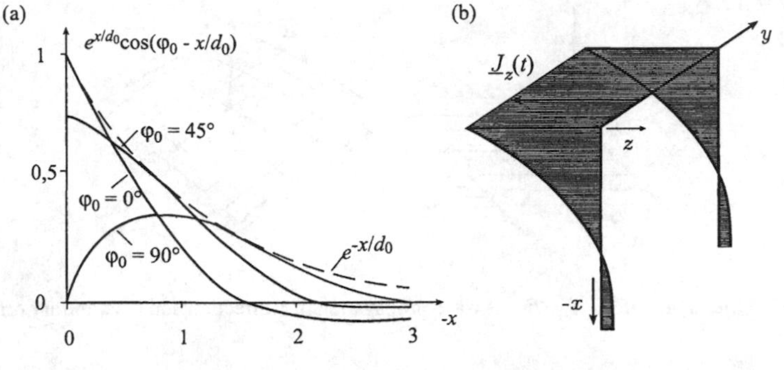

Figure 6.6: Skin effect: (a) $x\ \underline{E}_{z1}(x, 0, 0)$ for $\varphi = \omega t + \varphi_0$ and (b) current distribution in the conductor.

order d_0 we can describe the current by a surface current. From (6.35a), (6.35b) and (6.35c) we obtain

$$\underline{\mathcal{E}} = Z_{F1}\,\underline{H}_{y0}\exp(\chi_1 x - \gamma z)\left(\frac{Z_{F1}}{Z_{F0}}\,\mathrm{d}x + \mathrm{d}z\right), \tag{6.43}$$

$$\underline{\mathcal{H}} = \underline{H}_{y0}\exp(\chi_1 x - \gamma z)\,\mathrm{d}y\,. \tag{6.44}$$

Corresponding to (2.57) we obtain for the conductor region $x \geq 0$ the relation between the current density $\underline{\mathcal{J}}$ and the electric field intensity $\underline{\mathcal{E}}$

$$\underline{\mathcal{J}} = \sigma_1 \star \underline{\mathcal{E}}\,. \tag{6.45}$$

Inserting (6.43) into (6.45), considering $|Z_{F1}| \ll Z_{F0}$ and integrating $\underline{\mathcal{J}}$ over x over the interval $(-\infty, 0)$ we obtain the surface current density form

$$\underline{\mathcal{J}}_A(u, v) = \int_{-\infty}^{0} \mathcal{J}(u, v, n) = \frac{\sigma_1 Z_{F1}}{\chi_1}\,\underline{\mathcal{H}}_t\Big|_{x=0} = \underline{\mathcal{H}}_t\Big|_{x=0}. \tag{6.46}$$

At the surface the magnetic field only exhibits a tangential component. Therefore we have set $\underline{\mathcal{H}}\big|_{x=0} = \underline{\mathcal{H}}_t\big|_{x=0}$. Since in a metallic conductor $|Z_{F1}| \ll Z_{F0}$ is fulfilled we may neglect inside the conductor the electric field component normal to its surface and obtain from (6.43) the approximation

$$\underline{\mathcal{E}} \cong \underline{\mathcal{E}}_t = Z_{F1}\,\underline{H}_{y0}\exp(\chi_1 x - \gamma z)\,\mathrm{d}z\,. \tag{6.47}$$

Figure 6.6(b) shows the current distribution under the surface of the metal. Using the twist operator (2.170) we can express the relation between the tangential electric and

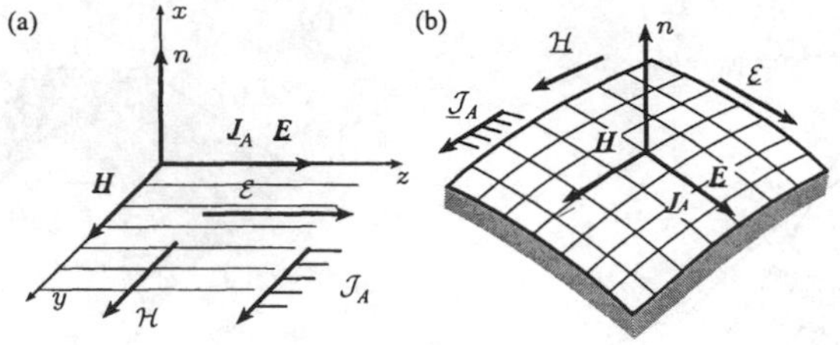

Figure 6.7: Orientation of $\boldsymbol{H}$, $\boldsymbol{J}_A$, $\boldsymbol{E}$: (a) wave propagation in z-direction and (b) arbitrary orientation.

magnetic fields (6.44), (6.47) on the conductor surface by

$$\underline{\mathcal{E}}_t = Z_{F1} \perp_n \underline{\mathcal{H}}_t \,,$$
$$\underline{\mathcal{H}}_t = -Z_{F1}^{-1} \perp_n \underline{\mathcal{E}}_t \,. \tag{6.48}$$

The magnitude of the surface current density is equal to the magnitude of the tangential magnetic field at the surface. The surface current field is orthogonal to the tangential magnetic field at the surface. As we can see from Figure 6.7, the surface normal vector $\boldsymbol{n}$, the magnetic field $\boldsymbol{H}$, and the surface current density $\boldsymbol{J}_A$ form a positive-oriented orthogonal tripod. On a conductor surface the tangential magnetic field and the surface current density have equal amplitude and are mutually orthogonal. The surface current field lines are obtained by rotating the magnetic field lines counter-clockwise by 90°.

We now consider the relation between the surface current density and the electric field in the conductor. The vectors $\boldsymbol{n}$, $\boldsymbol{H}$ and $\boldsymbol{E}$ form the positive-oriented orthogonal tripod depicted in Figure 6.7. Equation (6.48) gives the so-called *practical boundary condition*. We can apply this practical boundary condition also to curved surfaces, if the radius of curvature is large compared to the skin depth. In this case we assume the surface to be replaced in every point by its tangential plane. From (6.46) and (6.48) we obtain

$$\underline{\mathcal{E}}_t = Z_{F1} \perp_n \underline{\mathcal{J}}_A \,,$$
$$\underline{\mathcal{J}}_A = -Z_{F1}^{-1} \perp_n \underline{\mathcal{E}}_t \,. \tag{6.49}$$

The relation between the surface current density $\boldsymbol{J}_A$ and the tangential electric field $\boldsymbol{E}_t$ on the surface is given by the wave impedance Z_{F1} of the conductor.

Let us consider the rectangular area element shown in Figure 6.8. We assume that the longer side l of the rectangular area element is parallel to the direction of the surface current. We assume a surface current flowing in the z-direction. The surface

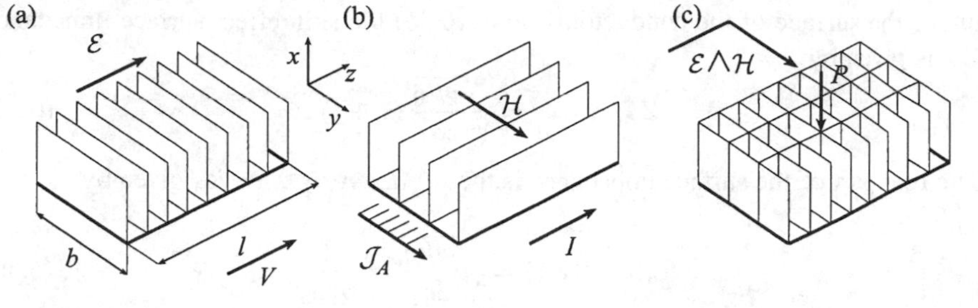

Figure 6.8: Conductor surface element: (a) electric field, (b) magnetic field, (c) Poynting field.

current form $\underline{\mathcal{J}}_A$ and the tangential electric and magnetic field forms $\underline{\mathcal{E}}_t$ and $\underline{\mathcal{H}}_t$ are given by

$$\underline{\mathcal{J}}_A = \underline{J}_{Az}\,\mathrm{d}y\,, \quad \underline{\mathcal{E}}_t = \underline{E}_z\,\mathrm{d}z\,, \quad \underline{\mathcal{H}}_t = \underline{H}_y\,\mathrm{d}y\,. \tag{6.50}$$

With (6.48) and (6.49) we obtain

$$\underline{\mathcal{E}}_t = Z_{F1}\underline{J}_{Az}\,\mathrm{d}z\,, \qquad \underline{\mathcal{H}}_t = \underline{J}_{Az}\,\mathrm{d}y\,. \tag{6.51}$$

The surface current flowing through this area element in the z-direction is given by

$$\underline{I} = \int_{y_1}^{y_2} \underline{\mathcal{J}}_A = b\underline{J}_{Az}\,. \tag{6.52}$$

The voltage across the longer side of the area element is given by

$$\underline{V} = \int_{z_1}^{z_2} \underline{\mathcal{E}}_t = l\,\underline{E}_{z1} = l\,Z_{F1}\,\underline{J}_{Az}\,. \tag{6.53}$$

From this we obtain the impedance Z of the rectangular area element of length l and width b. The impedance of the rectangular area element oriented in parallel to the surface currents only depends on the ratio side lengths l and b and the wave impedance Z_F of the material,

$$Z = \frac{\underline{U}}{\underline{I}} = \frac{l}{b} Z_{F1}\,. \tag{6.54}$$

A square area element exhibits an area impedance equal to the wave impedance $Z_{F1}F$ independently from its size. At low frequencies we may define for any thin conducting layer with a thickness d a *surface impedance* $Z_A = 1/\sigma d$. If the skin effect occurs, also for thick conductive layers the current only is flowing within a thin layer under the surface of the conductor. Due to (6.36) the skin effect surface impedance Z_A is given by

$$Z_A = Z_{F1} = \sqrt{\frac{\omega\mu_0}{2\sigma_1}}\,(1+\mathrm{j})\,. \tag{6.55}$$

The real part of the surface impedance is the *surface resistance* R_A given by

$$R_A = \Re\{Z_A\} = \sqrt{\frac{\omega\mu_0}{2\sigma_1}} = \frac{1}{\sigma_1 d_0}\,. \tag{6.56}$$

The surface resistance $\frac{1}{\sigma_1 d_0}$ is equal to the surface resistance of a conducting layer with thickness d_0 and the conductivity σ_1. The imaginary part of the surface X_A of the surface impedance is equal to its real part. The imaginary part of the surface impedance corresponds to a surface inductance originating from the penetration of the magnetic field into the metal.

$$X_A = R_A\,. \tag{6.57}$$

If we are only interested in the electromagnetic field in the free space or in the dielectric material, but not in the field distribution inside the conductor, the practical boundary condition (6.48) allows us to simplify the field computation considerably. We solve the field equations in free space or in the dielectric material, respectively, and match the electric and magnetic field intensities together at the boundary surface of the conductor using the practical boundary conditions. Let us consider for example the surface wave along a plane. From (6.13d) to (6.13f) we obtain

$$\frac{\underline{E}_{x2}}{\underline{H}_{y2}} = \frac{\gamma}{\mathrm{j}\omega\varepsilon_2}\,, \tag{6.58}$$

$$\frac{\underline{E}_{z2}}{\underline{H}_{y2}} = \frac{\chi_2}{\mathrm{j}\omega\varepsilon_2}\,. \tag{6.59}$$

From the practical boundary conditions (6.48) we obtain

$$\frac{\underline{E}_{z2}}{\underline{H}_{y2}} = Z_{F1}\,. \tag{6.60}$$

From (6.4), (6.6) and (6.58) – (6.60) we obtain γ and χ_2, which are coincident with (6.25) and (6.28). Thereby we have determined the electromagnetic field in region 2 completely.

The results of this section may also be applied to a wire with a circular cross-section, if the cross-sectional radius is considerably larger than d_0. In this case small

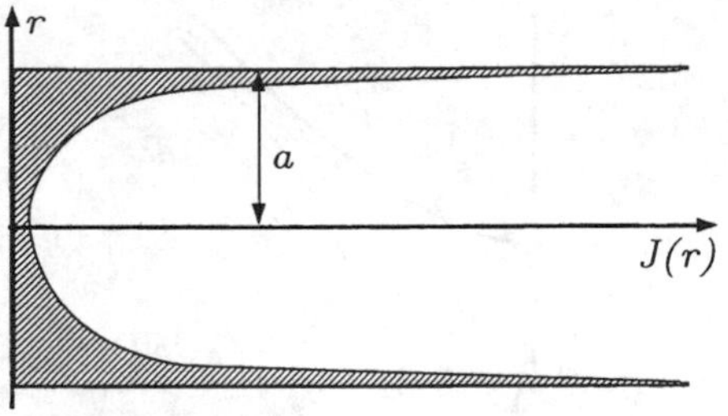

Figure 6.9: Current distribution in conductor with circular cross-section.

surface elements of the wire can be approximated by their tangential plane and the practical boundary conditions may be applied. Due to the symmetry of a conductor with circular cross-section, the current distribution over the circumference of the conductor is uniform. In Figure 6.9 the current distribution of the current in the circular conductor for $d_0 \ll a$ is shown. With the surface impedance Z_A according to (6.55) we obtain the impedance of a conductor of length l.

$$Z = R + \mathrm{j}X = \frac{l}{2\pi\sigma_1 d_0 a}(1 + \mathrm{j}) . \tag{6.61}$$

The impedance Z exhibits an inductive imaginary part X, which is due to the penetration of the magnetic field under the conductor surface. This inductive part, however, in general can be neglected in comparison with the inductance due to the magnetic field outside the conductor. At high frequencies due to the small penetration depths, the real part of the impedance Z may be much larger than the DC resistance

$$R_0 = \frac{l}{a^2\pi}\frac{1}{\sigma_1} . \tag{6.62}$$

Figure 6.10 shows the results of an exact computation of R and X for the conductor with a circular cross-section.

6.3 SURFACE CURRENT LOSSES

Surface currents in a conductor yield power loss by conversion of electromagnetic energy into thermal energy. This power loss is the reason for the attenuation of the electromagnetic surface wave. Let us now compute the surface current losses per unit of area. Since the electromagnetic field under the conductor surface decays exponentially with a very small penetration depth d_0, we can assume that the electromagnetic

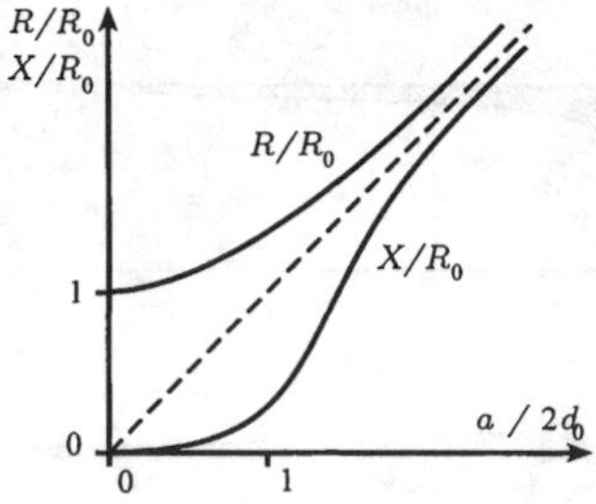

Figure 6.10: Exact values of R and X for a conductor with a circular cross-section.

energy flowing through the surface into the conductor is converted there directly into heat. The power loss P_L in the conductor consequently is equal to the active power flowing into the conductor through the surface. From (4.40) we obtain

$$P_L = -\Re\left\{\int_A \mathcal{T}\right\} . \tag{6.63}$$

The negative sign in (6.63) occurs since the area normal vector $\boldsymbol{n}$ is directed towards the conductor outside, whereas we are computing the power flowing into the conductor. From (4.21) it follows that

$$P_L = -\frac{1}{2}\Re\left\{\int_A \underline{\mathcal{E}}_t \wedge \underline{\mathcal{H}}_t^*\right\} . \tag{6.64}$$

With (6.48) we obtain

$$P_L = -\frac{1}{2}\Re\left\{Z_{F1}\int_A \left(\perp_n \underline{\mathcal{H}}_t\right) \wedge \underline{\mathcal{H}}_t^*\right\} . \tag{6.65}$$

In an orthogonal coordinate system (u, v, n) with the coordinates u and v tangential to the surface and n normal to the surface and the corresponding basis forms s_1, s_2 and n, the magnetic field form is

$$\underline{\mathcal{H}} = \underline{H}_u s_1 + \underline{H}_v s_2 . \tag{6.66}$$

and we obtain

$$\left(\perp_n \underline{\mathcal{H}}_t\right) \wedge \underline{\mathcal{H}}_t^* = \left(|\underline{H}_u|^2 + |\underline{H}_v|^2\right) s_1 \wedge s_2 = |\underline{\boldsymbol{H}}_t|^2\, s_1 \wedge s_2 \tag{6.67}$$

and from this

$$P_L = \frac{1}{2}\int_A R_A |\underline{\boldsymbol{H}}_t|^2 s_1 \wedge s_2 . \tag{6.68}$$

Since the magnetic field due to (2.122) is free of divergence and since the magnetic field disappears inside the conductor, the magnetic field exhibits only a tangential component at the conductor surface. We therefore may set $\underline{H} = \underline{H}_t$. If the magnetic field at the conductor surface is known, (6.65) and (6.68) respectively allow us to compute the conductor losses. This allows a further simplification of the field computation, which we will apply in the following as far as possible. The field distribution in the free space or in the quasi-dielectric, respectively, does not change considerably, if a metallic conductor is replaced by an ideal conductor. Therefore we can proceed in the field computation in two steps:

1. The field is computed for an ideal conductor ($\sigma = \infty$).
2. From the tangential component of the magnetic field at the conductor surface, the losses are computed using (6.68).

If we are using this method, we have to consider the following: If we let $\sigma \to \infty$, this does not influence the normal components of the electric field and the tangential component of the magnetic field considerably, whereas the tangential component of the electric field at the conductor surface is going to zero. However, this does not cause essential changes in the field distribution, since the tangential component of the electric field in the case of metallic conductors is by orders of magnitude smaller than the normal component.

Let us now consider as an example the power loss in the area element according to Figure 6.8. Inserting (6.50) in (4.17) we obtain the complex Poynting's form

$$\mathcal{T} = -\frac{1}{2}\underline{E}_z\underline{H}_y^{\star}\,\mathrm{d}y \wedge \mathrm{d}z\,. \tag{6.69}$$

With (6.51) it follows that

$$\mathcal{T}_x\Big|_{x=0} = -\frac{1}{2}Z_A|\underline{J}_{Az}^2|\,\mathrm{d}y \wedge \mathrm{d}z\,. \tag{6.70}$$

The loss of power flowing into the conductor per unit of area is given by

$$\Re\{\mathcal{T}_x\Big|_{x=0}\} = -\frac{1}{2}R_A|\underline{J}_{Az}^2|. \tag{6.71}$$

With (6.63) the loss power flowing into the area element according to Figure 6.8 is given by

$$P_L = \frac{1}{2}blR_A|\underline{J}_{Az}^2|\,. \tag{6.72}$$

Similar to (6.54) the real part of the impedance of the area element is given by

$$R = \frac{l}{b}R_A\,. \tag{6.73}$$

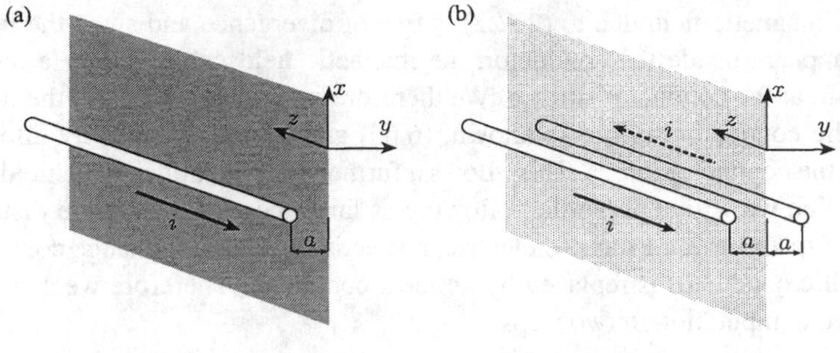

Figure 6.11: (a) Wire in distance a parallel to a conducting wall, (b) Mirror wire.

With (6.52) we obtain from (6.72) the power loss

$$P_v = \frac{1}{2} R |I^2| \tag{6.74}$$

flowing into the area element. This is exactly the power loss occurring in a resistor R, if a current I is impressed.

6.4 INDUCED SURFACE CURRENTS

A static magnetic field can penetrate a non-magnetic conductor without any perturbation. Contrary to this a high-frequency electromagnetic field induces surface currents on the conductor, which shield the electromagnetic field from inside the conductor. If the skin penetration depth d_0 is small in comparison with the linear dimensions of the conductor, we can assume the electromagnetic field to be completely shielded from inside the conductor. If the alternating magnetic field is known at the conductor surface, the surface current density may be computed from (6.48). However, we have to consider that the induced surface currents contribute to the electromagnetic field too.

In the following example we consider a straight circular conductor in parallel to a plane conducting surface, as depicted in Figure 6.11(a). We assume the circular conductor as well as the conducting plane to have infinite extension. The distance between the circular conductor and the conducting plane is a. The plane conducting surface is positioned in the plane $y = 0$. The circular conductor is oriented in parallel to the z-axis. We assume that the radius of the circular conductor may be neglected in comparison to the distance a. We first compute the magnetic field generated by a current I flowing through the conductor in the negative z-direction without assuming the presence of the conducting plane. Due to the symmetry, the magnetic field lines in this case are concentric circles with the circular conductor in the center.

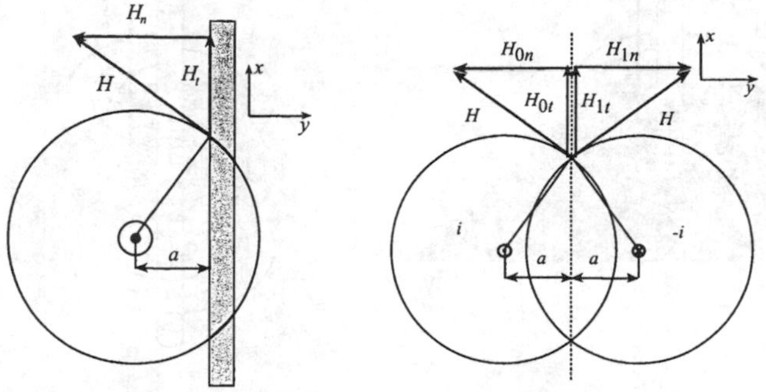

Figure 6.12: (a) Magnetic field around a straight wire in free space and (b) introduction of a mirror conductor.

We compute the magnetic field due to the current i flowing through a straight wire in free space. From Ampère's law we obtain the magnitude of the magnetic field $\underline{H}_0$ at a distance r from the axis of the wire given by

$$\underline{H}_0 = \frac{i}{2r\pi} \,. \tag{6.75}$$

At the plane $y = 0$ the tangential component of the magnetic field, $\underline{H}_{0t}$ is given by

$$\underline{H}_{0t} = \frac{a}{r}\underline{H}_0 = \frac{i}{2\pi}\frac{a}{a^2 + x^2} \,, \tag{6.76}$$

as can be concluded directly from the similarity of the two triangles $\{r, a, x\}$ and $\{\boldsymbol{H}, \boldsymbol{H}_t, \boldsymbol{H}_n\}$ in Figure 6.12(a). If we now insert the conducting plane at $y = 0$, the magnetic field $\boldsymbol{H}_0$ induces a current in the conducting wall shielding the magnetic field from inside the conductor. For symmetry reasons all the induced wall current flows parallel to the z-axis. The magnetic field $\boldsymbol{H}_1$ generated by the induced wall current exhibits tangential components that are antisymmetric with respect to the plane $y = 0$. Since the magnetic field $\boldsymbol{H}_1$, which is generated by the wall currents, is compensating the primary field $\boldsymbol{H}_0$ inside the conductor, it follows that outside the conductor the tangential components of $\boldsymbol{H}_1$ and $\boldsymbol{H}_0$ exhibit equal magnitude and equal sign. From this we obtain the surface current density

$$\underline{J}_{Az} = \underline{H}_t = 2\underline{H}_{0t} = \frac{i}{\pi}\frac{a}{a^2 + x^2} \,. \tag{6.77}$$

In Figure 6.12(b) a mirror conductor is introduced. The wall at $y = 0$ is removed and an image conductor at $y = a$ is inserted. In the image conductor a current $-i$ with

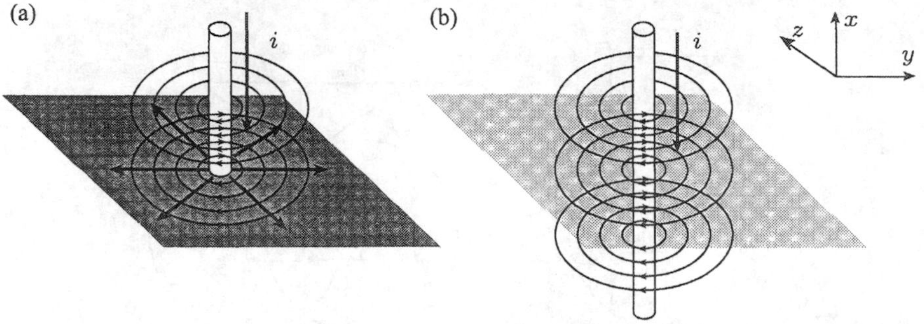

Figure 6.13: (a) Wire normal to a conducting plane and (b) Mirror wire.

the same magnitude and opposite direction is flowing. In the half-space $y < 0$ this image conductor produces the same magnetic field as the shielding currents in $y = 0$. Figure 6.11 illustrates the *mirror principle*.

We obtain the total current induced in the wall, i_{ind} by integrating the area current density induced in the wall over x:

$$i_{\text{ind}} = \int_{-\infty}^{+\infty} \underline{J}_{Az} \mathrm{d}x = i \, . \tag{6.78}$$

The total current induced in the wall has the same magnitude as the current flowing in the wire and exhibits opposite direction. We also can use the mirror principle to compute the total magnetic field due to the current in the wire and the current in the wall. If we remove again the wall in the plane $y = 0$ and arrange a mirror wire at $y = a$ with a current $-i$ impressed, the total magnetic field can be computed as the superposition of the field contributions from both wires.

Figure 6.13 illustrates the application of the mirror principle to a wire normal to a conducting plane. In the arrangement of a conducting plane and a straight wire shown in Figure 6.13(a), positioned normal to the conducting plane according to Figure 6.13(b) the conducting plane is replaced by the mirror image of the line in the region $x < 0$. In the case of the wire normal to the plane the direction of the current in the mirror wire is the same as in the original wire.

We may further generalize the mirror principle to wires of arbitrary shape. We can replace any arrangement of conductors and an infinitely extended wall by the original arrangement of conductors plus the mirror image of this arrangement. The mirror plane is the plane of symmetry. The currents in the mirror conductor arrangement have opposite components in parallel to the symmetry plane and identical components normal to the symmetry plane. If charges are mirrored at a conducting wall the mirror charges have the opposite sign. The mirror currents and the mirror charges are the sources of mirror electromagnetic fields. With respect to the plane of symmetry the tangential components of the magnetic field and the normal component of the electric field are keeping their sign, whereas the tangential components of the electric field and the normal component of the magnetic field are changing their sign.

If we use an ideally conducting magnetic wall as the mirror, the mirror currents have the same tangential components and opposite normal components, and the mirror charges have the same sign as the original charges. Furthermore the tangential components of the electric field and the normal component of the magnetic field are keeping their sign, whereas the tangential components of the magnetic field and the normal component of the electric field are changing their sign.

REFERENCES

[1] J. Zenneck, "Über die Fortpflanzung ebener elektromagnetischer Wellen längs einer ebenen Leiterfläche und ihre Beziehung zur Drahtlosen Telegraphie," *Annalen der Physik*, vol. 23, pp. 846–866, 2000.

[2] J. A. Stratton, *Electromagnetic Theory*. New York: McGraw-Hill, 1941.

[3] R. F. Harrington, *Time Harmonic Electromagnetic Fields*. New York: McGraw-Hill, 1961.

[4] L. Felsen and N. Marcuvitz, *Radiation and Scattering of Waves*. Englewood Cliffs, NJ: Prentice Hall, 1972.

[5] C. A. Balanis, *Advanced Engineering Electromagnetics*. New York: John Wiley & Sons, 1989.

Chapter 7

Transmission Lines

7.1 THE PRINCIPLES OF TRANSMISSION LINES

A *transmission line* allows the guidance of electromagnetic energy along a certain path. Usually the radiation of energy from the line into the environment and also the excitation of waves on the line by environmental fields should be avoided. We use the expression *waveguide* to denote various structures that have the purpose to guide electromagnetic energy. Waveguides are treated in detail in [1–5]. We therefore introduce the following definition: *A waveguide is a structure consisting of various materials that can guide electromagnetic waves along a given curve in space.* A homogeneous cylindric waveguide has the shape of a generalized homogeneous cylinder. A generalized cylinder is a surface that is generated by moving a closed curve in parallel to a line. A homogeneous cylinder is invariant with respect to translations in the direction of the generating line.

Figure 7.1 shows examples of various homogeneous cylindric waveguides. The *parallel wire line*, Figure 7.1(a), and the *coaxial line*, Figure 7.1(b), exhibit a two-fold connected cross-section. These lines have no cut-off frequency and can be used for all frequencies from DC upwards. The *cut-off frequency* of a certain type of a guided wave is the frequency below which this type of wave cannot propagate. In the coaxial line the electromagnetic field is confined within the region between the inner conductor and the outer conductor. The coaxial line is completely shielded. In the parallel wire line the electromagnetic field is not confined within a finite transverse cross-section. At higher frequencies the coaxial line is preferred compared to the parallel wire line, since the coaxial line exhibits no radiation losses. Hollow-pipe waveguides as the *rectangular waveguide* shown in Figure 7.1(c) and the *circular waveguide* shown in Figure 7.1(d) exhibit only one connected metallic boundary. The region inside the waveguide is either empty or filled with dielectric material. Hollow-pipe waveguides exhibit a lower cut-off frequency, since there is no electrostatic field solution for the region inside the hollow waveguide. In a hollow waveguide waves

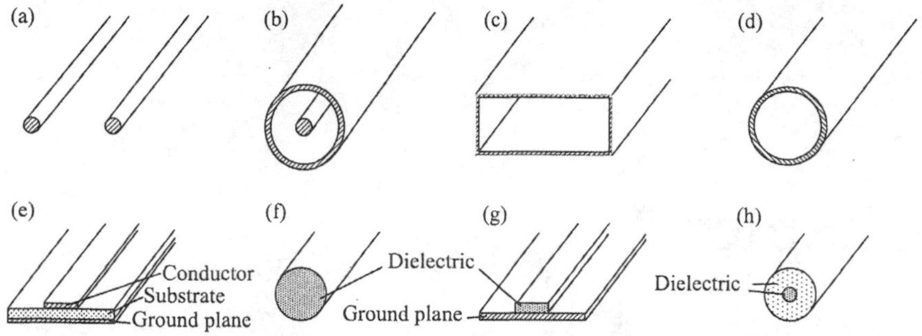

Figure 7.1: (a) Parallel wire line, (b) coaxial line, (c) rectangular waveguide, (d) circular waveguide, (e) microstrip line (f) circular dielectric waveguide, (g) image line and (h) optical fiber.

can only propagate if half the wavelength of a plane wave is smaller than the largest cross-sectional dimension of the hollow waveguide. By filling the hollow waveguide with the dielectric material, the lower cut-off frequency can be moved to lower values. This, however, will increase the line losses. In comparison with coaxial lines, hollow waveguides exhibit lower conductive losses and are preferred at frequencies above 10 GHz compared with coaxial lines. In order to transmit high power, hollow waveguides are also used at lower frequencies below 1 GHz. At millimeter-wave frequencies and higher frequencies, hollow waveguides also exhibit considerable conductor losses. Therefore, for submillimeter waves, infrared waves and in the optical wavelength region dielectric waveguides are preferred. Figure 7.1(e) shows a *microstrip line*, consisting of a metallic strip on an insulating substrate with metallic ground plane. The microstrip line is a planar structure and can be fabricated using printed-circuit techniques. Figure 7.1(f) shows a *circular dielectric waveguide*. The *dielectric waveguide* exhibits a higher permittivity than the surrounding dielectric material or free space, respectively. The electromagnetic field is not strictly confined within the dielectric waveguide, but decays exponentially in a transverse direction outside the core region of the dielectric waveguide. The *image line* depicted in Figure 7.1(g) is a dielectric waveguide mounted on conducting substrate. The electromagnetic field is mirrored on the substrate. Compared with the dielectric waveguide in Figure 7.1(f), the image line provides a mounting of the waveguide without perturbing the electromagnetic field. For very short wavelengths, especially in the optical region, a dielectric waveguide as depicted in Figure 7.1(h) is advantageous. The core of this waveguide with higher permittivity is embedded in a dielectric material with lower permittivity. In the outer region of the waveguide, the electromagnetic field is rapidly decaying, and at the boundary of the outer region the electromagnetic field has decayed sufficiently so that the field is not perturbed by the mounting of the waveguide. Also a cladding deposited onto the outer region will not increase the attenuation of

Table 7.1: Classification of Waveguide Modes

Mode Type	*Longitudinal Electric Field*	*Longitudinal Magnetic Field*
TEM or L modes (Lecher modes, transverse electromagnetic modes)	$\underline{E}_z = 0$	$\underline{H}_z = 0$
TE or H modes (transverse electric modes)	$\underline{E}_z = 0$	$\underline{H}_z \neq 0$
TM or E modes (transverse magnetic modes)	$\underline{E}_z \neq 0$	$\underline{H}_z = 0$
Hybrid modes	$\underline{E}_z \neq 0$	$\underline{H}_z \neq 0$

the electromagnetic wave. Circular dielectric waveguides with core and cladding are realized as *optical fibers*.

We now consider a homogeneous cylindric waveguide. We assume that the waveguide is oriented along the z-axis of our coordinate system. We seek a solution to the Maxwell's equations, which exhibits the following form:

$$\underline{\boldsymbol{E}}(x, y, z) = \underline{\boldsymbol{E}}_0(x, y)\, \mathrm{e}^{\pm\gamma z}, \tag{7.1a}$$

$$\underline{\boldsymbol{H}}(x, y, z) = \underline{\boldsymbol{H}}_0(x, y)\, \mathrm{e}^{\pm\gamma z}. \tag{7.1b}$$

Due to the translational invariance of the waveguide geometry in the z-direction we can decompose the expressions for the field intensities into the exponential factor $\mathrm{e}^{\pm\gamma z}$ and the transverse functions $\underline{\boldsymbol{E}}_0(x, y)$ and $\underline{\boldsymbol{H}}_0(x, y)$, respectively. We will see that there is an infinite number of solutions existing with distinct transverse field structures $\underline{\boldsymbol{E}}_0(x, y)$ and $\underline{\boldsymbol{H}}_0(x, y)$, respectively. Each of these solutions is named *mode*. We can subdivide the modes into *propagating modes* and *evanescent modes*. In the case of a lossless waveguide, the propagating mode exhibits an imaginary propagation coefficient $\gamma = \mathrm{j}\beta$, whereas the evanescent mode exhibits a real propagation coefficient $\gamma = \alpha$. If a mode exhibits a cut-off frequency, below this cut-off frequency the mode will be an evanescent mode, and above the cut-off frequency the mode will be a propagating mode. The number of existing propagating modes increases with increasing frequency. If below some cut-off frequency there exists only one propagating mode, this propagating mode is called the *fundamental mode* of the waveguide. We also may classify the modes with respect to the occurring field components. Table 7.1 summarizes the classification scheme of waveguides. The *transverse electromagnetic modes* or *Lecher modes* exhibit only transverse field components. The *transverse electric modes* exhibit no electric field components in the longitudinal direction, whereas the *transverse magnetic mode* exhibits no magnetic field components in the longitudinal direction. The *hybrid modes* have electric longitudinal field components as well as magnetic longitudinal field components.

7.2 PHASE AND GROUP VELOCITY

For a propagating mode we obtain from (7.1a) for a wave propagating in the positive z-direction the electric field intensity

$$\boldsymbol{E}(\boldsymbol{x}, t) = \Re\{\underline{\boldsymbol{E}}_0(x, y)\, \mathrm{e}^{\mathrm{j}(\omega t - \beta z)}\}\,. \tag{7.2}$$

The velocity by which a plane of constant phase is propagating, is called the *phase velocity*. We obtain the phase velocity by setting the exponential term in (7.2) constant

$$c = \frac{\omega}{\beta}\,. \tag{7.3}$$

A harmonic electromagnetic wave exhibits a phase velocity, which in general depends on frequency. The frequency dependence of the phase velocity may be caused by the geometric properties of the waveguides as well as by the frequency dependence of the permittivity and permeability of the material filling the waveguide. A wave packet as depicted in Figure 7.2 may be considered a superposition of harmonic waves. The electric field of such a wave packet may be described by

$$\boldsymbol{E}(\boldsymbol{x}, t) = \Re\left\{\int_{\omega_0-\Delta\omega}^{\omega_0+\Delta\omega} \underline{\boldsymbol{E}}_0(x, y, \omega)\, \mathrm{e}^{\mathrm{j}(\omega t - \beta(\omega) z)}\, \mathrm{d}\omega\right\}. \tag{7.4}$$

We assume the phase coefficient $\beta(\omega)$ to be frequency-dependent. Considering a narrow band wave packet, the phase coefficient $\beta(\omega)$ may be expanded in a certain frequency interval around the center frequency ω_0 into a Taylor series.

$$\omega t - \beta(\omega) z = \omega_0 t - \beta(\omega_0) z + (\omega - \omega_0)\left(t - \left.\frac{\mathrm{d}\beta}{\mathrm{d}\omega}\right|_{\omega_0} z\right). \tag{7.5}$$

After inserting of (7.5) into (7.4), we obtain

$$\boldsymbol{E}(\boldsymbol{x}, t) = \Re\left\{\mathrm{e}^{\mathrm{j}(\omega_0 t - \beta(\omega_0) z)} \int_{\omega_0-\Delta\omega}^{\omega_0+\Delta\omega} \underline{\boldsymbol{E}}_0(x, y, \omega)\, \mathrm{e}^{\mathrm{j}(\omega-\omega_0)\left(t - \left.\frac{\mathrm{d}\beta}{\mathrm{d}\omega}\right|_{\omega_0} z\right)}\, \mathrm{d}\omega\right\}. \tag{7.6}$$

The exponential term describes a harmonic wave propagating in the z-direction with an angular frequency ω_0. This harmonic wave propagates according to (7.3) with a

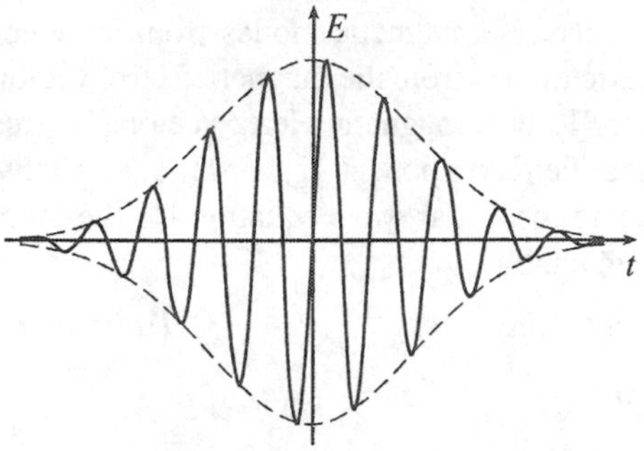

Figure 7.2: A periodic wave train.

phase velocity c. The integral in (7.6) describes the envelope of the wave. Setting $t - (\mathrm{d}\beta/\mathrm{d}\omega)z$ constant we obtain the velocity v_g of propagation of the envelope,

$$v_g = \left(\frac{\mathrm{d}\beta}{\mathrm{d}\omega}\right)^{-1} . \tag{7.7}$$

This velocity v_g is called the *group velocity*. Energy and information propagate with the group velocity. The phase velocity is only a virtual velocity of the phase planes and does not determine the velocity of transmission of energy or information, respectively.

7.3 THE FIELD COMPONENTS

In this section we discuss the way to evaluate the electromagnetic field components in linear waveguides. Without any restriction in generality we assume the waveguides to be oriented in the z-direction. Furthermore, we assume that the cross-section of the waveguide may be subdivided into subsections and that each of these subsections is homogeneously filled by conducting dielectric material or by free space. One strategy to solve the Maxwell's equations for such structures is to seek the partial solutions for each homogeneously filled subsection and to match the solution along the boundaries. If the metallic conductors are considered to be ideally conducting, we do not need to solve the field equation inside the metallic regions. By superposition of partial solutions the boundary conditions may be fulfilled. The mathematical effort may be reduced considerably, if we can choose a cylindric coordinate system in which boundary surfaces of the waveguide may be defined by setting constant one coordinate. In this case a single partial solution may already represent the field distribution of a mode.

We may derive transverse magnetic modes from the electric Hertz vector $\underline{\mathbf{\Pi}}_e$ and the transverse electric modes from the magnetic Hertz vector $\underline{\mathbf{\Pi}}_m$. We can choose an electric Hertz vector $\underline{\mathbf{\Pi}}_e$ or a magnetic Hertz vector $\underline{\mathbf{\Pi}}_m$, respectively, which exhibits only a longitudinal field component $\underline{\Pi}_{ez}$ or $\underline{\Pi}_{mz}$, respectively. The electric and magnetic Hertz forms and the scalar wave equation for the longitudinal components of the Hertz forms are given by:

Transverse Magnetic Field

$$\underline{H}_z = 0$$

$$\underline{\Pi}_e(\boldsymbol{x}) = \underline{\Pi}_{ez}(\boldsymbol{x})\,\mathrm{d}z\,, \tag{7.8a}$$

$$\Delta\underline{\Pi}_{ez} - \underline{\gamma}_{M0}^2\underline{\Pi}_{ez} = 0\,, \tag{7.9a}$$

Transverse Electric Field

$$\underline{E}_z = 0$$

$$\underline{\Pi}_m(\boldsymbol{x}) = \underline{\Pi}_{mz}(\boldsymbol{x})\,\mathrm{d}z\,, \tag{7.8b}$$

$$\Delta\underline{\Pi}_{mz} - \underline{\gamma}_{M0}^2\underline{\Pi}_{mz} = 0\,, \tag{7.9b}$$

$$\underline{\gamma}_{M0}^2 = -\omega^2\,\underline{\mu}\,\underline{\varepsilon}\,, \tag{7.10}$$

$$\underline{\mathcal{E}} = -\tilde{\mathrm{d}}\,\mathrm{d}\,\underline{\Pi}_e = \star\,\mathrm{d}\star\,\mathrm{d}\,\underline{\Pi}_e\,, \tag{7.11a}$$

$$\underline{\mathcal{H}} = \mathrm{j}\omega\underline{\varepsilon}\star\,\mathrm{d}\,\underline{\Pi}_e\,, \tag{7.12a}$$

$$\underline{\mathcal{H}} = -\tilde{\mathrm{d}}\,\mathrm{d}\,\underline{\Pi}_m = \star\,\mathrm{d}\star\,\mathrm{d}\,\underline{\Pi}_m\,, \tag{7.11b}$$

$$\underline{\mathcal{E}} = -\mathrm{j}\omega\underline{\mu}\star\,\mathrm{d}\,\underline{\Pi}_m\,. \tag{7.12b}$$

Transverse Magnetic Field

$$\underline{H}_z = 0$$

$$\underline{E}_x = \frac{\partial^2\underline{\Pi}_{ez}}{\partial x\partial z}\,, \tag{7.13a}$$

$$\underline{E}_y = \frac{\partial^2\underline{\Pi}_{ez}}{\partial y\partial z}\,, \tag{7.14a}$$

$$\underline{E}_z = \frac{\partial^2\underline{\Pi}_{ez}}{\partial z^2} - \underline{\gamma}_{M0}^2\underline{\Pi}_{ez}\,, \tag{7.15a}$$

$$\underline{H}_x = \mathrm{j}\omega\underline{\varepsilon}\,\frac{\partial\underline{\Pi}_{ez}}{\partial y} \tag{7.16a}$$

$$\underline{H}_y = -\mathrm{j}\omega\underline{\varepsilon}\,\frac{\partial\underline{\Pi}_{ez}}{\partial x}\,, \tag{7.17a}$$

$$\underline{H}_z = 0\,, \tag{7.18a}$$

Transverse Electric Field

$$\underline{E}_z = 0$$

$$\underline{H}_x = \frac{\partial^2\underline{\Pi}_{mz}}{\partial x\partial z}\,, \tag{7.13b}$$

$$\underline{H}_y = \frac{\partial^2\underline{\Pi}_{mz}}{\partial y\partial z}\,, \tag{7.14b}$$

$$\underline{H}_z = \frac{\partial^2\underline{\Pi}_{mz}}{\partial z^2} - \underline{\gamma}_{M0}^2\underline{\Pi}_{mz}\,, \tag{7.15b}$$

$$\underline{E}_x = -\mathrm{j}\omega\underline{\mu}\,\frac{\partial\underline{\Pi}_{mz}}{\partial y}\,, \tag{7.16b}$$

$$\underline{E}_y = \mathrm{j}\omega\underline{\mu}\,\frac{\partial\underline{\Pi}_{mz}}{\partial x}\,, \tag{7.17b}$$

$$\underline{E}_z = 0\,. \tag{7.18b}$$

For hybrid modes a linear combination of a longitudinal electric Hertz vector according to (7.8a) and a magnetic Hertz vector according to (7.8b) may be chosen.

7.4 WAVEGUIDES FOR TRANSVERSE ELECTROMAGNETIC WAVES

Transverse electromagnetic (TEM) *waves* exhibit no field components in the direction of propagation. Choosing the z-direction as the direction of propagation, we obtain

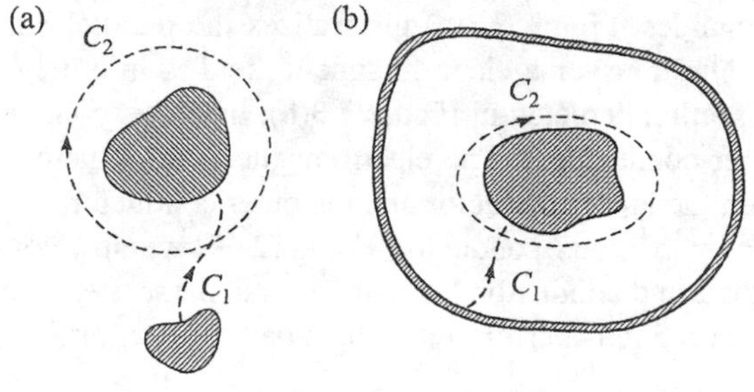

Figure 7.3: Paths of integration C_1 and C_2 for (a) parallel wire line and (b) shielded line.

$\underline{E}_z = 0$ and $\underline{H}_z = 0$. Therefore we can derive the transverse electromagnetic wave from (7.8a) as well from (7.8b). We are choosing the formulation (7.8a). In this case the Helmholtz equation (7.9a) is valid and the field components in the cartesian coordinate system are given by (7.13a) – (7.18a). With $\underline{E}_z = 0$ we obtain from (7.15a):

$$\frac{\partial^2 \underline{\Pi}_{ez}}{\partial z^2} - \underline{\gamma}_{M0}^2 \underline{\Pi}_{ez} = 0\,. \tag{7.19}$$

This equation is satisfied by

$$\underline{\Pi}_{ez}(\boldsymbol{x}) = \underline{\Pi}_{e0}^{(+)}(x, y)\, \mathrm{e}^{-\underline{\gamma}_{M0} z} + \underline{\Pi}_{e0}^{(-)}(x, y)\, \mathrm{e}^{\underline{\gamma}_{M0} z}\,. \tag{7.20}$$

For lossless lines we obtain $\underline{\gamma}_{M0} = \mathrm{j}\beta_{M0}$. Due to (7.10) the phase velocity of the TEM wave is equal to the phase velocity of the plane electromagnetic wave. In the following we assume ideal conductors in lossless media. From (7.9a) and (7.19), we obtain

$$\frac{\partial^2 \underline{\Pi}_{e0}}{\partial x^2} + \frac{\partial^2 \underline{\Pi}_{e0}}{\partial y^2} = 0\,, \tag{7.21}$$

which holds for $\underline{\Pi}_{e0}^{(+)}(x, y)$ as well $\underline{\Pi}_{e0}^{(-)}(x, y)$. This equation is the two-dimensional *Laplace equation* known from electrostatics. Due to (7.13a), (7.14a), (7.16a) and (7.17a) the two-dimensional Laplace equation must also be satisfied by the components $\underline{E}_x$, $\underline{E}_y$, $\underline{H}_x$ and $\underline{H}_y$. The transverse field distribution therefore corresponds to the field distribution of the static two-dimensional problem. Furthermore from the validity of the two-dimensional Laplace equation for the transverse field components it follows that in a waveguide bounded by a connected conductor a transverse electromagnetic wave cannot occur. Transverse electromagnetic waves only may occur, if the waveguide cross-section exhibits at least two unconnected conductors. Figure 7.3 shows schematically the cross-sections of two transverse electromagnetic waveguides. Figure 7.3(a) generalizes the parallel wire line according to Figure 7.1(a). The transverse electromagnetic field is infinitely extended. In the waveguide cross-section depicted in Figure 7.3(b), the outer conductor 1 completely surrounds the inner conductor 2. The electromagnetic field therefore is bounded to the region between the inner conductor and the outer conductor and is shielded from the outside region by the outer conductor. Using (2.59) we now compute the voltage between conductor 2 and conductor 1 in a transverse plane, i.e., in a plane normal to the z-axis. The integration is performed along a path of integration C_1. We obtain

$$\underline{V}_{21}(z) = -\int_{C_1} \underline{\mathcal{E}} \cdot \tag{7.22}$$

For any transverse plane, i.e., for a certain z we obtain a voltage $\underline{V}_{21}(z)$, which is independent from the choice of the path of integration. The independence from the path of integration is due to the circumstance that in the transverse electromagnetic mode no longitudinal magnetic field occurs and any closed contour of integration yields a zero contribution to the path integral (7.22). We therefore may define for a certain transverse plane a voltage between the two conductors 2 and 1. The voltage $\underline{V}_{21}(z)$, however, depends on the longitudinal coordinate z. In the following we shall use the simplified notation $\underline{V}(z)$ instead of $\underline{V}_{21}(z)$. We now determine the current $\underline{I}_2$ flowing through conductor 2. To do this we insert (2.63) into (2.50) and integrate over the closed path C_2 surrounding the conductor 2. Since the transverse electromagnetic mode exhibits no longitudinal electric flux, the integral of the magnetic field over the closed path C_2 is equal to the current flowing through the inner conductor in positive z-direction. We obtain

$$\underline{I}_2(z) = \oint_{C_2} \underline{\mathcal{H}} \cdot \tag{7.23}$$

In the following we write $\underline{I}(z)$ instead of $\underline{I}_2(z)$. If the waveguide cross-section exhibits only two unconnected conductors, the currents flowing into these conductors are of equal magnitude and opposite direction. $\underline{E}_z = 0$ is only valid for ideal conductors with $\sigma = \infty$. In the case of finite conductivity there also occurs a small z component of the electric field. However, in most cases we need not consider the finite conductivity σ of the metallic surfaces for computing the electromagnetic field distribution of the modes. Electric losses and conductor losses may be calculated on the base of field distributions of lossless lines. Furthermore we restrict our considerations to electromagnetic waves propagating in the positive z-direction. From (7.13a), (7.14a), (7.16a) and (7.17a) we obtain for the transverse field components in the lossless medium:

$$\underline{E}_x = -\mathrm{j}\beta_{M0}\frac{\partial \underline{\Pi}_{ez}}{\partial x}\,, \qquad \underline{H}_x = \mathrm{j}\omega\varepsilon\frac{\partial \underline{\Pi}_{ez}}{\partial y}\,, \tag{7.24}$$

$$\underline{E}_y = -\mathrm{j}\beta_{M0}\frac{\partial \underline{\Pi}_{ez}}{\partial y}\,, \qquad\qquad \underline{H}_y = -\mathrm{j}\omega\varepsilon\frac{\partial \underline{\Pi}_{ez}}{\partial x}\,. \tag{7.25}$$

From (2.78) and (7.10) it follows that

$$\underline{E}_x = Z_F \underline{H}_y, \quad \underline{E}_y = -Z_F \underline{H}_x\,. \tag{7.26}$$

These equations may be generalized in vector notation as follows

$$\underline{\mathcal{E}} = -Z_F \star (\mathrm{d}z \wedge \underline{\mathcal{H}})\,, \tag{7.27a}$$

$$\underline{\mathcal{H}} = \frac{1}{Z_F} \star (\mathrm{d}z \wedge \underline{\mathcal{E}})\,. \tag{7.27b}$$

The electromagnetic field distribution is completely described by specifying $\underline{I}(z)$ and $\underline{V}(z)$. We can express the complex field intensities $\underline{\boldsymbol{E}}(\boldsymbol{x})$ and $\underline{\boldsymbol{H}}(\boldsymbol{x})$ as product of the complex scalar amplitudes $\underline{V}(z)$ and $\underline{I}(z)$ with normalized real field vectors $\mathbf{e}(x, y)$ and $\mathbf{h}(x, y)$:

$$\underline{\boldsymbol{E}}(\boldsymbol{x}) = \underline{V}(z)\mathbf{e}(x, y)\,, \tag{7.28a}$$

$$\underline{\boldsymbol{H}}(\boldsymbol{x}) = \underline{I}(z)\mathbf{h}(x, y)\,. \tag{7.28b}$$

We call $\mathbf{e}(x, y)$ the *electric structure function* and $\mathbf{h}(x, y)$ the *magnetic structure function* of the TEM mode . The structure functions are given by

$$\mathbf{e}(x, y) = \left[e_x(x, y), e_y(x, y)\right]^T\,, \tag{7.29a}$$

$$\mathbf{h}(x, y) = \left[h_x(x, y), h_y(x, y)\right]^T \tag{7.29b}$$

where $e_x(x, y)$, $e_y(x, y)$, $h_x(x, y)$ and $h_y(x, y)$ are the components of the structure functions. The corresponding differential forms are the *electric structure form* $\mathsf{e}(x, y)$ and the *magnetic structure form* $\mathsf{h}(x, y)$, given by

$$\mathsf{e}(x, y) = e_x(x, y)\,\mathrm{d}x + e_y(x, y)\,\mathrm{d}y\,, \tag{7.30a}$$

$$\mathsf{h}(x, y) = h_x(x, y)\,\mathrm{d}x + h_y(x, y)\,\mathrm{d}y\,. \tag{7.30b}$$

From (7.22), (7.23), (7.28a) and (7.28b) we obtain

$$-\int_{C_1} \mathsf{e}(x, y) = 1\,, \tag{7.31a}$$

$$\oint_{C_2} \mathsf{h}(x, y) = 1\,. \tag{7.31b}$$

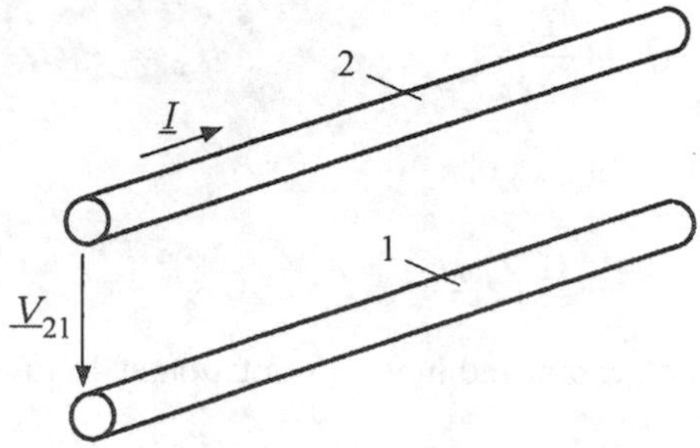

Figure 7.4: Arrows for $\underline{V}(z)$ and $\underline{V}(z)$ in the TEM line.

Due to the uniqueness of the solution of the two-dimensional Laplace equation (7.21) the field vectors in distinct transverse planes only differ by a factor independent from x and y. The normalized field vectors e and h are independent from z. The lossless TEM waveguide may be characterized by the *capacitance per unit of length* C' and *inductance per unit of length* L' . Let us consider the *charge per unit of length* Q' on conductor 2. We determine the charge $Q'\Delta z$ stored within a section of length Δz by

$$Q'\Delta z = \varepsilon \int_{A_2} \star\, \underline{\mathcal{E}}\,. \tag{7.32}$$

The volume of integration V is the cylinder shown in Figure 7.5. The side surface of the cylinder is generated by the curve C_2. Due to the transverse character of the electric field we only have to integrate over the side surface of the cylinder.

Inserting (7.27a) into (7.32) yields

$$Q'\Delta z = -\varepsilon Z_F \int_{A_2} \mathrm{d}z \wedge \underline{\mathcal{H}} = \varepsilon Z_F \int_{A_2} \underline{\mathcal{H}} \wedge \mathrm{d}z\,, \tag{7.33a}$$

$$Q' = \varepsilon Z_F \int_{C_2} \underline{\mathcal{H}} = \varepsilon Z_F \underline{I}\,. \tag{7.33b}$$

Inserting (7.28a) into (7.32) we obtain

$$Q'\Delta z = \underline{V}\,\varepsilon \int_{A_2} \star\, \mathsf{e}\,. \tag{7.34}$$

Inserting (7.30a) yields

$$\begin{aligned} Q'\Delta z &= \underline{V}\,\varepsilon \int_{A_2} \left(e_x\, \mathrm{d}y \wedge \mathrm{d}z + e_y\, \mathrm{d}z \wedge \mathrm{d}x\right) \\ &= \underline{V}\,\varepsilon \int_{A_2} \left(e_x\, \mathrm{d}y - e_y\, \mathrm{d}x\right) \wedge dz\,. \end{aligned} \tag{7.35}$$

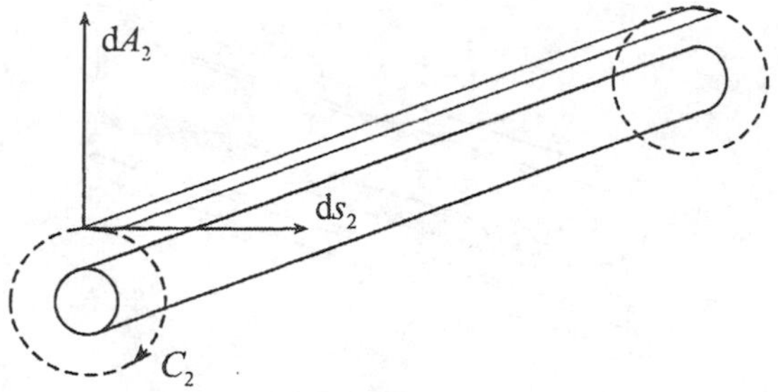

Figure 7.5: Volume for integration to determine the charge.

The capacitance per unit of length, C' is given by

$$C' = \frac{Q'}{\underline{V}} = \varepsilon \int_{A_2} \left(e_x \,\mathrm{d}y - e_y \,\mathrm{d}x \right) . \tag{7.36}$$

If the electromagnetic wave is propagating only in the positive z-direction, the ratio of voltage and current due to (7.33b) and (7.36) is given by

$$Z_0 = \frac{\underline{V}}{\underline{I}} = \frac{Q'}{C'\underline{I}} = \frac{Z_F \,\varepsilon}{C'} . \tag{7.37}$$

Z_0 is the *characteristic impedance* of the transmission line. The characteristic impedance is the impedance of a line with infinite length. With (2.56) and (2.78) we obtain

$$Z_0 = \frac{1}{cC'} . \tag{7.38}$$

The characteristic impedance depends on the phase velocity and on the capacity per unit of length. The phase velocity c of the transverse electromagnetic wave is equal to the phase velocity of the plane electromagnetic wave propagating in the same dielectric medium that is filling the space between the conductors in the transverse electromagnetic waveguide. The capacitance C' per unit of length corresponds to the electrostatic capacitance per unit of length. We now determine the inductance per unit of length. We consider the line segment of length Δz. Through the space between both lines a *magnetic flux per unit of length* $\underline{\Phi}'$ is flowing. From (2.145) we obtain

$$\underline{\Phi}' \Delta z = \mu \int_{A_1} \star\, \underline{\mathcal{H}} . \tag{7.39}$$

The integration is performed over the area A_1, which is obtained by parallel transla-

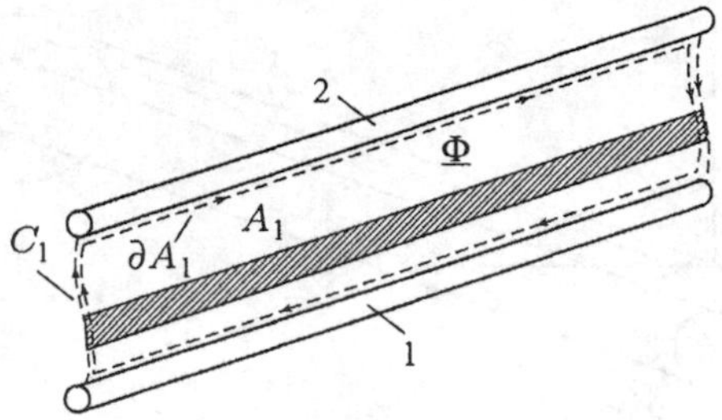

Figure 7.6: Area of integration to determine the magnetic flux $\underline{\Phi}'\Delta z$.

tion of the curve C_1 in z-direction by a distance Δz (Figure 7.6). With (7.27b) we obtain

$$\underline{\Phi}'\Delta z = \frac{\mu}{Z_F}\int_{A_1} \mathrm{d}z \wedge \underline{\mathcal{E}} = -\frac{\mu}{Z_F}\int_{A_1} \underline{\mathcal{E}} \wedge \mathrm{d}z\,. \tag{7.40}$$

This yields directly

$$\underline{\Phi}' = -\frac{\mu}{Z_F}\int_{C_1} \underline{\mathcal{E}}\,. \tag{7.41}$$

With (7.22) we obtain

$$\underline{\Phi}' = \frac{\mu}{Z_F}\underline{V}\,. \tag{7.42}$$

The inductance L' per unit of length is defined by

$$L' = \frac{\underline{\Phi}'}{\underline{I}}\,. \tag{7.43}$$

From (7.28b) and (7.39) we obtain

$$\begin{aligned} \underline{\Phi}'\Delta z &= \underline{I}\,\mu\int_{A_1} \left(h_x\,\mathrm{d}y \wedge \mathrm{d}z + h_y\,\mathrm{d}z \wedge \mathrm{d}x\right) \\ &= \underline{I}\,\mu\int_{A_2} \left(h_x\,\mathrm{d}y - h_y\,\mathrm{d}x\right) \wedge dz\,. \end{aligned} \tag{7.44}$$

This yields directly

$$\underline{\Phi}' = \underline{I}\,\mu\int_{A_2} \left(h_x\,\mathrm{d}y - h_y\,\mathrm{d}x\right)\,. \tag{7.45}$$

With (7.43) we obtain

$$L' = \mu\int_{A_2} \left(h_x\,\mathrm{d}y - h_y\,\mathrm{d}x\right)\,. \tag{7.46}$$

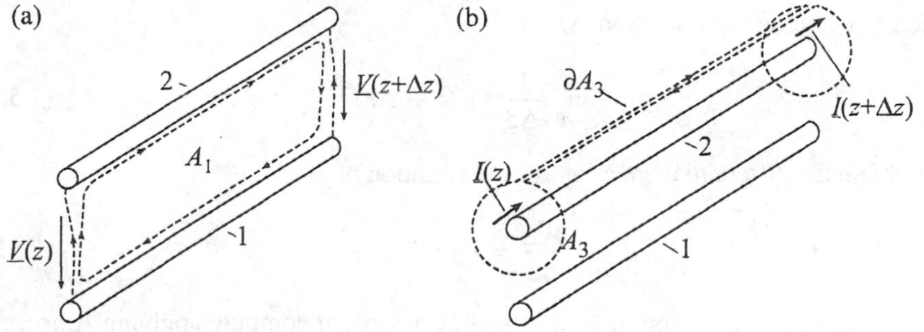

Figure 7.7: (a) Application of Faraday's law., (b) Application of Ampère's law.

Using (7.37), (7.42), (7.43), (2.74) and (2.78) we can express the characteristic impedance Z_0 of the transmission line by the phase velocity c of the electromagnetic wave and the inductance per unit of length L'.

$$Z_0 = \frac{\underline{V}}{\underline{\Phi}'} L' = \frac{Z_F}{\mu} L' = c\, L' . \tag{7.47}$$

Furthermore we obtain from (2.74), (7.38) and (7.47)

$$Z_0 = \sqrt{\frac{L'}{C'}} , \tag{7.48}$$

$$c = \frac{1}{\sqrt{L'C'}} = \frac{1}{\sqrt{\varepsilon\mu}} . \tag{7.49}$$

Let us now consider wave propagation in the positive z-direction and in negative z-direction as well, and let us derive the transmission line equations for $\underline{V}$ and $\underline{I}$. To investigate the variation of $\underline{V}(z)$ with z we evaluate the difference of the voltages $\underline{V}(z)$ and $\underline{V}(z+\Delta z)$ in Figure 7.7(a). Since the tangential components of the electric field intensity vanishes at the conductor surface, the difference of the voltages $\underline{V}(z)$ and $\underline{V}(z+\Delta z)$ is given by the circulation integral

$$\underline{V}(z+\Delta z) - \underline{V}(z) = \oint_{\partial A_1} \underline{\mathcal{E}} = \int_{A_1} \mathrm{d}\underline{\mathcal{E}} . \tag{7.50}$$

Using Faraday's law (2.134) and (7.39) and (7.43) we obtain

$$\underline{V}(z+\Delta z) - \underline{V}(z) = -\mathrm{j}\omega\mu \int_{A_1} \star\, \underline{\mathcal{H}} = -\mathrm{j}\omega \underline{\Phi}' \Delta z = -\mathrm{j}\omega L' I \Delta z . \tag{7.51}$$

After performing the transition $\Delta z \to 0$,

$$\frac{\mathrm{d}\underline{V}}{\mathrm{d}z} = \lim_{\Delta z \to 0} \frac{1}{\Delta z} \left[\underline{V}(z + \Delta z) - \underline{V}(z) \right] , \tag{7.52}$$

we obtain the differential equation for the variation of $\underline{V}(z)$:

$$\frac{\mathrm{d}\underline{V}}{\mathrm{d}z} = -\mathrm{j}\omega L' \underline{I} . \tag{7.53}$$

The difference of currents $\underline{I}(z + \Delta z)$ and $\underline{I}(z)$, we can compute applying Ampère's law (2.50), and integrating over the cut cylindric surface in Figure 7.7(b). The circulation integral along ∂A_3 is composed of the circulation integrals in the transverse planes at z and at $z + \Delta z$. These circulation integrals yield the current contributions $\underline{I}(z)$ and $-\underline{I}(z + \Delta z)$. The contributions of the path integrals parallel to the z-axis from z to $z + \Delta z$ compensate each other and therefore give no contribution to the circulation. Therefore we obtain

$$\underline{I}(z + \Delta z) - \underline{I}(z) = -\oint_{\partial A_3} \underline{\mathcal{H}} = -\int_{A_3} \mathrm{d}\underline{\mathcal{H}} . \tag{7.54}$$

In the area integral we can replace the cut surface A_3 by the uncut surface A_2 according to Figure 7.5. Since the area A_2 is not penetrated by a conduction current, Ampere's law, (2.133), together with (2.144) and (7.32) yields

$$\underline{I}(z + \Delta z) - \underline{I}(z) = -\mathrm{j}\omega\varepsilon \int_{A_3} \star\, \underline{\mathcal{E}} = -\mathrm{j}\omega Q' \,\Delta z = -\mathrm{j}\omega C' \,\underline{V}\,\Delta z . \tag{7.55}$$

With the transition $\Delta z \to 0$ we obtain

$$\frac{\mathrm{d}\underline{I}}{\mathrm{d}z} = \lim_{\Delta z \to 0} \frac{1}{\Delta z} (\underline{I}(z + \Delta z) - \underline{I}(z)) \tag{7.56}$$

and therewith it follows from (7.55) that

$$\frac{\mathrm{d}\underline{I}}{\mathrm{d}z} = -\mathrm{j}\omega C' \underline{V} . \tag{7.57}$$

From (7.53) and (7.57) we can eliminate $\underline{V}$ or $\underline{I}$, respectively, and we obtain the *telegrapher's equation*:

$$\frac{\mathrm{d}^2\underline{I}}{\mathrm{d}z^2} + \beta^2 \underline{I} = 0 , \tag{7.58a}$$

$$\frac{\mathrm{d}^2\underline{V}}{\mathrm{d}z^2} + \beta^2 \underline{V} = 0 , \tag{7.58b}$$

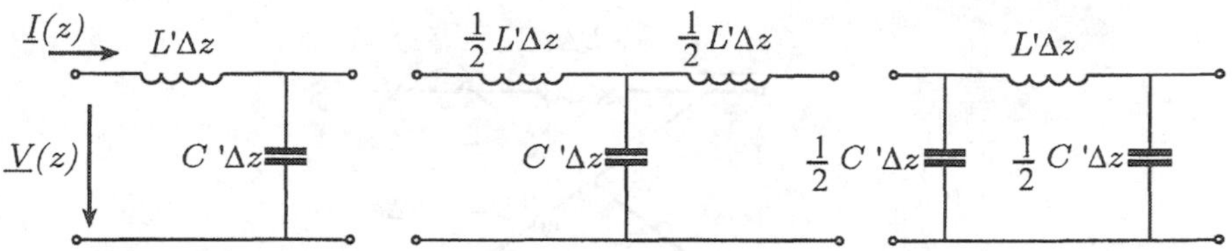

Figure 7.8: Equivalent circuits for a lossless TEM line of length Δz: (a) elementary equivalent circuit, (b) T-equivalent circuit and (c) Π-equivalent circuit.

with

$$\beta = \beta_{M0} = \omega\sqrt{L'C'} \,. \tag{7.59}$$

Equations (7.52), (7.57), (7.58a) and (7.58b) are the *transmission line equations* for the lossless TEM line and describe the z dependence of $\underline{V}$ and $\underline{I}$. The phase coefficient β of the TEM mode is identical with the phase coefficient β_{M0} of the plane wave propagating in the same dielectric material. For short line segments we can use equivalent circuits with concentrated elements. Figure 7.8 shows three of these equivalent circuits for lossless TEM lines of length Δz. The equivalent circuits provide a good approximation, if Δz is very small in comparison with the wavelength. Let us consider the equivalent circuit according to Figure 7.8(a). As we can see easily, the relation of the four amplitudes and both pairs of nodes of the equivalent circuit is given by

$$\underline{V}(z + \Delta z) = \underline{V}(z) - \mathrm{j}\omega L'\Delta z\,\underline{I}(z)\,, \tag{7.60a}$$
$$\underline{I}(z + \Delta z) = \underline{I}(z) - \mathrm{j}\omega C'\Delta z\underline{V}(z)\,. \tag{7.60b}$$

Performing the transition $\Delta z \to 0$, we obtain (7.52) and (7.57). This means that we can approximate a line segment with arbitrary accuracy, if we are choosing a sufficiently small Δz. If we have to model longer line segments by this equivalent circuit with high accuracy, we can cascade several of these equivalent circuits. The T-equivalent circuit according to Figure 7.8(b) and the Π-equivalent circuit in Figure 7.8(c) are obtained by connecting two equivalent circuits according to Figure 7.8(a).

We can improve the equivalent circuit for the line segment by replacing (7.60a) and (7.60b) by a more accurate integration scheme:

$$\underline{V}(z + \Delta z) = \underline{V}(z) - \mathrm{j}\omega L'\Delta z\,\frac{1}{2}\left[\underline{I}(z) + \underline{I}(z + \Delta z)\right]\,, \tag{7.61a}$$

$$\underline{I}(z + \Delta z) = \underline{I}(z) - \mathrm{j}\omega C'\Delta z\frac{1}{2}\left[\underline{V}(z) + \underline{V}(z + \Delta z)\right]\,. \tag{7.61b}$$

From these equations we obtain the admittance representation

$$\begin{pmatrix} \underline{I}(z) \\ -\underline{I}(z + \Delta z) \end{pmatrix} = \begin{pmatrix} \underline{Y}_1 & \underline{Y}_2 \\ \underline{Y}_2 & \underline{Y}_1 \end{pmatrix} \begin{pmatrix} \underline{V}(z) \\ \underline{V}(z + \Delta z) \end{pmatrix} \tag{7.62}$$

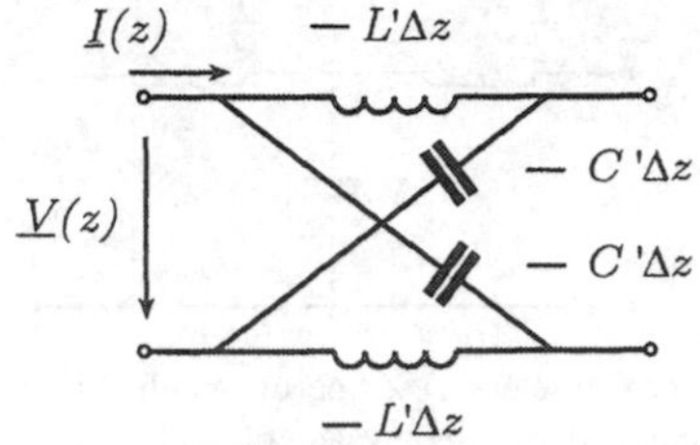

Figure 7.9: Equivalent circuit for a lossless TEM line of length Δz.

with

$$\underline{Y}_1 = \frac{j\omega\Delta z\, C'}{4} + \frac{1}{j\omega\Delta z\, L'}, \quad \underline{Y}_2 = \frac{j\omega\Delta z\, C'}{4} - \frac{1}{j\omega\Delta z\, L'}. \tag{7.63}$$

The equivalent circuit is shown in Figure 7.9.

Up to now we have considered lossless TEM waveguides only. In the real waveguide, however, losses occur due to the finite conductivity of the metallic conductors. Due to the skin effect these losses are increasing with frequency. If the line is filled by dielectric, we also have to consider the dielectric losses. The skin effect losses as well as the dielectric losses increase with increasing frequency. Let us first compute the losses in the dielectric material. If the dielectric material has an ohmic conductivity $\sigma \neq 0$, due to (2.57) a conduction current is flowing in the dielectric material. The conductive current $\underline{I}'_L \Delta z$ flowing within the length interval Δz from conductor 2 to conductor 1 is given by

$$\underline{I}'_L \Delta z = \sigma \int_{A_2} \star\, \underline{\mathcal{E}}\,. \tag{7.64}$$

$\underline{I}'_L$ is the dielectric loss current per unit of length. The integration is performed over the area A_2 in Figure 7.5. Inserting (7.32) and (7.36) we obtain the conductive loss current $\underline{I}'_L$ per unit of length:

$$\underline{I}'_L = \frac{\sigma}{\varepsilon}\underline{Q}' = \frac{\sigma}{\varepsilon}C'\underline{V}\,. \tag{7.65}$$

We define the *conductance per unit of length* G' as the ratio of the current flowing per unit of length from conductor 2 to conductor 1 and the voltage between conductor 2 and conductor 1:

$$G' = \frac{\sigma}{\varepsilon}C' = \omega C' \tan\delta_e\,. \tag{7.66}$$

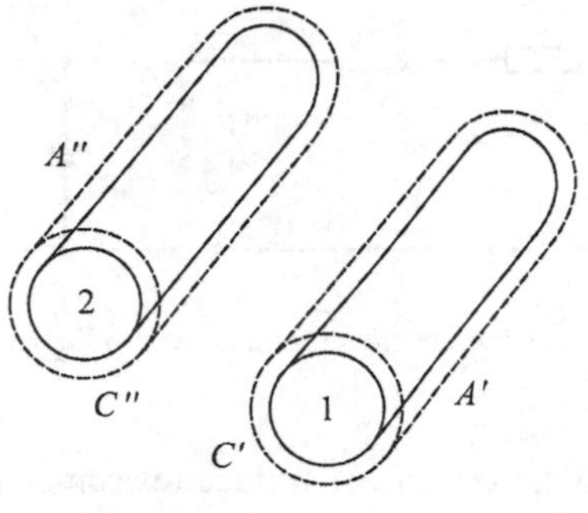

Figure 7.10: Area of integration to determine the skin effect losses.

If a loss current $\underline{I}'_L$ is flowing from conductor 2 to conductor 1 per unit of length, we have to modify (7.57) as follows:

$$\frac{\mathrm{d}\underline{I}}{\mathrm{d}z} = -\mathrm{j}\omega C'\underline{V} - \underline{I}'_L\,. \tag{7.67}$$

With (7.65) and (7.66) we obtain

$$\frac{\mathrm{d}\underline{I}}{\mathrm{d}z} = -(\mathrm{j}\omega C' + G')\underline{V}. \tag{7.68}$$

The dielectric losses are considered by the loss conductance per unit of length G'.

To compute the skin effect losses in conductors 1 and 2 we use (6.68) and integrate over the areas A' and A'' in Figure 7.10. We introduce a cylindric coordinate system (n, u, z) with the transverse coordinates n and u and the longitudinal coordinate z. The coordinates u and z are tangential to the conductor surface and the coordinate n is normal to the conductor surface. The corresponding basis differentials are n, s_1, $\mathrm{d}z$. With (6.68) we obtain the skin effect losses $\Delta z\, P'_L$ within a line segment of length Δz

$$P'_L \Delta z = \frac{1}{2}\int_{A'+A''} R_A |\underline{\boldsymbol{H}}_t|^2\,\mathrm{d}z \wedge s_1 = \frac{1}{2}\Delta z \int_{C'+C''} R_A |\underline{\boldsymbol{H}}_t|^2 s_1\,, \tag{7.69}$$

where R_A is the skin effect surface resistance defined in (6.56). This yields

$$P'_L = \frac{1}{2}\int_{C'+C''} R_A |\underline{\boldsymbol{H}}_t|^2 s_1\,. \tag{7.70}$$

The skin effect losses per unit of length are proportional to the square of the line current, i.e., the skin effect losses may be expressed by a resistance per unit of length

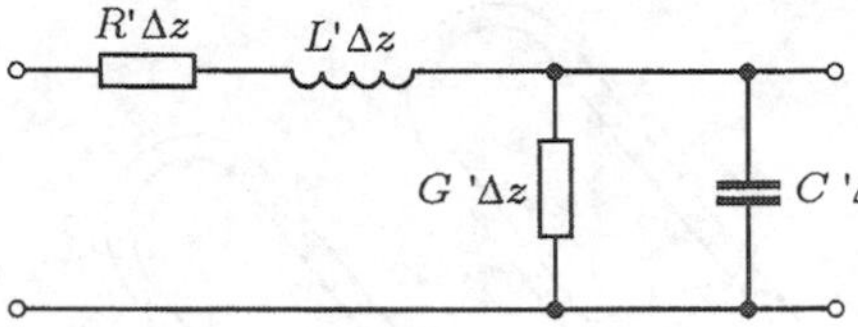

Figure 7.11: Equivalent circuit for a lossy TEM line of length Δz.

R'. The power loss due to the skin effect surface resistance is given by

$$P_L' = \frac{1}{2} R' |\underline{I}|^2 . \tag{7.71}$$

Using (7.23), (7.70) and (7.71) we can express the skin effect resistance per unit of length, R', by

$$R' = \frac{\oint_{C'+C''} R_A |\underline{\boldsymbol{H}}_t|^2 s_1}{\left(\oint_{C''} |\underline{\boldsymbol{H}}_t| s_1\right)^2} . \tag{7.72}$$

Due to the resistance per unit of length, R', the voltage decreases per unit of length by $R'\underline{I}$. Therefore we have to modify (7.52) as follows:

$$\frac{\mathrm{d}\underline{V}}{\mathrm{d}z} = -(\mathrm{j}\omega L' + R')\underline{I} . \tag{7.73}$$

For a line segment of length Δz of a lossy TEM waveguide, we obtain the equivalent circuit Figure 7.11 We note that due to the conductor losses also a very small longitudinal electric field component occurs. Nevertheless, we still call this waveguide mode transverse-electromagnetic. We can again eliminate $\underline{V}(z)$ or $\underline{I}(z)$, respectively, from (7.68) and (7.73) and obtain the equations

$$\frac{\mathrm{d}^2}{\mathrm{d}z^2}\underline{V} - \gamma^2 \underline{V} = 0 , \tag{7.74a}$$

$$\frac{\mathrm{d}^2}{\mathrm{d}z^2}\underline{I} - \gamma^2 \underline{I} = 0 . \tag{7.74b}$$

Figure 7.12 shows the modification of the equivalent circuit depicted in 7.9 for the lossy transmission line. The admittances $\underline{Y}_1$ and $\underline{Y}_2$ in 7.62 are given by

$$\underline{Y}_1 = \frac{\Delta z\, \underline{Y}'}{4} + \frac{1}{\Delta z\, \underline{Z}'} , \quad \underline{Y}_2 = \frac{\Delta z\, \underline{Y}'}{4} - \frac{1}{\Delta z\, \underline{Z}'} . \tag{7.75}$$

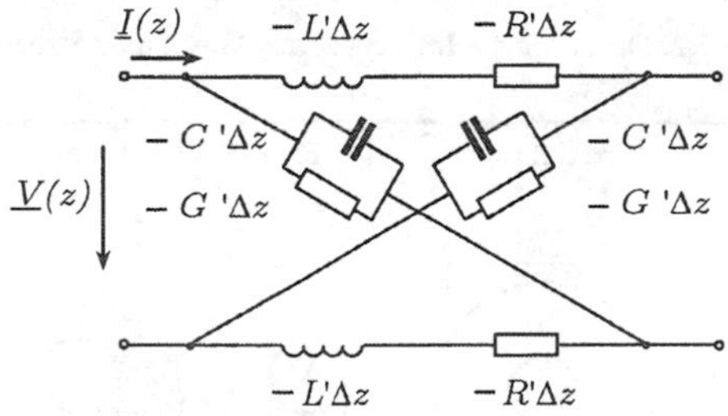

Figure 7.12: Equivalent circuit for a lossy TEM line of length Δz.

with

$$\underline{Z}' = \mathrm{j}\omega L' + R' \,, \quad \underline{Y}' = \mathrm{j}\omega C' + G' \,. \tag{7.76}$$

With the complex propagation coefficient Gamma given by

$$\gamma = \sqrt{(\mathrm{j}\omega C' + G')(\mathrm{j}\omega L' + R')} \,. \tag{7.77}$$

The solution of (7.74a) and (7.74b) is given by

$$\underline{V}(z) = \underline{V}_0^{(+)}\,\mathrm{e}^{-\gamma z} + \underline{V}_0^{(-)}\,\mathrm{e}^{+\gamma z} \,, \tag{7.78a}$$

$$\underline{I}(z) = \underline{I}_0^{(+)}\,\mathrm{e}^{-\gamma z} + \underline{I}_0^{(-)}\,\mathrm{e}^{+\gamma z} \,. \tag{7.78b}$$

If the electromagnetic wave propagates only in one direction, the ratio of current and voltage due to (7.68) and (7.73) is given by

$$\underline{V}(z) = \pm Z_0 \underline{I}(z) \,, \tag{7.79}$$

where the complex characteristic impedance Z_0 is given by

$$Z_0 = \sqrt{\frac{\mathrm{j}\omega L' + R'}{\mathrm{j}\omega C' + G'}} \,. \tag{7.80}$$

The positive sign in (7.79) is valid, if the electromagnetic wave is propagating in the positive z-direction, whereas the negative sign corresponds to a wave propagating in the negative z-direction. For small losses, i.e., $G' \ll \omega C'$ and $R' \ll \omega L'$, we obtain the following approximation for the propagation coefficient and the characteristic

Table 7.2: Data of Some Important TEM Waveguide Structures.

	Coaxial Line	Parallel Wire Line	Parallel Plate Line $a \ll b$
	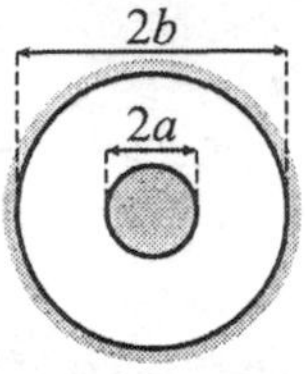	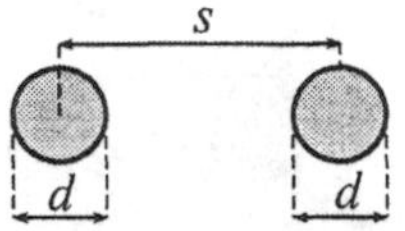	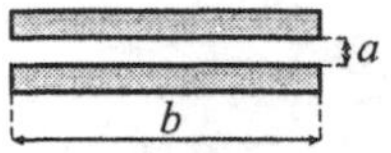
Capacitance per unit of length C'	$\dfrac{2\pi\varepsilon}{\ln\frac{b}{a}}$	$\dfrac{\pi\varepsilon}{\operatorname{arcosh}\frac{s}{d}}$	$\dfrac{\varepsilon b}{a}$
Inductance per unit of length L'	$\dfrac{\mu}{2\pi}\ln\dfrac{b}{a}$	$\dfrac{\mu}{\pi}\operatorname{arcosh}\dfrac{s}{d}$	$\dfrac{\mu a}{b}$
Conductance per unit of length G'	$\dfrac{2\pi\sigma}{\ln\frac{b}{a}}$	$\dfrac{\pi\sigma}{\operatorname{arcosh}\frac{s}{d}}$	$\dfrac{\sigma b}{a}$
Resistance per unit of length R'	$\dfrac{R_A}{2\pi}\left(\dfrac{1}{a}+\dfrac{1}{b}\right)$	$\dfrac{2R_A}{\pi d}\dfrac{s}{\sqrt{s^2-d^2}}$	$\dfrac{2R_A}{d}$
Characteristic impedance Z_w (lossless case)	$\dfrac{Z_F}{2\pi}\ln\dfrac{b}{a}$	$\dfrac{Z_F}{\pi}\operatorname{arcosh}\dfrac{s}{d}$	$Z_F\dfrac{a}{b}$

impedance:

$$\gamma \cong \omega\sqrt{L'C'}\left[\mathrm{j}+\frac{1}{2\omega}\left(\frac{G'}{C'}+\frac{R'}{L'}\right)\right], \tag{7.81a}$$

$$Z_0 \cong \sqrt{\frac{L'}{C'}}\left[1-\frac{\mathrm{j}}{2\omega}\left(\frac{R'}{L'}-\frac{G'}{C'}\right)\right]. \tag{7.81b}$$

Table 7.2 summarizes the data of some important TEM waveguide structures.

7.5 RECTANGULAR WAVEGUIDES

Figure 7.13 shows a waveguide with a rectangular cross-section and the inner dimensions a and b. The waveguide may be empty inside or filled with dielectric material. We first consider the lossless case and assume the waveguide walls to be perfectly

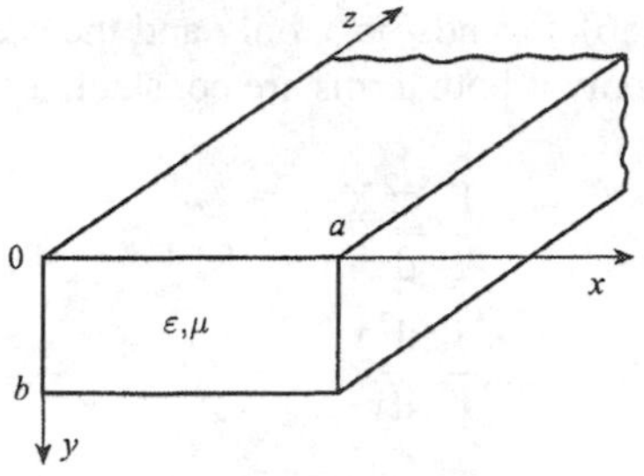

Figure 7.13: Waveguide with rectangular cross-section.

conductive and the inner region to be either empty or filled with a homogeneous isotropic lossless dielectric.

7.5.1 Transverse Electric Waves

For the transverse electric (TE or H) modes we choose the magnetic Hertz vector according to (7.8b). In the lossless case the Helmholtz equation (7.1b) is given by

$$\Delta \underline{\Pi}_{mz} + \beta_{M0}^2 \underline{\Pi}_{mz} = 0\,, \tag{7.82}$$

where the phase coefficient β_{M0} is given by

$$\beta_{M0} = \omega\sqrt{\varepsilon\mu} = \beta_0\sqrt{\varepsilon_r\mu_r}\,. \tag{7.83}$$

For a wave propagating in the positive z-direction we choose

$$\underline{\Pi}_{mz}(\mathbf{x}) = \underline{\Pi}_{m0}(x, y)\,\mathrm{e}^{-\mathrm{j}\beta z}\,. \tag{7.84}$$

Furthermore for $\underline{\Pi}_{m0}(x, y)$ we choose

$$\underline{\Pi}_{m0}(x, y) = \underline{X}(x)\,\underline{Y}(y) \tag{7.85}$$

and obtain from (7.82), (7.83) and (7.84)

$$\frac{1}{\underline{X}}\frac{\mathrm{d}^2\underline{X}}{\mathrm{d}x^2} + \frac{1}{\underline{Y}}\frac{\mathrm{d}^2\underline{Y}}{\mathrm{d}y^2} + (\beta_{M0}^2 - \beta^2) = 0\,. \tag{7.86}$$

Since the first term in (7.86) depends on x only and the second term on y only, this equation can be satisfied only if both terms are constant. Therefore the equations

$$\frac{1}{\underline{X}} \frac{\mathrm{d}^2 \underline{X}}{\mathrm{d}x^2} = -\beta_x^2 \,, \tag{7.87a}$$

$$\frac{1}{\underline{Y}} \frac{\mathrm{d}^2 \underline{Y}}{\mathrm{d}y^2} = -\beta_y^2 \tag{7.87b}$$

must be satisfied and β_x and β_y have to fulfill the condition

$$\beta_x^2 + \beta_y^2 = \beta_{M0}^2 - \beta^2 \,. \tag{7.88}$$

The general solutions for $\underline{X}(x)$ and $\underline{Y}(y)$ are given by

$$\underline{X}(x) = \underline{A}' \sin \beta_x x + \underline{B}' \cos \beta_x x \,, \tag{7.89a}$$

$$\underline{Y}(y) = \underline{C}' \sin \beta_y y + \underline{D}' \cos \beta_y y \,. \tag{7.89b}$$

Since we have assumed ideally conducting waveguide walls, the tangential electric field components must vanish at the waveguide walls. From (7.16b), (7.85), (7.89a) and (7.89b) we obtain $\underline{C}' = 0$, if $\underline{E}_x$ has to vanish at $y = 0$. Furthermore it follows

$$\beta_y = \frac{n\pi}{b} \quad \text{for} \quad n = 0, 1, 2 \ldots \,, \tag{7.90a}$$

if $\underline{E}_x \big|_{y=b}$ has to be satisfied. In the same way we obtain from (7.17b) $\underline{A}' = 0$ and

$$\beta_x = \frac{m\pi}{a} \quad \text{for} \quad m = 0, 1, 2 \ldots \,, \tag{7.90b}$$

if $\underline{E}_y$ has to vanish at $y = 0$ and $y = a$. From (7.85) and (7.89a) to (7.90b) we obtain therefore

$$\underline{\Pi}_{m0}(x, y) = \underline{A} \cos \frac{m\pi x}{a} \cos \frac{n\pi y}{b} \,, \tag{7.91}$$

where $\underline{A}$ is a complex amplitude not yet determined. The quantities β_x and β_y are determined by the cross-sectional dimensions a and b of the rectangular waveguide and by the positive integer numbers m and n. The waveguide modes are indicated by m and n. The transverse electric mode belonging to the indices m and n is named TE_{mn} mode or H_{mn} mode, respectively. From (7.91) and (7.84) we obtain for the wave propagating in the positive z-direction

$$\underline{\Pi}_{mz}(\boldsymbol{x}) = \underline{A} \cos \frac{m\pi x}{a} \cos \frac{n\pi y}{b} \, \mathrm{e}^{-\mathrm{j}\beta z} \,. \tag{7.92}$$

After inserting into (7.82) we obtain

$$\beta^2 = \beta_{M0}^2 - \beta_{Mc}^2 \tag{7.93}$$

with

$$\beta_{Mc} = \sqrt{\beta_x^2 + \beta_y^2} = \sqrt{\left(\frac{m\pi}{a}\right)^2 + \left(\frac{n\pi}{b}\right)^2}\,. \tag{7.94}$$

The relation between the wavelength of the plane wave in free-space, λ_0, and the phase coefficient of the plane wave in the medium, β_{M0}, is given by

$$\lambda_0 = \frac{2\pi}{\beta_0} = \sqrt{\varepsilon_r \mu_r}\cdot\frac{2\pi}{\beta_{M0}}\,. \tag{7.95}$$

The waveguide wavelength λ_H is given by

$$\lambda_H = \frac{2\pi}{\beta}\,. \tag{7.96}$$

We define the so-called *cut-off wavelength* λ_c by

$$\lambda_c = \sqrt{\varepsilon_r \mu_r}\,\frac{2\pi}{\beta_{Mc}} = \frac{2\sqrt{\varepsilon_r \mu_r}}{\sqrt{\left(\frac{m}{a}\right)^2 + \left(\frac{n}{b}\right)^2}}\,. \tag{7.97}$$

From (7.93) to (7.97) it follows

$$\lambda_H = \frac{1}{\sqrt{\varepsilon_r \mu_r}}\,\frac{\lambda_0}{\sqrt{1 - \left(\frac{\lambda_0}{\lambda_c}\right)^2}}\,. \tag{7.98}$$

A waveguide mode can only propagate, if the free-space wavelength λ_0 is smaller than the cut-off wavelength λ_c. The cut-off wavelength λ_c depends on the mode and decreases for increasing m and n. The *cut-off frequency*

$$f_c = \frac{c_0}{\lambda_c}\,. \tag{7.99}$$

is the frequency corresponding to the cut-off wavelength. For $a > b$ the fundamental mode, i.e., the mode with the largest cut-off wavelength, is the TE_{10} mode. For the TE_{10} mode we obtain

$$\lambda_c = 2a\sqrt{\varepsilon_r \mu_r}\,. \tag{7.100}$$

If in a rectangular waveguide a side ratio $a/b = 2/1$ is chosen in the frequency interval between the cut-off frequency and the double cut-off frequency, only the fundamental TE_{10} mode can propagate. The modes TE_{20} and TE_{01} have twice the cut-off frequency of the fundamental mode and all other modes have higher cut-off frequencies. From (7.3), (7.95), (7.96) and (7.98) we obtain with $c_0 = \omega/\beta_0$ the phase velocity c_H of the guided wave

$$c_H = \frac{1}{\sqrt{\varepsilon_r \mu_r}} \frac{c_0}{\sqrt{1 - \left(\frac{\lambda_0}{\lambda_c}\right)^2}} . \tag{7.101}$$

The group velocity of the waveguide wave is computed from (7.7) under the assumption that the permittivity of the dielectric material filling the waveguide is frequency-independent. From (7.93) we obtain

$$2\beta \frac{\mathrm{d}\beta}{\mathrm{d}\omega} = 2\varepsilon_r \mu_r \beta_0 \frac{\mathrm{d}\beta_0}{\mathrm{d}\omega} = 2\varepsilon_r \mu_r \frac{\omega}{c_0^2} \tag{7.102}$$

and from that using (7.7) and (7.101) the group velocity v_{gH} of the waveguide wave.

$$v_{gH} = \left(\frac{\mathrm{d}\beta}{\mathrm{d}\omega}\right)^{-1} = \frac{c_0^2}{\varepsilon_r \mu_r} \frac{\beta}{\omega} = \frac{c_0}{\sqrt{\varepsilon_r \mu_r}} \sqrt{1 - \left(\frac{\lambda_0}{\lambda_c}\right)^2} . \tag{7.103}$$

Figure 7.14 shows the wavelength dependence of the phase velocity and the group velocity of the waveguide wave. For small wavelengths the phase and group velocity approach the velocity of the plane wave propagating in the waveguide medium. If the wavelength approaches the cut-off wavelength, the waveguide group velocity approaches zero, and the waveguide phase velocity goes to infinity. The group velocity in the waveguide never can exceed the propagation velocity of the plane wave in the waveguide medium. We obtain the field components of the TE_{mn} mode from (7.13b) to (7.18b) and (7.92).

$$\underline{E}_x = \mathrm{j}\omega\mu \left(\frac{n\pi}{b}\right) \underline{A} \cos\frac{m\pi x}{a} \sin\frac{n\pi y}{b} \mathrm{e}^{-j\beta z} , \tag{7.104a}$$

$$\underline{E}_y = -\mathrm{j}\omega\mu \left(\frac{m\pi}{a}\right) \underline{A} \sin\frac{m\pi x}{a} \cos\frac{n\pi y}{b} \mathrm{e}^{-j\beta z} , \tag{7.104b}$$

$$\underline{E}_z = 0 , \tag{7.104c}$$

$$\underline{H}_x = \mathrm{j}\beta \left(\frac{m\pi}{a}\right) \underline{A} \sin\frac{m\pi x}{a} \cos\frac{n\pi y}{b} \mathrm{e}^{-\mathrm{j}\beta z} , \tag{7.104d}$$

$$\underline{H}_y = \mathrm{j}\beta \left(\frac{n\pi}{b}\right) \underline{A} \cos\frac{m\pi x}{a} \sin\frac{n\pi y}{b} \mathrm{e}^{-\mathrm{j}\beta z} , \tag{7.104e}$$

$$\underline{H}_z = \beta_{Mc}^2 \underline{A} \cos\frac{m\pi x}{a} \cos\frac{n\pi y}{b} \mathrm{e}^{-\mathrm{j}\beta z} . \tag{7.104f}$$

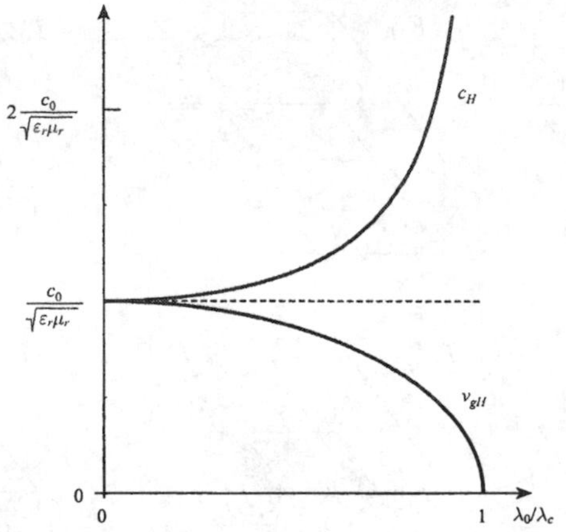

Figure 7.14: Wavelength dependence of c_H and v_{gH}.

For the wave propagating in the negative z-direction we have to replace $\mathrm{e}^{-\mathrm{j}\beta z}$ by $\mathrm{e}^{\mathrm{j}\beta z}$ and due to the partial derivation with respect to z in (7.13b) and (7.14b) the sign of (7.104d) and (7.104e) must be changed. Figure 7.15 shows the field distribution of some waveguide modes of waveguides with a rectangular cross-section. From (7.104a), (7.104b), (7.104d) and (7.104e) and using (2.78), (7.95), (7.96) and (7.98) we obtain for the wave propagating in the positive z-direction

$$\frac{\underline{E}_x}{\underline{H}_y} = -\frac{\underline{E}_y}{\underline{H}_x} = \frac{\omega\mu}{\beta} = \frac{Z_F}{\sqrt{1-\left(\frac{\lambda_0}{\lambda_c}\right)^2}}\,. \tag{7.105}$$

We define the *wave impedance* Z_{TE} of the TE mode by

$$Z_{\mathrm{TE}} = \frac{Z_F}{\sqrt{1-\left(\frac{\lambda_0}{\lambda_c}\right)^2}}\,. \tag{7.106}$$

In the general notation of (7.105) we obtain with (2.170) for the waveguide wave propagating in the positive z-direction

$$\underline{\mathcal{E}}_{tr} = -Z_{\mathrm{TE}} \perp_z \underline{\mathcal{H}}_{tr}\,. \tag{7.107}$$

$\underline{\mathcal{E}}_{tr}$ and $\underline{\mathcal{H}}_{tr}$ are the transverse electric and the magnetic field forms

$$\underline{\mathcal{E}}_{tr} = \underline{E}_x\,\mathrm{d}x + \underline{E}_y\,\mathrm{d}y\,, \tag{7.108a}$$

$$\underline{\mathcal{H}}_{tr} = \underline{H}_x\,\mathrm{d}x + \underline{H}_y\,\mathrm{d}y\,. \tag{7.108b}$$

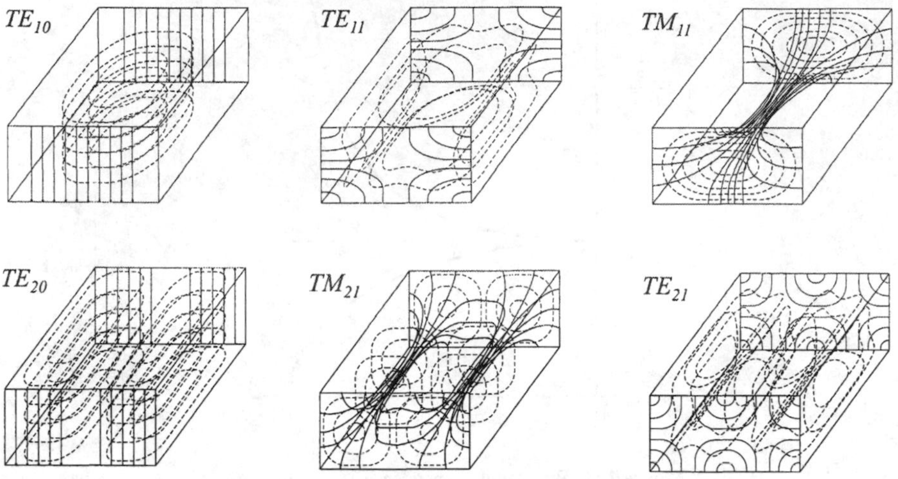

Figure 7.15: Field distribution of some waveguide modes of waveguides with a rectangular cross-section.

For the wave propagating in the negative z-direction we have to replace Z_{TE} in (7.107) by minus $-Z_{\mathrm{TE}}$. For the TE_{10} mode we obtain from (7.104a) to (7.104f):

$$\underline{E}_x = 0\,, \tag{7.109a}$$

$$\underline{E}_y = -\mathrm{j}\omega\mu\frac{\pi}{a}\underline{A}\sin\frac{\pi x}{a}\,\mathrm{e}^{-\mathrm{j}\beta z}\,, \tag{7.109b}$$

$$\underline{E}_z = 0\,, \tag{7.109c}$$

$$\underline{H}_x = \mathrm{j}\beta\frac{\pi}{a}\underline{A}\sin\frac{\pi x}{a}\,\mathrm{e}^{-\mathrm{j}\beta z}\,, \tag{7.109d}$$

$$\underline{H}_y = 0\,, \tag{7.109e}$$

$$\underline{H}_z = \beta_{Mc}^2\underline{A}\cos\frac{\pi x}{a}\,\mathrm{e}^{-\mathrm{j}\beta z}\,. \tag{7.109f}$$

Waveguides with the side ratio $a/b = 2/1$ have a maximum monomode frequency range from f_c to $2f_c$ and are in general used in a frequency region from $1.25f_c$ to $1.9f_c$. In Table 7.3 the data of standardized waveguides with a rectangular cross-section are summarized. The frequency regions of standardized waveguides are overlapping.

7.5.2 Transverse Magnetic Waves

To investigate transverse magnetic field types we start with a formulation according to (7.8a). The derivation is performed as in the preceding section. Equations (7.93)

Table 7.3: Data of Standardized Rectangular Waveguides.

Type	Frequency Range $TE_{10}-mode$ $1.25 \cdot f_c$ until $1.9 \cdot f_c$ [GHz]	Dimensions [mm] *a*	*b*	Attenuation of $TE_{10}-mode$ for $1.5 \cdot f_c$ [GHz]	$\sigma_{cu} = 5.8 \cdot 10^7$ S/m α [dB/m] Theor.	max accept.
R 3	0.32 – 0.49	584.2	292.1	0.385	0.00078	0.001
R 4	0.35 – 0.53	533.4	266.7	0.422	0.00090	0.0012
R 5	0.41 – 0.62	457.2	228.6	0.49	0.00113	0.0015
R 6	0.49 – 0.75	381.0	190.5	0.59	0.00149	0.002
R 8	0.64 – 0.98	292.1	146.05	0.77	0.00221	0.003
R 9	0.76 – 1.15	247.65	123.82	0.91	0.00283	0.004
R 12	0.96 – 1.46	195.58	97.79	1.15	0.00405	0.005
R 14	1.13 – 1.73	165.1	82.55	1.36	0.00522	0.007
R 18	1.45 – 2.2	129.54	64.77	1.74	0.00748	0.01
R 22	1.72 – 2.61	109.22	54.61	2.06	0.00967	0.013
R 26	2.17 – 3.3	86.36	43.18	2.6	0.0138	0.018
R 32	2.6 – 3.95	72.14	34.04	3.12	0.0188	0.024
R 40	3.22 – 4.90	58.17	29.08	3.87	0.0249	0.032
R 48	3.94 – 5.99	47.549	22.149	4.73	0.0354	0.046
R 58	4.64 – 7.05	40.386	20.193	5.57	0.0430	0.056
R 70	5.38 – 8.17	34.849	15.799	6.45	0.0575	0.075
R 84	6.57 – 9.99	28.499	12.624	7.89	0.0791	0.103
R100	8.2 – 12.5	22.860	10.160	9.84	0.110	0.143
R120	9.84 – 15.0	19.050	9.525	11.8	0.133	
R140	11.9 – 18.0	15.799	7.899	14.2	0.176	
R180	14.5 – 22.0	12.954	6.477	17.4	0.236	
R220	17.6 – 26.7	10.668	4.318	21.1	0.368	
R260	21.7 – 33.0	8.636	4.318	26.0	0.436	
R320	26.3 – 40.0	7.112	3.556	31.6	0.538	
R400	32.9 – 50.1	5.690	2.845	39.5	0.815	
R500	39.2 – 59.6	4.775	2.388	47.1	1.058	
R620	49.8 – 75.8	3.759	1.880	59.8	1.52	
R740	60.5 – 91.9	3.0998	1.5494	72.6	2.02	
R900	73.8 – 112	2.5400	1.2700	88.5	2.73	
R1200	92.2 – 140	2.0320	1.0160	110.7	3.81	
R1400	113 – 173	1.651	0.8255	136.2	5.21	
R1800	145 – 220	1.2954	0.6477	173.6	7.49	
R2200	172 – 261	1.0922	0.5461	205.9	9.68	
R26001	217 – 330	0.8636	0.4318	260.2	13.76	

– (7.99) and (7.101) – (7.103) are also valid for TM modes. These modes are named TM_{mn} or E_{mn} modes. Instead of (7.92) we obtain using (7.13a) and (7.14a) for the waveguide wave propagating in the positive z-direction

$$\underline{\Pi}_{ez}(\boldsymbol{x}) = \underline{B} \sin \frac{m\pi x}{a} \sin \frac{n\pi y}{b} \mathrm{e}^{-\mathrm{j}\beta z} . \tag{7.110}$$

Using (7.13a) to (7.18a) we obtain the field components of the TM_{mn} modes:

$$\underline{E}_x = -\mathrm{j}\beta \left(\frac{m\pi}{a}\right) \underline{B} \cos \frac{m\pi x}{a} \sin \frac{n\pi y}{b} \mathrm{e}^{-\mathrm{j}\beta z} , \tag{7.111a}$$

$$\underline{E}_y = -\mathrm{j}\beta \left(\frac{n\pi}{b}\right) \underline{B} \sin \frac{m\pi x}{a} \cos \frac{n\pi y}{b} \mathrm{e}^{-\mathrm{j}\beta z} , \tag{7.111b}$$

$$\underline{E}_z = \beta_{Mc}^2 \underline{B} \sin \frac{m\pi x}{a} \sin \frac{n\pi y}{b} \mathrm{e}^{-\mathrm{j}\beta z} , \tag{7.111c}$$

$$\underline{H}_x = \mathrm{j}\omega\varepsilon \left(\frac{n\pi}{b}\right) \underline{B} \sin \frac{m\pi x}{a} \cos \frac{n\pi y}{b} \mathrm{e}^{-\mathrm{j}\beta z} , \tag{7.111d}$$

$$\underline{H}_y = -\mathrm{j}\omega\varepsilon \left(\frac{m\pi}{a}\right) \underline{B} \cos \frac{m\pi x}{a} \sin \frac{n\pi y}{b} \mathrm{e}^{-\mathrm{j}\beta z} , \tag{7.111e}$$

$$\underline{H}_z = 0 . \tag{7.111f}$$

The field distribution of the TM_{11} and TM_{21} modes are shown in Figure 7.15. From (7.111a), (7.111b), (7.111d) and (7.111e) we obtain with (2.78), (7.95), (7.96) and (7.98) for the wave propagating in the positive z-direction:

$$\frac{\underline{E}_x}{\underline{H}_y} = -\frac{\underline{E}_y}{\underline{H}_x} = \frac{\beta}{\omega\mu} = Z_F \sqrt{1 - \left(\frac{\lambda_0}{\lambda_c}\right)^2} . \tag{7.112}$$

We introduce the *wave impedance* Z_{TM} for the transverse magnetic modes

$$Z_{\text{TM}} = Z_F \sqrt{1 - \left(\frac{\lambda_0}{\lambda_c}\right)^2} . \tag{7.113}$$

In the general notation of (7.112) we obtain with (2.170) for the transverse magnetic wave propagating in the positive z-direction:

$$\underline{\mathcal{E}}_{tr} = -Z_{\text{TM}} \perp_z \underline{\mathcal{H}}_{tr} . \tag{7.114}$$

7.5.3 Power Flow in the Waveguide

The active power P flowing through the waveguide is obtained from (4.40) as the real part of the integral of the complex Poynting vector $\underline{\boldsymbol{T}}$ over the waveguide cross-section $A|_z$ at the longitudinal coordinate z:

$$P(z) = \Re\{\int_{A|_z} \mathcal{T}(x)\} = \Re\{\int_{A|_z} T_z(x)\,\mathrm{d}x \wedge \mathrm{d}y\} = \Re\{\int_0^b \left(\int_0^a T_z(x)\,\mathrm{d}x\right)\mathrm{d}y\}. \tag{7.115}$$

In the lossless waveguide the active power is independent from the longitudinal coordinate z.

Let us first compute the power transmitted by the TE$_{mn}$ mode. From (4.21) and (7.106) we obtain for the electromagnetic wave propagating in the positive z-direction

$$\underline{T}_z = \frac{1}{2}(\underline{E}_x\underline{H}_y^\star - \underline{E}_y\underline{H}_x^\star) = \frac{1}{2}Z_{\mathrm{TE}}(|\underline{H}_y|^2 + |\underline{H}_x|^2). \tag{7.116}$$

We now insert (7.104d), (7.104e) and (7.116) into (7.115). Using

$$\int_0^a \int_0^b \sin^2\frac{m\pi x}{a}\cos^2\frac{n\pi y}{b}\,\mathrm{d}x\,\mathrm{d}y = \frac{ab}{4}(1-\delta_{m0})(1+\delta_{n0}) \tag{7.117}$$

where δ_{mn} is the Kronecker symbol defined in (A.8) we obtain the power P transmitted through the waveguide

$$P = \frac{ab}{8}Z_{\mathrm{TE}}\beta^2|A|^2\left[\left(\frac{m\pi}{a}\right)^2(1-\delta_{m0})(1+\delta_{n0}) + \left(\frac{n\pi}{b}\right)^2(1+\delta_{m0})(1-\delta_{n0})\right]. \tag{7.118}$$

From this we obtain

$$P_{mn}^{\mathrm{TE}} = \begin{cases} \frac{ab}{8}Z_{\mathrm{TE}mn}\beta^2\beta_{Mc}^2|A_{mn}|^2 & \text{for} \quad m \neq 0, n \neq 0 \\ \frac{ab}{4}Z_{\mathrm{TE}mn}\beta^2\beta_{Mc}^2|A_{mn}|^2 & \text{for} \quad n = 0 \text{ or } m = 0 \end{cases}. \tag{7.119}$$

Transverse magnetic modes only exist for $m \neq 0, n \neq 0$.

For the TM modes evaluation of the integral (7.115) under consideration of (7.111a), (7.111b), (7.112) and (7.113) yields

$$P_{mn}^{\mathrm{TM}} = \frac{ab}{8Z_{\mathrm{TM}mn}}\beta^2\beta_{Mc}^2|B_{mn}|^2. \tag{7.120}$$

7.5.4 Orthogonality of the Waveguide Modes

Under general excitation conditions several modes may propagate in a waveguide. The electromagnetic field is represented by

$$\underline{\mathcal{E}}(x) = \sum_{m,n}\left[\underline{\mathcal{E}}_{mn}^{\mathrm{TE}}(x) + \underline{\mathcal{E}}_{mn}^{\mathrm{TM}}(x)\right], \tag{7.121a}$$

$$\underline{\mathcal{H}}(x) = \sum_{m,n}\left[\underline{\mathcal{H}}_{mn}^{\mathrm{TE}}(x) + \underline{\mathcal{H}}_{mn}^{\mathrm{TM}}(x)\right]. \tag{7.121b}$$

The summation is performed over all propagating modes. Inserting these modal expansions of $\underline{\mathcal{E}}(\boldsymbol{x})$ and $\underline{\mathcal{H}}(\boldsymbol{x})$ into (4.21) and (4.40) we obtain the active power P flowing through the waveguide cross-section $A|_z$ at the longitudinal coordinate z as

$$\begin{aligned} P(z) &= \frac{1}{2}\Re\left\{\int_{A|_z} \underline{\mathcal{E}}(\boldsymbol{x}) \wedge \underline{\mathcal{H}}^*(\boldsymbol{x})\right\} \\ &= \frac{1}{2} \sum_{\substack{m,n,M \\ m',n',M'}} \Re\left\{\int_{A|_z} \underline{\mathcal{E}}_{mn}^{M}(\boldsymbol{x}) \wedge \underline{\mathcal{H}}_{m'n'}^{M'*}(\boldsymbol{x})\right\} . \end{aligned} \tag{7.122}$$

where the superscript M may assume TE or TM. With (7.104a), (7.104b), (7.104d), (7.104e), (7.111a), (7.111b), (7.111d), (7.111e), (7.119) and (7.120) we obtain the *orthogonality* relationships

$$\frac{1}{2}\Re\left\{\int_{A|_z} \underline{\mathcal{E}}_{mn}^{\mathrm{TE}}(\boldsymbol{x}) \wedge \underline{\mathcal{H}}_{m'n'}^{\mathrm{TE}}(\boldsymbol{x})\right\} = \delta_{mm'}\delta_{nn'}P_{mn}^{\mathrm{TE}}, \tag{7.123a}$$

$$\frac{1}{2}\Re\left\{\int_{A|_z} \underline{\mathcal{E}}_{mn}^{\mathrm{TM}}(\boldsymbol{x}) \wedge \underline{\mathcal{H}}_{m'n'}^{\mathrm{TM}}(\boldsymbol{x})\right\} = \delta_{mm'}\delta_{nn'}P_{mn}^{\mathrm{TM}}, \tag{7.123b}$$

$$\frac{1}{2}\Re\left\{\int_{A|_z} \underline{\mathcal{E}}_{mn}^{\mathrm{TE}}(\boldsymbol{x}) \wedge \underline{\mathcal{H}}_{m'n'}^{\mathrm{TM}}(\boldsymbol{x})\right\} = 0, \tag{7.123c}$$

$$\frac{1}{2}\Re\left\{\int_{A|_z} \underline{\mathcal{E}}_{mn}^{\mathrm{TM}}(\boldsymbol{x}) \wedge \underline{\mathcal{H}}_{m'n'}^{\mathrm{TE}}(\boldsymbol{x})\right\} = 0, \tag{7.123d}$$

where δ_{mn} is the Kronecker delta function, defined in (A.8). The orthogonality of functions is discussed in section 12.2. Inserting these equations into (7.122) yields

$$\begin{aligned} P(z) &= \frac{1}{2}\sum_{m,n} \Re\left\{\int_{A|_z} \underline{\mathcal{E}}_{mn}^{\mathrm{TE}}(\boldsymbol{x}) \wedge \underline{\mathcal{H}}_{mn}^{\mathrm{TE}}(\boldsymbol{x})\right\} \\ &+ \frac{1}{2}\sum_{m,n} \Re\left\{\int_{A|_z} \underline{\mathcal{E}}_{mn}^{\mathrm{TM}}(\boldsymbol{x}) \wedge \underline{\mathcal{H}}_{mn}^{\mathrm{TM}}(\boldsymbol{x})\right\} . \end{aligned} \tag{7.124}$$

7.5.5 Generalized Currents and Voltages in Waveguides

Section 7.4 shows that for TEM waves the transverse complex field vectors $\underline{\boldsymbol{E}}$ and $\underline{\boldsymbol{H}}$ may be represented as products of scalar *complex amplitudes* and real vector *structure functions*. The complex scalar amplitudes only depend on the longitudinal coordinate z, whereas the structure function only depends on the transverse coordinates x and y. This decomposition of the transverse electric and magnetic field functions into a real structure function and a scalar complex amplitude has been very useful, since the vector structure function only depends on the mode, and hence is independent of the excitation of the waveguide. The excitation of the waveguide only determines the scalar amplitude. Since in the waveguide longitudinal field components also exist, the integrals defining voltage and currents now are dependent on the paths of integration. It is, however, also possible to introduce complex scalar amplitudes for hollow waveguides. One possibility may be to introduce scalar amplitudes as path integrals of the complex field intensities $\underline{\boldsymbol{E}}$ and $\underline{\boldsymbol{H}}$ over specified paths of integration. Another way to introduce integral field quantities is to define complex amplitudes via area integrals of the field quantities. We demonstrate this method for the TE_{10} mode. We introduce a generalized complex current $\underline{I}(z)$ and a generalized complex voltage $\underline{V}(z)$ for the electromagnetic wave propagating in the positive z-direction. We impose the condition that the generalized voltage $\underline{V}(z)$ and the generalized current $\underline{I}(z)$ are related in analogy to (7.37) via the wave impedance Z_{TE} of the waveguide

$$\underline{V}(z) = Z_{\mathrm{TE}}\underline{I}(z)\,. \tag{7.125}$$

Furthermore we want to introduce the generalized voltage $\underline{V}(z)$ and the generalized current $\underline{I}(z)$ such that the transmitted active power can be described in the usual way by

$$P(z) = \frac{1}{2}\Re\{\underline{V}(z)\underline{I}(z)^{\star}\}\,. \tag{7.126}$$

With (7.125) it follows that the wave propagating in the positive z-direction is carrying the active power

$$P(z) = \frac{1}{2}Z_{\mathrm{TE}}|\underline{I}(z)|^2\,. \tag{7.127}$$

Using (7.119) and (7.127) we now can express the complex amplitude $\underline{A}$ by the generalized current $\underline{I}(z)$, and we obtain

$$\underline{A} = \frac{\mathrm{j}}{\beta_{Mc}}\sqrt{\frac{2}{ab}}\,I(z=0)\,, \tag{7.128}$$

where we have to impose a condition for the choice of the phase of $\underline{I}(z)$. Using (7.125) and (7.128) we can express the transverse field components of the TE_{10} mode in (7.109b) and (7.109d) by the generalized voltage $\underline{V}(z)$ and the generalized current $\underline{I}(z)$. The phase of $\underline{I}(z)$ in (7.127) has been chosen such that $\underline{V}(z)$ and $\underline{\boldsymbol{E}}(x, y, z)$ both are in phase. For the wave propagating in positive z-direction the z dependence of $\underline{V}(z)$ and $\underline{I}(z)$ is given by an exponential factor $\mathrm{e}^{-\mathrm{j}\beta z}$, for a wave propagating in the negative z-direction the z dependence of $\underline{V}(z)$ and $\underline{I}(z)$ is given by the exponential factor $\mathrm{e}^{\mathrm{j}\beta z}$. In this case (7.125) has to be replaced by

$$\underline{V}(z) = -Z_{\mathrm{TE}}\underline{I}(z)\,. \tag{7.129}$$

Now we allow wave propagation in both directions and use the equations

$$\underline{E}_y = \sqrt{\frac{2}{ab}}\underline{V}(z)\sin\frac{\pi x}{a}\,, \tag{7.130a}$$

$$\underline{H}_x = -\sqrt{\frac{2}{ab}}\underline{I}(z)\sin\frac{\pi x}{a}\,, \tag{7.130b}$$

with not yet determined $\underline{V}(z)$ and $\underline{I}(z)$. From Faraday's law (2.51) it follows that

$$\frac{\partial \underline{E}_y}{\partial x} = -\mathrm{j}\omega\mu\underline{H}_z\,. \tag{7.131}$$

We now can express the longitudinal component $\underline{H}_z$ of the magnetic field by the generalized voltage $\underline{V}(z)$

$$\underline{H}_z = \frac{\mathrm{j}}{\omega\mu}\frac{\pi}{a}\sqrt{\frac{2}{ab}}\underline{V}(z)\cos\frac{\pi x}{a}\,. \tag{7.132}$$

Furthermore it follows from (2.51)

$$\frac{\partial \underline{E}_y}{\partial z} = +\mathrm{j}\omega\mu\underline{H}_x\,. \tag{7.133}$$

Inserting (7.130a) and (7.130b) we obtain

$$\frac{\mathrm{d}\underline{V}}{\mathrm{d}z} = -\mathrm{j}\omega\mu\underline{I}\,. \tag{7.134}$$

From Ampere's law, (2.50), it follows that

$$\frac{\partial \underline{H}_x}{\partial z} - \frac{\partial \underline{H}_z}{\partial x} = \mathrm{j}\omega\varepsilon\underline{E}_y\,. \tag{7.135}$$

After inserting of (7.130a), (7.130b) and (7.132) we obtain

$$\frac{\mathrm{d}\underline{I}}{\mathrm{d}z} = -\mathrm{j}\left[\omega\varepsilon - \frac{1}{\omega\mu}\left(\frac{\pi}{a}\right)^2\right]\underline{V}\,. \tag{7.136}$$

Equations (7.134) and (7.136) correspond with the line equations

$$\frac{\mathrm{d}\underline{V}}{\mathrm{d}z} = -\mathrm{j}\omega L_s'\underline{I}\,, \tag{7.137a}$$

$$\frac{\mathrm{d}\underline{I}}{\mathrm{d}z} = -\mathrm{j}\left(\omega C_p' - \frac{1}{\omega L_p''}\right)\underline{V}\,. \tag{7.137b}$$

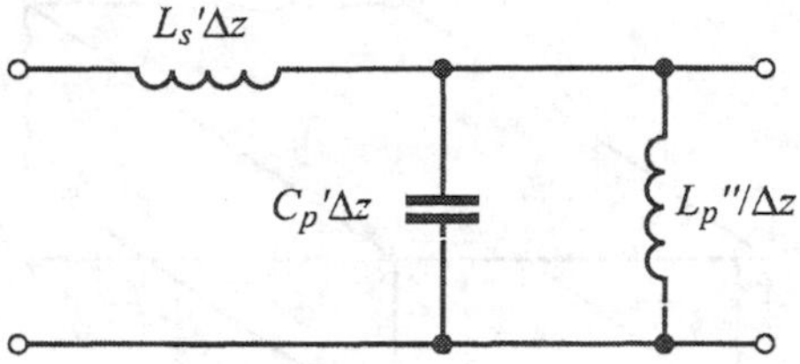

Figure 7.16: Equivalent circuit of a waveguide segment of length Δz excited in the TE_{10} mode.

The series inductance per unit of length L_s', the parallel inductance per unit of length C_p' and the parallel capacitance per unit of length L_p'' are defined in the following way

$$L_s' = \mu = \frac{Z_F}{c}, \tag{7.138a}$$

$$C_p' = \varepsilon = \frac{1}{Z_F c}, \tag{7.138b}$$

$$L_p'' = \mu \left(\frac{a}{\pi}\right)^2 = \frac{Z_F}{c\beta_{Mc}^2}. \tag{7.138c}$$

From the line equations (7.137a) and (7.137b) we obtain for a short line segment of length Δz the equivalent circuit as depicted in Figure 7.16. Due to the parallel inductance $L_p''/\Delta z$ the equivalent circuit exhibits high pass behavior. This corresponds to the high pass property of the waveguide.

7.5.6 Attenuation Due to Conductor Losses

Due to the conductor losses the guided wave is attenuated. If α is the attenuation coefficient of the guided wave, and considering that the transmitted active power P_z is proportional to the absolute square of the amplitude of the guided wave, for an attenuated wave propagating in the positive z-direction the relation

$$P(z) = P_0\,\mathrm{e}^{-2\alpha z} \tag{7.139}$$

must hold. From this it follows that

$$\frac{\mathrm{d}P}{\mathrm{d}z} = -2\alpha P \tag{7.140}$$

and therefrom we obtain

$$\alpha = -\frac{1}{2P}\frac{\mathrm{d}P}{\mathrm{d}z}. \tag{7.141}$$

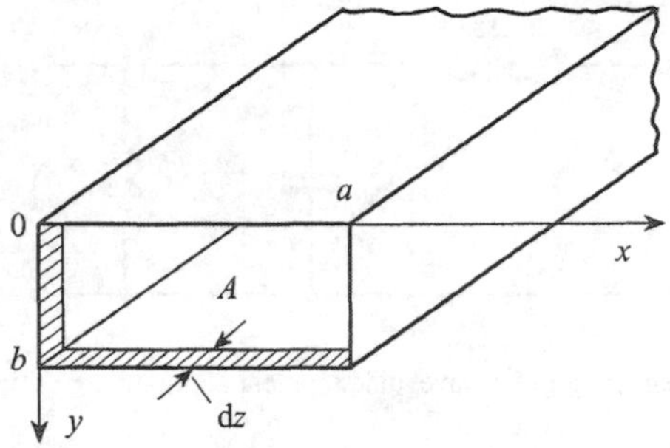

Figure 7.17: The wall currents in the rectangular waveguide.

Therefore we can compute the attenuation coefficients α, if we know the transmitted power $P(z)$ and the power loss per unit of length $P'_v = -\frac{\mathrm{d}P}{\mathrm{d}z}$. Initially we consider the field solution for the lossless waveguide and compute the power losses due to the wall currents using (6.68). We integrate (6.68) over an infinitesimal wall surface element of length $\mathrm{d}z$ according to Figure 7.17. The power loss in this area element is equal to $-\,\mathrm{d}P$, and we obtain

$$P'_L = -\frac{\mathrm{d}P}{\mathrm{d}z} = \frac{1}{2}\oint_{\partial A} R_A |H_t|^2 \,\mathrm{d}s\,. \tag{7.142}$$

First we compute the power losses due to the wall currents for the TE$_{mn}$ mode of the rectangular waveguide. From (7.142) we obtain

$$P'_L = R_A \int_0^a [|\underline{H}_x|^2 + |\underline{H}_z|^2]\,\mathrm{d}x\Big|_{y=0} + R_A \int_0^b [|\underline{H}_y|^2 + |\underline{H}_z|^2]\,\mathrm{d}y\Big|_{x=0}\,. \tag{7.143}$$

With (7.104d) to (7.104f) we obtain

$$[|\underline{H}_x|^2 + |\underline{H}_z|^2]\Big|_{y=0} = \beta^2\left(\frac{m\pi}{a}\right)^2 |A|^2 \sin^2\frac{m\pi x}{a} + \beta_{Mc}^4 |A|^2 \cos^2\frac{m\pi x}{a}\,, \tag{7.144}$$

$$[|\underline{H}_y|^2 + |\underline{H}_z|^2]\Big|_{x=0} = \beta^2\left(\frac{n\pi}{b}\right)^2 |A|^2 \sin^2\frac{n\pi y}{b} + \beta_{Mc}^4 |A|^2 \cos^2\frac{n\pi y}{b}\,. \tag{7.145}$$

From this it follows with (7.143) for the power losses due to the wall currents per unit of length

$$P'_L = \frac{1}{2} R_A |A|^2 \left[\beta^2 \left(\frac{m\pi}{a} \right)^2 (1 - \delta_{m0}) a + \right.$$
$$\left. + \beta^4_{Mc} (1 + \delta_{m0}) a + \beta^2 \left(\frac{n\pi}{b} \right)^2 (1 - \delta_{n0}) b + \beta^4_{Mc} (1 + \delta_{n0}) b \right] . \quad (7.146)$$

Inserting (7.118) and (7.146) into (7.141) we obtain the attenuation coefficient

$$\frac{P'_L}{2P} = \frac{2R_A \left\{ \beta^2 \left[\left(\frac{m\pi}{a}\right)^2 a + \left(\frac{n\pi}{b}\right)^2 b \right] + \beta^4_{Mc} [(1 + \delta_{m0}) a + (1 + \delta_{n0}) b] \right\}}{ab Z_{\mathrm{TE}} \beta^2 \left[\left(\frac{m\pi}{a}\right)^2 (1 + \delta_{n0}) + \left(\frac{n\pi}{b}\right)^2 (1 + \delta_{m0}) \right]} . \quad (7.147)$$

Using (7.93) – (7.98) and (7.106) we obtain

$$\alpha = 2R_A \frac{\left(\frac{m^2}{a} + \frac{n^2}{b}\right) \left(1 - \frac{\lambda_0^2}{\lambda_c^2}\right) + \left(\frac{m^2}{a^2} + \frac{n^2}{b^2}\right) [(1 + \delta_{m0}) a + (1 + \delta_{n0}) b] \frac{\lambda_0^2}{\lambda_c^2}}{ab Z_F \sqrt{1 - \left(\frac{\lambda_0}{\lambda_c}\right)^2} \left[\left(\frac{m}{a}\right)^2 (1 + \delta_{n0}) + \left(\frac{n}{b}\right)^2 (1 + \delta_{m0}) \right]} . \quad (7.148)$$

For the TE_{10} mode it follows with $\delta_{m0} = 1$, $\delta_{n0} = 0$:

$$\alpha = \frac{2R_A}{b Z_F \sqrt{1 - \left(\frac{\lambda_0}{\lambda_c}\right)^2}} \left(\frac{1}{2} + \frac{b}{a} \frac{\lambda_0^2}{\lambda_c^2} \right) . \quad (7.149)$$

For the TE_{10} mode the theoretic losses and the maximum losses allowed due to the DIN standard are summarized in Table 7.3. To compute the wall current losses for the TM_{10} modes we obtain from (7.142)

$$P'_L = R_A \int_0^a |\underline{H}_x|^2 \, dx \bigg|_{y=0} + R_A \int_0^b |\underline{H}_y|^2 \, dy \bigg|_{x=0} . \quad (7.150)$$

After inserting (7.111d) and (7.111e) and applying (7.94), (7.112) and (7.113) we obtain

$$-\frac{dP}{dz} = \frac{1}{2} R_A |B|^2 \left[\left(\frac{n\pi}{b} \right)^2 a + \left(\frac{m\pi}{a} \right)^2 b \right] \frac{\beta^2}{Z^2_{\mathrm{TM}}} . \quad (7.151)$$

With (7.120) we obtain the attenuation coefficient of the TM_{mn} mode

$$\alpha = \frac{2R_A}{ab Z_{\mathrm{TM}}} \frac{\left(\frac{m}{a}\right)^2 b + \left(\frac{n}{b}\right)^2 a}{\left(\frac{m}{a}\right)^2 + \left(\frac{n}{b}\right)^2} . \quad (7.152)$$

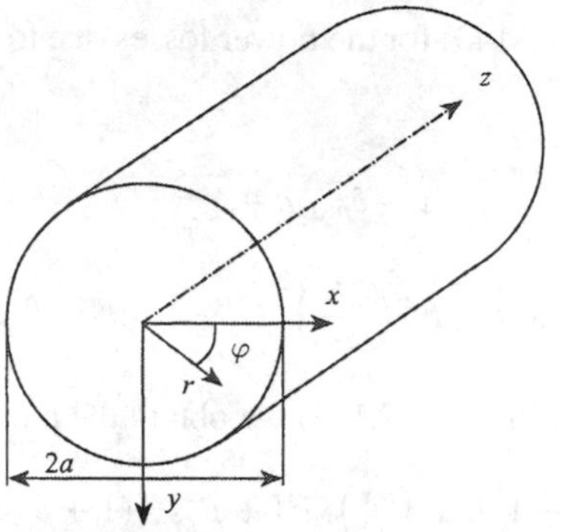

Figure 7.18: Circular cylindric waveguide.

7.6 CIRCULAR CYLINDRIC WAVEGUIDES

7.6.1 The Circular Waveguide Modes

Figure 7.18 shows a circular waveguide with inner diameter $2a$. We investigate the lossless circular waveguide with a perfectly conducting wall and free-space inner region. To investigate the TM and TE modes of the circular cylindric waveguide we derive the fields either from an electric Hertz form $\underline{\Pi}_e$ or a magnetic Hertz form $\underline{\Pi}_m$ exhibiting only a z component.

$$\underline{\Pi}_e = \underline{\Pi}_{ez}\,\mathrm{d}z\,, \tag{7.153}$$

$$\underline{\Pi}_m = \underline{\Pi}_{mz}\,\mathrm{d}z\,. \tag{7.154}$$

For both cases the Helmholtz equation (3.34) has the following form:

$$\Delta\underline{\Pi}_{iz} + \beta_0^2\underline{\Pi}_{iz} = 0 \qquad i = e, m\,, \tag{7.155}$$

with

$$\beta_0^2 = \omega^2\varepsilon_0\mu_0\,. \tag{7.156}$$

With (A.145) and (A.146) we obtain for circular cylindric coordinates

$$\frac{\partial^2\underline{\Pi}_{iz}}{\partial r^2} + \frac{1}{r}\frac{\partial\underline{\Pi}_{iz}}{\partial r} + \frac{1}{r^2}\frac{\partial^2\underline{\Pi}_{iz}}{\partial\varphi^2} + \frac{\partial^2\underline{\Pi}_{iz}}{\partial z^2} + \beta_0^2\underline{\Pi}_{iz} = 0\,. \tag{7.157}$$

We seek solutions for waves propagating in the positive z-direction and choose the separation formulation

$$\underline{\Pi}_{iz}(r,\varphi,z) = R(r)\,f(\varphi)\,\mathrm{e}^{-j\beta z}\,. \tag{7.158}$$

From this it follows that

$$f\frac{\mathrm{d}^2R}{\mathrm{d}r^2}+f\frac{1}{r}\frac{\mathrm{d}R}{\mathrm{d}r}+\frac{R}{r^2}\frac{\mathrm{d}^2f}{\mathrm{d}\varphi^2}+(\beta_0^2-\beta^2)Rf=0\,. \tag{7.159}$$

We introduce the parameter k_c given by

$$k_c^2=\beta_0^2-\beta^2 \tag{7.160}$$

and obtain

$$r^2\frac{1}{R}\frac{\mathrm{d}^2R}{\mathrm{d}r^2}+r\frac{1}{R}\frac{\mathrm{d}R}{\mathrm{d}r}+\frac{1}{f}\frac{\mathrm{d}^2f}{\mathrm{d}\varphi^2}+r^2k_c^2=0\,. \tag{7.161}$$

The first two terms of this equation are dependent on r only, whereas the third term only depends on φ. This equation therefore can be fulfilled only, if the sum of the first two terms and the last term as well as the third term each are independently constant. Therefore we set the third term equal $-n^2$ and obtain

$$\frac{\mathrm{d}^2f}{\mathrm{d}\varphi^2}+n^2f=0\,. \tag{7.162}$$

The solution of this equation is

$$f=A'\sin n\varphi+B'\cos n\varphi\,. \tag{7.163}$$

Since $f(\varphi)$ is periodic with 2π, the parameter n must be an integer. Furthermore $A'=0$ without loss of generality, since both solutions in (7.163) only are distinguished by a rotation of the field distribution around the z-axis by 90°. For $R(r)$ we obtain from (7.161) the differential equation

$$r^2\frac{\mathrm{d}^2R}{\mathrm{d}r^2}+r\frac{\mathrm{d}R}{\mathrm{d}r}+(r^2k_c^2-n^2)R=0\,. \tag{7.164}$$

This is Bessel's differential equation. The solution is given by

$$R(r)=C'J_n(k_cr)+D'Y_n(k_cr) \tag{7.165}$$

where $J_n(x)$ and $Y_n(x)$ are the ordinary Bessel functions of the nth order and first and second kind. Since $Y_n(x)$ for $x=0$ exhibits a singularity, $D'=0$ must be valid in (7.165). Up to now the treatment of the Helmholtz equation is the same for the electric and magnetic Hertz forms $\underline{\Pi}_e$ and $\underline{\Pi}_m$.

We now investigate the transverse magnetic and transverse electric field modes. From (3.35), (3.36), (3.46), (3.47) and (7.158), (7.163), (7.165) we obtain
TM modes:

$$\underline{\Pi}_e(r,\varphi,z) = B J_n(k_c r)\cos n\varphi\, \mathrm{e}^{-j\beta z}\,\mathrm{d}z\,, \tag{7.166}$$

$$\underline{\mathcal{E}} = \mathrm{d}\tilde{\mathrm{d}}\,\underline{\Pi}_e + \beta_0^2\,\underline{\Pi}_e\,, \tag{7.167}$$

$$\underline{\mathcal{H}} = \mathrm{j}\omega\underline{\varepsilon} \star \mathrm{d}\,\underline{\Pi}_e\,. \tag{7.168}$$

TE modes:

$$\underline{\Pi}_m(r,\varphi,z) = A J_n(k_c r)\cos n\varphi\, \mathrm{e}^{-j\beta z}\,\mathrm{d}z\,, \tag{7.169}$$

$$\underline{\mathcal{H}} = \mathrm{d}\tilde{\mathrm{d}}\,\underline{\Pi}_m + \beta_0^2\,\underline{\Pi}_m\,, \tag{7.170}$$

$$\underline{\mathcal{E}} = -\mathrm{j}\omega\underline{\mu} \star \mathrm{d}\,\underline{\Pi}_m\,. \tag{7.171}$$

A and B are arbitrary complex amplitudes. For circular cylindric coordinates we obtain for the formulation (7.153)

$$\mathrm{d}\,\underline{\Pi}_e = \frac{1}{r}\frac{\partial \underline{\Pi}_{ez}}{\partial\varphi}\,\mathrm{d}r - \frac{\partial \underline{\Pi}_{ez}}{\partial r}\,r\,\mathrm{d}\varphi\,, \tag{7.172}$$

$$\mathrm{d}\tilde{\mathrm{d}}\,\underline{\Pi}_e = \frac{\partial^2 \underline{\Pi}_{ez}}{\partial r\partial z}\,\mathrm{d}r + \frac{1}{r}\frac{\partial^2 \underline{\Pi}_{ez}}{\partial\varphi\partial z}\,r\,\mathrm{d}\varphi + \frac{\partial^2 \underline{\Pi}_{ez}}{\partial z^2}\,\mathrm{d}z\,. \tag{7.173}$$

With (7.160) we obtain

$$\mathrm{d}\tilde{\mathrm{d}}\,\underline{\Pi}_e + \beta_0^2\,\underline{\Pi}_e = \frac{\partial^2 \underline{\Pi}_{ez}}{\partial r\partial z}\,\mathrm{d}r + \frac{1}{r}\frac{\partial^2 \underline{\Pi}_{ez}}{\partial\varphi\partial z}\,r\,\mathrm{d}\varphi + k_c^2\,\underline{\Pi}_{ez}\,\mathrm{d}z\,. \tag{7.174}$$

With this we obtain from (7.166) to (7.171) the field components summarized in (7.175) – (7.186).
TM modes:

$$\underline{E}_r = -j\beta k_c B J_n'(k_c r)\cos n\varphi\,\mathrm{e}^{-j\beta z}\,, \tag{7.175}$$

$$\underline{E}_\varphi = \mathrm{j}\beta\frac{n}{r} B J_n(k_c r)\sin n\varphi\,\mathrm{e}^{-\mathrm{j}\beta z}\,, \tag{7.176}$$

$$\underline{E}_z = k_c^2 B J_n(k_c r)\cos n\varphi\,\mathrm{e}^{-\mathrm{j}\beta z}\,, \tag{7.177}$$

$$\underline{H}_r = -\mathrm{j}\omega\varepsilon_0\frac{n}{r} B J_n(k_c r)\sin n\varphi\,\mathrm{e}^{-\mathrm{j}\beta z}\,, \tag{7.178}$$

$$\underline{H}_\varphi = -\mathrm{j}\omega\varepsilon_0 k_c B J_n'(k_c r)\cos n\varphi\,\mathrm{e}^{-\mathrm{j}\beta z}\,, \tag{7.179}$$

$$\underline{H}_z = 0\,. \tag{7.180}$$

TE modes:

$$\underline{H}_r = -\mathrm{j}\beta k_c A J_n'(k_c r) \cos n\varphi \, \mathrm{e}^{-\mathrm{j}\beta z} , \tag{7.181}$$

$$\underline{H}_\varphi = \mathrm{j}\beta \frac{n}{r} A J_n(k_c r) \sin n\varphi \, \mathrm{e}^{-\mathrm{j}\beta z} , \tag{7.182}$$

$$\underline{H}_z = k_c^2 A J_n(k_c r) \cos n\varphi \, \mathrm{e}^{-\mathrm{j}\beta z} , \tag{7.183}$$

$$\underline{E}_r = \mathrm{j}\omega\mu_0 \frac{n}{r} A J_n(k_c r) \sin n\varphi \, \mathrm{e}^{-\mathrm{j}\beta z} , \tag{7.184}$$

$$\underline{E}_\varphi = \mathrm{j}\omega\mu_0 k_c A J_n'(k_c r) \cos n\varphi \, \mathrm{e}^{-\mathrm{j}\beta z} , \tag{7.185}$$

$$\underline{E}_z = 0 . \tag{7.186}$$

With J_n' we denote the derivative of the ordinary Bessel function of first kind with respect to the argument. At the ideally conducting waveguide wall the tangential component of the electric field must vanish at $r = a$, i.e., $\underline{E}_\varphi\big|_{r=a} = 0$ and $\underline{E}_z\big|_{r=a} = 0$. From this it follows for the TM- and TE-modes:

TM modes

$$J_n(k_c a) = 0 , \tag{7.187a}$$

$$J_n'(k_c a) = 0 , \tag{7.188a}$$

TE modes

$$(k_c)_{nm} = \frac{\xi_{nm}}{a} , \tag{7.187b}$$

$$(k_c)_{nm} = \frac{\xi_{nm}'}{a} . \tag{7.188b}$$

For the transverse magnetic modes the Bessel function J_n for $r = a$ must be zero, whereas for transverse electric modes the derivative of the Bessel function J_n' must be zero at $r = a$. The mth zero of J_n is denoted with ξ_{nm}, the mth zero of J_n' is denoted with ξ_{nm}'. The lowest order zeros ξ_{nm} and ξ_{nm}' are listed in Table 7.4.

The k_c are determined by the conditions (7.187a) and (7.188a) respectively. According to (7.160) we obtain the phase coefficient β of the waveguide wave.

$$\beta = \sqrt{\beta_0^2 - k_c^2} . \tag{7.189}$$

With the free-space wavelength $\lambda_0 = 2\pi/\beta_0$, the waveguide wavelength $\lambda_H = 2\pi/\beta$ and the cut-off wavelength λ_c defined via

$$\lambda_c = \frac{2\pi}{k_c} \tag{7.190}$$

it follows that

$$\lambda_H = \frac{\lambda_0}{\sqrt{1 - \left(\frac{\lambda_0}{\lambda_c}\right)^2}} . \tag{7.191}$$

Table 7.4: The mth Roots for $J_n(x)$ and mth Roots for $J_n'(x)$

ξ_{nm}:

n	$m=1$	$m=2$	$m=3$
0	2.405	5.520	8.654
1	3.832	7.016	10.173
2	5.135	8.417	11.62

ξ'_{nm}:

n	$m=1$	$m=2$	$m=3$
0	3.832	7.016	10.173
1	1.84	5.33	8.54
2	3.054	6.706	9.969

This equation is identical with (7.98) for the guide with rectangular cross-section. A waveguide mode only can propagate, if $\lambda_0 < \lambda_c$. From (7.187b), (7.188b) and (7.190) we obtain the cut-off wavelength
TM modes:

$$(\lambda_c)_{mn} = \frac{2\pi a}{\xi_{nm}}, \tag{7.192}$$

TE modes:

$$(\lambda_c)_{mn} = \frac{2\pi a}{\xi'_{nm}}. \tag{7.193}$$

The wave impedance Z_{TM} and Z_{TE} follow from (7.175) – (7.185) and (7.189) – (7.191).
TM modes:

$$Z_{\text{TM}} = \frac{\underline{E}_r}{\underline{H}_\varphi} = -\frac{\underline{E}_\varphi}{\underline{H}_r} = \frac{\beta}{\omega\varepsilon_0} = Z_{F0}\sqrt{1-\left(\frac{\lambda_0}{\lambda_c}\right)^2}, \tag{7.194}$$

TE modes:

$$Z_{\text{TE}} = \frac{\underline{E}_r}{\underline{H}_\varphi} = -\frac{\underline{E}_\varphi}{\underline{H}_r} = \frac{\omega\mu_0}{\beta} = \frac{Z_{F0}}{\sqrt{1-\left(\frac{\lambda_0}{\lambda_c}\right)^2}}. \tag{7.195}$$

Therefore (7.106) and (7.113) are valid also for circular waveguides. Furthermore the generalization (7.107) and (7.114) is valid, since $\boldsymbol{e}_r$, $\boldsymbol{e}_\varphi$ and $\boldsymbol{e}_z$ form a right-handed orthonormal frame.

The field lines of the most important modes are shown in Figure 7.19. The data of some standardized waveguides are presented in Table 7.5. A relative frequency range for single-mode operation in the case of cylindric waveguides is smaller than that for rectangular waveguides. For circular waveguides the TE_{0m} modes are of special interest since the attenuation of these modes decreases with increasing frequency.

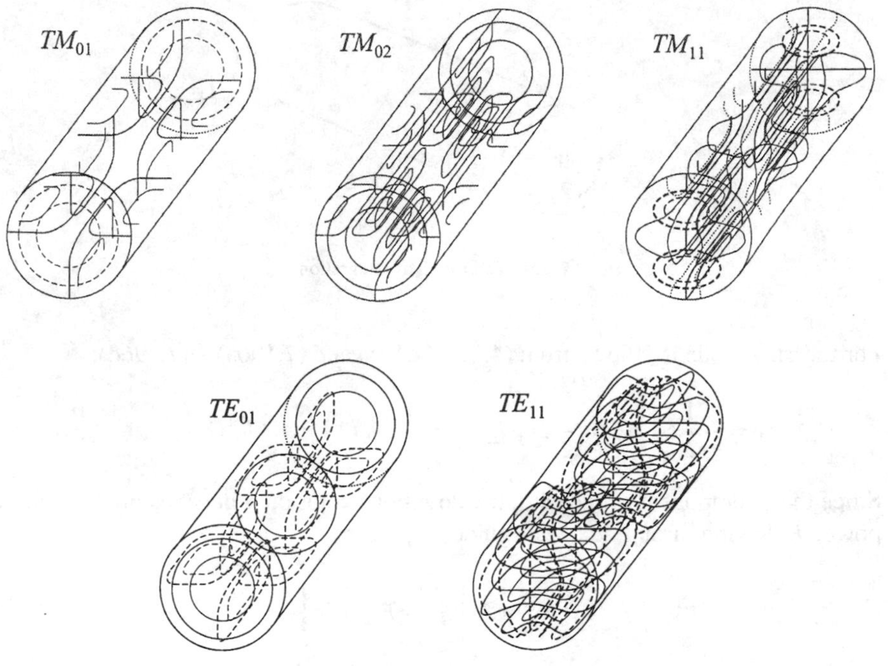

Figure 7.19: Field distribution of some waveguide modes of waveguides with a circular cross-section.

From (7.181) – (7.186) we obtain for the TE_{0m} modes the following field distribution:

$$\underline{E}_r = 0\,, \tag{7.196a}$$

$$\underline{E}_\varphi = \mathrm{j}\omega\mu_0 k_c A J_0'(k_c r)\,\mathrm{e}^{-\mathrm{j}\beta z}\,, \tag{7.196b}$$

$$\underline{E}_z = 0\,, \tag{7.196c}$$

$$\underline{H}_r = -\mathrm{j}\beta k_c A J_0'(k_c r)\,\mathrm{e}^{-\mathrm{j}\beta z}\,, \tag{7.196d}$$

$$\underline{H}_\varphi = 0\,, \tag{7.196e}$$

$$\underline{H}_z = k_c^2 A J_0(k_c r)\,\mathrm{e}^{-\mathrm{j}\beta z}\,. \tag{7.196f}$$

Figure 7.19 shows the field distribution of some waveguide modes of waveguides with a rectangular cross-section.

7.6.2 Power Flow and Attenuation in the TE_{01} Mode

We determine the active power flowing through the waveguide as the real part of the integral of the complex Poynting differential form over the waveguide cross-section.

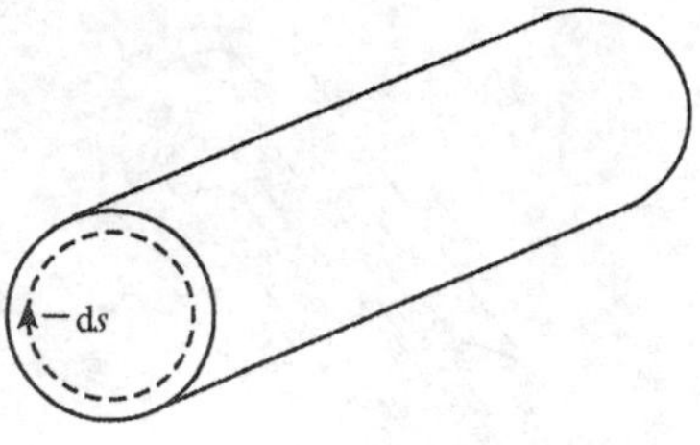

Figure 7.20: Path of integration.

For the TE$_{01}$ mode it follows from (4.21), (7.195) and (7.196a) – (7.196d):

$$T_z = -\frac{1}{2}\underline{E}_\varphi \underline{H}_r^\star = \frac{1}{2} Z_{\text{TE}} |\underline{H}_r|^2 = \frac{1}{2} Z_{\text{TE}} \beta^2 k_c^2 |A|^2 J_0'^2(k_c r)\,. \tag{7.197}$$

Since the z component of the complex Poynting vector only depends on r, the active power P flowing through the waveguide is given by

$$P = 2\pi \Re \left\{ \int_0^a r T_z(r)\, \mathrm{d}r \right\}\,. \tag{7.198}$$

After insertion of (7.197) we obtain

$$P = \pi Z_{\text{TE}} \beta^2 k_c^2 |A|^2 \int_0^a r J_0'^2(k_c r)\, \mathrm{d}r\,. \tag{7.199}$$

With

$$\int_0^{x_0} x J_1^2(\alpha x)\, \mathrm{d}x = \int_0^{x_0} x J_0^{2'}(\alpha x)\, \mathrm{d}x = \begin{cases} \frac{x_0^2}{2} J_1^2(\alpha x_0) & \text{for} \quad J_0(\alpha x_0) = 0 \\ \frac{x_0^2}{2} J_0^2(\alpha x_0) & \text{for} \quad J_0'(\alpha x_0) = 0 \\ & \text{and} \quad J_1(\alpha x_0) = 0 \end{cases} \tag{7.200}$$

we obtain from this

$$P = \frac{\pi}{2} Z_{\text{TE}} \beta^2 |A|^2 (k_c a)^2 J_0^2(k_c a)\,. \tag{7.201}$$

With $\xi_{01}' = 3.83$ it follows that

$$\frac{\pi}{2}(k_c a)^2 J_0^2(k_c a) = 3.73 \tag{7.202}$$

and with this

$$P = 3.73 Z_{\text{TE}} \beta^2 |A|^2\,. \tag{7.203}$$

Table 7.5: Data of Standardized Circular Waveguides.

Type	Cut-off Frequencies TE11 f_c [GHz]	TM01 f_c [GHz]	Dimension $2a$ [mm]	Attenuation of TE11 mode $1.2 \cdot f_c$ [GHz]	$\sigma_{cu} = 5.8 \cdot 10^7$ S/m α [dB/m] Theor.	max accept.
C 3.3	0.27	0.35	647.9	0.325	0.00067	0.0009
C 6.2	0.51	0.66	345.1	0.611	0.00174	0.0023
C 12	0.96	1.25	183.77	1.147	0.00447	0.0058
C 25	2.1	2.74	83.6	2.521	0.014	0.018
C 48	3.95	5.16	44.45	4.74	0.0375	0.049
C 104	8.68	11.3	20.244	10.42	0.122	0.15
C 255	21.1	27.5	8.33	25.31	0.462	–
C 495	40.2	52.5	4.369	48.26	1.211	–
C 890	73.6	96.1	2.388	88.3	3.011	–

The conductor losses per unit of length P'_L are calculated with (7.142). The integration is performed over the boundary ∂A of the waveguide cross-section A according to Figure 7.20. Since $\underline{H}_z$ is independent from φ it follows that

$$P'_L = \frac{1}{2} R_A \oint_{\partial A} |\underline{H}_z|^2 \, \mathrm{d}s = \pi a R_A |\underline{H}_z|^2 \Big|_{r=a} . \tag{7.204}$$

After inserting (7.196f) we obtain the current losses per unit of length

$$P'_L = \pi a R_A k_c^4 |A|^2 J_0^2(k_c a) . \tag{7.205}$$

The attenuation coefficient alpha follows from (7.141):

$$\alpha = \frac{-1}{2P}\frac{\mathrm{d}P}{\mathrm{d}z} = \frac{1}{2}\frac{P'_L}{P} = \frac{R_A k_c^2}{Z_{\mathrm{TE}} \beta^2 a} . \tag{7.206}$$

Using (7.160), (7.191), (7.195), $\lambda_c = 2\pi/k_c$ and $\lambda_0 = 2\pi/\beta_0$ we obtain from this

$$\alpha = \frac{R_A}{Z_{F0} a} \frac{\left(\frac{\lambda_0}{\lambda_c}\right)^2}{\sqrt{1 - \left(\frac{\lambda_0}{\lambda_c}\right)^2}} . \tag{7.207}$$

We see that for $\lambda_0 \ll \lambda_c$ the attenuation coefficient α is proportional to $\lambda_0^{3/2}$. The decrease of attenuation with increasing frequency in the TE01 mode has the following reason. Since according to (7.196e) the φ component of the magnetic field vanishes, the wall currents and therewith the wall current losses are only due to the longitudinal magnetic field component $\underline{H}_z$. The the ratio of the longitudinal magnetic field component $\underline{H}_z$ to the transverse magnetic field decreases according to $1/f$. The transmitted power is proportional to the square of the magnitude of the transverse field. The wall current losses are proportional to the square of the magnitude longitudinal field. For frequency-independent surface resistance R_A therefore the losses would decrease with increasing frequency according to $1/f^2$. Taking into consideration the skin effect, according to (6.56) the surface resistance R_A increases with $f^{1/2}$, and considering both effects we obtain a total frequency dependence of the losses proportional to $1/f^{3/2}$.

This holds for all TE_{0n} modes of the circular waveguide. Since in all that cases according to (7.196e) we obtain $\underline{H}_\varphi = 0$. The TE_{01} mode is the most interesting mode here, since it exhibits the lowest cut-off frequency. However, if this mode will be used for low-attenuation signal transmission it has to be considered that the TE_{01} mode is not the fundamental mode. Also, if the mode is used at higher frequencies, a number of higher order modes may also be excited at these frequencies in principle, and therefore it is necessary to take care that only the TE_{01} mode will be excited. Inhomogeneities and perturbations in the waveguide have to be avoided in order to obtain mode conversion. It has to be considered that the TM_{11} mode exhibits the same cut-off frequency as the TE_{01} mode. Therefore both modes have the same waveguide wavelength at the same frequency. This is called *mode degeneration*. Mode degeneration may yield to mode coupling even in the case of small inhomogeneities of the waveguide.

7.7 DIELECTRIC WAVEGUIDES AND OPTICAL FIBERS

7.7.1 Homogeneous Planar Dielectric Waveguides

The simplest dielectric waveguide is the *planar dielectric waveguide* depicted in Figure 7.21. The planar dielectric waveguide consists of a dielectric slab of thickness h and permittivity ε_2 enclosed by two regions with lower permittivities ε_1 and ε_3 respectively [3, 5, 6]. In Figure 7.21 the slab region is called region 2. Regions 1 and 3 either are filled by a dielectric or are free-space regions. For $\varepsilon_2 > \varepsilon_1, \varepsilon_3$ the slab can guide electromagnetic waves. In the following we assume a symmetric dielectric waveguide with same permittivity ε_1 in regions 1 and 3. We assume in all regions nonmagnetic media with permittivity μ_0. The TM-modes may be derived from an electric Hertz form exhibiting only a z component $\underline{\Pi}_{ez}\,\mathrm{d}z$ whereas the TE-modes may

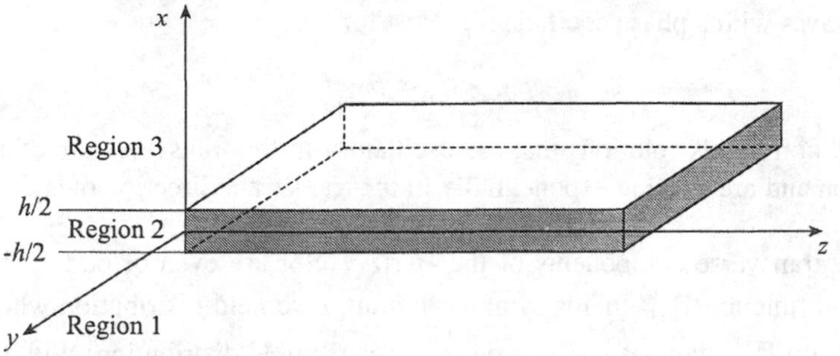

Figure 7.21: The planar dielectric waveguide.

be derived using a magnetic Hertz form $\underline{\Pi}_{mz}\,\mathrm{d}z$.

$$\underline{\Pi}_{ei} = \underline{\Pi}_{ezi}\,\mathrm{d}z \quad \text{for TM modes}\,, \tag{7.208a}$$
$$\underline{\Pi}_{mi} = \underline{\Pi}_{mzi}\,\mathrm{d}z \quad \text{for TE modes}\,. \tag{7.208b}$$

For the TM-modes the Helmholtz equation in medium i is given by

$$\Delta\underline{\Pi}_{ei} + \beta_{0i}^2\underline{\Pi}_{ei} = 0 \quad i = 1, 2 \tag{7.209}$$

where the phase coefficients β_{01} and β_{02} for transverse electromagnetic waves in media 1 and 2 are

$$\beta_{0i} = \omega\sqrt{\varepsilon_i \mu_i} \quad i = 1, 2\,. \tag{7.210}$$

Regions 1 and 3 are filled by medium 1, whereas region 2 is filled by medium 2. We seek for solutions of the Helmholtz equation guiding waves propagating in the z-direction. We assume uniform field distribution in y-direction and therefore $\partial/\partial y = 0$. This yields the two-dimensional Helmholtz equation

$$\frac{\partial^2\underline{\Pi}_{ezi}}{\partial x^2} + \frac{\partial^2\underline{\Pi}_{ezi}}{\partial z^2} + \beta_{0i}{}^2 = 0\,. \tag{7.211}$$

Assuming wave propagation in the positive z-direction we choose

$$\underline{\Pi}_{ezi} = X_i(x)\,\mathrm{e}^{-\mathrm{j}\beta z}\,.$$

Inserting this into (7.211) gives

$$\frac{\mathrm{d}^2 X_i}{\mathrm{d}x^2} + (\beta_{0i}^2 - \beta^2)X_i = 0\,. \tag{7.212}$$

Guided waves with a phase coefficient β exist for

$$\beta_{01} < \beta < \beta_{02} \,. \tag{7.213}$$

This condition yields solutions that are oscillatory in the transverse direction in the slab region and are varying exponentially in the transverse direction outside the slab region.

The transverse components of the Hertz vector are even or odd functions of x. The odd function $\underline{\Pi}_{zi}^{(o)}$ yields symmetric transverse field distribution whereas the even function $\underline{\Pi}_{zi}^{(e)}$ yields anti-symmetric transverse field distribution. With the transverse phase coefficient β_{x2} for region 2 and the transverse attenuation coefficient for regions 1 and 3, given by

$$\beta_{x2} = \sqrt{\beta_{02}^2 - \beta^2}\,, \tag{7.214a}$$

$$\alpha_{x1}^2 = \sqrt{\beta^2 - \beta_{01}^2} \tag{7.214b}$$

we obtain the odd solutions

$$\underline{X}_1^{(o)}(x) = B \sin \beta_x x \qquad \text{for } |x| \leq h/2\,, \tag{7.215a}$$

$$\underline{X}_1^{(o)}(x) = \pm D\,\mathrm{e}^{-\alpha_x |x|} \qquad \text{for } |x| \geq h/2 \tag{7.215b}$$

and the even solutions

$$\underline{X}_1^{(e)}(x) = B \cos \beta_x x \qquad \text{for } |x| \leq h/2\,, \tag{7.216a}$$

$$\underline{X}_1^{(e)}(x) = D\,\mathrm{e}^{-\alpha_x |x|} \qquad \text{for } |x| \geq h/2\,. \tag{7.216b}$$

The expression for the TM-mode therefore is

$$\Pi_z^{(ez)} = \begin{cases} \underline{\Pi}_{ez1}^{(o)} = D\,\mathrm{e}^{\alpha_x x}\,\mathrm{e}^{-\mathrm{j}\beta z} & \text{for } x \leq -h/2 \\ \underline{\Pi}_{ez2}^{(o)} = B \sin \beta_x x\,\mathrm{e}^{-\mathrm{j}\beta z} & \text{for } |x| \leq h/2\,, \\ \underline{\Pi}_{ez3}^{(o)} = -D\,\mathrm{e}^{-\alpha_x x}\,\mathrm{e}^{-\mathrm{j}\beta z} & \text{for } x \geq h/2 \end{cases} \tag{7.217}$$

$$\Pi_z^{(ez)} = \begin{cases} \underline{\Pi}_{ez1}^{(e)} = D\,\mathrm{e}^{\alpha_x x}\,\mathrm{e}^{-\mathrm{j}\beta z} & \text{for } x \leq -h/2 \\ \underline{\Pi}_{ez2}^{(e)} = B \cos \beta_x x\,\mathrm{e}^{-\mathrm{j}\beta z} & \text{for } |x| \leq h/2\,. \\ \underline{\Pi}_{ez3}^{(e)} = D\,\mathrm{e}^{-\alpha_x x}\,\mathrm{e}^{-\mathrm{j}\beta z} & \text{for } x \geq h/2 \end{cases} \tag{7.218}$$

Inserting (7.217) into (6.12a) to (6.12f) we obtain the field components

$$\underline{E}_x = \begin{cases} \mathrm{j}\alpha_x \beta D\,\mathrm{e}^{-\alpha_x x}\,\mathrm{e}^{-\mathrm{j}\beta z} & \text{for } x < -a \\ -\mathrm{j}\beta_x \beta B \cos \beta_x x\,\mathrm{e}^{-\mathrm{j}\beta z} & \text{for } |x| < a\,, \\ \mathrm{j}\alpha_x \beta D\,\mathrm{e}^{\alpha_x x}\,\mathrm{e}^{-\mathrm{j}\beta z} & \text{for } x > a \end{cases} \tag{7.219}$$

$$\underline{E}_z = \begin{cases} \alpha_x^2 D\, \mathrm{e}^{-\alpha_x x}\, \mathrm{e}^{-\mathrm{j}\beta z} & \text{for} \quad x < -a \\ \beta_x^2 B \sin \beta_x x\, \mathrm{e}^{-\mathrm{j}\beta z} & \text{for} \quad |x| < a \\ -\alpha_x^2 D\, \mathrm{e}^{\alpha_x x}\, \mathrm{e}^{-\mathrm{j}\beta z} & \text{for} \quad x > a \end{cases} , \tag{7.220}$$

$$\underline{H}_y = \begin{cases} \mathrm{j}\omega\varepsilon_1 \alpha_x D\, \mathrm{e}^{-\alpha_x x}\, \mathrm{e}^{-j\beta z} & \text{for} \quad x < -a \\ -\mathrm{j}\omega\varepsilon_2 \beta_x B \cos \beta_x x\, \mathrm{e}^{-j\beta z} & \text{for} \quad |x| < a \\ \mathrm{j}\omega\varepsilon_1 \alpha_x D\, \mathrm{e}^{-\alpha_x x}\, \mathrm{e}^{-j\beta z} & \text{for} \quad x > a \end{cases} . \tag{7.221}$$

For simplicity of notation we choose

$$u = \beta_x h/2 , \tag{7.222a}$$

$$v = \alpha_x h/2 . \tag{7.222b}$$

Continuity of $\underline{E}_z$ and $\underline{H}_y$ at $x \pm h/2$ requires

$$v^2\, \mathrm{e}^{-v}\, D + u^2 \sin u\, B = 0 , \tag{7.223a}$$

$$\varepsilon_1 v\, \mathrm{e}^{-v}\, D + \varepsilon_2 u \cos u\, B = 0 . \tag{7.223b}$$

This set of homogenous equations only has solutions with non-vanishing B and D if the determinant of the coefficients is zero. This condition yields

$$v = \frac{\varepsilon_1}{\varepsilon_2}\, u \tan u \quad \text{for odd TM mode} . \tag{7.224}$$

From (7.214a), (7.214b), (7.222a) and (7.222b) we obtain

$$u^2 + v^2 = w^2 . \tag{7.225}$$

with the *normalized frequency* w defined by

$$w = \frac{h}{2}\sqrt{\beta_{02}^2 - \beta_{01}^2} = \frac{\omega h}{2c_0}\sqrt{\varepsilon_{r2} - \varepsilon_{r1}} = \frac{\omega h}{2c_0}\sqrt{n_2^2 - n_1^2} \tag{7.226}$$

where n_2 is the refractive index of the slab medium and n_1 is the refractive index of the medium in region 1 and 3. This, coupled with (7.224) is the *characteristic equation* for determining the phase coefficient β and the cutoff frequency of the odd TM modes of the symmetric planar waveguide.

Even TM-modes are obtained by inserting (7.217) into (6.12a) to (6.12f) and following the same procedure as above. The characteristic equation for even TM-modes is

$$v = -\frac{\varepsilon_1}{\varepsilon_2}\, u \cot u \quad \text{for even TM mode} , \tag{7.227}$$

together with (7.225).

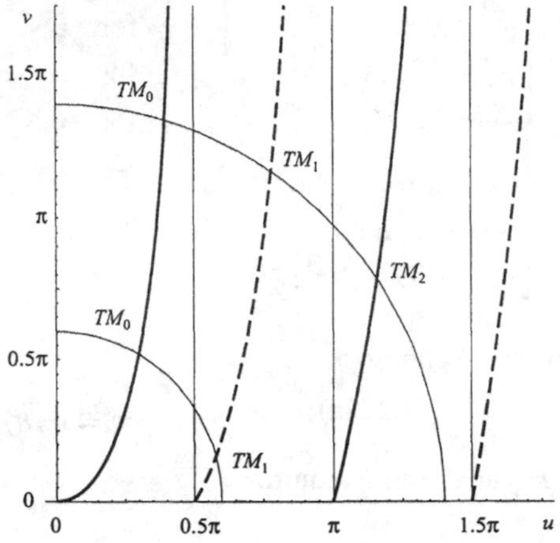

Figure 7.22: Graphical solution for eigenvalues for TM modes.

The characteristic equations for even and odd TM modes can be solved graphically. This is accomplished by the diagram shown in Figure 7.22 where the abscissa represents u and the ordinate v. The full curves are the plot of (7.224) and the dashed curves represent (7.227). Both plots are drawn for $\varepsilon_2/\varepsilon_1 = 2.5$. The plots of (7.225) are circles with the radius w. The intersections of the circles with the curves representing (7.224) and (7.227) determine the modes that propagate unattenuated in the dielectric slab waveguide.

TE modes of the planar dielectric waveguide may be derived using (7.208b). There is a complete duality between TM and TE modes. In the coordinate system of Figure 7.21 the TE mode propagating in the z-direction will exhibit an electric field in the y-direction and magnetic field components in the x- and z-direction. Applying the principle of duality to (7.224) and (7.227) we obtain the characteristic equations for the TE modes given by

$$v = \frac{\mu_1}{\mu_2} u \tan u \qquad \text{for odd TE mode ,} \tag{7.228a}$$

$$v = -\frac{\mu_1}{\mu_2} u \cot u \qquad \text{for even TE mode .} \tag{7.228b}$$

together with (7.225). For nonmagnetic media we obtain $\mu_1/\mu_2 = 1$.

The cutoff-frequencies for TE_n and TM_n modes are identical for the same n. The cut-off frequency TE_0 mode as well as the TM_0 mode is zero. This means the TE_0 and TM_0 propagate unattenuated no matter how thin the slab is. However, if the slab is very thin the field extends wide into the regions 1 and 3 and becomes the TEM wave with infinite extension in x-direction when the frequency becomes zero. In this limiting case the wave is not guided any more by the slab.

The normalized cutoff frequency w_{cn} of the TE_n and TM_n modes is given by

$$w_{cn} = \frac{n\pi}{2} . \tag{7.229}$$

With (7.226) we obtain the cutoff frequency

$$f_{cn} = \frac{nc_0}{2h\sqrt{\varepsilon_{r2} - \varepsilon_{r1}}} . \tag{7.230}$$

In the design of planar circuits for higher frequencies one has to take care that higher order surface waves cannot propagate. The lowest higher order modes are the TE_1 and TM_1 modes. In case of a silicon substrate with $\varepsilon_r = 12.6$ the cutoff-frequency of TE_1 and TM_1 modes is

$$f_{c1} = \frac{c_0}{2h\sqrt{\varepsilon_{r1} - 1}} = \frac{42.1\,\text{GHz}}{h/[\text{mm}]} . \tag{7.231}$$

7.7.2 Dielectric Slab with Single-Sided Metallization

Figure 7.23 shows the dielectric slab of thickness h with single sided metallization. The slab has the permittivity ε_{r2}. The region above the slab ($x > h$) exhibits a permittivity ε_{r1}. The dielectric plate with metallization on one side carries surface waves of TM type as well as such of TE type. We can obtain the field solutions for the dielectric slab with single-sided metallization directly from the solutions for the slab without metallization by application of the mirror principle. At the plane $x = 0$ odd TM modes and the even TE modes of the dielectric slab waveguide without metallization exhibit an electric field normal to this plane and a magnetic field tangential to this plane. Therefore these field solutions remain valid in either half-space $x \geq 0$ or $x \leq 0$ if an ideally conducting sheet is inserted in the plane $x = 0$.

In the case of the TE wave the current lines are normal to the direction of propagation, whereas the TM wave exhibits current lines in direction of propagation. The phase coefficients for the TM waves and TE waves respectively are determined via the eigenvalue equations

$$v = \frac{\varepsilon_{r1}}{\varepsilon_{r2}} u \tan u \qquad \text{for TM modes,} \tag{7.232a}$$

$$v = -u \cot u \qquad \text{for TE, modes} \tag{7.232b}$$

where the parameters u and v are given by

$$\begin{aligned} u &= h\sqrt{\beta_{02}^2 - \beta^2} , \\ v &= h\sqrt{\beta^2 - \beta_{01}^2} . \end{aligned} \tag{7.233}$$

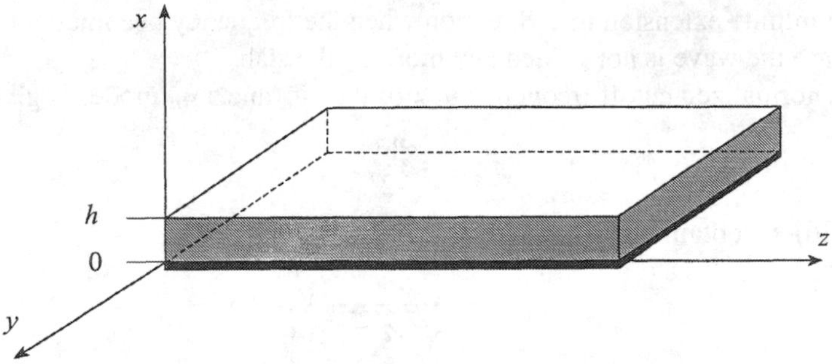

Figure 7.23: The planar dielectric waveguide.

The normalized frequency w is given by

$$w = \sqrt{u^2 + v^2} = h\sqrt{\beta_{02}^2 - \beta_{01}^2} . \qquad (7.234)$$

The TM$_0$ mode has no lower cut-off frequency. For the higher modes the normalized cut-off frequencies are given by

$$v_{c,\mathrm{TM}n} = n\pi \qquad \text{for TM modes}, \qquad (7.235)$$

$$v_{c,\mathrm{TE}n} = (n + \frac{1}{2})\pi \qquad \text{for TE modes}. \qquad (7.236)$$

The lowest higher order mode is the TE$_0$ mode for which the cutoff-frequency on silicon substrate is given by

$$f_{c,\mathrm{TE}0} = \frac{c_0}{4h\sqrt{\varepsilon_{r1} - 1}} = \frac{21.0\,\mathrm{GHz}}{h/[\mathrm{mm}]} . \qquad (7.237)$$

7.7.3 Circular Dielectric Waveguides with Step Index Profile

The circular cylindric dielectric waveguide plays an important role as optical fibers [7–10]. Figure 7.24(a) shows the circular cylindric dielectric waveguide. The core region $r \leq a$ (region 1) is filled with a homogeneous isotropic dielectric with a permittivity ε_1. The outer region $r > a$ (region 2) also is homogeneous and isotropic and has a permittivity ε_2. The outer region either is free space ($\varepsilon_2 = \varepsilon_0$) or filled with a dielectric. In order to guide a wave $\varepsilon_2 < \varepsilon_1$ must be valid. The dielectric waveguide is an open waveguide. This means that the electromagnetic field is not confined within the waveguide, but is extended also in the outer region. Therefore we have to solve the field equations in both regions $r \leq a$ and $r > a$. This is done by seeking

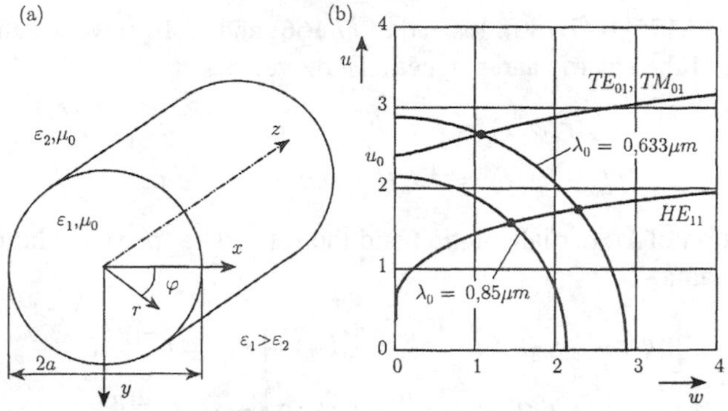

Figure 7.24: (a) Dielectric waveguide and (b) graphic solution of the eigenvalue equation.

the general solutions in both regions assuming either region to be infinitely extended and then by matching these partial solutions along the boundary. Since the boundary surface between the core region and outer region is a circular cylinder, the circular cylindric coordinate system is most suitable to solve the problem. The circular cylindric dielectric waveguide exhibits also hybrid modes. These are modes exhibiting electric as well as magnetic longitudinal components. Therefore in general we have to superimpose solutions derived from a longitudinal electric hertz vector field $\underline{\Pi}_{ez}$ and solutions derived from a longitudinal magnetic hertz vector field $\underline{\Pi}_{mz}$, if we want to use Hertz vector fields exhibiting z components only. For the core region 1 with $r \leq a$ the solutions (7.175) – (7.186) of the field equations are valid. We now have to consider that the core region is filled with a dielectric and therefore (7.160) has to be replaced with

$$k_c^2 = \beta_{01}^2 - \beta^2 . \tag{7.238}$$

With the phase coefficient β_{01} of the plane wave in a medium with permittivity ε_1 given by

$$\beta_{01} = \omega\sqrt{\mu_0 \varepsilon_1} = \sqrt{\varepsilon_{r1}}\beta_0 . \tag{7.239}$$

The occurrence of hybrid modes follows from the circumstance that the boundary conditions at $r = a$ can be generally fulfilled only if transverse electric *and* transverse magnetic partial solutions are superimposed. Comparing the field components of the transverse magnetic partial solutions (7.175) – (7.180) with the field components of the transverse electric partial solutions (7.181) – (7.186) we see that we have to rotate one of the two partial solutions by 90° around the z-axis in order to match the TE partial solutions and TM partial solutions. We rotate the TM partial solutions by 90° using the solution $\sin n\varphi$ instead of the solution $\cos n\varphi$ of (7.163). The corresponding solutions are obtained by the substitutions $\cos n\varphi \rightarrow \sin n\varphi$ and $\sin n\varphi \rightarrow -\cos n\varphi$

in (7.166) and (7.175) – (7.180). Instead of (7.166) and (7.169) we use the following z components of the electric and magnetic Hertz vectors

$$\underline{\Pi}_e(r, \varphi, z) = B J_n(k_c r) \sin n\varphi \, \mathrm{e}^{-\mathrm{j}\beta z} \, \mathrm{d}z \,, \tag{7.240a}$$

$$\underline{\Pi}_m(r, \varphi, z) = A J_n(k_c r) \cos n\varphi \, \mathrm{e}^{-\mathrm{j}\beta z} \, \mathrm{d}z \,. \tag{7.240b}$$

By superposition of TE partial solution and the rotated TM partial solution (7.175) – (7.180) we obtain:

$$\underline{E}_r = \mathrm{j}\left[\omega\mu_0 \frac{n}{r} A J_n(k_c r) - \beta k_c B J_n'(k_c r)\right] \sin n\varphi \, \mathrm{e}^{-\mathrm{j}\beta z} \,, \tag{7.241a}$$

$$\underline{E}_\varphi = \mathrm{j}\left[\omega\mu_0 k_c A J_n'(k_c r) - \beta \frac{n}{r} B J_n(k_c r)\right] \cos n\varphi \, \mathrm{e}^{-\mathrm{j}\beta z} \,, \tag{7.241b}$$

$$\underline{E}_z = k_c^2 B J_n(k_c r) \sin n\varphi \, \mathrm{e}^{-\mathrm{j}\beta z} \,, \tag{7.241c}$$

$$\underline{H}_r = -\mathrm{j}\left[\beta k_c A J_n'(k_c r) - \omega\varepsilon_1 \frac{n}{r} B J_n(k_c r)\right] \cos n\varphi \, \mathrm{e}^{-\mathrm{j}\beta z} \,, \tag{7.241d}$$

$$\underline{H}_\varphi = \mathrm{j}\left[\beta \frac{n}{r} A J_n(k_c r) - \omega\varepsilon_1 k_c B J_n'(k_c r)\right] \sin n\varphi \, \mathrm{e}^{-\mathrm{j}\beta z} \,, \tag{7.241e}$$

$$\underline{H}_z = k_c^2 A J_n(k_c r) \cos n\varphi \, \mathrm{e}^{-\mathrm{j}\beta z} \,. \tag{7.241f}$$

For region 2 we are seeking solutions decaying stronger than exponentially with $r \to \infty$. Such solutions describe guided waves. It exists for $\beta_{02} < \beta$, where β_{02} is the phase coefficient for a plane wave in the medium of region 2, given by

$$\beta_{02} = \omega\sqrt{\mu_0 \varepsilon_2} = \sqrt{\varepsilon_2}\beta_0 \,. \tag{7.242}$$

We introduce the parameter κ_c, given by

$$\kappa_c^2 = \beta^2 - \beta_{02}^2 \tag{7.243}$$

and obtain for the subregion 2 instead of (7.164) the Bessel differential equation

$$r^2 \frac{\mathrm{d}^2 R}{\mathrm{d}r^2} + r \frac{\mathrm{d}R}{\mathrm{d}r} - (r^2 \kappa_c^2 + n^2) R = 0 \,. \tag{7.244}$$

The solutions are the modified Bessel functions $K_n(x)$ and $I_n(x)$. The solution $K_n(x)$ for $x \to \infty$ goes to 0 stronger than exponentially, whereas the solution $I_n(x)$ is not bounded for $x \to \infty$. We obtain the solutions for subregion 2 from the solutions for subregion 1,(7.241a) – (7.241f), by performing the following substitutions

$$J_n(k_c r) \to K_n(\kappa_c r) \,, \tag{7.245a}$$

$$k_c J_n'(k_c r) \to \kappa_c K_n'(\kappa_c r) \,, \tag{7.245b}$$

$$\varepsilon_1 \to \varepsilon_2 \,, \tag{7.245c}$$

$$k_c^2 \to -\kappa_c^2 \,. \tag{7.245d}$$

and obtain:

$$E_r = \mathrm{j}\left[\omega\mu_0 \frac{n}{r} C K_n(\kappa_c r) - \beta\kappa_c D K_n'(\kappa_c r)\right] \sin n\varphi\, \mathrm{e}^{-\mathrm{j}\beta z}, \tag{7.246a}$$

$$E_\varphi = \mathrm{j}\left[\omega\mu_0\kappa_c C K_n'(\kappa_c r) - \beta \frac{n}{r} D K_n(\kappa_c r)\right] \cos n\varphi\, \mathrm{e}^{-\mathrm{j}\beta z}, \tag{7.246b}$$

$$E_z = -\kappa_c^2 D K_n(\kappa_c r) \sin n\varphi\, \mathrm{e}^{-\mathrm{j}\beta z}, \tag{7.246c}$$

$$\underline{H}_r = -\mathrm{j}\left[\beta\kappa_c C K_n'(\kappa_c r) - \omega\varepsilon_2 \frac{n}{r} D K_n(\kappa_c r)\right] \cos n\varphi\, \mathrm{e}^{-\mathrm{j}\beta z}, \tag{7.246d}$$

$$\underline{H}_\varphi = \mathrm{j}\left[\beta \frac{n}{r} C K_n(\kappa_c r) - \omega\varepsilon_2\kappa_c D K_n'(\kappa_c r)\right] \sin n\varphi\, \mathrm{e}^{-\mathrm{j}\beta z}, \tag{7.246e}$$

$$\underline{H}_z = -\kappa_c^2 C K_n(\kappa_c r) \cos n\varphi\, \mathrm{e}^{-\mathrm{j}\beta z}. \tag{7.246f}$$

We now introduce the normalized parameters u and v by

$$u = k_c a = a\sqrt{\beta_{01}^2 - \beta^2}, \tag{7.247a}$$

$$v = \kappa_c a = a\sqrt{\beta^2 - \beta_{02}^2}. \tag{7.247b}$$

With (7.238) and (7.243) we obtain therefrom

$$u^2 + v^2 = a^2(\beta_{01}^2 - \beta_{02}^2), \tag{7.248a}$$

$$u^2 + v^2 = a^2\beta_0^2(\varepsilon_{r1} - \varepsilon_{r2}). \tag{7.248b}$$

The partial solutions for subregions 1 and 2 must fulfill the boundary conditions (2.175) and (2.179) at $r = a$. The tangential components E_φ, E_z, $\underline{H}_\varphi$ and $\underline{H}_z$ of the fields at the boundary surface must be continuous at $r = a$. From (7.241b), (7.241c), (7.241e), (7.241f), (7.246b), (7.246c), (7.246e) and (7.246f) we obtain therewith

$$\omega\mu_0 u J_n'(u) A - \beta n J_n(u) B = \omega\mu_0 v K_n'(v) C - \beta n K_n(v) D, \tag{7.249a}$$

$$u^2 J_n(u) B = -v^2 K_n(v) D, \tag{7.249b}$$

$$\beta n J_n(u) A - \omega\varepsilon_1 u J_n'(u) B = \beta n K_n(v) C - \omega\varepsilon_2 v K_n'(v) D, \tag{7.249c}$$

$$u^2 J_n(u) A = -v^2 K_n(v) C. \tag{7.249d}$$

Using (7.249b) and (7.249d) we can express the complex amplitudes C and D by A and B, and after inserting into (7.249a) and (7.249c) we obtain

$$\omega\mu_0 \left(\frac{1}{u}\frac{J_n'(u)}{J_n(u)} + \frac{1}{v}\frac{K_n'(v)}{K_n(v)}\right) A - \beta n \left(\frac{1}{u^2} + \frac{1}{v^2}\right) B = 0, \tag{7.250a}$$

$$\beta n \left(\frac{1}{u^2} + \frac{1}{v^2}\right) A - \omega\varepsilon_0 \left(\frac{\varepsilon_{r1}}{u}\frac{J_n'(u)}{J_n(u)} + \frac{\varepsilon_{r2}}{v}\frac{K_n'(v)}{K_n(v)}\right) B = 0. \tag{7.250b}$$

This homogeneous system of equations for A and B only has a solution, if the coefficient determinant vanishes. This yields the eigenvalue equation

$$\frac{\omega^2}{c_0^2}\left[\frac{1}{u}\frac{J_n'(u)}{J_n(u)}+\frac{1}{v}\frac{K_n'(v)}{K_n(v)}\right]\left[\frac{\varepsilon_{r1}}{u}\frac{J_n'(u)}{J_n(u)}+\frac{\varepsilon_{r2}}{v}\frac{K_n'(v)}{K_n(v)}\right]-\beta^2 n^2\left(\frac{1}{u^2}+\frac{1}{v^2}\right)^2=0\,. \tag{7.251}$$

We eliminate the unknown phase coefficient β. From (7.247a) and (7.247b) we obtain

$$\beta^2\left[\frac{1}{u^2}+\frac{1}{v^2}\right]=\left[\frac{\beta_{01}^2}{u^2}+\frac{\beta_{02}^2}{v^2}\right]=\frac{\omega^2}{c_0^2}\left[\frac{\varepsilon_{r1}}{u^2}+\frac{\varepsilon_{r2}}{v^2}\right] \tag{7.252}$$

and with this from (7.251) the eigenvalue equation in the following form:

$$\left[\frac{1}{u}\frac{J_n'(u)}{J_n(u)}+\frac{1}{v}\frac{K_n'(v)}{K_n(v)}\right]\left[\frac{\varepsilon_{r1}}{u}\frac{J_n'(u)}{J_n(u)}+\frac{\varepsilon_{r2}}{v}\frac{K_n'(v)}{K_n(v)}\right]=n^2\left[\frac{\varepsilon_{r1}}{u^2}+\frac{\varepsilon_{r2}}{v^2}\right]\left[\frac{1}{u^2}+\frac{1}{v^2}\right]. \tag{7.253}$$

From (7.248b) and (7.253) the normalized parameters u and v and from this the quantities β, k_c and κ_c may be determined. The solution of the eigenvalue equation only may be obtained by numeric or graphic methods. To obtain the graphic solution we introduce the functions

$$\xi_n(u)=\frac{1}{u}\frac{J_n'(u)}{J_n(u)}\,, \tag{7.254}$$

$$\eta_n(v)=\frac{1}{v}\frac{K_n'(v)}{K_n(v)} \tag{7.255}$$

and obtain the eigenvalue equations

$$(\xi_n(u)+\eta_n(v))(\varepsilon_{r1}\xi_n(u)+\varepsilon_{r2}\eta_n(v))=n^2\left(\frac{\varepsilon_{r1}}{u^2}+\frac{\varepsilon_{r2}}{v^2}\right)\left(\frac{1}{u^2}+\frac{1}{v^2}\right), \tag{7.256}$$

$$u^2+v^2=a^2\beta_0^2(\varepsilon_{r1}-\varepsilon_{r2})\,. \tag{7.257}$$

These equations are solved graphically by drawing the curves given by both equations in the uv-plane. The solutions are given by the intersections of the curves. The curve given by (7.256) only depends on the parameter n and the geometric parameters of the waveguide. By (7.257) circles in uv-plane are defined, the radius of which is proportional to the frequency. The graphic solution is depicted in Figure7.24(b). In (7.256) a family of curves belongs to each n. According to (7.241a) to (7.241f) the φ dependence of the field components is given by $\sin n\varphi$ and $\cos n\varphi$ respectively. The parameter n therefore determines the number of node planes in the field. Furthermore the circular cylindric dielectric waveguide exhibits a number of node cylinder

Table 7.6: Parameter u_0 at the Cut-off Wavelength λ_c

	$\mathrm{HE_{nm}}$ – Modes ($\mathrm{TM_{0m}}$ for n = 0)	$\mathrm{EH_{nm}}$ – Modes ($\mathrm{TE_{0m}}$ for n = 0)
$n = 0$	$J_0(u_0) = 0$	$J_0(u_0) = 0$
$n = 1$	$J_1(u_0) = 0, \; (u_0 \neq 0)$	$J_1(u_0) = 0, \; (u_0 \neq 0)$
$n \geq 2$	$\frac{J_{n-2}(u_0)}{J_{n-1}(u_0)} = \frac{\varepsilon_{r1}-\varepsilon_{r2}}{\varepsilon_{r1}+\varepsilon_{r2}}$	$J_n(u_0) = 0, \; (u_0 \neq 0)$

surfaces. Modes are marked with indices n and m where n is the number of node planes and m is the number of node cylinders for a certain mode. To each pair of indices two field types are assigned. For $n = 0$ these two field types are transverse electric modes and the transverse magnetic modes. For $n \neq 0$ the modes are of hybrid type exhibiting longitudinal electric as well longitudinal magnetic field components. These field types are HE_{nm} and EH_{nm} modes respectively depending on whether the transverse field structure is similar to the TE_{nm} modes or the TM_{nm} modes of the circular cylindric hollow waveguide.

Let us first consider the modes with $n = 0$. These modes exhibit rotational symmetry. For $n = 0$ the right side of the eigenvalue equations (7.253) and (7.256) respectively vanish. These equation are fulfilled, if one of the expressions in brackets on the left side vanishes. For the TE_{0m} waves we obtain

$$\xi_0(u) + \eta_0(v) = 0\,, \tag{7.258}$$

whereas for the TM_{0m} waves

$$\varepsilon_{r1}\xi_0(u) + \varepsilon_{r2}\eta_0(w) = 0 \tag{7.259}$$

is valid. The assignment of these equations for $n = 0$ to the transverse electric and transverse magnetic fields respectively becomes obvious by comparison with (7.250a) and (7.250b). For $n = 0$ both equations are uncoupled. The transverse electric field with amplitude A and the transverse magnetic field with amplitude B therefore are not coupled.

In Figure 7.24(b) solution curves of (7.256) are drawn for the HE_{11} mode, the H_{01} mode and the E_{01} mode. The solution curve for the HE_{11} mode originates at $u = 0$, $v = 0$. Therefore for circles according to (7.257) with an arbitrarily small radius a point of intersection exists, i.e., the HE_{11} mode exhibits no lower cut-off frequency and can propagate for arbitrarily small frequencies. The field distribution

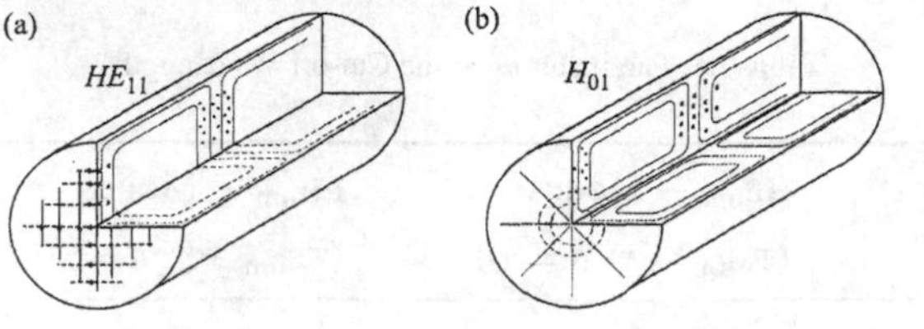

Figure 7.25: Field lines of the (a) HE_{11} mode and (b) the TE_{01} mode.

of the HE_{11} mode is depicted in Figure 7.25(a). For all other modes the curves of solution of (7.256) exhibit $v = 0$ for $u > 0$. The values u_0 of u for which the curve of solution of the specific modes are going through $v = 0$ are summarized in Table 7.6. These values u_0 determine the lowest values of β_0 according to (7.257) and the largest values of the free-space wavelength λ_0 for which the mode can propagate. The cut-off wavelengths λ_c of the modes are given by

$$\lambda_c = \frac{2\pi a}{u_0} \sqrt{\varepsilon_{r1} - \varepsilon_{r2}} \,. \tag{7.260}$$

With the exception of the HE_{11} mode which has no cut-off frequency, the TE_{01} mode exhibits the lowest cut-off frequency. According to Table 7.6 and (7.260) it is given by

$$\lambda_c^{(TE_{01})} = \frac{2\pi a}{2.405} \sqrt{\varepsilon_{r1} - \varepsilon_{r2}} \,. \tag{7.261}$$

Figure 7.24(b) shows the solution curves for the HE_{11} and the TE_{01} modes for $\sqrt{\varepsilon_{r1}} = 1.5$ and $\sqrt{\varepsilon_{r2}} = 1.4955$. The square roots of the relative permittivities are equal to the refractive indices. For optical fibers such small differences of the refractive indices between core and cladding region are usual. Such small differences may be realized by different doping of the quartz material. Small differences between core and cladding refractive indices allow for the realization of optical fibers with a core diameter of up to 10 μm; in single-mode operation for less than 1 μm wavelength. Figure 7.24(b) shows the graphic solution of the eigenvalue equations for a circular cylindric dielectric waveguide with a core diameter of $2a$ =5 μm and core and cladding refractive indices of 1.5 and 1.4955 respectively. From (7.261) we obtain a cut-off wavelength $\lambda_c = 0.758$ μm.

Figure 7.25 depicts the field distribution of the TE_{01} mode. For the small difference between ε_{r1} and ε_{r2}, (7.258) and (7.259) are nearly identical. Therefore the solution curves for the TE_{01} and the TM_{01} modes cannot be distinguished. Figure 7.26 shows the dispersion diagram of the TE_{01} mode. The HE_{11} mode exhibits no lower cut-off frequency; however, it is not suitable to guide waves of an arbitrarily low frequency. The lower the frequency becomes, the less the electromagnetic field is guided by the core of the fiber and the more it is spread in the cladding of the fiber or in free

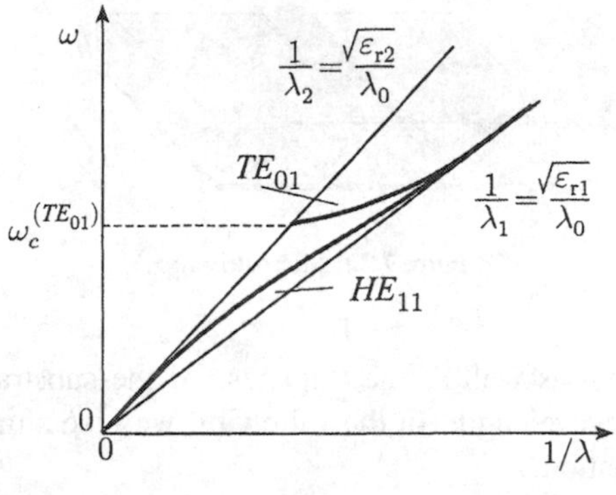

Figure 7.26: Dispersion diagram for the HE_{11} and TE_{01} modes.

space in the case of small perturbations of the waveguide or even small bending. The optical waveguide has to be designed such that the electromagnetic field is rapidly decaying in the cladding region. Otherwise the electromagnetic energy will be attenuated by lossy material on the surface of the cladding or it will be scattered.

7.8 PLANAR TRANSMISSION LINES

Planar transmission lines are formed by metallized plane dielectric plates [11–14]. Planar transmission lines play an important role as the basic line structures in hybrid and monolithic integrated circuits.

7.8.1 The Microstrip Line

The *microstrip line* is depicted in Figure 7.27. In the microstrip line the electromagnetic field is not confined to the substrate, but spreads over the free space. The microstrip line is an *inhomogeneous waveguide* since the transverse dielectric region is non-homogeneous. Due to this circumstance the fundamental mode is not a TEM mode, but a so-called *quasi* TEM *mode*. The quasi TEM mode approaches a TEM mode if the frequency goes to zero.

The analysis of microstrip lines only can be performed using numerical methods. Accurate methods of modelling (integral equation method, spectral domain method, partial wave synthesis, finite difference method, transmission line matrix method) in some cases may require a high numerical effort. The easiest approximation is the *quasi-static approximation*. In the quasi static approximation the transmission line wave is approximated by a TEM wave and the transmission line properties are calculated from the electrostatic capacitance. This approximation can be used if the

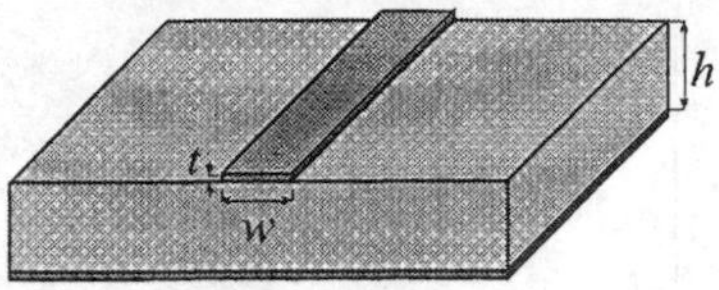

Figure 7.27: Microstrip line.

transmission line width as well as the thickness of the substrate are very small in comparison with the wavelength. In the following we give simple formulae for the quasi-static approximation.

Let C_a be the capacitance per unit of length of the microstrip line for the case in which the dielectric is replaced by free space and let C be the capacitance per unit of length with dielectric. The wave impedance Z_0 and the phase coefficient β are given by

$$Z_0 = 1/c\sqrt{CC_a}\ , \tag{7.262}$$

$$\beta = k_0(C/C_a)^{1/2} = k_0\sqrt{\varepsilon_{re}}\ . \tag{7.263}$$

with

$$\varepsilon_{re} = (\lambda_0/\lambda_c)^2 = C/C_a \tag{7.264}$$

where λ_0 is the free-space wavelength and λ_c is the wavelength of the guided wave. Closed-form expressions for Z_0 and ε_{re} have been given by Wheeler [15,16], Schneider [17] and Hammerstad [18].

The maximum frequency up to which a microstrip line may be used is limited by the losses increasing with frequency. Especially due to the excitation of substrate waves. The cut-off frequency f_{cs} beyond which a strong coupling between the quasi-TEM mode of the microstrip line and the surface wave of lowest order occurs is given by

$$f_{cs} = \frac{150}{\pi h}\sqrt{\frac{2}{\varepsilon_{re}-1}}\tan^{-1}\varepsilon_r\ \text{GHz mm} \tag{7.265}$$

with the cut-off frequency f_{cs} in GHz and the height h in millimeters. From this it follows for a GaAs substrate with ε_r=12,9 that for a frequency of 100 GHz the maximum thickness of the substrate should not exceed 0.3 mm.

Furthermore the maximum substrate thickness is limited by the radiation losses excited at transmission line discontinuities. For a $\lambda/2$ resonator the Q factor due to radiation approximately is given by

$$Q_r = \frac{3\varepsilon_r Z_0 \lambda_0^2}{32\,\eta_0 h^2}\ . \tag{7.266}$$

At 100 GHz a GaAs substrate therefore must be thinner than 0.125 mm to achieve a $Q > 100$.

In the following the approximate formula for the computation of the parameters of a microstrip line are summarized [13]. The approximation formula for the characteristic impedance is given by

$$Z_0 = \begin{cases} \frac{\eta_0}{2\pi\sqrt{\varepsilon_{re}}} \ln\left(\frac{8h}{W'} + 0.25\frac{W'}{h}\right) & \text{for } W/h \leq 1\,, \\ \frac{\eta_0}{\sqrt{\varepsilon_{re}}}\left[\frac{W'}{h} + 1.393 + 0.667\ln\left(\frac{W'}{h} + 1.444\right)\right]^{-1} & \text{for } W/h \geq 1\,, \end{cases} \tag{7.267}$$

with

$$\frac{W'}{h} = \begin{cases} \frac{W}{h} + \frac{1.25}{\pi}\frac{t}{h}\left(1 + \ln\frac{4\pi W}{t}\right) & \text{for } W/h \leq 1\,, \\ \frac{W'}{h} = \frac{W}{h} + \frac{1.25}{\pi}\frac{t}{h}\left(1 + \ln\frac{2h}{t}\right) & \text{for } W/h \geq 1\,. \end{cases} \tag{7.268}$$

The approximation formula for the effective permittivity is given by

$$\varepsilon_{re} = \frac{\varepsilon_r + 1}{2} + \frac{\varepsilon_r - 1}{2} F(W/h) - \frac{\varepsilon_r - 1}{4.6} - \frac{t/h}{\sqrt{W/h}} \tag{7.269}$$

with

$$F(W/h) = \begin{cases} (1 + 12h/W)^{-1/2} + 0.04(1 - W/h)^2 & \text{for } W/h \leq 1\,, \\ (1 + 12h/W)^{-1/2} & \text{for } W/h \geq 1\,. \end{cases} \tag{7.270}$$

The attenuation coefficient is given by

$$\alpha_c = \begin{cases} 1.38\frac{R_s}{hZ_0}\frac{32-(W'/h)^2}{32+(W'/h)^2}\Lambda\,, \\ 6.1 \times 10^{-5}\frac{R_s Z_0 \varepsilon_{re}}{h}\left[W'/h + \frac{0.667W'/h}{W'/h+1.444}\right]\Lambda\,, \end{cases} \tag{7.271}$$

with

$$\Lambda = \begin{cases} 1 + \frac{h}{W'}\left(1 + \frac{1.25t}{\pi W} + \frac{1.25}{\pi}\ln\frac{4\pi W}{t}\right)\,, \\ 1 + \frac{h}{W'}\left(1 - \frac{1.25t}{\pi h} + \frac{1.25}{\pi}\ln\frac{2h}{t}\right)\,. \end{cases} \tag{7.272}$$

The microstrip line is the most common planar transmission line structure. The full metallization of the bottom side of the substrate facilitates the mounting. The insertion of circuit elements in series can be done without difficulty; however, parallel circuiting of circuit elements requires either wire holes or the realization of short circuits via $\lambda/4$ lines. The latter solution is possible within a narrow band only.

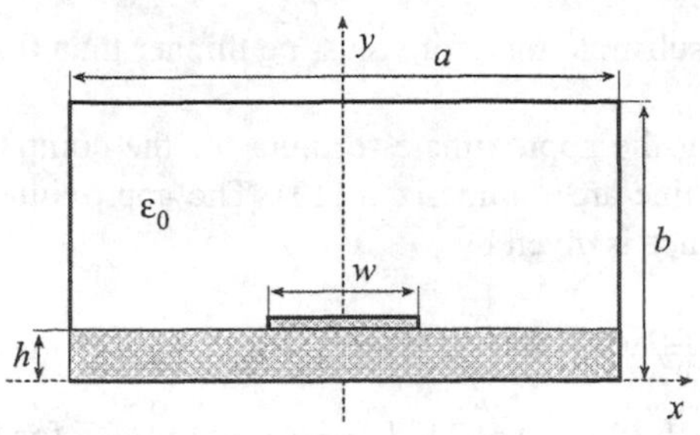

Figure 7.28: Cross-section of the shielded microstrip line.

7.8.2 Quasistatic Approximation for the Microstrip Line

Figure 7.28 shows a cross-sectional view of the shielded microstrip line. The microstrip line exhibits a conductor strip of width w on a dielectric substrate of height h. The permittivity of the substrate is ε_1. The ground plane of the substrate of a microstrip line usually is fully metallized. The thickness of the strip conductor is neglected. For computational reasons we have embedded the microstrip line in a waveguide of width a and height h. We assume a and h to be sufficiently large, so that the influence of the embedding on the characteristics of the microstrip line may be neglected.

We apply a *quasi-static approximation* [19] to analyze the shielded microstrip line where we assume that in subregions with constant ε the transverse electric field is governed by the two-dimensional *Laplace equation*

$$\frac{\partial^2 \Phi}{\partial x^2} + \frac{\partial^2 \Phi}{\partial y^2} = 0 . \tag{7.273}$$

The boundary conditions for the electric potential $\Phi(x, y)$ are

$$\Phi\left(-\frac{a}{2}, y\right) = \Phi\left(\frac{a}{2}, y\right) = \Phi(x, 0) = \Phi(x, b) . \tag{7.274}$$

We expand $\Phi(x, y)$ into products of the basis functions $\varphi_n(x)$ and $\psi_n(y)$

$$\Phi(x, y) = \sum_n c_n \varphi_n(x) \psi_n(y) . \tag{7.275}$$

As the basis functions $\varphi_n(x)$ we choose

$$\varphi_n(x) = \cos \frac{n\pi x}{a} \quad \text{with} \quad n = 1, 3, 5, \ldots \tag{7.276}$$

Presupposing that each term in the sum at the right-hand side of (7.275) fulfills the Laplace equation (7.273) it follows that

$$\psi_n(y) = \sinh \frac{n\pi y}{a} \quad \text{for} \quad 0 \leq y \leq h \tag{7.277}$$

and we obtain for the region $0 \leq y \leq h$ the series expansion

$$\Phi_1(x, y) = \sum_{n=1,3,5,\ldots} c_n \cos \frac{n\pi x}{a} \sinh \frac{n\pi y}{a} \quad \text{for} \quad 0 \leq y \leq h\,. \tag{7.278}$$

In a similar way we obtain for $h \leq y \leq b$ the series expansion

$$\Phi_0(x, y) = \sum_{n=1,3,5,\ldots} d_n \cos \frac{n\pi x}{a} \sinh \frac{n\pi (y-b)}{a} \quad \text{for} \quad h \leq y \leq b\,. \tag{7.279}$$

On the boundary surface $y = h$ the area charge density ρ_A is given by

$$\rho_A = \sum_{n=1,3,5,\ldots} r_n \cos \frac{n\pi x}{a}\,. \tag{7.280}$$

A surface charge density $\rho_A \neq 0$ only exists on the conductor, that means in the region $|x| \leq w/2$. This condition is enforced by extending the Fourier integral for determination of the coefficients r_n only over the interval $[-w/2, w/2]$:

$$r_n = \frac{2}{a} \int_{-w/2}^{w/2} \rho_A(x) \cos \frac{n\pi x}{a}\,\mathrm{d}x\,. \tag{7.281}$$

On the boundary surface $x = h$ the continuity condition

$$\Phi_1(x, h) = \Phi_0(x, h) \tag{7.282}$$

must be fulfilled. From this it follows that

$$d_n \sinh \frac{n\pi (h-b)}{a} = c_n \sinh \frac{n\pi h}{a}\,. \tag{7.283}$$

From the boundary condition for the electric flux density it follows that

$$-\varepsilon_0 \frac{\partial}{\partial y} \Phi_0(x, y) + \varepsilon_1 \frac{\partial}{\partial y} \Phi_1(x, y) = \rho_A \Big|_{y=h} \tag{7.284}$$

and from this

$$\frac{n\pi}{a}\left(-\varepsilon_0 d_n \cosh\frac{n\pi(h-b)}{a} + \varepsilon_1 c_n \cosh\frac{n\pi h}{a}\right) = r_n\,. \tag{7.285}$$

From (7.279), (7.281), (7.283) and (7.285) we finally obtain

$$\Phi(x,h) = \sum_{n=1,3,5,\ldots} g_n \cos\frac{n\pi x}{a}\int_{-w/2}^{w/2} \rho_A(x')\cos\frac{n\pi x'}{a}\,\mathrm{d}x' \tag{7.286}$$

with

$$g_n = \frac{2}{n\pi}\,\frac{1}{\varepsilon_1 \coth\frac{n\pi h}{a} - \varepsilon_0 \coth\frac{n\pi(h-b)}{a}}\,. \tag{7.287}$$

In (7.286) the integration is only performed over the interval $\left[-w/2, w/2\right]$ since the integrand vanishes outside this interval. We put this equation into the form

$$V_0 = \Phi(x,h) = \int_{-w/2}^{w/2} G(x,x')\rho_A(x')\,\mathrm{d}x' \quad \text{for} \quad |x| \le \frac{w}{2}\,, \tag{7.288}$$

where V_0 is the potential of the conductor strip and the *Green's function* for this problem $G(x,x')$ is given by

$$G(x,x') = \sum_{n=1,3,5,\ldots} g_n \cos\frac{n\pi x}{a}\cos\frac{n\pi x'}{a}\,. \tag{7.289}$$

For $b \to \infty$ we obtain from (7.287)

$$g_n = \frac{2}{n\pi}\,\frac{1}{\varepsilon_1 \coth\frac{n\pi h}{a} + \varepsilon_0}\,. \tag{7.290}$$

The charge per unit of lengths q' is

$$q' = \int_{-w/2}^{w/2} \rho_A(x)\,\mathrm{d}x\,. \tag{7.291}$$

The capacitance per unit of lengths C' of the microstrip line is

$$C' = \frac{q'}{V_0}\,. \tag{7.292}$$

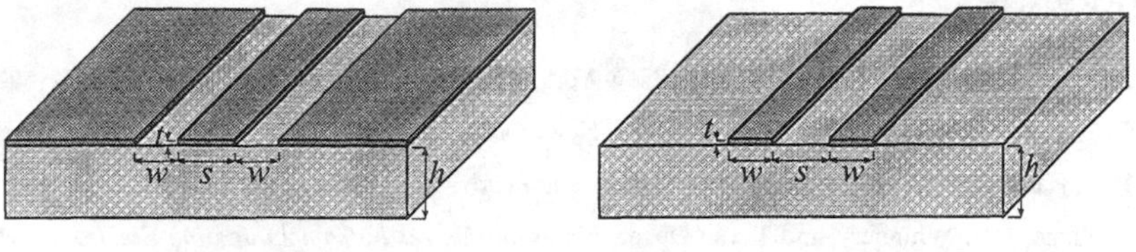

Figure 7.29: (a) Coplanar waveguide and (b) coplanar stripline.

7.8.3 Coplanar Line and Coplanar Strip Line

Figure 7.29 shows the coplanar line and the coplanar strip line. On coplanar lines and coplanar strip lines quasi TEM modes may propagate. The approximation formula for the characteristic impedance of a coplanar waveguide is

$$Z_0 = \frac{Z_{F0}}{4\sqrt{\varepsilon_{re}}} \frac{K(k')}{K(k)} \tag{7.293}$$

and the approximation formula for the effective relative permittivity of a coplanar waveguide is

$$\varepsilon_{re} = 1 + \frac{\varepsilon_r - 1}{2} \frac{K(k')K(k_1)}{K(k)K(k_1')} \tag{7.294}$$

with $k = a/b$; $a = s/2$, $b = s/2 + w$ and $k_1 = \sinh(\pi a/2h)/\sinh(\pi b/2h)$. The functions $K(k)$ and $K'(k)$ are elliptic integrals. The ratio $K(k)/K'(k)$ is given by the following approximation formula

$$\frac{K(k)}{K'(k)} = \begin{cases} \left[\frac{1}{\pi}\ln\left(2\,\frac{1+\sqrt{k}}{1-\sqrt{k}}\right)\right]^{-1} & \text{for } 0 \le k \le 0,7 \\ \frac{1}{\pi}\ln\left(2\,\frac{1+\sqrt{k}}{1-\sqrt{k}}\right) & \text{for } 0,7 \le k \le 1 \end{cases} . \tag{7.295}$$

The approximation formula for the characteristic impedance of a coplanar stripline is

$$Z_0 = \frac{Z_{F0}}{\sqrt{\varepsilon_{re}}} \frac{K(k)}{K(k')} . \tag{7.296}$$

The approximation formula for the effective permittivity of a coplanar stripline is also given by (7.294)

REFERENCES

[1] J. A. Stratton, *Electromagnetic Theory*. New York: McGraw-Hill, 1941.

[2] R. E. Collin, *Field Theory of Guided Waves*. New York: IEEE Press, 1991.

[3] R. F. Harrington, *Time Harmonic Electromagnetic Fields*. New York: McGraw-Hill, 1961.

[4] S. Ramo, J. R. Whinnery, and T. van Duzer, *Fields and Waves in Communication Electronics*. New York: John Wiley & Sons, 1965.

[5] C. A. Balanis, *Advanced Engineering Electromagnetics*. New York: John Wiley & Sons, 1989.

[6] H. Kogelnik, "Theory of dielectric waveguides," in *Integrated Optics* (T. Tamir, ed.), no. 7 in Topics in Applied Optics, pp. 13–81, Berlin Heidelberg New York: Springer, 1979.

[7] H. Unger, *Planar Optical Waveguides and Fibers*. Oxford: Clarendon Press, 1977.

[8] D. Marcuse, *Light Transmission Optics*. New York: Van Nostrand Reinhold, 1982.

[9] J. A. Buck, *Fundamentals of Optical Fibers*. New York: John Wiley & Sons, 1995.

[10] K. Chang, *Handbook of Microwave and Optical Components, Vol. 3, Optical Components*. New York: John Wiley & Sons, 1990.

[11] K. Gupta, R. Garg, I. Bahl, and P. Bhartia, *Microstrip Lines and Slotlines*. Boston: Artech House, 1996.

[12] B. C. Wadell, *Transmission Line Design Handbook*. Boston: Artech House, 1991.

[13] K. Chang, *Handbook of Microwave and Optical Components, Vol. 1, Microwave Passive and Antenna Components*. New York: John Wiley & Sons, 1989.

[14] R. Hoffmann, *Integrierte Mikrowellenschaltungen*. Berlin: Springer, 1983.

[15] H. Wheeler, "Transmission line properties of parallel strips separated by a dielectric sheet," *IEEE Trans. Microwave Theory Techn.*, vol. 13, pp. 172–185, Mar. 1965.

[16] H. Wheeler, "Transmission line properties of a strip on a dielectric sheet on a plane," *IEEE Trans. Microwave Theory Techn.*, vol. 25, pp. 631–647, Aug. 1977.

[17] M. Schneider, "Microstrip lines for microwave integrated circuits," *Bell System Tech. J.*, vol. 1969, pp. 1422–1444, 1969.

[18] E. Hammerstad, "Accurate models for microstrip computer-aided design," *1980 Int. Microwave Symposium Digest*, pp. 407–409, June 1980.

[19] R. C. Booton, *Computational Methods for Electromagnetics and Microwaves*. New York: John Wiley & Sons, 1992.

Chapter 8

The Transmission Line Equations

8.1 THE TRANSMISSION LINE CONCEPT

In this chapter we are concerned with the longitudinal variations of the wave amplitudes on a transmission line. We assume the transmission line to be excited in a certain mode. The transverse field distribution is determined by the excited mode. In longitudinal direction the spatial variation of the field is governed by the *transmission line equations* (7.58a) and (7.58b). Transmission line theory is presented in [1–3].

In our treatment of the TEM waveguide we have observed that the transverse field distribution is only determined by the geometry of the waveguide. The state of a transmission line is described completely by the scalar quantities $\underline{V}(z)$ and $\underline{I}(z)$, respectively. Current and voltage are governed by the line equations (7.68) and (7.73). In the same way we can describe the TE_{10} mode of a rectangular waveguide by the transmission line equations (7.137a) and (7.137b), if we are introducing *generalized currents* and *generalized voltages* to describe the electromagnetic wave in the waveguide. If we are choosing a certain mode in a waveguide, the specific state of excitation also is given by the generalized voltage and the generalized current, which depend on the longitudinal coordinate z only. We now can formulate the line equations in a more general form, which is valid for TEM modes in two-conductor waveguides

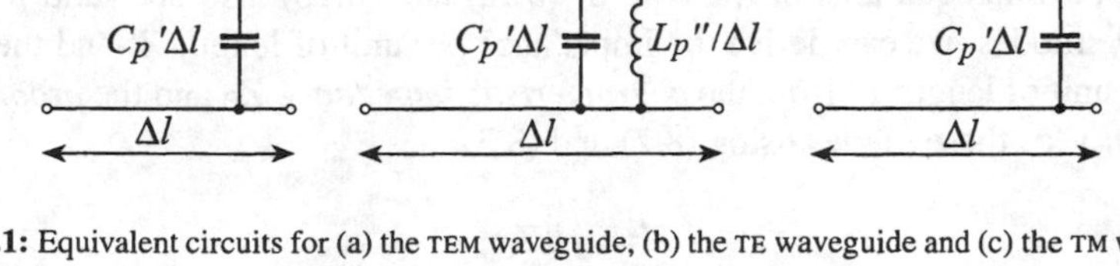

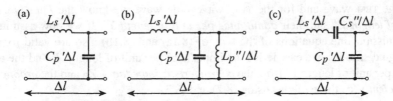

Figure 8.1: Equivalent circuits for (a) the TEM waveguide, (b) the TE waveguide and (c) the TM waveguide.

Table 8.1: Transmission Line Parameters of Lossless Waveguides

Mode	*Impedance per unit of length* Z'	*Admittance per unit of length* Y'	*Characteristic impedance* Z_0	*Phase coefficient* β
TEM	$\mathrm{j}\omega L'_s$	$\mathrm{j}\omega C'_p$	$\sqrt{\frac{L'_s}{C'_p}}$	$\omega\sqrt{L'_s C'_p}$
TE	$\mathrm{j}\omega L'_s$	$\mathrm{j}\left(\omega C'_p - \frac{1}{\omega L'_s}\right)$	$Z_H = \frac{Z_F}{\sqrt{1-\left(\frac{\lambda_0}{\lambda_c}\right)^2}}$	$\frac{\omega}{c}\sqrt{1-\left(\frac{\lambda_0}{\lambda_c}\right)^2}$
	$\mathrm{j}\frac{\omega Z_F}{c}$	$\mathrm{j}\frac{\omega}{cZ_F}\left(1-\frac{\omega_c^2}{\omega^2}\right)$		
TM	$\mathrm{j}\left(\omega L'_s - \frac{1}{\omega C'_p}\right)$	$\mathrm{j}\omega C'_p$	$Z_E = Z_F\sqrt{1-\left(\frac{\lambda_0}{\lambda_c}\right)^2}$	$\frac{\omega}{c}\sqrt{1-\left(\frac{\lambda_0}{\lambda_c}\right)^2}$
	$\mathrm{j}\frac{\omega Z_F}{c}\left(1-\frac{\omega_c^2}{\omega^2}\right)$	$\mathrm{j}\frac{\omega}{cZ_F}$		

as well as for TE and TM modes:

$$\frac{\mathrm{d}\underline{V}}{\mathrm{d}z} = -Z'\underline{I}, \tag{8.1a}$$

$$\frac{\mathrm{d}\underline{I}}{\mathrm{d}z} = -Y'\underline{V}. \tag{8.1b}$$

The *characteristic impedance* Z_0 is given by

$$Z_0 = \sqrt{\frac{Z'}{Y'}}. \tag{8.2}$$

The propagation coefficient $\underline{\gamma}$ is given by

$$\gamma = \alpha + \mathrm{j}\beta = \sqrt{Z'Y'}. \tag{8.3}$$

For the TEM wave and for the TE_{10} waveguide wave we know the *impedance per unit of length* Z' and their *admittance per unit of length* $\underline{Y'}$. If we now assume that transmission line equations of the type of (8.1a) and (8.1b) also are valid for other waveguide modes, we can derive the impedance per unit of length Z' and the admittance per unit of length Y' from the *characteristic impedance* Z_0 and the *propagation coefficient* γ of these modes using (8.2) and (8.3).

$$Z' = \mathrm{j}\beta Z_0, \tag{8.4a}$$

$$Y' = \frac{\mathrm{j}\beta}{Z_0}\,. \tag{8.4b}$$

These quantities are summarized in Table 8.1 for the lossless line. Using the impedance per unit of length Z' and the admittance per unit of length Y' we can specify the equivalent circuits according to Figure 8.1 for short line segments of length Δl. The equivalent circuits according to Figure 8.1(a) and 8.1(b) are the equivalent circuits already derived for the TEM waveguide mode and for the rectangular waveguide mode. The equivalent circuits according to Figure 8.1(b) and 8.1(c) are valid in general for the TE_{mn} modes and TM_{mn} modes of waveguides, respectively. In Figure 8.1(b) and 8.1(c) we have applied for the first time the network concept to a waveguide segment where current and voltage are not defined as usual by line integrals over magnetic and electric fields respectively. This extension of the network concept will be very useful in the following. It allows not only a simple description of waveguide circuits, but also the application of network theory to waveguide circuits.

8.2 GENERALIZED VOLTAGES AND CURRENTS

Like voltages and currents generalized voltages and generalized currents are integral field quantities. For these integral field quantities different definitions are used in literature. For transverse electromagnetic waves the definition of current and voltage as usual by line integrals makes sense. We also want to introduce for waveguide modes definitions of generalized voltages and generalized currents, which are independent from the waveguide modes. For the H_{10} mode of the rectangular waveguide we have introduced the generalized voltage $\underline{V}$ and the generalized current $\underline{I}$ such that they are interrelated with the transmitted active power according to (7.126). We want to keep this definition in the following. Since the transmitted active power depends on the transverse components of the electromagnetic field, it makes sense to define the generalized currents and voltages as integrals of the transverse field intensities of the waveguide modes.

We subdivide the field in the waveguide into transverse and longitudinal components. The transverse field components are described by the differential forms $\underline{\mathcal{E}}_{tr}$ and $\underline{\mathcal{H}}_{tr}$ respectively and the longitudinal components are described by the differential forms $\underline{\mathcal{E}}_l$ and $\underline{\mathcal{H}}_l$ respectively. For the xy-plane as the transverse plane and z-direction as the longitudinal direction, we obtain

$$\underline{\mathcal{E}}(\boldsymbol{x}) = \underline{\mathcal{E}}_{tr}(\boldsymbol{x}) + \underline{\mathcal{E}}_l(\boldsymbol{x})\,, \tag{8.5a}$$
$$\underline{\mathcal{H}}(\boldsymbol{x}) = \underline{\mathcal{H}}_{tr}(\boldsymbol{x}) + \underline{\mathcal{H}}_l(\boldsymbol{x})\,. \tag{8.5b}$$

with

$$\underline{\mathcal{E}}_{tr}(\boldsymbol{x}) = \underline{E}_x(\boldsymbol{x})\,dx + \underline{E}_y(\boldsymbol{x})\,dy\,, \tag{8.6a}$$
$$\underline{\mathcal{H}}_{tr}(\boldsymbol{x}) = \underline{H}_x(\boldsymbol{x})\,dx + \underline{H}_y(\boldsymbol{x})\,dy\,, \tag{8.6b}$$
$$\underline{\mathcal{E}}_l(\boldsymbol{x}) = \underline{E}_z(\boldsymbol{x})\,dz\,, \tag{8.6c}$$
$$\underline{\mathcal{H}}_l(\boldsymbol{x}) = \underline{H}_z(\boldsymbol{x})\,dz\,. \tag{8.6d}$$

We now split up the transverse field intensities as in (7.28a) and (7.28b) into the *complex amplitudes* $\underline{V}(z)$ and $\underline{I}(z)$ and the real normalized *structure functions* $\mathbf{e}(x, y)$ and $\mathbf{h}(x, y)$. For the TEM mode $\mathbf{e}(x, y)$ and $\mathbf{h}(x, y)$ have been normalized in Section 4.4 such that $\underline{V}(z)$ and $\underline{I}(z)$ are defined via the path integral from conductor 1 to conductor 2 and the circulation integral around conductor 2.

In (7.28a) and (7.28b) we introduced the structure functions $\mathbf{e}$ and $\mathbf{h}$. Introducing structure functions for the transverse field components we generalize the definition of the *structure functions* for TEM, TE and TM waves:

$$\underline{\boldsymbol{E}}_{tr}(\boldsymbol{x}) = \underline{V}(z)\mathbf{e}(x, y)\,, \tag{8.7a}$$
$$\underline{\boldsymbol{H}}_{tr}(\boldsymbol{x}) = \underline{I}(z)\mathbf{h}(x, y)\,. \tag{8.7b}$$

In (7.28a) and (7.28b) the differential forms e and h were introduced:

$$e(x, y) = e_x(x, y)\,dx + e_y(x, y)\,dy\,, \tag{8.8a}$$
$$h(x, y) = h_x(x, y)\,dx + h_y(x, y)\,dy\,. \tag{8.8b}$$

From (8.7a) and (8.7b) we obtain the corresponding relation for the differential forms:

$$\underline{\mathcal{E}}_{tr}(\boldsymbol{x}) = \underline{V}(z)e(x, y)\,, \tag{8.9a}$$
$$\underline{\mathcal{H}}_{tr}(\boldsymbol{x}) = \underline{I}(z)h(x, y)\,. \tag{8.9b}$$

Let us first consider the TEM modes. For the TEM mode the field intensities are coincident with the transverse field intensities. We can normalize the structure functions via path integrals with the normalization already given in (7.31a) and (7.31b).

Figure 8.2 shows the relation of the direction of the arrows for current and voltage and the direction of field lines and the Poynting vector. For the waveguide the introduction of integral field quantities can only be performed, if the path of integration is specified. The disadvantage of this method, however, is that we have to specify a path of integration for every waveguide mode, and a definition of a path of integration independently from the mode is not possible. We therefore will perform the normalization on the basis of the area integral over the absolute squares of the

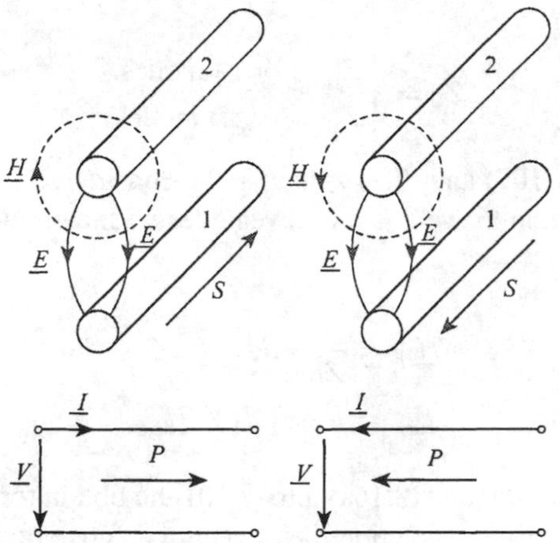

Figure 8.2: Relation of arrows and field orientation in the TEM waveguide.

transverse field quantities. The active power transmitted by the waveguide is given by

$$P = \frac{1}{2}\Re\left\{\int_A \mathcal{E}_{tr}(x,y) \wedge \mathcal{H}^*_{tr}(x,y)\right\}. \tag{8.10}$$

The area integral is performed over the cross-sectional area A of the waveguide. Inserting (8.9a) and (8.9b) yields

$$P = \frac{1}{2}\Re\{\underline{V}(z)\,\underline{I}^*(z)\} \int_A e(x,y) \wedge h(x,y)\}. \tag{8.11}$$

From (7.31a) and (7.31b) we obtain

$$\int_A e(x,y) \wedge h(x,y) = -\int_{C_1} e(x,y) \oint_{C_2} h(x,y) = 1 \tag{8.12}$$

and with this from (8.11)

$$P = \frac{1}{2}\Re\left\{\underline{V}(z)\,\underline{I}(z)^*\right\}. \tag{8.13}$$

To consider rectangular and circular hollow waveguides we summarize (7.105), (7.106), (7.112) and (7.113), which have been derived for waveguides with a rectangular cross-section in

$$\frac{\underline{E}_x}{\underline{H}_y} = -\frac{\underline{E}_y}{\underline{H}_x} = Z_0\,. \tag{8.14}$$

Furthermore

$$Z_0 = \begin{cases} Z_{\mathrm{H}} & \text{for TE modes} \\ Z_{\mathrm{E}} & \text{for TM modes} \end{cases} \tag{8.15}$$

is valid. Since also (7.107) and (7.114) have to be considered for waveguides with a rectangular cross-section as well as for waveguides with a circular cross-section, we obtain

$$\underline{\mathcal{H}}_{tr} = \frac{1}{Z_0} \star dz \wedge \underline{\mathcal{E}}_{tr}\,, \tag{8.16a}$$

$$\underline{\mathcal{E}}_{tr} = -Z_0 \star dz \wedge \underline{\mathcal{H}}_{tr}\,. \tag{8.16b}$$

We now are inserting (8.9a) and (8.16a) into (8.10) and obtain for the waveguide wave propagating in the positive z-direction the transmitted active power

$$P = \frac{1}{2Z_0}|\underline{V}|^2 \int_A \mathrm{e} \wedge \star (dz \wedge \mathrm{e})\,. \tag{8.17}$$

This yields

$$P = \frac{1}{2Z_0}|\underline{V}|^2 \int_A \left(e_x^2 + e_y^2\right)\, dx \wedge dy\,. \tag{8.18}$$

We now specify the structure function $\mathbf{e}$ to be real and normalize it according to

$$\int_A \left(e_x^2 + e_y^2\right)\, dx \wedge dy = 1\,. \tag{8.19}$$

Inserting (8.8b) and (8.16b) into (8.10) we obtain for the active power P transmitted in the positive z-direction

$$P = \frac{Z_0}{2}|\underline{I}|^2 \int_A \left(h_x^2 + h_y^2\right)\, dx \wedge dy\,. \tag{8.20}$$

We now normalize the real structure function $\mathbf{h}$ according to

$$\int_A \left(h_x^2 + h_y^2\right)\, dx \wedge dy = 1\,. \tag{8.21}$$

The differential forms of the electric and magnetic structure functions are related via

$$h = \star\, (\mathrm{d}z \wedge \mathrm{e})\,, \tag{8.22a}$$

$$\mathrm{e} = -\star\, (\mathrm{d}z \wedge h)\,. \tag{8.22b}$$

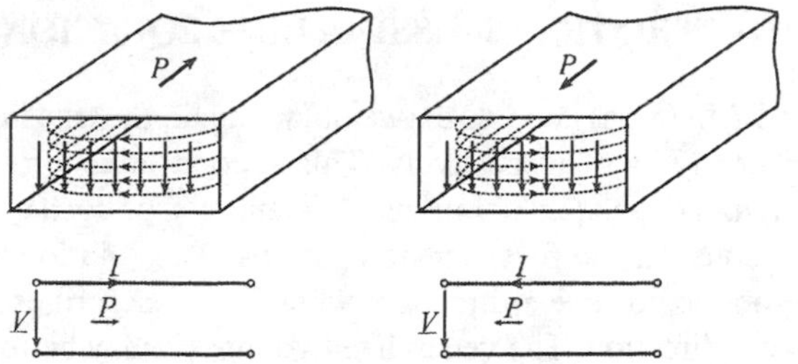

Figure 8.3: Relation of arrows and field orientation in a hollow waveguide.

From (8.22a), (8.22b) and (8.20) it follows that

$$\int_{A_k} e_k \wedge h_k = 1 \, . \tag{8.23}$$

Application of this normalization to the H_{10} mode of the rectangular waveguide yields

$$\underline{\mathcal{E}}_{tr} = \sqrt{\frac{2}{ab}}\, \underline{V} \sin \frac{\pi x}{a} dy \, . \tag{8.24a}$$

$$\underline{\mathcal{H}}_{tr} = \star dz \wedge \underline{\mathcal{E}}_{tr} = -\sqrt{\frac{2}{ab}}\, \underline{I} \sin \frac{\pi x}{a} dy \, . \tag{8.24b}$$

Concerning the structure functions we have only decided that they are real and that their absolute square is normalized. We may choose the sign of $\underline{V}$ and $\underline{I}$, and we have to decide how the signs of $\underline{V}$ and $\underline{I}$ shall be related to the direction of the electric and magnetic fields. Figure 8.3 shows the assignment chosen in the following for the relation between the arrow of the generalized voltage $\underline{V}$ and the generalized current $\underline{I}$ respectively and the direction of the field quantities in the H_{10} mode.

We note that for TEM modes the currents and voltages are defined in the conventional way via the line integrals over $\mathcal{H}$ and $\mathcal{E}$. Therefore the ratio of voltage and current differs from the ratio of transverse electric and magnetic field. The characteristic impedance for a TEM transmission line wave therefore is not identical with the wave impedance of the corresponding field. For TE and TM modes however we have defined generalized voltages and currents in such a way that the ratio of generalized voltage and generalized current is identical with the ratio of the transverse electric and magnetic fields. For the modes of hollow waveguides the characteristic impedance therefore is identical with the wave impedance.

8.3 SOLUTION OF THE TRANSMISSION LINE EQUATIONS

Eliminating $\underline{V}$ and $\underline{I}$ in (8.1a) and (8.1b), we obtain again the transmission line equation, see (7.74a) and (7.74a), respectively. This second order ordinary differential equation has the two independent solutions $e^{-\gamma z}$ and $e^{\gamma z}$ or $\cosh \gamma z$ and $\sinh \gamma z$, respectively. Let us consider the first pair of solutions. The solution $e^{-\gamma z}$ describes a wave propagating in the positive z-direction, whereas $e^{\gamma z}$ describes a wave propagating in the negative z-direction. The generalized voltage V may be represented by

$$\underline{V}(z) = \underline{V}^{(+)} \mathrm{e}^{-\gamma z} + \underline{V}^{(-)} \mathrm{e}^{\gamma z}, \tag{8.25}$$

where $\underline{V}^{(+)}$ and $\underline{V}^{(-)}$ are the complex amplitudes of the wave propagating in the positive and negative z-direction, respectively. Introducing (8.25) into (8.1a) we obtain

$$\underline{I}(z) = \frac{1}{Z_0}(\underline{V}^{(+)} \mathrm{e}^{-\gamma z} - \underline{V}^{(-)} \mathrm{e}^{\gamma z}). \tag{8.26}$$

In the same way as $\underline{V}$, the generalized current $\underline{I}$ may be represented by

$$\underline{I}(z) = \underline{I}^{(+)} \mathrm{e}^{-\gamma z} + \underline{I}^{(-)} \mathrm{e}^{\gamma z} \tag{8.27}$$

where $\underline{I}^{(+)}$ and $\underline{I}^{(-)}$ are the complex amplitudes of the electromagnetic waves propagating in the positive and negative set direction respectively. Comparing (8.26) and (8.27) we obtain

$$\underline{I}^{(+)} = \frac{\underline{V}^{(+)}}{Z_0}, \tag{8.28a}$$

$$\underline{I}^{(-)} = -\frac{\underline{V}^{(-)}}{Z_0}. \tag{8.28b}$$

If an electromagnetic wave is propagating only in one direction the ratio between voltage and current is given by Z_0 and is independent from z. If we know voltage and current at a certain point of the line, e.g., at $z = 0$, and want to compute voltage and current at any other point of the line it is useful to represent the solution of the transmission line equation in terms of $\cosh \gamma z$ and $\sinh \gamma z$. For the voltage $\underline{V}$ we obtain the general solution

$$\underline{V}(z) = \underline{A} \cosh \gamma z + \underline{B} \sinh \gamma z \tag{8.29}$$

where $\underline{A}$ and $\underline{B}$ are complex amplitudes. With (8.1a) we obtain

$$\underline{I}(z) = -\frac{1}{Z_0}(\underline{A} \sinh \gamma z + \underline{B} \cosh \gamma z). \tag{8.30}$$

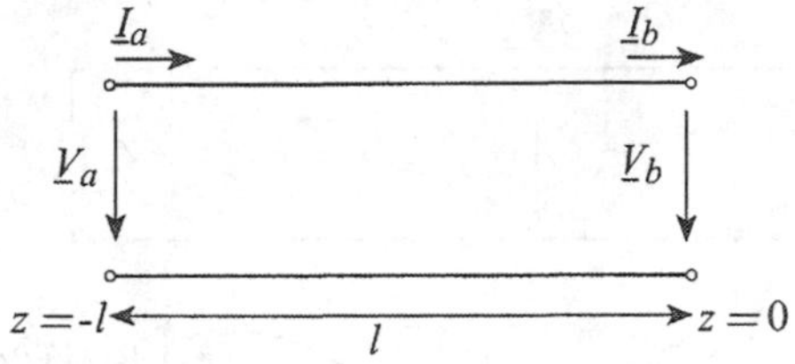

Figure 8.4: Line segment of length l.

At $z = 0$ voltage and current are given by

$$\underline{V}(z=0) = \underline{A}\,, \tag{8.31a}$$

$$\underline{I}(z=0) = -\frac{\underline{B}}{Z_0}\,. \tag{8.31b}$$

We therefore can express $\underline{A}$ and $\underline{B}$ by $\underline{V}$ and $\underline{I}$ and obtain

$$\underline{V}(z) = \underline{V}(z=0)\cosh\gamma z - Z_0\underline{I}(z=0)\sinh\gamma z\,, \tag{8.32a}$$

$$\underline{I}(z) = -\frac{\underline{V}(z=0)}{Z_0}\sinh\gamma z + \underline{I}(z=0)\cosh\gamma z\,. \tag{8.32b}$$

Since we may choose the point $z = 0$ at any position of the line we may calculate from $\underline{V}(0)$ and $\underline{I}(0)$, the voltage $\underline{V}(z)$ and the current $\underline{I}(z)$ at any other position z on the line. Let us consider a line segment of length l as depicted in Figure 8.4. Let $\underline{V}_a$ and $\underline{I}_a$ be voltage and current at the input of the line segment and $\underline{V}_b$ and $\underline{I}_b$ the voltage and the current at the output of the line segment. We can compute $\underline{V}_a$ and $\underline{I}_a$ as a function of $\underline{V}_b$ and $\underline{I}_b$. From (8.32a) and (8.32b) choosing $z = 0$ at the output, at the line input we then obtain $z = -l$.

$$\underline{V}_a = \underline{V}_b\cosh\gamma l + Z_0\underline{I}_b\sinh\gamma l\,, \tag{8.33a}$$

$$\underline{I}_a = \frac{\underline{V}_b}{Z_0}\sinh\gamma l + \underline{I}_b\cosh\gamma l\,. \tag{8.33b}$$

The hyperbolic functions with the complex argument γl may be represented by

$$\sinh\gamma l = \sinh\alpha l\cos\beta l + \mathrm{j}\cosh\alpha l\sin\beta l\,, \tag{8.34a}$$

$$\cosh\gamma l = \cosh\alpha l\cos\beta l + \mathrm{j}\sinh\alpha l\sin\beta l\,. \tag{8.34b}$$

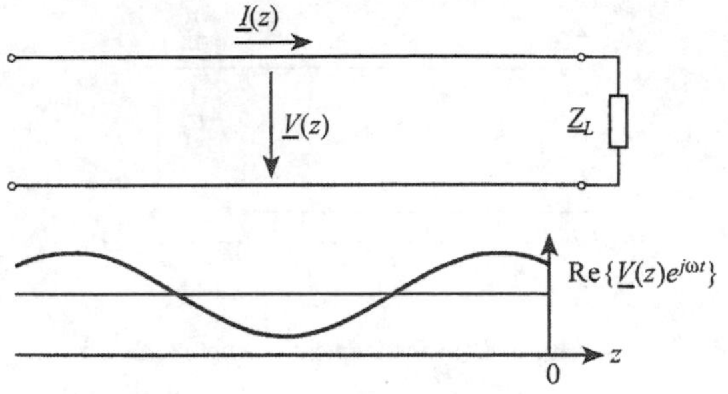

Figure 8.5: Line terminated with Z_L.

For the lossless line with the characteristic impedance Z_0 and the phase coefficient α we obtain

$$\underline{V}_a = \underline{V}_b \cos\beta l + \mathrm{j}\, Z_0 \underline{I}_b \sin\beta l\,, \tag{8.35a}$$

$$\underline{I}_a = \mathrm{j}\frac{\underline{V}_b}{Z_0} \sin\beta l + \underline{I}_b \cos\beta l\,. \tag{8.35b}$$

Considering a line terminated with an impedance Z_L at the end of the line at $z = 0$ according to Figure 8.5 we obtain the relation

$$\underline{V}_b = Z_L \underline{I}_b\,, \tag{8.36a}$$

$$Z_a = \frac{\underline{V}_a}{\underline{I}_a}\,. \tag{8.36b}$$

After inserting (8.33a) and (8.33b) we obtain

$$\frac{Z_a}{Z_0} = \frac{Z_L + Z_0 \tanh\gamma l}{Z_0 + Z_L \tanh\gamma l}\,. \tag{8.37}$$

For complex γ we use

$$\tanh\gamma l = \frac{\tanh\alpha l + \mathrm{j}\tan\beta l}{1 + \mathrm{j}\tanh\alpha l \tan\beta l}\,. \tag{8.38}$$

For the lossless line we therefore obtain

$$\frac{Z_a}{Z_0} = \frac{Z_L + \mathrm{j}\, Z_0 \tan\beta l}{Z_0 + \mathrm{j}\, Z_L \tan\beta l}\,. \tag{8.39}$$

8.4 WAVE AMPLITUDES

We already have shown that a transmission line wave of a certain mode may be described by two complex amplitudes $\underline{V}(z)$ and $\underline{I}(z)$. If the wave is propagating in one direction only, it may be completely described either by the voltage amplitude $\underline{V}(z)$ or the current amplitude $\underline{I}(z)$, since for a wave propagating in the positive z-direction the voltage to current ratio is given by $\underline{V}(z)/\underline{I}(z) = Z_0$ and for a wave propagating in the negative z-direction the voltage to current ratio is given by $\underline{V}(z)/\underline{I}(z) = -Z_0$. Instead of voltage $\underline{V}(z)$ and current $\underline{I}(z)$ we may also use the amplitudes of the electromagnetic waves propagating in the positive and negative z-direction, respectively. In the following we assume the line to be lossless and the characteristic impedance to be real and define the *wave amplitudes* $\underline{a}(z)$ and $\underline{b}(z)$ via

$$\underline{a}(z) = \frac{1}{2\sqrt{2Z_0}}[\underline{V}(z) + Z_0\underline{I}(z)], \tag{8.40a}$$

$$\underline{b}(z) = \frac{1}{2\sqrt{2Z_0}}[\underline{V}(z) - Z_0\underline{I}(z)]. \tag{8.40b}$$

After inserting (8.25) and (8.26), we obtain

$$\underline{a}(z) = \frac{1}{\sqrt{2Z_0}}\underline{V}^{(+)}\,\mathrm{e}^{-\mathrm{j}\beta z}, \tag{8.41a}$$

$$\underline{b}(z) = \frac{1}{\sqrt{2Z_0}}\underline{V}^{(-)}\,\mathrm{e}^{\mathrm{j}\beta z}. \tag{8.41b}$$

These equations show that $\underline{a}(z)$ describes the wave propagating in the positive z-direction, whereas $\underline{b}(z)$ describes a wave propagating in the negative z-direction. The voltage $\underline{V}(z)$ and the current $\underline{I}(z)$ may be expressed by the wave amplitudes $\underline{a}(z)$ and $\underline{b}(z)$ as follows

$$\underline{V}(z) = \sqrt{2Z_0}[\underline{a}(z) + \underline{b}(z)], \tag{8.42a}$$

$$\underline{I}(z) = \sqrt{\frac{2}{Z_0}}[\underline{a}(z) - \underline{b}(z)]. \tag{8.42b}$$

From (8.1a), (8.1b), (8.4a), (8.4b), (8.40a) and (8.40b) we obtain the transmission line equations for the wave amplitudes $\underline{a}(z)$ and $\underline{b}(z)$

$$\frac{\mathrm{d}\underline{a}}{\mathrm{d}z} = -\mathrm{j}\beta\underline{a}, \tag{8.43a}$$

$$\frac{\mathrm{d}\underline{b}}{\mathrm{d}z} = \mathrm{j}\beta\underline{b}. \tag{8.43b}$$

The transmission line equations for the wave amplitudes $\underline{a}(z)$ and $\underline{b}(z)$ are not coupled. The transmission line waves propagating in the positive and negative z-direction are propagating independently. The solutions of (8.43a) and (8.43b) are given by

$$\underline{a}(z) = \underline{a}(z = 0)\, \mathrm{e}^{-\mathrm{j}\beta z}\,, \tag{8.44a}$$

$$\underline{b}(z) = \underline{b}(z = 0)\, \mathrm{e}^{\mathrm{j}\beta z}\,. \tag{8.44b}$$

The active power transmitted through the line in the positive z-direction is given by

$$P(z) = \frac{1}{2}\Re\{\underline{V}(z)\underline{I}^{\star}(z)\} = \Re\{|\underline{a}(z)|^2 - |\underline{b}(z)|^2 + \underline{a}^{\star}(z)\underline{b}(z) - \underline{a}(z)\underline{b}^{\star}(z)\}\,. \tag{8.45}$$

From this it follows

$$P(z) = |\underline{a}(z)|^2 - |\underline{b}(z)|^2\,. \tag{8.46}$$

The first term $|\underline{a}(z)|^2$ describes the power transmitted by the wave $\underline{a}(z)$ in the positive z-direction, whereas the second term $|\underline{b}(z)|^2$ describes the power transmitted by the wave $\underline{b}(z)$ in the negative z-direction. We have normalized $\underline{a}(z)$ and $\underline{b}(z)$ such that their absolute squares describe the active power carried by the waves. We have defined the wave amplitudes only for lossless lines. This definition may also be used for lines with small losses where the complex characteristic impedance of the transmission line can be approximated by a real characteristic impedance Z_0. In (8.41a), (8.41b) and (8.43a) – (8.44b) we have to replace $\mathrm{j}\beta$ with γ. This approximation is justified since the characteristic impedance of a transmission line with weak losses only differs slightly from the characteristic impedance of a lossless line. In the treatment of waveguides and waveguide circuits the use of wave amplitudes $\underline{a}(z)$ and $\underline{b}(z)$ is more common than the use of generalized voltages and generalized currents, with the reason being that for waveguides wave amplitudes are physically descriptive and, furthermore, may be measured directly. Numerous simple design methods for microwave circuits are based on the wave amplitude description. The introduction of generalized voltages and currents, however, allows the application of common network theoretic design methods. Therefore, one always will choose the representations of the integral field quantities, which are better suited for either measurement or analysis and change the representation, if necessary.

8.5 REFLECTION COEFFICIENT AND SMITH CHART

The lossless line terminated at the end exhibits an input impedance as described by (8.39). The input impedance gives the ratio of generalized voltage $\underline{V}$ and generalized current $\underline{I}$ at the input of the line. We also may describe the impedance by the ratio of

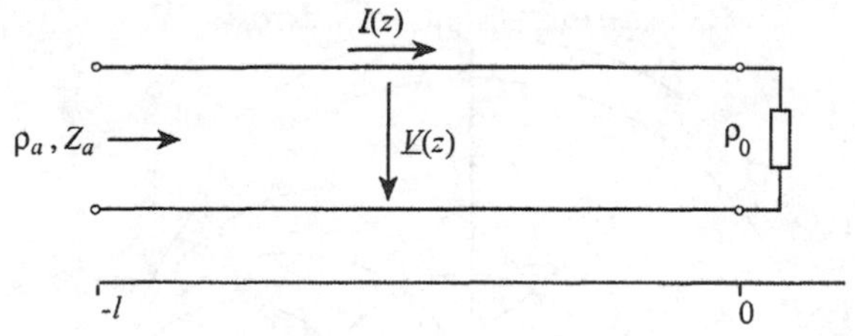

Figure 8.6: Line terminated with ρ_0.

the complex amplitudes of the incident wave and the reflected wave. We name this ratio the *reflection coefficient* ρ defined by

$$\rho(z) = \frac{\underline{b}(z)}{\underline{a}(z)} . \tag{8.47}$$

Let us assume the reflection coefficient ρ at $z = 0$ is given by

$$\rho_0 = \frac{\underline{b}(0)}{\underline{a}(0)} . \tag{8.48}$$

With (8.44a) and (8.44b) we obtain

$$\rho(z) = \rho_0 \, e^{2j\beta z} . \tag{8.49}$$

For a line segment of length l according to Figure 8.6 that is terminated at the end with ρ_0, we obtain from (8.49) the reflection coefficient ρ_i at the line input

$$\rho_a = \rho(z = -l) = \rho_0 \, e^{-2j\beta l} . \tag{8.50}$$

From (8.42a) and (8.42b) we obtain the input impedance Z_a of the terminated line

$$Z_a = \frac{\underline{V}_a}{\underline{I}_a} = Z_0 \frac{1 + \rho_a}{1 - \rho_a} . \tag{8.51}$$

From this we obtain the following relations between impedance and reflection coefficient

$$\frac{Z}{Z_0} = \frac{1 + \rho}{1 - \rho} , \tag{8.52}$$

$$\rho = \frac{Z - Z_0}{Z + Z_0} . \tag{8.53}$$

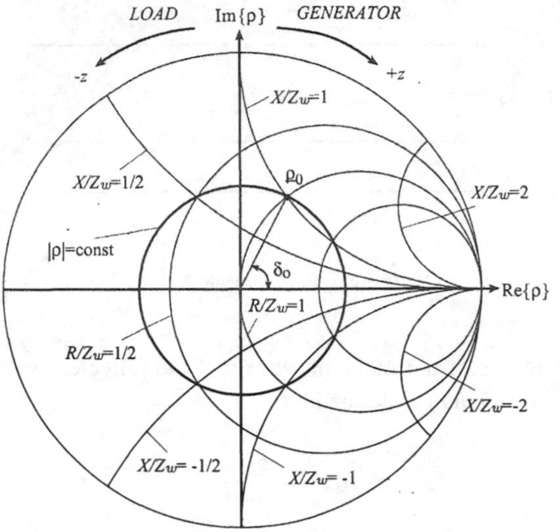

Figure 8.7: The Smith chart.

The description of the terminated lines by reflection coefficient at the input is equivalent to the description by the input impedance. Comparing (8.39) and (8.49) we see that the relation between the reflection coefficient and length of the line has a much simpler form than the relation between line impedance and line length. If the reflection coefficient ρ is given at a certain point of the line, and if we propagate along the line in the positive or negative z-direction, the reflection coefficient ρ is moving on a circle with center zero in the complex ρ plane. The graphical representation of the impedance transformation properties of a line is given by the *Smith chart* [4], depicted in Figure 8.7. The point with

$$\rho_0 = |\rho_0|\, e^{j\delta_0} \tag{8.54}$$

is marked in the diagram. Increasing the distance from the line termination means to move in the negative z-direction. According to (8.49) and (8.50) in this case we move clockwise on a circle ρ = const. Moving by half the line wavelength $\lambda/2$ corresponds to a complete rotation of 360° in the Smith chart. Mapping the coordinate grid of the Z plane into the ρ plane yields a circular coordinate grid as depicted in Figure 8.7. Any rational transformation in the complex plane is mapping circles into circles. Straight lines also belong to the set of circles, since a straight line may be considered a circle through the infinite point. It is useful to normalize the mapping between the impedance and reflections coefficient with respect to the characteristic impedance Z_0. We obtain the orthogonal circular coordinate grid for X = const. and R = const., respectively. This diagram with the Z coordinate grid drawn in the ρ plane is called the *Smith chart.* The Smith chart allows one in an easy way to determine the input impedance of a line terminated at the end and the graphic solution of impedance matching problems.

If the line is terminated with the wave impedance Z_0, i.e., $Z_L = Z_0$, we obtain from (8.53) the input reflection coefficient $\rho_0 = 0$. If the line is terminated with its characteristic impedance no wave will be reflected from its end. Due to (8.49) we have $\rho = 0$ over the whole line length and therefore also at the input of the line no wave is reflected. In the case of a non-reflecting terminated line we have power matching. For $\rho_0 \neq 0$ the magnitudes $|\underline{V}(z)|$ and $|\underline{I}(z)|$ depend on z. To determine the z dependence of the magnitude of the voltage, we insert (8.49) and (8.54) into (8.42a) and obtain

$$|\underline{V}(z)| = |\underline{a}(0)|\sqrt{2Z_0[1 + |\rho_0|^2 + 2|\rho_0|\cos(2\beta z + \delta_0)]}\,. \tag{8.55}$$

In the same way we obtain from (8.42b) the absolute value of the current

$$|\underline{I}(z)| = |\underline{a}(0)|\sqrt{\frac{2}{Z_0}[1 + |\rho_0|^2 - 2|\rho_0|\cos(2\beta z + \delta_0)]}\,. \tag{8.56}$$

The z dependence of $|\underline{V}(z)|$ and $|I(z)|$ is demonstrated in Figure 8.8. The magnitude of the voltage $|\underline{V}(z)|$ assumes its maximum value $|\underline{V}(z)|_{\text{max}}$ at z_a and its minimum value $|\underline{V}(z)|_{\text{min}}$ at z_b. From (8.55) and (8.56) we obtain for the maximum and minimum magnitudes of $|\underline{V}(z)|$ and $|\underline{I}(z)|$:

$$V_{\text{max}} = Z_0 I_{\text{max}} = \sqrt{2Z_0}|\underline{a}(0)|(1 + |\rho_0|)\,, \tag{8.57a}$$

$$V_{\text{min}} = Z_0 I_{\text{min}} = \sqrt{2Z_0}|\underline{a}(0)|(1 - |\rho_0|)\,. \tag{8.57b}$$

We define the *standing wave ratio* (SWR) as

$$s = \frac{V_{\text{max}}}{V_{\text{min}}} = \frac{I_{\text{max}}}{I_{\text{min}}}\,. \tag{8.58}$$

The standing wave ratio s is a real number such that $1 \leq s \leq \infty$. This quantity also is called *voltage standing wave ratio* (VSWR). With (8.52), (8.57a) and (8.57b) it follows that

$$s = \frac{Z(z_a)}{Z_0} = \frac{Z_0}{Z(z_b)}\,. \tag{8.59}$$

From measuring the z dependence of the magnitude of the voltage $|\underline{V}(z)|$ with a potential probe we can determine the reflection coefficient. From the ratio of the maximum magnitude and the minimum magnitude of the voltage we can determine

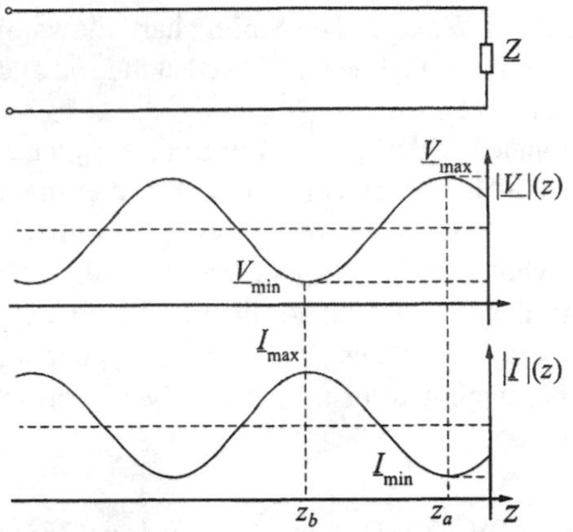

Figure 8.8: Variation of $|\underline{V}(z)|$ and $|\underline{I}(z)|$ along the transmission line.

the magnitude of the reflection coefficient, from the position z_a or z_b, respectively; using (8.55) we can determine δ_0 and thereby the phase ρ_0. This determination may also be performed using the Smith chart since z_a and z_b are positioned on the $X = 0$ axis of the Smith chart. The right intersection of the circle $\rho =$ const. with the $X = 0$ axis yields Z_a, whereas the left intersection of the circle $\rho =$ const. with the $X = 0$ axis yields Z_b. From the Smith chart we may compute not only the complex impedance Z, but also the complex admittance Y. The *normalized impedance* $\underline{z}$ and the *normalized admittance* $\underline{y}$ are given by

$$\underline{z} = \frac{Z}{Z_0}, \quad (8.60a) \qquad \underline{y} = YZ_0 . \quad (8.60b)$$

The normalized impedance $\underline{z}$ and the normalized admittance $\underline{y}$ are related by

$$\underline{y} = \underline{z}^{-1} . \tag{8.61}$$

The relation between the reflection coefficient ρ and the normalized impedance $\underline{z}$ and the normalized $\underline{y}$, respectively, according to (8.53), (8.60b) and (8.61) is given by

$$\rho = \frac{\underline{z} - 1}{\underline{z} + 1}, \tag{8.62a}$$

$$-\rho = \frac{\underline{y} - 1}{\underline{y} + 1} . \tag{8.62b}$$

We can use the coordinate grid of the Smith chart for the normalized impedance $\underline{z}$ as well as for the normalized admittance $\underline{y}$. If we consider that transforming from the

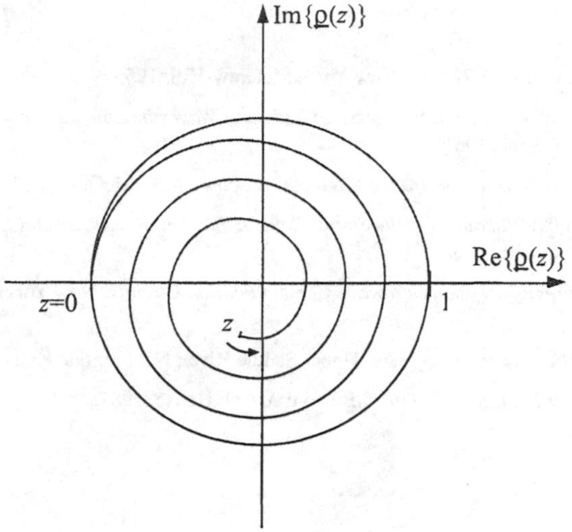

Figure 8.9: Reflection coefficient of a lossy line terminated with ($\rho_0 = -1$).

normalized impedance to the normalized admittance means the change of sign of the reflection coefficient and therefore mirroring at ρ the origin of the Smith chart.

The Smith chart provides information about circuits containing transmission line elements and is a valuable tool for microwave circuit design [5, 6]. Up to now we have assumed the transmission line to be lossless. If the transmission line is lossy, however, we have to replace (8.50) with

$$\rho_a = \rho_0 \, \mathrm{e}^{-2\gamma l} \, . \tag{8.63}$$

Propagating on a lossy line from the load towards the generator, we are moving in the Smith chart on a logarithmic spiral instead of a circle (Figure 8.9). For a long lossy line the input impedance converges to the characteristic impedance of the line also if the line is not terminated with its characteristic impedance. With a lossy line we cannot perform extreme impedance transformations, since normalized impedances with very large magnitudes as well as normalized impedances with very small magnitudes are located in the Smith chart close to the point $\rho = \pm 1$. With the spiral we cannot transform to such points. A short circuit, for example, cannot be transformed with a lossy $\lambda/4$ line into an open circuit with $\rho = 1$, but is transformed due to (8.63) into a finite impedance with $\rho = \mathrm{e}^{-\alpha\lambda/2}$. A detailed treatment of lossy transmission lines is given in [7].

REFERENCES

[1] R. King, *Transmission-Line Theory*. New York: McGraw-Hill, 1955.

[2] S. Ramo, J. R. Whinnery, and T. van Duzer, *Fields and Waves in Communication Electronics*. New York: John Wiley & Sons, 1965.

[3] R. E. Collin, *Foundations of Microwave Engineering*. New York: McGraw-Hill, 1992.

[4] P. Smith, *Electronic Applications of the Smith Chart in Waveguide, Circuit and Components Analysis*. New York: McGraw-Hill, 1969.

[5] D. Misra, *Radio-Frequency and Microwave Communication Circuits*. New York: John Wiley & Sons, 2001.

[6] M. Radmanesh, *Microwave Electronics*. Upper Saddle River, NJ: Prentice Hall PTR, 2001.

[7] F. E. Gardiol, *Lossy Transmission Lines*. Boston: Artech House, 1987.

Chapter 9

Resonant Circuits and Resonators

Resonant circuits and *resonators* are important circuit elements in radio-frequency engineering. A resonant circuit is formed by a capacitor and an inductor connected in parallel or in series. As components in active and passive circuits, resonant circuits and resonators allow a selective transmission or blocking of signals. In oscillators these components serve as the frequency-determining elements.

9.1 THE LINEAR PASSIVE ONE-PORT

A circuit element with a single port is called a *one-port*. Figure 9.1 shows the schematic drawing and the equivalent circuit of a one-port. Incident and reflected waves are related to a given transverse plane of reference. The signal at the port is described either by the complex amplitudes $\underline{a}$ and $\underline{b}$ of the incident and scattered wave or by the voltage $\underline{V}$ and current $\underline{I}$. If the relation between $\underline{V}$ and $\underline{I}$ or $\underline{a}$ and $\underline{b}$ respectively is known, the one-port is defined as a circuit element. If the relation between $\underline{V}$ and $\underline{I}$ or $\underline{a}$ and $\underline{b}$ respectively is linear, the one-port is named a linear one-port. The complex power P_c flowing into the one-port is given by

$$P_c = -\int_A \mathcal{T} \, . \tag{9.1}$$

The integration is performed over the cross-sectional area A of the waveguide port in the plane of reference. With (4.21), (8.9a), (8.9b) and (8.22a) we obtain a complex power P_c flowing into the one-port:

$$P_c = \frac{1}{2}\underline{V}\,\underline{I}^* \, . \tag{9.2}$$

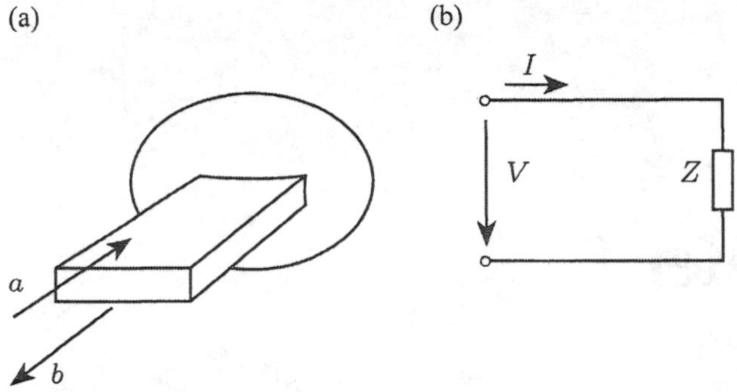

Figure 9.1: One-port: (a) schematic drawing and (b) equivalent circuit.

The real part of the complex power P_c is the active power P. The imaginary part is the reactive power P_r:

$$P_c = P + \mathrm{j} P_r \,. \tag{9.3}$$

Using (8.42a) and (8.42b) we express the complex power P_c by wave amplitudes a and b:

$$P_c = |a|^2 - |b|^2 + a^* b - a b^* \,. \tag{9.4}$$

For $P = 0$ the one-port is lossless. A lossless one-port also is called a reactive one-port. For the passive one-port we obtain $|\rho| \leq 1$. For the reactive one-port $|\rho| = 1$ is valid. For the source-free one-port we obtain from (4.39) and (9.1)

$$P_c = P + 2\mathrm{j}\omega(W_m - W_e) \,, \tag{9.5}$$

where W_m is the average stored magnetic energy and W_e is the average stored electric energy. The momentary values of W_e and W_m oscillate between zero and their maximum value with the double oscillation frequency. For $W_e = W_m$ within a quarter of the period of oscillation the stored magnetic energy is completely transformed in electric energy and vice versa. For $W_m \neq W_e$ a periodical energy exchange occurs also with an external circuit driving the resonant circuit. In this case reactive power is flowing through the port. The ratio of voltage and current or generalized voltage and generalized current is given by the complex impedance Z or the complex admittance Y respectively. It follows from (9.2) that

$$P_c = \frac{1}{2} Z |\underline{I}|^2 = \frac{1}{2} Y^* |\underline{V}|^2 \,. \tag{9.6}$$

With (9.5) we obtain

$$Z = \frac{P + 2j\omega(W_m - W_e)}{\frac{1}{2}|\underline{I}|^2}, \quad (9.7a) \qquad Y = \frac{P + 2j\omega(W_e - W_m)}{\frac{1}{2}|\underline{V}|^2}. \quad (9.7b)$$

The reactance X and the susceptance B are

$$X = \frac{4\omega}{|\underline{I}|^2}(W_m - W_e), \quad (9.8a) \qquad B = \frac{4\omega}{|\underline{V}|^2}(W_e - W_m). \quad (9.8b)$$

9.2 THE REACTANCE THEOREM

Foster's reactance theorem [1–3] imposes a condition on the frequency dependence of a reactance. From the complex Maxwell equations (2.144) and (2.145) we obtain for real permittivity, real permeability and no sources

$$\mathrm{d}\underline{\mathcal{H}} = \mathrm{j}\omega\varepsilon \star \underline{\mathcal{E}}, \quad (9.9a)$$

$$\mathrm{d}\underline{\mathcal{E}} = -\mathrm{j}\omega\mu \star \underline{\mathcal{H}}. \quad (9.9b)$$

The partial derivative of these equations with respect to ω is given by

$$\mathrm{d}\frac{\partial \underline{\mathcal{H}}}{\partial\omega} = \mathrm{j}\varepsilon \star \underline{\mathcal{E}} + \mathrm{j}\omega\varepsilon \star \frac{\partial \underline{\mathcal{E}}}{\partial\omega}, \quad (9.10a)$$

$$\mathrm{d}\frac{\partial \underline{\mathcal{E}}}{\partial\omega} = -\mathrm{j}\mu \star \underline{\mathcal{H}} - \mathrm{j}\omega\mu \star \frac{\partial \underline{\mathcal{H}}}{\partial\omega}. \quad (9.10b)$$

We now compute the following expressions

$$\mathrm{d}\left(\frac{\partial \underline{\mathcal{H}}}{\partial\omega} \wedge \underline{\mathcal{E}}^*\right) = \mathrm{j}\varepsilon\,(\star\,\underline{\mathcal{E}}) \wedge \underline{\mathcal{E}}^* + \mathrm{j}\omega\varepsilon\left(\star\frac{\partial \underline{\mathcal{E}}}{\partial\omega}\right) \wedge \underline{\mathcal{E}}^* - \mathrm{j}\omega\mu\frac{\partial \underline{\mathcal{H}}}{\partial\omega} \wedge (\star\,\underline{\mathcal{H}}^*), \quad (9.11a)$$

$$\mathrm{d}\left(\frac{\partial \underline{\mathcal{E}}}{\partial\omega} \wedge \underline{\mathcal{H}}^*\right) = -\mathrm{j}\mu\,(\star\,\underline{\mathcal{H}}) \wedge \underline{\mathcal{H}}^* - \mathrm{j}\omega\mu\left(\star\frac{\partial \underline{\mathcal{H}}}{\partial\omega}\right) \wedge \underline{\mathcal{H}}^* + \mathrm{j}\omega\varepsilon\frac{\partial \underline{\mathcal{E}}}{\partial\omega} \wedge (\star\,\underline{\mathcal{E}}^*). \quad (9.11b)$$

From these equations and (4.28) and (4.29) we obtain

$$\mathrm{d}\left(\frac{\partial \underline{\mathcal{E}}}{\partial\omega} \wedge \underline{\mathcal{H}}^*\right) - \mathrm{d}\left(\frac{\partial \underline{\mathcal{H}}}{\partial\omega} \wedge \underline{\mathcal{E}}^*\right) = -8\mathrm{j}\left(\overline{\mathcal{W}}_e + \overline{\mathcal{W}}_m\right). \quad (9.12)$$

Integrating this equation over a volume V with the boundary ∂V and applying Stokes' theorem (A.88) yields

$$\oint_{\partial V}\left(\frac{\partial \underline{\mathcal{E}}}{\partial\omega} \wedge \underline{\mathcal{H}}^* - \frac{\partial \underline{\mathcal{H}}}{\partial\omega} \wedge \underline{\mathcal{E}}^*\right) = -8\mathrm{j}\int_V \left(\overline{\mathcal{W}}_e + \overline{\mathcal{W}}_m\right). \quad (9.13)$$

We apply this relation to the one-port depicted in Figure 9.1(a). The boundary surface ∂V is assumed to enclose the complete one-port. The reference plane A of the waveguide defining the port is assumed to be a part of the boundary ∂V. Therefore on the left-hand side of (9.13) we only need to perform the integration over the cross-sectional area A of the waveguide. We consider that the orientation of the boundary surface ∂V of the volume V is outwards whereas the cross-sectional area A of the waveguides is oriented inwards. Therefore we have to change the sign on the left-hand side of (9.13) when we are replacing ∂V by A_i. Since A is a transverse plane, $\underline{\mathcal{E}}$ and $\underline{\mathcal{H}}$ may be replaced by their transverse components $\underline{\mathcal{E}}_{tr}$ and $\underline{\mathcal{H}}_{tr}$ as introduced in (8.6a) and (8.6b). From (8.9a) and (8.9b) we obtain

$$\underline{\mathcal{E}}_{tr}(\boldsymbol{x}) = \underline{V}\, e(u, v) , \tag{9.14a}$$

$$\underline{\mathcal{H}}_{tr}(\boldsymbol{x}) = \underline{I}\, h(u, v) , \tag{9.14b}$$

$$\frac{\partial}{\partial \omega}\underline{\mathcal{E}}_{tr}(\boldsymbol{x}) = \frac{\mathrm{d}\underline{V}}{\mathrm{d}\omega}\, e(u, v) , \tag{9.14c}$$

$$\frac{\partial}{\partial \omega}\underline{\mathcal{H}}_{tr}(\boldsymbol{x}) = \frac{\mathrm{d}\underline{I}}{\mathrm{d}\omega}\, h(u, v) \tag{9.14d}$$

where u, v are the transverse coordinates in the port plane. From (9.14a) to (9.14d) and (8.23) it follows that

$$\oint_{\partial V} \frac{\partial \underline{\mathcal{E}}}{\partial \omega} \wedge \underline{\mathcal{H}}^* = -\oint_A \frac{\partial \underline{\mathcal{E}}_{tr}}{\partial \omega} \wedge \underline{\mathcal{H}}_{tr}^* = -\frac{\mathrm{d}\underline{V}}{\mathrm{d}\omega}\underline{I}^* , \tag{9.15a}$$

$$\oint_{\partial V} \frac{\partial \underline{\mathcal{H}}}{\partial \omega} \wedge \underline{\mathcal{E}}^* = -\oint_A \frac{\partial \underline{\mathcal{H}}_{tr}}{\partial \omega} \wedge \underline{\mathcal{E}}_{tr}^* = \frac{\mathrm{d}\underline{I}}{\mathrm{d}\omega}\underline{V}^* . \tag{9.15b}$$

Inserting this into (9.13) we obtain

$$\frac{\mathrm{d}\underline{V}}{\mathrm{d}\omega}\underline{I}^* + \frac{\mathrm{d}\underline{I}}{\mathrm{d}\omega}\underline{V}^* = 8\mathrm{j} \int_V \left(\overline{\mathcal{W}}_e + \overline{\mathcal{W}}_m\right) = 8\mathrm{j}\,(W_e + W_m) . \tag{9.16}$$

From

$$\underline{V} = \mathrm{j}X\underline{I}, \qquad \underline{I} = \mathrm{j}B\underline{V} \tag{9.17}$$

we obtain

$$\frac{\mathrm{d}\underline{V}}{\mathrm{d}\omega} = \mathrm{j}\frac{\mathrm{d}X}{\mathrm{d}\omega}\underline{I}\Big|_{\underline{I}=\mathrm{const.}} , \qquad \frac{\mathrm{d}\underline{I}}{\mathrm{d}\omega} = \mathrm{j}\frac{\mathrm{d}B}{\mathrm{d}\omega}\underline{V}\Big|_{\underline{V}=\mathrm{const.}} . \tag{9.18}$$

and

$$\frac{\mathrm{d}X}{\mathrm{d}\omega}|\underline{I}|^2 = \frac{\mathrm{d}B}{\mathrm{d}\omega}|\underline{V}|^2 . \tag{9.19}$$

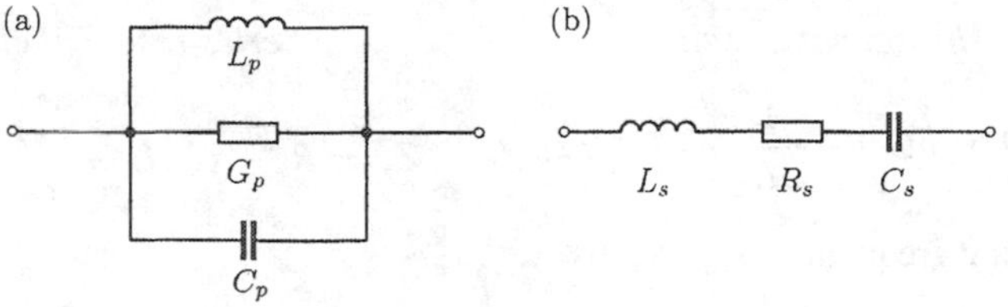

Figure 9.2: Resonant circuit (a) parallel circuit, (b) series circuit.

Inserting (9.18) and (9.19) into (9.16) yields

$$\frac{\mathrm{d}X}{\mathrm{d}\omega}|\underline{I}|^2 = 4\,(W_e + W_m)\ , \tag{9.20a}$$

$$\frac{\mathrm{d}B}{\mathrm{d}\omega}|V|^2 = 4\,(W_e + W_m)\ . \tag{9.20b}$$

Together with (9.8a) and (9.8b) this yields

$$W_e = \frac{1}{8}\left(\frac{\mathrm{d}X}{\mathrm{d}\omega} - \frac{X}{\omega}\right)|\underline{I}|^2 = \frac{1}{8}\left(\frac{\mathrm{d}B}{\mathrm{d}\omega} + \frac{B}{\omega}\right)|\underline{V}|^2\,, \tag{9.21a}$$

$$W_m = \frac{1}{8}\left(\frac{\mathrm{d}X}{\mathrm{d}\omega} + \frac{X}{\omega}\right)|\underline{I}|^2 = \frac{1}{8}\left(\frac{\mathrm{d}B}{\mathrm{d}\omega} - \frac{B}{\omega}\right)|\underline{V}|^2\,. \tag{9.21b}$$

Since W_e and W_m are non-negative it follows for passive lossless one-ports

$$\frac{dX}{d\omega} > \left|\frac{X}{\omega}\right|\,, \tag{9.22a}$$

$$\frac{dB}{d\omega} > \left|\frac{B}{\omega}\right|\,. \tag{9.22b}$$

This is called *Foster's reactance theorem.*

9.3 RESONANT CIRCUITS

In Figure 9.2 the *parallel resonant circuit* and the *series resonant circuit* are depicted. Parallel and serial are mutually dual. Therefore we will treat them in the following in parallel. The admittance Y_p of the parallel resonant circuit and the impedance Z_s of the series resonant circuit are given by

Parallel resonant circuit *Series resonant circuit*

$$Y_p = G_p + \mathrm{j}(\omega C_p - \frac{1}{\omega L_p}), \quad (9.23a) \qquad Z_s = R_s + \mathrm{j}(\omega L_s - \frac{1}{\omega C_s}). \quad (9.23b)$$

At the resonant frequency ω_0 given by

Parallel resonant circuit *Series resonant circuit*

$$\omega_0 = \frac{1}{\sqrt{L_p C_p}}, \quad (9.24a) \qquad \omega_0 = \frac{1}{\sqrt{L_s C_s}} \quad (9.24b)$$

the admittance Y_p and Z_s respectively are real.

Parallel resonant circuit

For constant impressed current the parallel resonant circuit exhibits a voltage maximum at the resonant frequency.

Series resonant circuit

For impressed voltage a series resonant circuit exhibits a current maximum at the resonant frequency.

The complex power P_c flowing into the resonant circuit and the active power P are given by

Parallel resonant circuit *Series resonant circuit*

$$P_c = \frac{1}{2} Y_p^* |\underline{V}|^2, \quad (9.25a) \qquad P_c = \frac{1}{2} Z_s |\underline{I}|^2, \quad (9.25b)$$

$$P = \frac{1}{2} G_p |\underline{V}|^2, \quad (9.26a) \qquad P = \frac{1}{2} R_s |\underline{I}|^2. \quad (9.26b)$$

The average electric energy W_e stored in the capacitor and the average magnetic energy W_m stored in the inductor for both cases are given by

$$W_e = \frac{1}{4} C |\underline{V}|^2, \tag{9.27}$$

$$W_m = \frac{1}{4} L |\underline{I}|^2. \tag{9.28}$$

For resonance we obtain in both cases

$$W_e = W_m. \tag{9.29}$$

The total stored energy W_{tot} is given by

$$W_{\mathrm{tot}} = W_e + W_m = \frac{1}{2} C |\underline{V}|^2 = \frac{1}{2} L |\underline{I}|^2. \tag{9.30}$$

The quality Q of a resonant circuit is defined by

$$Q = 2\pi\,\frac{\text{Stored energy}}{\text{Energy dissipated per period}}\,. \tag{9.31}$$

The *energy dissipated per period* is given by

$$\frac{2\pi P}{\omega_0} = \text{Energy dissipated per period} \tag{9.32}$$

With this we obtain

Parallel resonant circuit | *Series resonant circuit*

$$Q = \frac{\omega_0 C_p}{G_p} = \frac{1}{\omega_0 L_p G_p}\,, \tag{9.33a}$$

$$Q = \frac{\omega_0 L_s}{R_s} = \frac{1}{\omega_0 C_s R_s}\,. \tag{9.33b}$$

In many cases it is useful to introduce the so-called detuning parameter v

$$v = \left(\frac{\omega}{\omega_0} - \frac{\omega_0}{\omega}\right). \tag{9.34}$$

For small detuning v the following approximate formula

$$v = 2\frac{\omega - \omega_0}{\omega_0} \quad \text{for} \quad \frac{\omega - \omega_0}{\omega_0} \ll 1. \tag{9.35}$$

is valid. With (9.24a), (9.24b), (9.33a), (9.33b) and (9.34) we obtain the normalized representation

Parallel resonant circuit | *Series resonant circuit*

$$Y_p = G_p(1 + \mathrm{j}Qv)\,, \tag{9.36a}$$

$$Z_s = R_s(1 + \mathrm{j}Qv)\,. \tag{9.36b}$$

In Figure 9.3 the admittance and impedance curves of the parallel and series resonant circuits are depicted in the Y plane as well as in the Z plane. The 3 dB cut-off frequencies ω_+ and ω_- are the frequencies for which the voltage across the parallel resonant circuit for the impressed current of constant amplitude decreases by a factor $1/\sqrt{2}$ compared with the resonant case. For the series resonant circuit with impressed voltage of constant amplitude the current is reduced by a factor $1/\sqrt{2}$ compared with the resonant case. The 3 dB cut-off frequency ω_+ is given by $v = 1/Q$ and the cut-off frequency ω_- is given by $v = -1/Q$. With (9.34) it follows that

$$\frac{\omega_+}{\omega_0} = \frac{1}{2Q} + \sqrt{\left(\frac{1}{2Q}\right)^2 + 1}, \tag{9.37}$$

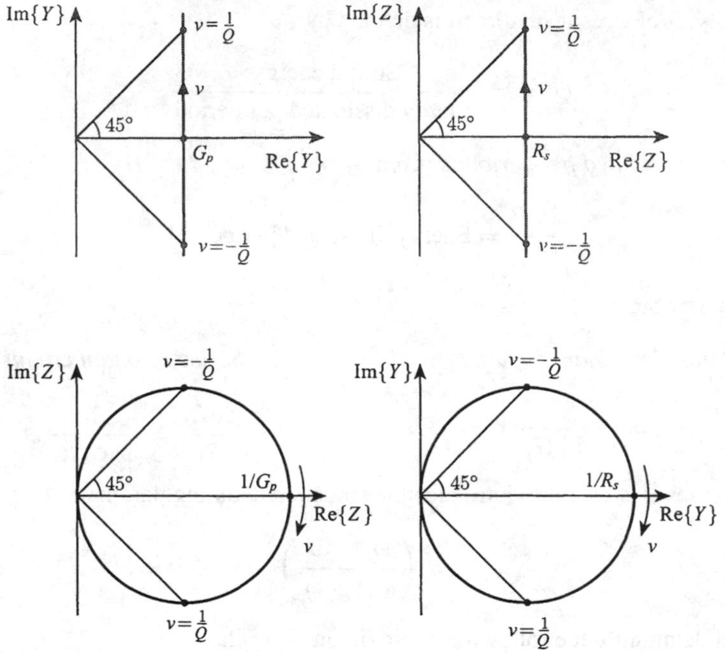

Figure 9.3: Impedance and admittance plots of series and parallel resonant circuits in the Y-plane and Z-plane.

$$\frac{\omega_-}{\omega_0} = -\frac{1}{2Q} + \sqrt{\left(\frac{1}{2Q}\right)^2 + 1}. \tag{9.38}$$

The 3 dB bandwidth $\Delta\omega$ is given by the difference

$$\Delta\omega = \omega_+ - \omega_- . \tag{9.39}$$

With (9.37) and (9.38) we obtain

$$\Delta\omega = \frac{\omega_0}{Q} . \tag{9.40}$$

The *relative bandwidth* $\Delta\omega/\omega_0$ is equal to the reciprocal quality factor. In some cases in literature also the parameter *damping* d is used

$$\frac{\Delta\omega}{\omega_0} = \frac{1}{Q} = d . \tag{9.41}$$

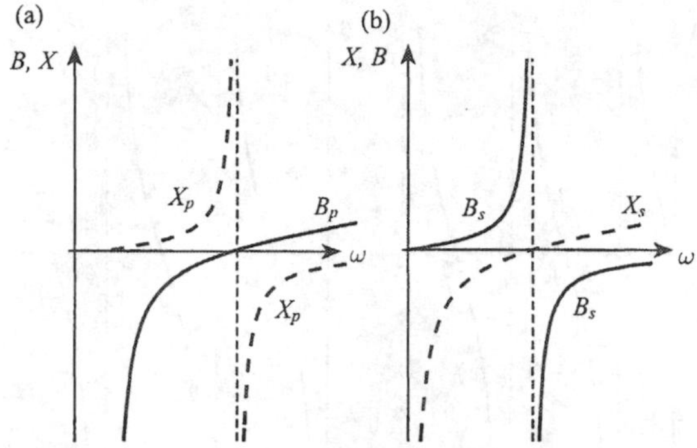

Figure 9.4: (a) Reactance X and susceptance B of the lossless parallel resonant circuit and (b) the lossless series resonant circuit.

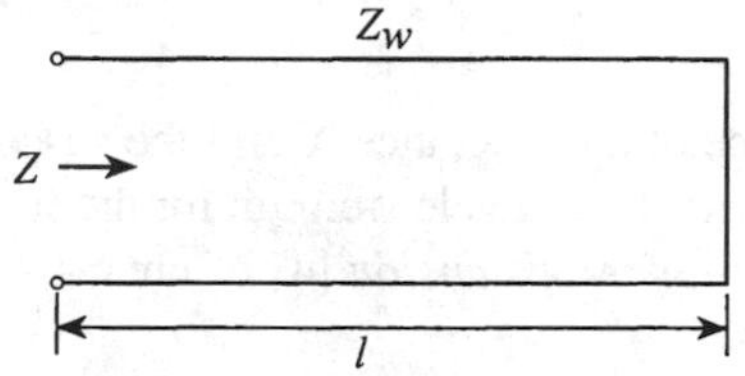

Figure 9.5: Transmission line of length l short-circuited at one end.

For lossless resonant circuits $G_p = 0$ and $R_s = 0$ respectively is valid. In Figure 9.4 the frequency dependence of reactance X and the susceptance B according to (9.23a) and (9.23b) is depicted.

9.4 THE TRANSMISSION LINE RESONATOR

We consider a lossless transmission line of length l with characteristic impedance Z_0, which is short-circuited at one end as depicted in Figure 9.5. This short-circuited transmission line is a reactive one-port. From (8.39) we obtain the input impedance Z and the input admittance Y:

$$Z = \mathrm{j}X = \mathrm{j}Z_0 \tan\frac{\omega l}{c}, \quad (9.42a) \qquad Y = \mathrm{j}B = -\frac{\mathrm{j}}{Z_0}\cot\frac{\omega l}{c}. \quad (9.42b)$$

We introduce the angular frequency $\omega_1 = \pi\frac{c}{l}$ and obtain

$$Z = \mathrm{j}X = \mathrm{j}Z_0 \tan\pi\frac{\omega}{\omega_1}, \quad (9.43a) \qquad Y = \mathrm{j}B = -\frac{\mathrm{j}}{Z_0}\cot\pi\frac{\omega}{\omega_1}. \quad (9.43b)$$

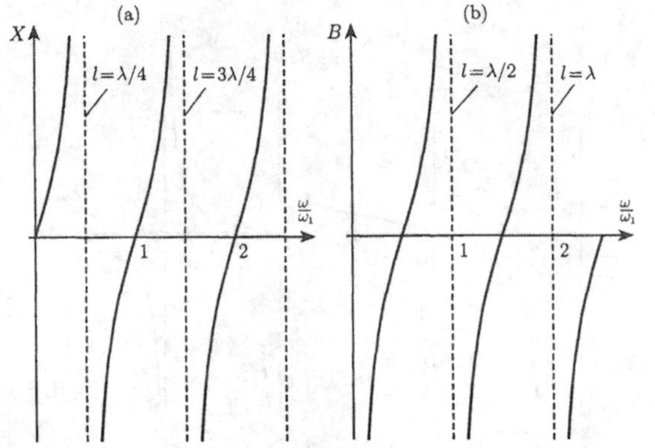

Figure 9.6: (a) Frequency dependence of the reactance X and (b) the susceptance B of the short-circuited transmission line.

The frequency dependence of the reactance X and the susceptance B are depicted in Figure 9.6. In order to obtain the equivalent circuit for the short-circuited transmission line we perform a *Mittag-Leffler expansion* [4] of $\tan \pi \frac{\omega}{\omega_1}$ and $\cot \pi \frac{\omega}{\omega_1}$ respectively and obtain

$$\tan \pi x = \frac{2x}{\pi} \sum_{n=1}^{\infty} \frac{1}{\left(n - \frac{1}{2}\right)^2 - x^2}, \quad (9.44a) \qquad \cot \pi x = \frac{1}{\pi x} + \frac{2x}{\pi} \sum_{n=1}^{\infty} \frac{1}{x^2 - n^2}. \quad (9.44b)$$

After inserting into (9.43a) and (9.43b) respectively it follows that

$$Z = \mathrm{j} Z_0 \frac{2\omega}{\pi \omega_1} \sum_{n=1}^{\infty} \frac{1}{\left(n - \frac{1}{2}\right)^2 - \left(\frac{\omega}{\omega_1}\right)^2}, \tag{9.45a}$$

$$Y = -\frac{\mathrm{j}}{Z_0} \left(\frac{\omega_1}{\pi \omega} + \frac{2\omega}{\pi \omega_1} \sum_{n=1}^{\infty} \frac{1}{\left(\frac{\omega}{\omega_1}\right)^2 - n^2} \right), \tag{9.45b}$$

and from this that

$$Z = \sum_{n=1}^{\infty} \frac{1}{\mathrm{j} \left[\omega \frac{\pi}{2\omega_1 Z_0} - \frac{1}{\omega} \frac{\left(n - \frac{1}{2}\right)^2 \pi \omega_1}{2 Z_0} \right]}, \tag{9.46a}$$

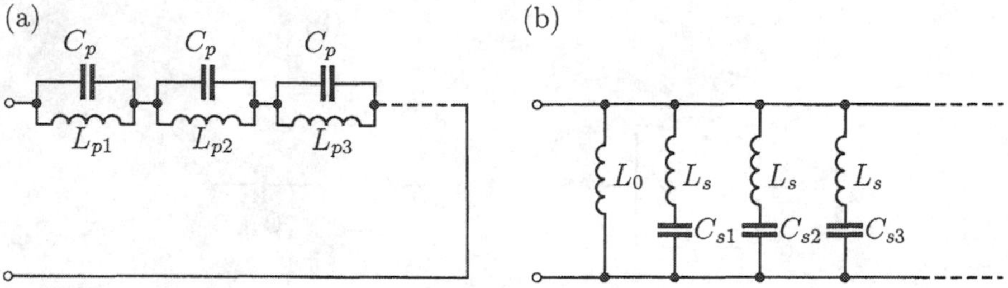

Figure 9.7: Equivalent circuits of the lossless transmission line resonator (a) according to (9.50a) and (b) according to (9.50b).

$$Y = \frac{1}{j\omega \frac{\pi Z_0}{\omega_1}} + \sum_{n=1}^{\infty} \frac{1}{j\left[\omega \frac{\pi Z_0}{2\omega_1} - \frac{1}{\omega}\frac{n^2\pi\omega_1 Z_0}{2}\right]}. \tag{9.46b}$$

We now introduce the following quantities

$$C_p = \frac{\pi}{2\omega_1 Z_0}, \tag{9.47a}$$

$$L_0 = \frac{\pi Z_0}{\omega_1}, \quad L_s = \frac{\pi Z_0}{2\omega_1}, \tag{9.47b}$$

$$L_{pn} = \frac{2Z_0}{\left(n - \frac{1}{2}\right)^2 \pi\omega_1}, \tag{9.48a}$$

$$C_{sn} = \frac{2}{n^2\pi\omega_1 Z_0}, \tag{9.48b}$$

$$\omega_{0pn} = \frac{1}{\sqrt{C_p L_{pn}}} = \left(n - \frac{1}{2}\right)\omega_1, \tag{9.49a}$$

$$\omega_{0sn} = \frac{1}{\sqrt{L_s C_{sn}}} = n\omega_1 \tag{9.49b}$$

and obtain with this

$$Z = \sum_{n=1}^{\infty} \frac{1}{j\left(\omega C_p - \frac{1}{\omega L_{pn}}\right)}, \tag{9.50a}$$

$$Y = \frac{1}{j\omega L_0} + \sum_{n=1}^{\infty} \frac{1}{j\left(\omega L_s - \frac{1}{\omega C_{sn}}\right)}. \tag{9.50b}$$

These *fractional expansion representations* are called the *Foster representations* [3, 5]. The *Foster representation of the first kind*, given by (9.50a) describes the series connection of an infinite number of parallel resonant circuits with resonance frequencies given by (9.49a), whereas the *Foster representation of the second kind*, (9.50b), describes the parallel connection of an infinite number of series resonant circuits and one inductance L_0 where the resonant frequencies of the series resonant circuits are given by (9.49b). The corresponding *equivalent circuits* are the *Foster equivalent circuit of the first kind* shown in Figure 9.7(a) and the *Foster equivalent circuit of the second kind* shown in Figure 9.7(b).

For lossy transmission lines we have to add loss resistors in the equivalent circuits. In the case of Figure 9.7 we have to add a loss conductor in parallel to each

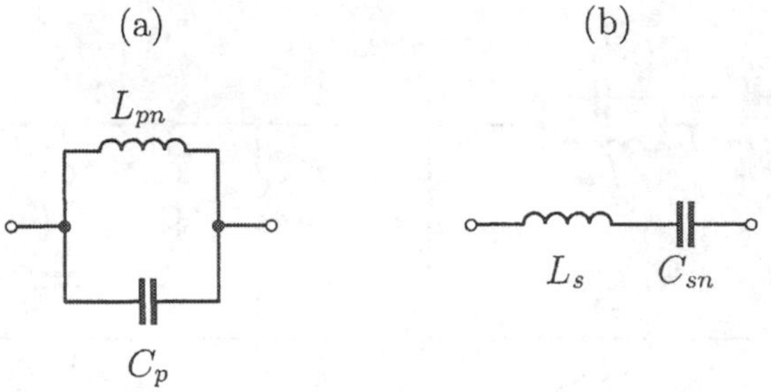

Figure 9.8: Equivalent circuits of the lossless transmission line resonator (a) near a parallel resonance and (b) near a series resonance.

parallel resonant circuit, and in the case of Figure 9.7(a) we have to add a loss resistor in series to each series resonant circuit. Considering a transmission line resonator at frequencies ω_{0pn} or ω_{0sn} in the neighborhood of one pole of the reactance function allows to neglect all poles with the exception of the pole under consideration. In this way the equivalent circuit may be reduced to a single resonant circuit describing the pole under consideration. Figure 9.8 shows the corresponding equivalent circuits consisting of a single parallel or series resonant circuit respectively. The short-circuited transmission line allows for the replacement of series resonant circuits as well as parallel resonant circuits. A short-circuited transmission line exhibits an infinite number of resonances; however, in general it is possible to design a circuit with transmission line resonators in such a way that only one dominant pole plays a role. Compared with lumped element resonant circuits realized with lumped elements, a resonator in general exhibits a much higher quality factor.

9.5 CAVITY RESONATORS

9.5.1 The Rectangular Cavity Resonator

We have seen that a transmission line short-circuited at the end is a resonator. The transmission line may be either a TEM transmission line or another type of transmission line, e.g., a waveguide. Such a transmission line segment terminated by short circuit or open circuit is a resonator. Resonators formed by segments of hollow waveguides are called cavity resonators.

Figure 9.9 shows a rectangular cavity resonator with side-lengths a, b and d. This cavity resonator consists of a waveguide segment of length d, terminated at both ends by conducting planes. The electromagnetic field in the resonator consists of a superposition of waves travelling in the positive and negative set direction exhibiting node planes at $z = 0$ and $z = d$. We obtain the boundary conditions $\boldsymbol{E}_{tr}\big|_{z=0} = 0$,

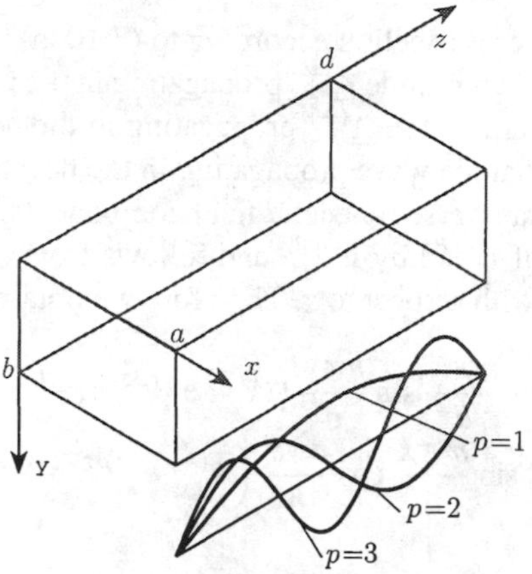

Figure 9.9: Rectangular cavity resonator.

$\boldsymbol{E}_{tr}\big|_{z=d} = 0$. To fulfill these boundary conditions the distance d must be an integer multiple of $\lambda_H/2$.

$$d = \frac{p}{2}\lambda_H, \quad \text{for } p = 0, 1, 2 \ldots . \tag{9.51}$$

Superimposing waves in the forward and backward direction yields nodes in distances $\lambda_H/2$. With (7.93) to (7.96) it follows that

$$\beta_{M0}^2 = \left(\frac{m\pi}{a}\right)^2 + \left(\frac{n\pi}{b}\right)^2 + \left(\frac{p\pi}{d}\right)^2 . \tag{9.52}$$

The resonant frequency ω_{mnp} and the corresponding free-space wavelength λ_{mnp} are given by

$$\omega_{mnp} = \frac{c_0}{\sqrt{\varepsilon_r \mu_r}} \sqrt{\left(\frac{m\pi}{a}\right)^2 + \left(\frac{n\pi}{b}\right)^2 + \left(\frac{p\pi}{d}\right)^2}, \tag{9.53}$$

$$\lambda_{mnp} = \frac{\sqrt{\varepsilon_r \mu_r}}{\sqrt{\left(\frac{m}{2a}\right)^2 + \left(\frac{n}{2b}\right)^2 + \left(\frac{p}{2d}\right)^2}}. \tag{9.54}$$

If the resonant mode originates from the TE_{mn} mode of the waveguide, the resonator mode is named the TE_{mnp} mode. If the resonator mode originates from the TM_{mn} mode of the waveguide, the resonator mode is named the TM_{mnp} mode.

9.5.1.1 The TE_{mnp} Modes of the Rectangular Cavity Resonator

To determine the field components of the TE_{mnp} modes of the rectangular cavity resonator we consider the field components of the TE_{mn} waveguide modes of the waveguide with a rectangular cross-section according to (7.104a) – (7.104f). We superimpose a waveguide wave of amplitude $\underline{A}^{(+)}$ propagating in the positive z-direction with a waveguide mode with amplitude $\underline{A}^{(-)}$ propagating in the negative z-direction. We take into consideration that the wave propagating in the negative z-direction exhibits the inverse sign of the transverse magnetic field components. This change of sign is due to the replacement of $\mathrm{e}^{-\mathrm{j}\beta z}$ by $\mathrm{e}^{+\mathrm{j}\beta z}$ and follows from (7.13b) and (7.14b) due to the partial derivation with respect to z. Therefore we obtain

$$\underline{E}_x = \mathrm{j}\omega\mu\left(\frac{n\pi}{b}\right)\cos\frac{m\pi x}{a}\sin\frac{n\pi y}{b}\left(A^{(+)}\,\mathrm{e}^{-\mathrm{j}\beta z} + A^{(-)}\,\mathrm{e}^{\mathrm{j}\beta z}\right), \tag{9.55a}$$

$$\underline{E}_y = -\mathrm{j}\omega\mu\left(\frac{m\pi}{a}\right)\sin\frac{m\pi x}{a}\cos\frac{n\pi y}{b}\left(A^{(+)}\,\mathrm{e}^{-\mathrm{j}\beta z} + A^{(-)}\,\mathrm{e}^{\mathrm{j}\beta z}\right), \tag{9.55b}$$

$$\underline{E}_z = 0\,, \tag{9.55c}$$

$$\underline{H}_x = \mathrm{j}\beta\left(\frac{m\pi}{a}\right)\sin\frac{m\pi x}{a}\cos\frac{n\pi y}{b}\left(A^{(+)}\,\mathrm{e}^{-\mathrm{j}\beta z} - A^{(-)}\,\mathrm{e}^{\mathrm{j}\beta z}\right), \tag{9.55d}$$

$$\underline{H}_y = \mathrm{j}\beta\left(\frac{n\pi}{b}\right)\cos\frac{m\pi x}{a}\sin\frac{n\pi y}{b}\left(A^{(+)}\,\mathrm{e}^{-\mathrm{j}\beta z} - A^{(-)}\,\mathrm{e}^{\mathrm{j}\beta z}\right), \tag{9.55e}$$

$$\underline{H}_z = \beta_{Mc}^2\cos\frac{m\pi x}{a}\cos\frac{n\pi y}{b}\left(A^{(+)}\,\mathrm{e}^{-\mathrm{j}\beta z} + A^{(-)}\,\mathrm{e}^{\mathrm{j}\beta z}\right). \tag{9.55f}$$

At $z = 0$ the transverse components of the electric field vanish, i.e., in this transverse plane $\underline{E}_x = 0$ and $\underline{E}_y = 0$ must be valid. From this we obtain

$$A^{(-)} = -A^{(+)}. \tag{9.56}$$

From this condition (9.51) follows. With (9.56) we obtain from (9.55a) – (9.55f)

$$\underline{E}_x = 2\beta Z_H\left(\frac{n\pi}{b}\right)A\cos\frac{m\pi x}{a}\sin\frac{n\pi y}{b}\sin\frac{p\pi z}{d}, \tag{9.57a}$$

$$\underline{E}_y = -2\beta Z_H\left(\frac{m\pi}{a}\right)A\sin\frac{m\pi x}{a}\cos\frac{n\pi y}{b}\sin\frac{p\pi z}{d}, \tag{9.57b}$$

$$\underline{E}_z = 0\,, \tag{9.57c}$$

$$\underline{H}_x = 2\mathrm{j}\beta\left(\frac{m\pi}{a}\right)A\sin\frac{m\pi x}{a}\cos\frac{n\pi y}{b}\cos\frac{p\pi z}{d}, \tag{9.57d}$$

$$\underline{H}_y = 2\mathrm{j}\beta\left(\frac{n\pi}{b}\right)A\cos\frac{m\pi x}{a}\sin\frac{n\pi y}{b}\cos\frac{p\pi z}{d}, \tag{9.57e}$$

$$\underline{H}_z = -2\mathrm{j}\beta_{Mc}^2 A\cos\frac{m\pi x}{a}\cos\frac{n\pi y}{b}\sin\frac{p\pi z}{d}. \tag{9.57f}$$

9.5.1.2 The TM_{mnp} Modes of the Rectangular Cavity Resonator

We obtain the TM_{mnp} modes of the rectangular cavity resonator from the TM_{mn} modes of the rectangular waveguide. Starting with (7.111a) – (7.111f) and considering that replacing $e^{-j\beta z}$ by $e^{+j\beta z}$ due to (7.13a) and (7.14a) yields a change of sign of $\underline{E}_x$ and $\underline{E}_y$, we obtain

$$\underline{E}_x = -j\beta\left(\frac{m\pi}{a}\right)\cos\frac{m\pi x}{a}\sin\frac{n\pi y}{b}\left(B^{(+)}e^{-j\beta z} - B^{(-)}e^{j\beta z}\right), \tag{9.58a}$$

$$\underline{E}_y = -j\beta\left(\frac{n\pi}{b}\right)\sin\frac{m\pi x}{a}\cos\frac{n\pi y}{b}\left(B^{(+)}e^{-j\beta z} - B^{(-)}e^{j\beta z}\right), \tag{9.58b}$$

$$\underline{E}_z = \beta_{Mc}^2\sin\frac{m\pi x}{a}\sin\frac{n\pi y}{b}\left(B^{(+)}e^{-j\beta z} + B^{(-)}e^{j\beta z}\right), \tag{9.58c}$$

$$\underline{H}_x = j\frac{\beta}{Z_E}\left(\frac{n\pi}{b}\right)\sin\frac{m\pi x}{a}\cos\frac{n\pi y}{b}\left(B^{(+)}e^{-j\beta z} + B^{(-)}e^{j\beta z}\right), \tag{9.58d}$$

$$\underline{H}_y = -j\frac{\beta}{Z_E}\left(\frac{m\pi}{a}\right)\cos\frac{m\pi x}{a}\sin\frac{n\pi y}{b}\left(B^{(+)}e^{-j\beta z} + B^{(-)}e^{j\beta z}\right), \tag{9.58e}$$

$$\underline{H}_z = 0. \tag{9.58f}$$

From the boundary conditions $\underline{E}_x\big|_{z=0} = 0$, $\underline{E}_y\big|_{z=0} = 0$ it follows that

$$B^{(+)} = B^{(-)} = B. \tag{9.59}$$

With this we obtain the fields component of the TM_{mnp} mode

$$\underline{E}_x = -2\beta\left(\frac{m\pi}{a}\right)B\cos\frac{m\pi x}{a}\sin\frac{n\pi y}{b}\sin\frac{p\pi z}{d}, \tag{9.60a}$$

$$\underline{E}_y = -2\beta\left(\frac{n\pi}{b}\right)B\sin\frac{m\pi x}{a}\cos\frac{n\pi y}{b}\sin\frac{p\pi z}{d}, \tag{9.60b}$$

$$\underline{E}_z = 2\beta_{Mc}^2 B\sin\frac{m\pi x}{a}\sin\frac{n\pi y}{b}\cos\frac{p\pi z}{d}, \tag{9.60c}$$

$$\underline{H}_x = 2j\frac{\beta}{Z_E}\left(\frac{n\pi}{b}\right)B\sin\frac{m\pi x}{a}\cos\frac{n\pi y}{b}\cos\frac{p\pi z}{d}, \tag{9.60d}$$

$$\underline{H}_y = -2j\frac{\beta}{Z_E}\left(\frac{m\pi}{a}\right)B\cos\frac{m\pi x}{a}\sin\frac{n\pi y}{b}\cos\frac{p\pi z}{d}, \tag{9.60e}$$

$$\underline{H}_z = 0. \tag{9.60f}$$

9.5.1.3 The Quality of Rectangular Cavity Resonators

The definition of the quality factor Q according to (9.31) has been so general that it may also be applied to the resonator. To every mode a quality factor Q_{mnp} may be assigned by

$$Q_{mnp} = \frac{\omega_{mnp}W_{mnp}}{P_{l\,mnp}}, \tag{9.61}$$

where W_{mnp} is the average energy stored in mode mnp and $P_{l\,mnp}$ is the energy of the mode dissipated per unit of time. Furthermore the quantities marked with mnp depend on whether we are considering the TM mode or the TE mode. With (2.34), (2.35), (4.3) and (4.4) the average stored energy W_{mnp} is given by

$$W_{mnp} = \frac{1}{4} \int_V \mathcal{E} \wedge \star\, \mathcal{E} + \mathcal{H} \wedge \star\, \mathcal{E} \,. \tag{9.62}$$

The integration is performed over the resonator volume V. According to (6.68), the power loss $P_{l\,mnp}$ is given by

$$P_{l\,mnp} = \frac{1}{2} \oint_{\partial V} R_A |\boldsymbol{H}_t|^2 s_1 \wedge s_2 \,. \tag{9.63}$$

The integration is performed over the boundary ∂V, i.e., the walls of the resonator. For the TE_{101} mode with the field components following from (9.57a) – (9.57f) we obtain

$$\underline{E}_y = -2\beta Z_H \left(\frac{\pi}{a}\right) A \sin \frac{\pi x}{a} \sin \frac{\pi z}{d} \,, \tag{9.64a}$$

$$\underline{H}_x = 2\mathrm{j}\beta \left(\frac{\pi}{a}\right) A \sin \frac{\pi x}{a} \cos \frac{\pi z}{d} \,, \tag{9.64b}$$

$$\underline{H}_z = -2\mathrm{j}\beta_{Mc}^2 A \cos \frac{\pi x}{a} \sin \frac{\pi z}{d} \,. \tag{9.64c}$$

The quality is given by

$$Q_{101} = \frac{\pi Z_F}{2R_A} \frac{b(a^2+d^2)^{\frac{3}{2}}}{ad(a^2+d^2)+2b(a^3+d^3)} \,. \tag{9.65}$$

The field lines of the TE_{101} mode of the rectangular resonator are depicted in Figure 9.10.

9.5.2 The Circular Cylindrical Cavity Resonator

The circular cylindric cavity resonator is formed by a segment of a circular cylindric waveguide terminated at both ends by a conducting plane. The resonator modes originate from the circular cylindric waveguide modes. The waveguide wavelength λ_H has to fulfill the condition (9.51). From the TE_{nm} modes and the TM_{nm} modes respectively of the circular cylindric waveguide we obtain the TE_{nmp} modes and the TM_{nmp} modes of the circular cylindric cavity resonator. The resonant frequencies following from (7.190), (7.191) and (9.51) are given.

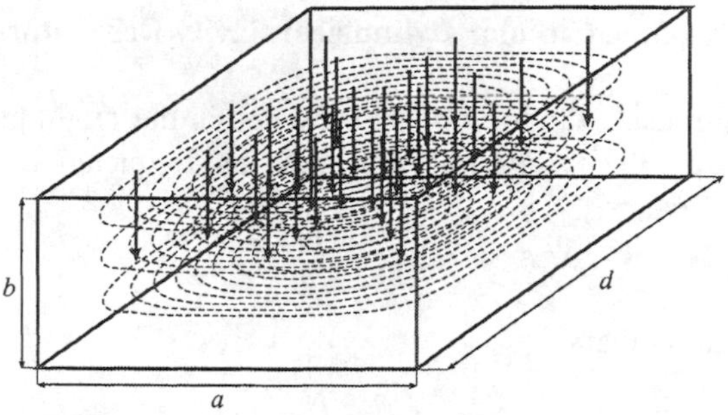

Figure 9.10: Field lines of the TE_{101} mode of the rectangular cavity resonator.

As an example we calculate the field of the TM_{010} mode of the circular cavity resonator. From (7.175) – (7.179) it follows that

$$\underline{E}_r = -\mathrm{j}\beta k_c J_0'(k_c r)\left(B^{(+)}\,\mathrm{e}^{-\mathrm{j}\beta z} - B^{(-)}\,\mathrm{e}^{\mathrm{j}\beta z}\right), \tag{9.66a}$$

$$\underline{E}_z = k_c^2 J_0(k_c r)\left(B^{(+)}\,\mathrm{e}^{-\mathrm{j}\beta z} + B^{(-)}\,\mathrm{e}^{\mathrm{j}\beta z}\right), \tag{9.66b}$$

$$\underline{H}_\varphi = -\mathrm{j}\omega\varepsilon_0 k_c J_0'(k_c r)\left(B^{(+)}\,\mathrm{e}^{-\mathrm{j}\beta z} + B^{(-)}\,\mathrm{e}^{\mathrm{j}\beta z}\right). \tag{9.66c}$$

The negative sign in (9.66a) originates from the partial derivate with respect to z in the first term on the right side of (7.174). For the TM_{010} mode we obtain $\beta = 0$. This corresponds to the operation of the waveguide at the cut-off wavelength. The radial electric field component $\underline{E}_r$ must vanish at $z = 0$ and $z = d$, and due to $\beta = 0$ vanishes everywhere. We obtain

$$B^{(+)} = B^{(-)} = B\,. \tag{9.67}$$

From (7.189) we obtain with $\beta = 0$ and (7.187b)

$$\beta_0 = \frac{2.405}{a}, \quad \lambda_0 = \frac{2\pi}{\beta_0} = 2.61a. \tag{9.68}$$

From (9.66a) to (9.66c) we obtain the two remaining field components $\underline{E}_z$ and $\underline{H}_\varphi$:

$$\underline{E}_z = 2B\beta_0^2 J_0(\beta_0 r), \tag{9.69a}$$

$$\underline{H}_\varphi = 2\mathrm{j}\,B\frac{\beta_0^2}{Z_{F0}} J_1(\beta_0 r). \tag{9.69b}$$

9.5.3 The Quality of the Circular Cylindrical Cavity Resonators

The quality of a circular cylindric resonator excited in the TM_{010} mode is computed in the following using (9.61) – (9.63). At the resonant frequency we obtain

$$W = 2W_e = \frac{\varepsilon_0}{2} \int_V |\underline{E}_z|^2 \mathrm{d}V. \tag{9.70}$$

With (9.69a) it follows that

$$W = 4\pi \varepsilon_0 |B|^2 \beta_0^4 d \int_0^a r J_0^2(\beta_0 r) \mathrm{d}r\,. \tag{9.71}$$

With (B.37) we obtain

$$W = 2\pi \varepsilon_0 |B|^2 \beta_0^4 a^2 d J_1^2(\beta_0 a)\,. \tag{9.72}$$

The power loss follows from (9.63):

$$P_l = \frac{1}{2} R_A \oint_{\partial V} |\underline{H}_\varphi|^2 \mathrm{d}A\,. \tag{9.73}$$

The integral has to be performed over the side wall and the top and bottom walls of the circular cavity resonator. We obtain

$$P_l = 2R_A |B|^2 \frac{\beta_0^4}{Z_{F0}^2} \left[2\pi a d J_1^2(\beta_0 a) + 4\pi \int_0^a r J_1^2(\beta_0 r) \mathrm{d}r \right], \tag{9.74}$$

where the first term corresponds to the integral over the side wall and the second term is the integral over the top and bottom walls. With (7.200) it follows that

$$P_l = 4\pi a (d + a) R_A |B|^2 \frac{\beta_0^4}{Z_{F0}^2} J_1^2(\beta_0 a)\,. \tag{9.75}$$

From (9.61), (9.72) and (9.75) we finally obtain

$$Q_0 = \frac{\varepsilon_0 \omega_0 a d}{2(d + a)} \frac{Z_{F0}^2}{R_A}\,. \tag{9.76}$$

With (7.187b) it follows that

$$\beta_0 a = \xi_{01} = 2.405 \tag{9.77}$$

and with $\varepsilon_0 \omega_0 Z_{F0} = \beta_0$ we obtain the quality factor

$$Q_0 = \frac{Z_{F0}}{R_A} \frac{\xi_{01}}{2\left(1 + \frac{a}{d}\right)}\,. \tag{9.78}$$

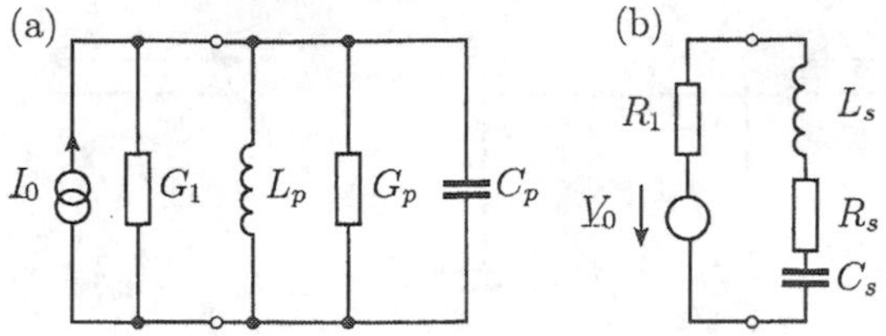

Figure 9.11: Resonant circuit connected with generator.

9.6 COUPLING OF RESONANT CIRCUITS AND RESONATORS

9.6.1 The Loaded Quality Factor

If a resonant circuit or a resonator is coupled to an external circuit, energy stored in the resonant circuit or the resonator will be exchanged with the external circuit. Figure 9.11 depicts the connection of a parallel resonant circuit and a series resonant circuit respectively with a generator. The generator consists of an impressed current source $\underline{I}_0$ and an inner conductance G_1 or of an impressed voltage source $\underline{V}_0$ and the inner resistance R_1. We have assumed the impedance of the generator to be real. This simplifies the following considerations; however, it does not impose restrictions since in the case of a reactive component of the generator impedance this can be easily compensated by detuning of the resonant circuit.

According to (9.33a) and (9.33b) the quality factor Q_0 of the resonant circuit not connected with the generator is given by

Parallel resonant circuit *Series resonant circuit*

$$Q_0 = \frac{\omega_0 C_p}{G_p}, \quad (9.79a) \qquad Q_0 = \frac{\omega_0 L_s}{R_s}. \quad (9.79b)$$

If the resonant circuit is connected with the generator, the resonant circuit also is damped by the external conductance G_p or the external resistance R_1 respectively.

Considering the generator as ideal current source or ideal voltage source respectively and G_1 or R_1 to be part of the lossy resonant circuit, we can introduce a loaded quality factor Q_L given by

Parallel resonant circuit *Series resonant circuit*

$$Q_L = \frac{\omega_0 C_p}{G_1 + G_p}, \quad (9.80a) \qquad Q_L = \frac{\omega_0 L_s}{R_1 + R_s}. \quad (9.80b)$$

We name Q_0 as the *unloaded quality factor* or *unloaded Q* and Q_L as the *loaded*

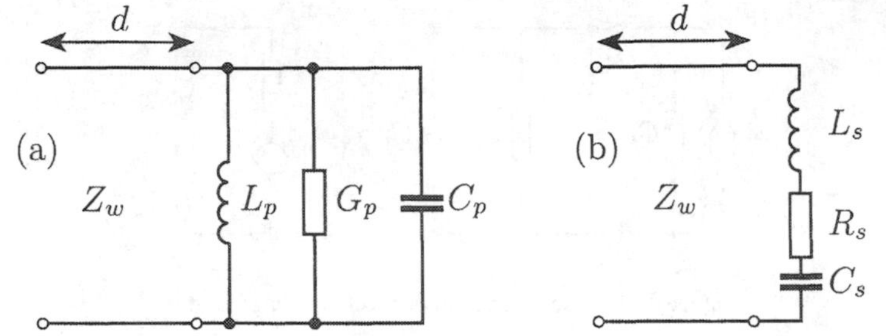

Figure 9.12: Transmission line terminated with resonant circuit.

quality factor or *loaded Q*. It is useful to introduce also a so-called *external quality factor* Q_{ext} given by

Parallel resonant circuit *Series resonant circuit*

$$Q_{ext} = \frac{\omega_0 C_p}{G_1}, \quad (9.81a) \qquad Q_{ext} = \frac{\omega_0 L_s}{R_1}. \quad (9.81b)$$

The external quality factor Q_{ext} is a measure for the damping resonant circuit by the external circuit. From (9.79a) – (9.81a) and (9.79b) – (9.81b) respectively it follows that

$$\frac{1}{Q_L} = \frac{1}{Q_0} + \frac{1}{Q_{ext}}. \quad (9.82)$$

The smaller Q_{ext} is, the larger the damping due to the coupling of the resonant circuit to an external circuit. For $Q_{ext} > Q_0$ the resonant circuit is *undercoupled*, for $Q_{ext} = Q_0$ we obtain *critical coupling*, and for $Q_{ext} < Q_0$ the resonant circuit is *overcoupled*.

9.6.2 Termination of a Transmission Line with a Resonant Circuit

We now consider a transmission line terminated either with a parallel resonant circuit or a series resonant circuit as depicted in Figure 9.12. The transmission line input is connected with a generator the impedance of which is equal to the wave impedance Z_0 of the transmission line. Therefore the resonant circuit also is terminated with the characteristic impedance Z_0. According to (9.81a) and (9.81b) we obtain

Parallel resonant circuit *Series resonant circuit*

$$Q_{ext} = \omega_0 C_p Z_0, \quad (9.83a) \qquad Q_{ext} = \frac{\omega_0 L_s}{Z_0}. \quad (9.83b)$$

Using (8.60a), (8.60b), (9.23a), (9.23b), (9.79a), (9.79b), (9.81a) and (9.81b) we normalize the admittance of the parallel resonant circuit and the impedance of the series resonant circuit with respect to the wave impedance and obtain

Parallel resonant circuit | *Series resonant circuit*

$$y_p = \frac{Q_{ext}}{Q_0}(1 + \mathrm{j}\,Q_0 v)\,, \qquad (9.84a) \qquad\qquad z_s = \frac{Q_{ext}}{Q_0}(1 + \mathrm{j}\,Q_0 v)\,. \qquad (9.84b)$$

With (8.62a) and (8.62b) we obtain the reflection coefficient

Parallel resonant circuit | *Series resonant circuit*

$$\rho_p = \frac{1 - \frac{Q_{ext}}{Q_0}(1 + \mathrm{j}\,Q_0 v)}{1 + \frac{Q_{ext}}{Q_0}(1 + \mathrm{j}\,Q_0 v)}\,, \qquad (9.85a) \qquad\qquad \rho_s = -\frac{1 - \frac{Q_{ext}}{Q_0}(1 + \mathrm{j}\,Q_0 v)}{1 + \frac{Q_{ext}}{Q_0}(1 + \mathrm{j}\,Q_0 v)}\,. \qquad (9.85b)$$

The rational functions (9.85a) and (9.85b) define circles in the ρ plane. In Figure 9.13 the curves of the parallel resonant circuit and the series resonant circuit are given for undercritial coupling, critical coupling and overcritical coupling.

In the case of crititcal coupling the impedance curve passes $\rho = 0$ at resonance. In the case of overcritical coupling the origin $\rho = 0$ is enclosed by the reflection factor curve, whereas for undercritical coupling $\rho = 0$ is not enclosed by the reflection factor curve. The 3 dB cut-off frequencies ω_- and ω_+ correspond to point on the reflection factor curve with $X = R$ and $X = -R$ of the Smith chart. We obtain these curves by drawing a circle with center at $\rho = -\mathrm{j}$ or $\rho = \mathrm{j}$ respectively and passing through $\rho = -1$ and $\rho = 1$.

To determine the input impedance of a transmission line of length d and characteristic impedance Z_0 terminated with a parallel resonant circuit we have to rotate the points of the ρ curve of the parallel resonant circuit according to Figure 9.12(a) in the Smith chart by $\pi d/\lambda$. Strictly speaking we have to rotate every point of the ρ curve by another angle since each point belongs to another frequency. If, however, the resonant circuit exhibits sufficiently high quality the essential part of the ρ curve belongs to a very small frequency interval. In this case by approximation we may rotate the ρ curve by an angle corresponding to the waveguide length d and the waveguide wavelength λ or λ_H respectively at the center frequency of the resonant circuit.

By comparing of Figure 9.13(a) and Figure 9.13(b) we see that the reflection factor curves of the parallel resonant circuit by a rotation over 180° in the Smith chart are transformed into the reflection factor curves of the series resonant circuit. Therefore a $\lambda/4$ transmission line terminated with a series resonant circuit behaves like a parallel resonant circuit and vice versa; a $\lambda/4$ transmission line terminated with a series resonant circuit behaves like a parallel resonant circuit.

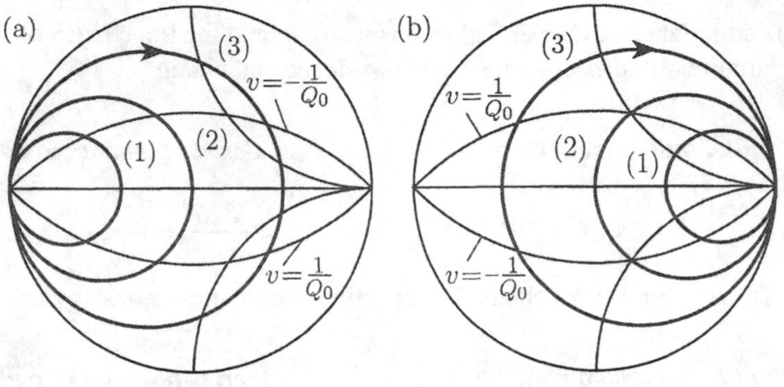

Figure 9.13: Curves (a) of the parallel resonant circuit and (b) of the series resonant circuit for (1) undercritical, (2) critical and (3) overcritical coupling.

9.6.3 Coupling of Cavity Resonators

Resonators may be coupled to external circuits, e. g. via coaxial lines, waveguides or other kinds of transmission lines. The coupling may be performed inductively over a loop or capacitively over a pin or by some combination of these methods. In Figure 9.14 various examples for the coupling of resonators to waveguides are depicted.

As an example we treat the inductive coupling of a coaxial line to the TM_{010} mode of the circular cylindric resonator. Figure 9.15 shows the inductive coupling of a coaxial line to a circular cylindric resonator. The inner conductor of the coaxial line inside the resonator forms a coupling loop enclosing an area A. The magnetic field in φ-direction intersects the loop normally. If at the resonant frequency of the TM_{010} mode a current with amplitude $\underline{I}$ is impressed, a magnetic field in the resonator is induced. On the other hand the alternating magnetic field of the excited TM_{010} mode induces a voltage in the loop. This voltage will assume a maximum value at the resonant frequency of the TM_{010} mode. Therefore we expect that the equivalent circuit of the cavity resonator excited in the TM_{010} mode via an inductive loop will be a parallel resonant circuit, if we assume that the transverse plane of reference is positioned in the side wall of the resonator. The voltage $\underline{V}$ induced in the inductive loop is given by

$$\underline{V} = \mathrm{j}\omega\mu_0 A\underline{H}_{\varphi} + \mathrm{j}\omega L\underline{I}\,, \tag{9.86}$$

where the first term at the right side of (9.86) describes the voltage induced due to Faraday's law (2.51) by the magnetic field of the TM_{010} mode in the coupling loop. The second term in (9.86) is due to the self-induction due to the inductance L of the coupling loop. The computation of L is more complicated since we have to consider all resonator modes for this. However, it is not necessary to know the value of L

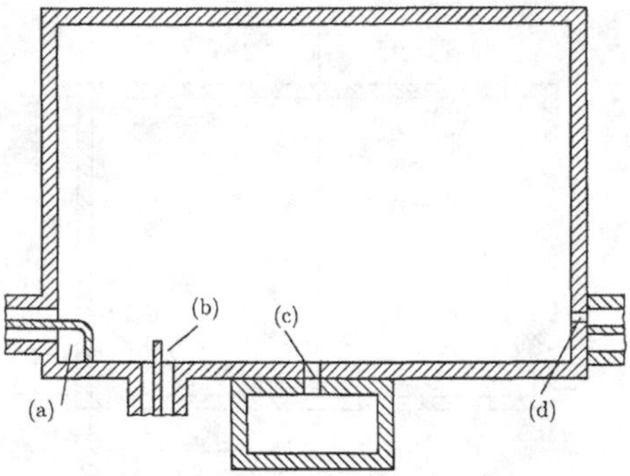

Figure 9.14: Coupling of a resonator to a waveguide with (a) inductive loop coupling, (b) capacitive pin coupling, (c) hole coupling and (d) inductive hole coupling.

since in the case of a high resonator Q factor the inductance L will cause only a minor detuning of the resonator. We can compensate the influence of L by a small change of frequency. The complex power P_c flowing into the resonator is given by

$$P_c = \frac{1}{2}\underline{V}\underline{I}^* = \frac{1}{2}\mathrm{j}\omega\mu_0 A\underline{H}_\varphi(a)\underline{I}^* + \frac{1}{2}\mathrm{j}\omega L|\underline{I}|^2 . \tag{9.87}$$

In the resonant case the complex power P_c is real and equal to the loss power P_l.

$$P_c = P_l = \frac{1}{2}\mathrm{j}\omega\mu_0 A\underline{H}_\varphi(a)\underline{I}^* \ldots \text{real}, \tag{9.88}$$

where the resonant case is defined by tuning for real input impedance. The contribution of the loop inductance L in this case is compensated by a small detuning of the resonator. Assuming that the power loss flowing into the cavity resonator only is due to the resonator wall losses, we obtain from (9.69b) and (9.75):

$$P_l = \pi a(d+a)R_A|\underline{H}_\varphi(a)|^2 \quad \text{for } \omega = \omega_0 . \tag{9.89}$$

From (9.88) and (9.89) it follows that

$$P_l = \frac{(\omega\mu_0 A)^2|\underline{I}|^2}{4\pi a(d+a)R_A} . \tag{9.90}$$

On the other hand according to Figure 9.2(a) we obtain for the parallel resonant circuit

$$P_l = \frac{|\underline{I}|^2}{2G_p} . \tag{9.91}$$

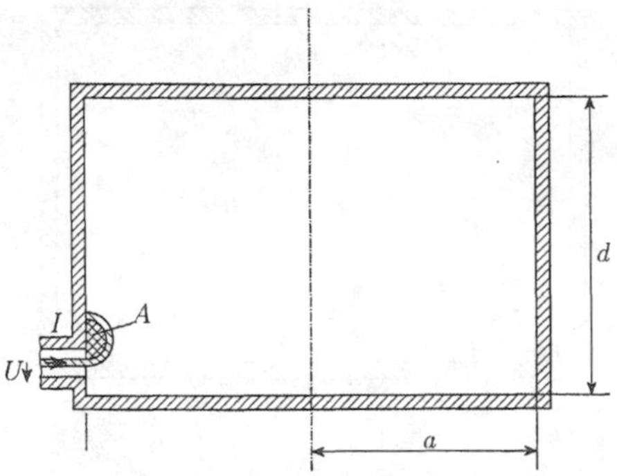

Figure 9.15: Inductive coupling of a coaxial line to a circular cylindric resonator.

From (9.90) and (9.91) we obtain the loss conductance G_p of the parallel resonant circuit according to Figure 9.2(a):

$$G_p = \frac{2\pi a(d+a)R_A}{(\omega\mu_0 A)^2}\,. \tag{9.92}$$

Since we already have computed the quality Q_0 for the TM_{010} mode of the circular cylindric resonator in (9.78) according to (9.36a), we obtain

$$Y = G_p(1 + \mathrm{j}\,Q_0 v)\,. \tag{9.93}$$

This is the admittance of the cavity resonator of excitation in the TM_{010} mode and refers to the plane of reference in the resonator wall. According to the made assumptions this admittance is only correct within a neighborhood of the resonant frequency of the TM_{010} mode, since only in this case the magnetic field contribution of the TM_{010} mode is the dominating magnetic flux contribution in the coupling loop.

REFERENCES

[1] R. Foster, "A reactance theorem," *Bell System Tech. J.*, vol. 3, pp. 259–267, 1924.

[2] H. Carlin and A. Giordano, *Network Theory*. Englewood Cliffs, NJ: Prentice Hall, 1964.

[3] R. F. Harrington, *Time Harmonic Electromagnetic Fields*. New York: McGraw-Hill, 1961.

[4] S. Hassani, *Mathematical Physics*. Berlin: Springer, 2002.

[5] V. Belevitch, *Classical network theory*. San Francisco, California: Holden-Day, 1968.

Chapter 10

Microwave Circuits

10.1 LINEAR MULTIPORTS

A general microwave circuit is a *multiport* [1–4]. A multiport is a circuit with a number of *ports*. In the network picture a port is constituted by a pair of external nodes where the currents flowing into the two nodes of a port have equal amplitude and opposite signs. In microwave circuits a multiport usually exhibits waveguide ports. The port is defined by the junction plane of the waveguide. It is assumed that at each port a single transverse mode is excited. If more than one mode is excited in the junction plane of the waveguide, we have to assign one port to every mode. In the following we assume that in every waveguide only a single transverse waveguide mode is excited. In this case the number of physical ports is identical with the number of ports in the abstract multiport scheme. If in some physical port of the multiport more than one waveguide mode may occur, we have to assign to each mode it's own port in

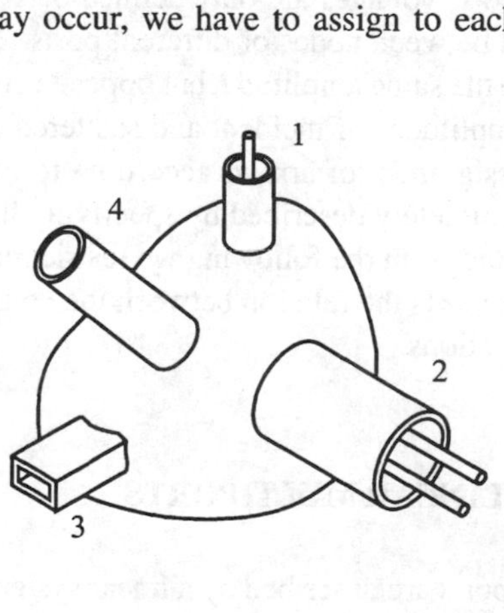

Figure 10.1: Multiport.

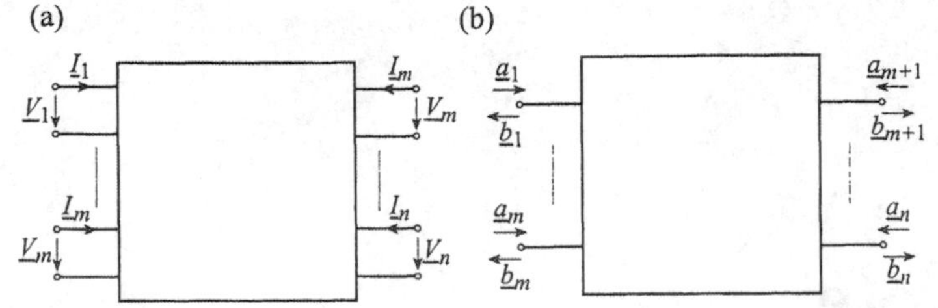

Figure 10.2: Multiport described by the amplitudes (a) $\underline{V}_k$, $\underline{I}_k$ and (b) $\underline{a}_k$, $\underline{b}_k$.

the abstract multiport scheme. In the following we assume that only one single mode may propagate through every physical port unless a different case is explicitly stated. Therefore to each physical waveguide port only a single port is assigned.

Figure 10.1 shows the schematic drawing of a multiport. The port either may be of a coaxial type, a waveguide type, or any other type. To each port a pair of complex amplitudes is assigned; these describe the incident and scattered electromagnetic waves completely. The complex amplitudes are defined for a certain plane of reference in the waveguide termination. To describe the state of the kth port ($k = 1 \dots n$) we may take the incident and scattered wave amplitudes $\underline{a}_k$ and $\underline{b}_k$ or the current $\underline{I}_k$ and the voltage $\underline{V}_k$. Current and voltage may be defined in a conventional way for TEM ports. For non-TEM waveguide ports, currents and voltages may be considered as generalized currents and voltages.

Figure 10.2 shows the schematic representations of a multiport. In Figure 10.2(a) to each port a pair of nodes is assigned. The arrows for the voltages $\underline{V}_k$ and currents $\underline{I}_k$ are drawn for every port. Voltages are only defined between the two nodes of the same port, but undefined between nodes of different ports. The currents flowing into the node of one port have the same amplitude, but opposite sign. If the signals are characterized by the wave amplitudes of incident and scattered electromagnetic waves a and b respectively, an assignment of arrows according to Figure 10.2(b) is possible. The multiport may be completely described by specifying the relations between incident and scattered amplitudes. In the following we restrict our consideration to linear multiports. In linear multiports the relation between the complex amplitudes is given by a system of linear equations.

10.2 SOURCE-FREE LINEAR MULTIPORTS

Source-free linear multiports are described by a linear system of equations [2,3,5]. In a multiport with n ports, also called an *n-port* the n port voltages $\underline{V}_1 \dots \underline{V}_n$ are related to the n port currents $\underline{I}_1 \dots \underline{I}_n$ via an n-dimensional linear systems of equations:

$$\begin{aligned}
\underline{V}_1 &= Z_{11}\underline{I}_1 + Z_{12}\underline{I}_2 + \dots Z_{1n}\underline{I}_n ,\\
\underline{V}_2 &= Z_{21}\underline{I}_1 + Z_{22}\underline{I}_2 + \dots Z_{2n}\underline{I}_n ,\\
&\vdots \\
\underline{V}_n &= Z_{n1}\underline{I}_1 + Z_{n2}\underline{I}_2 + \dots Z_{nn}\underline{I}_n ,
\end{aligned} \tag{10.1}$$

where Z_{mn} are impedances. These equations may be written in the following form

$$\underline{V}_i = \sum_{k=1}^{n} Z_{ik}\underline{I}_k \qquad \text{for} \qquad i = 1 \dots n . \tag{10.2}$$

Port voltages and port currents are described by n-dimensional vectors. An n-dimensional column vector is a matrix type $<m\times 1>$.

$$\underline{\boldsymbol{V}}_{<n\times 1>} = \begin{bmatrix} \underline{V}_1 \\ \vdots \\ \underline{V}_n \end{bmatrix} , \tag{10.3a}$$

$$\underline{\boldsymbol{I}}_{<n\times 1>} = \begin{bmatrix} \underline{I}_1 \\ \vdots \\ \underline{I}_n \end{bmatrix} . \tag{10.3b}$$

The impedances Z_{ik} may be summarized in a matrix of type $<n\times n>$ i.e. an nth order quadratic matrix

$$\boldsymbol{Z}_{<n\times n>} = (Z_{ik}) = \begin{bmatrix} Z_{11} & \dots & Z_{1n} \\ \vdots & \ddots & \vdots \\ Z_{n1} & \dots & Z_{nn} \end{bmatrix} . \tag{10.4}$$

The circuit equations (10.1) and (10.2) respectively can be written in matrix notation:

$$\underline{\boldsymbol{V}} = \boldsymbol{Z}\,\underline{\boldsymbol{I}} . \tag{10.5}$$

$\boldsymbol{Z}$ is called the *impedance matrix* of the multiport. The description of the multiport by (10.1), (10.2) and (10.5) is the ***Z**-representation* or *impedance representation* of the multiport equations. On the other hand, the *admittance matrix* $\boldsymbol{Y}$ and the ***Y**-representation* or *admittance representation* represent the port currents as a function of the port voltages.

$$\underline{\boldsymbol{I}} = \boldsymbol{Y}\,\underline{\boldsymbol{V}} . \tag{10.6}$$

From (10.6) and (10.5) it follows that

$$\underline{\boldsymbol{V}} = \boldsymbol{Z}\,\underline{\boldsymbol{I}} = \boldsymbol{Z}\,\boldsymbol{Y}\,\underline{\boldsymbol{V}} \tag{10.7}$$

and we obtain

$$\boldsymbol{Y} = \boldsymbol{Z}^{-1} . \tag{10.8}$$

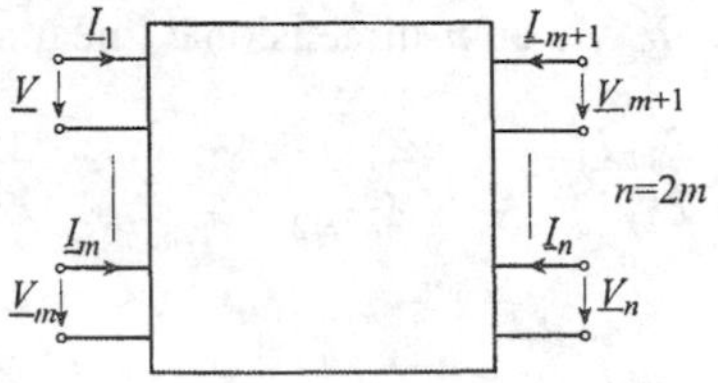

Figure 10.3: Port number symmetric multiport.

10.3 THE CHAIN MATRIX

The multiport in Figure 10.3 is *port-number symmetric*, i.e., it has the same number $m = n/2$ of input ports and output ports. In the *chain representation* of the circuit equations the currents and voltages of the m input ports are represented as functions of currents and voltages of the m outport ports. The input ports are numbered from 1 to m and the output ports are numbered from $m+1$ to n. The input amplitudes may be summarized in the m-dimensional vectors

$$\underline{\boldsymbol{V}}_{1<m\times 1>} = \left[\underline{V}_1, \underline{V}_2, \ldots \underline{V}_m\right]^T , \tag{10.9a}$$

$$\underline{\boldsymbol{I}}_{1<m\times 1>} = \left[\underline{I}_1, \underline{I}_2, \ldots \underline{I}_m\right]^T . \tag{10.9b}$$

To save space in the book we sometimes write column vectors as transposed row vectors. The output quantities are summarized in

$$\underline{\boldsymbol{V}}_{2<m\times 1>} = \left[\underline{V}_{m+1}, \underline{V}_{m+2}, \ldots \underline{V}_n\right]^T , \tag{10.9c}$$

$$\underline{\boldsymbol{I}}_{2<m\times 1>} = \left[\underline{I}_{m+1}, \underline{I}_{m+2}, \ldots \underline{I}_n\right]^T . \tag{10.9d}$$

The input quantities are represented in dependence of the output quantities by

$$\begin{aligned} \underline{\boldsymbol{V}}_1 &= \boldsymbol{A}_{11}\, \underline{\boldsymbol{V}}_2 + \boldsymbol{A}_{12}(-\underline{\boldsymbol{I}}_2) , \\ \underline{\boldsymbol{I}}_1 &= \boldsymbol{A}_{21}\, \underline{\boldsymbol{V}}_2 + \boldsymbol{A}_{22}(-\underline{\boldsymbol{I}}_2) . \end{aligned} \tag{10.10}$$

where $\boldsymbol{A}_{11<m\times m>} \ldots \boldsymbol{A}_{22<m\times m>}$ are quadratic mth order submatrices. These may be summarized in a supermatrix $\boldsymbol{A}_{<n\times n>}$, given by

$$\boldsymbol{A} = \begin{bmatrix} \boldsymbol{A}_{11} & \boldsymbol{A}_{12} \\ \boldsymbol{A}_{21} & \boldsymbol{A}_{22} \end{bmatrix} . \tag{10.11}$$

Summarizing the input quantities and the output quantities in n-dimensional vectors, we obtain

$$\begin{bmatrix} \underline{\boldsymbol{V}}_1 \\ \underline{\boldsymbol{I}}_1 \end{bmatrix} = \boldsymbol{A} \begin{bmatrix} \underline{\boldsymbol{V}}_2 \\ -\underline{\boldsymbol{I}}_2 \end{bmatrix} . \tag{10.12}$$

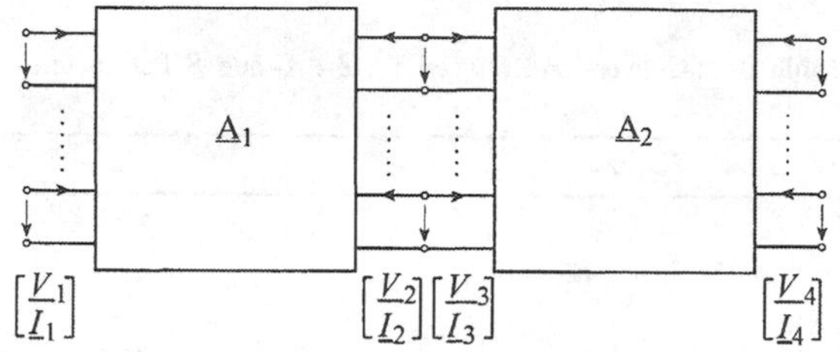

Figure 10.4: Cascading of two multiports.

The matrix $\boldsymbol{A}$ is called *chain matrix*. The chain representation is useful for the analysis of cascaded multiports. Figure 10.4 shows the cascading of two port-number symmetric multiports. The first multiport is described by the chain matrix $\boldsymbol{A}_1$, the second multiport by the chain matrix $\boldsymbol{A}_2$. The m output ports of the first multiport are connected with the m input ports of the second multiport. Both multiports are described in chain representation by

$$\begin{bmatrix} \underline{\boldsymbol{V}}_1 \\ \underline{\boldsymbol{I}}_1 \end{bmatrix} = \boldsymbol{A}_1 \begin{bmatrix} \underline{\boldsymbol{V}}_2 \\ -\underline{\boldsymbol{I}}_2 \end{bmatrix}, \tag{10.13}$$

$$\begin{bmatrix} \underline{\boldsymbol{V}}_3 \\ \underline{\boldsymbol{I}}_3 \end{bmatrix} = \boldsymbol{A}_2 \begin{bmatrix} \underline{\boldsymbol{V}}_4 \\ \underline{\boldsymbol{I}}_4 \end{bmatrix}. \tag{10.14}$$

For appropriate numbering of the output ports of the first multiport and the input ports of the second multiports we obtain

$$\begin{bmatrix} \underline{\boldsymbol{V}}_2 \\ -\underline{\boldsymbol{I}}_2 \end{bmatrix} = \begin{bmatrix} \underline{\boldsymbol{V}}_3 \\ \underline{\boldsymbol{I}}_3 \end{bmatrix}. \tag{10.15}$$

From (10.13) to (10.15) it follows that

$$\begin{bmatrix} \underline{\boldsymbol{V}}_1 \\ \underline{\boldsymbol{I}}_1 \end{bmatrix} = \boldsymbol{A}_1 \boldsymbol{A}_2 \begin{bmatrix} \underline{\boldsymbol{V}}_4 \\ -\underline{\boldsymbol{I}}_4 \end{bmatrix}. \tag{10.16}$$

The chain matrix

$$\boldsymbol{A} = \boldsymbol{A}_1 \boldsymbol{A}_2 \tag{10.17}$$

describes a multiport obtained by cascading of the multiports 1 and 2 in Figure 10.4. We obtain

$$\begin{bmatrix} \underline{\boldsymbol{V}}_1 \\ \underline{\boldsymbol{I}}_1 \end{bmatrix} = \boldsymbol{A} \begin{bmatrix} \underline{\boldsymbol{V}}_4 \\ -\underline{\boldsymbol{I}}_4 \end{bmatrix}. \tag{10.18}$$

To convert the $\boldsymbol{Z}$-representation into the $\boldsymbol{A}$-representation we first write (10.5) in the form

$$\begin{aligned} \underline{\boldsymbol{V}}_1 &= \boldsymbol{Z}_{11} \underline{\boldsymbol{I}}_1 + \boldsymbol{Z}_{12} \underline{\boldsymbol{I}}_2, \\ \underline{\boldsymbol{V}}_2 &= \boldsymbol{Z}_{21} \underline{\boldsymbol{I}}_1 + \boldsymbol{Z}_{22} \underline{\boldsymbol{I}}_2, \end{aligned} \tag{10.19}$$

Table 10.1: Conversion Between Y-, Z-, A- and B-Parameters

	Y	Z
Y	Y	Z^{-1}
Z	Y^{-1}	Z
A	$\begin{bmatrix} Y_{11} & -1 \\ Y_{21} & 0 \end{bmatrix}^{-1} \begin{bmatrix} Y_{12} & 0 \\ -Y_{22} & -1 \end{bmatrix}$	$\begin{bmatrix} -1 & \mathbf{Z}_{11} \\ 0 & \mathbf{Z}_{21} \end{bmatrix}^{-1} \begin{bmatrix} 0 & \mathbf{Z}_{12} \\ 1 & \mathbf{Z}_{22} \end{bmatrix}$
B	$\begin{bmatrix} -Y_{12} & 0 \\ -Y_{22} & 1 \end{bmatrix}^{-1} \begin{bmatrix} Y_{11} & 1 \\ Y_{21} & -0 \end{bmatrix}$	$\begin{bmatrix} 0 & \mathbf{Z}_{122} \\ -1 & \mathbf{Z}_{22} \end{bmatrix}^{-1} \begin{bmatrix} 1 & \mathbf{Z}_{11} \\ 0 & \mathbf{Z}_{21} \end{bmatrix}$

where $\underline{V}_1$, $\underline{I}_1$ are the input quantities and $\underline{V}_2$, $\underline{I}_2$ are the output quantities. The $\mathbf{Z}_{11} \ldots \mathbf{Z}_{22}$ are the four mth order quadratic submatrices of the matrix $\mathbf{Z}$. To compute $\boldsymbol{A}_{11}$ we set $\underline{I}_2 = 0$ in (10.10) and obtain

$$\underline{V}_1 = \boldsymbol{A}_{11}\, \underline{V}_2 \quad \text{for } \underline{I}_2 = 0\,. \tag{10.20}$$

From (10.19) it follows that

$$\underline{V}_1 = \mathbf{Z}_{11} \underline{I}_1 = \mathbf{Z}_{11}\, \mathbf{Z}_{21}^{-1}\, \underline{V}_2 \quad \text{for } \underline{I}_2 = 0\,. \tag{10.21}$$

From this we obtain

$$\boldsymbol{A}_{11} = \mathbf{Z}_{11}\, \mathbf{Z}_{21}^{-1} \quad \text{for } \underline{I}_2 = 0\,. \tag{10.22}$$

For $\boldsymbol{A}_{21}$ we obtain from (10.10)

$$\underline{I}_1 = \boldsymbol{A}_{21}\, \underline{V}_2 \quad \text{for } \underline{I}_2 = 0\,. \tag{10.23}$$

In the same way we obtain for $\underline{I}_2 = 0$ the submatrix

$$\boldsymbol{A}_{21} = \mathbf{Z}_{21}^{-1}\,. \tag{10.24}$$

To determine $\boldsymbol{A}_{12}$ and $\boldsymbol{A}_{22}$ we set $\underline{V}_2 = 0$ and obtain from (10.10) and (10.19)

$$\underline{I}_1 = -\mathbf{Z}_{21}^{-1}\, \mathbf{Z}_{22}\, \underline{I}_2 \quad \text{for } \underline{V}_2 = 0\,. \tag{10.25}$$

By comparison with (10.10) we obtain

$$\boldsymbol{A}_{22} = \mathbf{Z}_{21}^{-1}\, \mathbf{Z}_{22}\,. \tag{10.26}$$

Table 10.2: Conversion Between Y-, Z-, A- and B-Parameters

	A	B
Y	$\begin{bmatrix} 0 & A_{12} \\ 1 & A_{22} \end{bmatrix}^{-1} \begin{bmatrix} -1 & A_{11} \\ 0 & A_{21} \end{bmatrix}$	$\begin{bmatrix} B_{12} & 0 \\ B_{22} & 1 \end{bmatrix}^{-1} \begin{bmatrix} B_{11} & -1 \\ B_{21} & 0 \end{bmatrix}$
Z	$\begin{bmatrix} -1 & A_{11} \\ 0 & A_{21} \end{bmatrix}^{-1} \begin{bmatrix} 0 & A_{12} \\ 1 & A_{22} \end{bmatrix}$	$\begin{bmatrix} B_{11} & -1 \\ B_{21} & 0 \end{bmatrix}^{-1} \begin{bmatrix} B_{12} & 0 \\ B_{22} & -1 \end{bmatrix}$
A	A	$\begin{bmatrix} B_{11} & -B_{12} \\ B_{21} & -B_{22} \end{bmatrix}^{-1} \begin{bmatrix} 1 & 0 \\ 0 & -1 \end{bmatrix}$
B	$\begin{bmatrix} A_{11} & -A_{12} \\ A_{21} & -A_{22} \end{bmatrix}^{-1} \begin{bmatrix} 1 & 0 \\ 0 & -1 \end{bmatrix}$	B

From (10.19) it follows that

$$\underline{V}_1 = (-Z_{11}\, Z_{21}^{-1}\, Z_{22} + Z_{12})\, \underline{I}_2 \tag{10.27}$$

and therewith under comparison with (10.10) we obtain

$$A_{12} = -Z_{12} + Z_{11}\, Z_{21}^{-1}\, Z_{22} \quad \text{for } \underline{V}_2 = 0\,. \tag{10.28}$$

With this we can express the A matrix by the Z matrix in the following form:

$$A = \begin{bmatrix} Z_{11}\, Z_{21}^{-1} & -Z_{12} + Z_{11}\, Z_{21}^{-1}\, Z_{22} \\ Z_{21}^{-1} & Z_{21}^{-1}\, Z_{22} \end{bmatrix}. \tag{10.29}$$

The so-called *inverse chain matrix* B represents the output signals as a function of the input signals

$$\begin{bmatrix} \underline{V}_2 \\ \underline{I}_2 \end{bmatrix} = B \begin{bmatrix} \underline{V}_1 \\ -\underline{I}_1 \end{bmatrix}. \tag{10.30}$$

For the cascading of two multiports according to Figure 10.4 we obtain

$$B = B_2\, B_1\,. \tag{10.31}$$

We consider that in spite of the name inverse chain matrix $B \neq A^{-1}$. Conversion formulae between the Y-, Z-, A- and B-representation are given in Table 10.1 and Table 10.2.

10.4 THE SCATTERING MATRIX

The multiport according to Figure 10.2(b) is described by the wave amplitudes of the waves incident through the ports and scattered through the ports. The wave amplitudes $\underline{a}_k$ and $\underline{b}_k$ are related to the (generalized) voltages $\underline{V}_k$ and (generalized) currents $\underline{I}_k$ according to (8.40a) and (8.40b) via

$$\underline{a}_k = \frac{1}{2\sqrt{2}}\left(\frac{\underline{V}_k}{\sqrt{Z_{wk}}} + \sqrt{Z_{wk}}\underline{I}_k\right), \tag{10.32a}$$

$$\underline{b}_k = \frac{1}{2\sqrt{2}}\left(\frac{\underline{V}_k}{\sqrt{Z_{wk}}} - \sqrt{Z_{wk}}\underline{I}_k\right). \tag{10.32b}$$

We allow at various ports waveguide types of different kind, e.g., a multiport may contain coaxial ports as well as waveguide ports. Especially we have to consider that the wave impedances of the transmission lines of the various ports may be different. We summarize the wave amplitudes of the n incident waves $\underline{a}_1 \ldots \underline{a}_n$ and the n scattered waves $\underline{b}_1 \ldots \underline{b}_n$ in n-dimensional column vector

$$\underline{\boldsymbol{a}} = \left[\underline{a}_1, \underline{a}_2, \ldots \underline{a}_n\right]^T, \tag{10.33a}$$

$$\underline{\boldsymbol{b}} = \left[\underline{b}_1, \underline{b}_2, \ldots \underline{b}_n\right]^T. \tag{10.33b}$$

With the diagonal matrix $\boldsymbol{g}$ summarizing the square roots of the characteristic impedances,

$$\boldsymbol{g} = \text{diag}\left[\sqrt{Z_{w1}}, \sqrt{Z_{w2}}, \ldots \sqrt{Z_{wn}}\right], \tag{10.34}$$

We can write (10.32a) and (10.32b) in matrix notation:

$$\underline{\boldsymbol{a}} = \frac{1}{2\sqrt{2}}(\boldsymbol{g}^{-1}\,\underline{\boldsymbol{V}} + \boldsymbol{g}\,\underline{\boldsymbol{I}}), \tag{10.35a}$$

$$\underline{\boldsymbol{b}} = \frac{1}{2\sqrt{2}}(\boldsymbol{g}^{-1}\,\underline{\boldsymbol{V}} - \boldsymbol{g}\,\underline{\boldsymbol{I}}). \tag{10.35b}$$

If all ports exhibit the same characteristic impedance Z_0, we obtain

$$\boldsymbol{g} = \sqrt{Z_0}\mathbf{1}. \tag{10.36}$$

Multiplying (10.35a) and (10.35b) from the left with $\sqrt{2}\boldsymbol{g}$ and $\sqrt{2}\boldsymbol{g}^{-1}$ respectively and forming the sum and the difference, we obtain in analogy to (8.42a) and (8.42b)

$$\underline{\boldsymbol{V}} = \sqrt{2}\boldsymbol{g}\,(\underline{\boldsymbol{a}} + \underline{\boldsymbol{b}}), \tag{10.37a}$$

$$\underline{\boldsymbol{I}} = \sqrt{2}\boldsymbol{g}^{-1}\,(\underline{\boldsymbol{a}} - \underline{\boldsymbol{b}}). \tag{10.37b}$$

Table 10.3: Conversion Between Y-, Z- and S-Parameters

Y	$=$	$\left[q+Sq\right]^{-1}\left[q^{-1}-Sq^{-1}\right]$
Z	$=$	$\left[q^{-1}-Sq^{-1}\right]^{-1}\left[q+Sq\right]$
S	$=$	$\left[q^{-1}+Yq\right]^{-1}\left[q^{-1}-Yq\right]$
S	$=$	$\left[Zq^{-1}+q\right]^{-1}\left[Zq^{-1}-q\right]$

The linear system of equations

$$\underline{b} = \boldsymbol{S}\,\underline{a} \tag{10.38}$$

gives the dependence of $\underline{b}$ on $\underline{a}$. The matrix $\boldsymbol{S}$ is named the $\boldsymbol{S}$ matrix or *scattering matrix*. The representation of a multiport by the scattering matrix is the *scattering representation*. This name originates from the waves incident into the ports which are scattered inside the multiport and then leave the waves as scattered waves.

Conversion formulae between the $\boldsymbol{Y}$-, $\boldsymbol{Z}$- and $\boldsymbol{S}$-representations are given in Table 10.3. We show how the scattering matrix may be represented in terms of the impedance matrix. To do this we insert (10.37a) and (10.37b) into (10.5) and obtain

$$\boldsymbol{g}\,(\underline{a}+\underline{b}) = \boldsymbol{Z}\,\boldsymbol{g}^{-1}\,(\boldsymbol{a}-\underline{b})\,. \tag{10.39}$$

Multiplying this equation from the left with $\boldsymbol{g}^{-1}$ we obtain

$$(\boldsymbol{g}^{-1}\boldsymbol{Z}\,\boldsymbol{g}^{-1}+\boldsymbol{1})\,\underline{b} = (\boldsymbol{g}^{-1}\,\boldsymbol{Z}\,\boldsymbol{g}^{-1}-\boldsymbol{1})\,\underline{a}\,. \tag{10.40}$$

Multiplying from the left with the inverse of the bracketed expression on the right we obtain

$$\underline{b} = (\boldsymbol{g}^{-1}\boldsymbol{Z}\,\boldsymbol{g}^{-1}+\boldsymbol{1})^{-1}\,(\boldsymbol{g}^{-1}\,\boldsymbol{Z}\,\boldsymbol{g}^{-1}-\boldsymbol{1})\,\underline{a}\,. \tag{10.41}$$

Comparing this with (10.38) yields the representation of the scattering matrix via the $\boldsymbol{Z}$ matrix

$$\boldsymbol{S} = (\boldsymbol{g}^{-1}\,\boldsymbol{Z}\,\boldsymbol{g}^{-1}+\boldsymbol{1})^{-1}\,(\boldsymbol{g}^{-1}\,\boldsymbol{Z}\,\boldsymbol{g}^{-1}-\boldsymbol{1})\,. \tag{10.42}$$

We note that the bracketed expressions in (10.42) may be interchanged. The proof for this is as follows: The matrix $\boldsymbol{1}$ may be interchanged with every matrix, and therefore also with $\boldsymbol{g}^{-1}\,\boldsymbol{Z}\,\boldsymbol{g}^{-1}$. If two matrices may be interchanged, the sum of both matrices may also be interchanged with the difference of both matrices. Furthermore, the two matrices $\boldsymbol{A}$ and $\boldsymbol{B}^{-1}$ may be interchanged, if $\boldsymbol{A}$ and $\boldsymbol{B}$ are interchangeable. This may

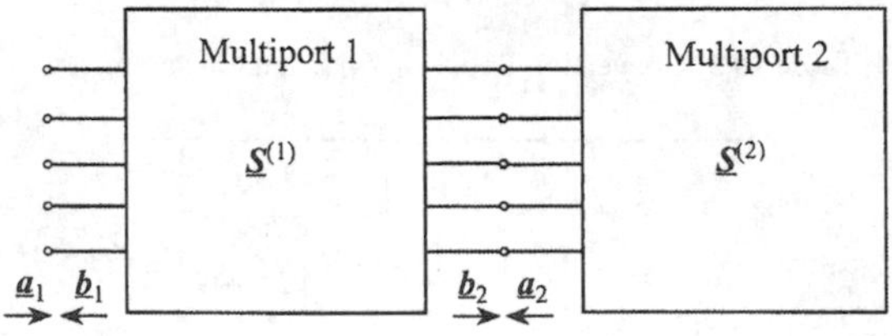

Figure 10.5: Connection of two multiports.

be demonstrated by multiplying the equation $\boldsymbol{A}\,\boldsymbol{B} = \boldsymbol{B}\,\boldsymbol{A}$ on both sides from the left and from the right with $\boldsymbol{B}^{-1}$.

We now consider the connection of two multiports according to Figure 10.5. The first multiport has a port number n_1; the second multiport has a port number $n_2 < n_1$. Every port of the second multiport is connected with a port of the first multiport. Therefore at the first multiport

$$n = n_1 - n_2 \tag{10.43}$$

ports remain unconnected. We number the ports of the first multiport such that the ports remaining unconnected have the numbers 1 to n. The multiports 1 and 2 are described by the scattering matrices $\boldsymbol{S}^{(1)}$ and $\boldsymbol{S}^{(2)}$. We summarize the wave amplitudes of the ports 1 to n of the first multiport in the vectors $\underline{\boldsymbol{a}}_1$ and $\underline{\boldsymbol{b}}_1$. The wave amplitudes of ports $n+1$ to n_1 are summarized in the column vectors $\underline{\boldsymbol{a}}_2$ and $\underline{\boldsymbol{b}}_2$. The waves incident into the second multiport are summarized in $\underline{\boldsymbol{b}}_2$; the waves flowing out from the ports of the second multiport are described by $\underline{\boldsymbol{a}}_2$. We assume that the connected ports of both multiports are of the same type. The circuit equations of both multiports are

$$\begin{bmatrix} \underline{\boldsymbol{b}}_1 \\ \underline{\boldsymbol{b}}_2 \end{bmatrix} = \boldsymbol{S}^{(1)} \begin{bmatrix} \underline{\boldsymbol{a}}_1 \\ \underline{\boldsymbol{a}}_2 \end{bmatrix}, \tag{10.44}$$

$$\underline{\boldsymbol{a}}_2 = \boldsymbol{S}^{(2)}\,\underline{\boldsymbol{b}}_2 . \tag{10.45}$$

Subdividing the scattering matrix of the first multiport into submatrices $\boldsymbol{S}_{11}$, $\boldsymbol{S}_{12}$, $\boldsymbol{S}_{21}$ and $\boldsymbol{S}_{22}$, we obtain from (10.44):

$$\underline{\boldsymbol{b}}_1 = \boldsymbol{S}_{11}^{(1)}\,\underline{\boldsymbol{a}}_1 + \boldsymbol{S}_{12}^{(1)}\,\underline{\boldsymbol{a}}_2 , \tag{10.46a}$$

$$\underline{\boldsymbol{b}}_2 = \boldsymbol{S}_{21}^{(1)}\,\underline{\boldsymbol{a}}_1 + \boldsymbol{S}_{22}^{(1)}\,\underline{\boldsymbol{a}}_2 . \tag{10.46b}$$

Inserting (10.46b) into (10.45) we obtain

$$\underline{\boldsymbol{a}}_2 = \boldsymbol{S}^{(2)}\,\boldsymbol{S}_{21}^{(1)}\,\underline{\boldsymbol{a}}_1 + \boldsymbol{S}^{(2)}\,\boldsymbol{S}_{22}^{(1)}\,\underline{\boldsymbol{a}}_2 . \tag{10.47}$$

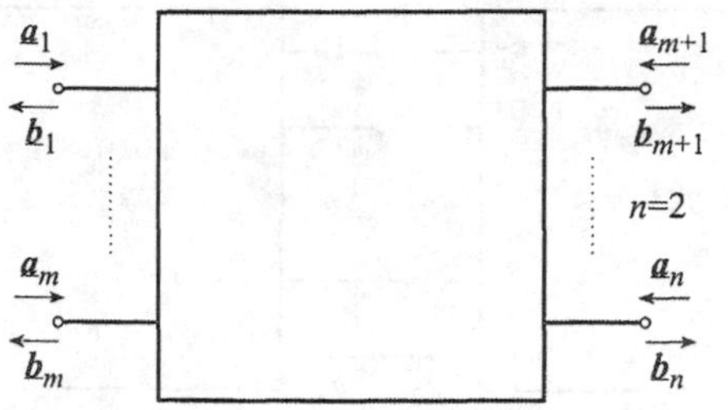

Figure 10.6: Port number symmetric multiport.

We bring this expression into the following form:

$$\underline{a}_2 = (\mathbf{1} - \boldsymbol{S}^{(2)}\, \boldsymbol{S}_{22}^{(1)})^{-1}\, \boldsymbol{S}^{(2)}\, \boldsymbol{S}_{21}^{(1)}\, \underline{a}_1\,. \tag{10.48}$$

After inserting into (10.46a), we obtain

$$\underline{b}_1 = \left[\boldsymbol{S}_{11}^{(1)} + \boldsymbol{S}_{12}^{(1)}\,(\mathbf{1} - \boldsymbol{S}^{(2)}\, \boldsymbol{S}_{22}^{(1)})^{-1}\, \boldsymbol{S}^{(2)}\, \boldsymbol{S}_{21}^{(1)}\right] \underline{a}_1\,. \tag{10.49}$$

For the scattering matrix $\boldsymbol{S}_{<n \times n>}$, which relates incident and scattered waves of the first n ports of the first multiport according to

$$\underline{b}_1 = \boldsymbol{S}_{<n \times n>}\, \underline{a}_1 \tag{10.50}$$

we obtain

$$\boldsymbol{S} = \boldsymbol{S}_{11}^{(1)} + \boldsymbol{S}_{12}^{(1)}\,(\mathbf{1} - \boldsymbol{S}^{(2)}\, \boldsymbol{S}_{22}^{(1)})^{-1}\, \boldsymbol{S}^{(2)}\, \boldsymbol{S}_{21}^{(1)}\,. \tag{10.51}$$

10.5 THE TRANSMISSION MATRIX

The *transmission matrix* may be defined for port-number symmetric multiports according to Figure 10.6. It represents the incident and scattered wave amplitudes of the input ports in dependence from the incident and scattered wave amplitudes of the output ports. This representation of a multiport is called the *transmission representation.*

$$\begin{bmatrix} \underline{b}_1 \\ \underline{a}_1 \end{bmatrix} = \boldsymbol{T} \begin{bmatrix} \underline{a}_2 \\ \underline{b}_2 \end{bmatrix}. \tag{10.52}$$

The transmission matrix for the wave amplitudes therefore corresponds to the chain matrix for (generalized) voltages and (generalized) currents. In the column vectors $\underline{a}_1$ and $\underline{b}_1$ the wave amplitudes of the m input ports and the vectors $\underline{a}_2$ and $\underline{b}_2$ the

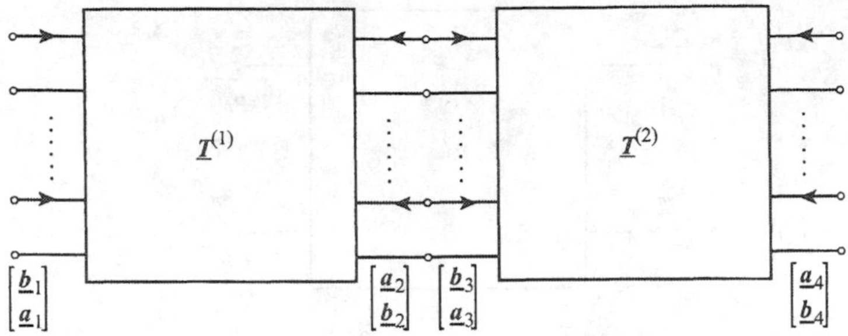

Figure 10.7: Cascading of two multiports.

wave amplitudes of the m output ports are summarized. The total number of ports is $n = 2m$.

The transmission matrix allows one to compute easily the cascading of two port-number symmetric multiports. For two cascaded multiports according to Figure 10.7 we obtain

$$\begin{bmatrix} \underline{b}_1 \\ \underline{a}_1 \end{bmatrix} = \boldsymbol{T}^{(1)} \begin{bmatrix} \underline{a}_2 \\ \underline{b}_2 \end{bmatrix}, \tag{10.53}$$

$$\begin{bmatrix} \underline{b}_3 \\ \underline{a}_3 \end{bmatrix} = \boldsymbol{T}^{(2)} \begin{bmatrix} \underline{a}_4 \\ \underline{b}_4 \end{bmatrix}, \tag{10.54}$$

where $\boldsymbol{T}^{(1)}$ and $\boldsymbol{T}^{(2)}$ are the transmission matrices of the multiports 1 and 2. By appropriately numbering the output ports of the first multiport and the input ports of the second multiport we obtain

$$\begin{bmatrix} \underline{a}_2 \\ \underline{b}_2 \end{bmatrix} = \begin{bmatrix} \underline{b}_3 \\ \underline{a}_3 \end{bmatrix}. \tag{10.55}$$

From (10.53) to (10.55) it follows that

$$\begin{bmatrix} \underline{b}_1 \\ \underline{a}_1 \end{bmatrix} = \boldsymbol{T} \begin{bmatrix} \underline{a}_4 \\ \underline{b}_4 \end{bmatrix}, \tag{10.56}$$

where the transmission matrix $\boldsymbol{T}$ of the multiport obtained by cascading multiport 1 and multiport 2 is given by

$$\boldsymbol{T} = \boldsymbol{T}^{(1)} \boldsymbol{T}^{(2)}. \tag{10.57}$$

The transformation between the scattering representation and the transmission representation is performed by the formulae summarized in Table 10.4.

Table 10.4: Conversion Between S- and T-Parameters

S	=	$\begin{bmatrix} -1 & T_{12} \\ 0 & T_{22} \end{bmatrix}^{-1} \begin{bmatrix} 0 & T_{11} \\ 1 & T_{21} \end{bmatrix}$
T	=	$\begin{bmatrix} -1 & S_{11} \\ 0 & S_{21} \end{bmatrix}^{-1} \begin{bmatrix} -S_{12} & 0 \\ -S_{22} & 1 \end{bmatrix}$

10.6 TELLEGEN'S THEOREM

We already have discussed the field form (4.49) of Tellegen's theorem in Section 4.2. Tellegen's theorem states fundamental relations between voltages and currents in a circuit. It is only based on topological relationships and is independent from the constitutive laws of the network [6–9].

To derive the *network form of Tellegen's theorem* from field form (4.49) of Tellegen's theorem we consider the set V_B of all boundary surfaces of the electromagnetic structure. The set of boundary surfaces is of zero volume. In the integral we can substitute the fields by the transverse field components since the normal field components give no contribution,

$$\int_{\partial V_B} \mathcal{E}'_{tr}(\boldsymbol{x}, t') \wedge \mathcal{H}''_{tr}(\boldsymbol{x}, t'') = 0 . \tag{10.58}$$

Each surface of V_B has two sides and the integral over the boundary ∂V_B is performed over both sides of every boundary surface. We mark each side of every part of the set of boundary surfaces by α or β.

We now expand the fields $\underline{\mathcal{E}}$ and $\underline{\mathcal{H}}$ at the boundary surfaces into sets of structure functions h_m^ξ with $\xi = \alpha, \beta$. The numbers of structure functions are N_α on side α and N_β on side β. The expansion coefficients are the generalized voltages V_m^ξ and the generalized currents I_m^ξ. Inserting (8.9a) and (8.9b) we obtain

$$\mathcal{E}_{tr}^\xi = \sum_m^{N_\xi} V_m^\xi e_m^\xi , \tag{10.59a}$$

$$\mathcal{H}_{tr}^\xi = \sum_m^{N_\xi} I_m^\xi h_m^\xi . \tag{10.59b}$$

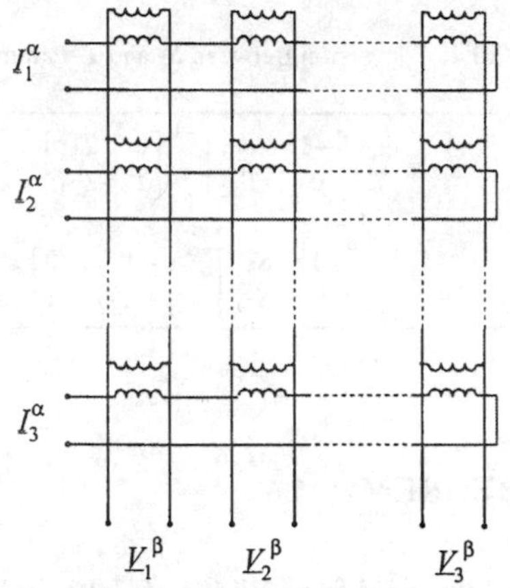

Figure 10.8: Canonical form of the connection network.

Inserting this expansions into (10.58) yields

$$\int_{\partial V_B} \mathcal{E}'(\boldsymbol{x}, t') \wedge \mathcal{H}''(\boldsymbol{x}, t'') = \sum_n^{N_\alpha} \sum_m^{N_\alpha} V_m^{\alpha'}(t') I_n^{\alpha''}(t'') \int_{\partial V_B} e_m^\alpha \wedge h_n^\alpha$$
$$+ \sum_n^{N_\beta} \sum_m^{N_\beta} V_m^{\beta'}(t') I_n^{\beta''}(t'') \int_{\partial V_B} e_m^\beta \wedge h_n^\beta = 0 . \quad (10.60)$$

We assume *orthonormal structure functions* as introduced in (8.23)

$$\int_{\partial V_B} e_m^\beta \wedge h_n^\beta = \delta_{mn} \quad (10.61)$$

and obtain

$$\sum_n^{N_\alpha} V_n^{\alpha'}(t') I_n^{\alpha''}(t'') + \sum_n^{N_\beta} V_n^{\beta'}(t') I_n^{\beta''}(t'') = 0 . \quad (10.62)$$

We summarize all voltage and currents in the vectors

$$\boldsymbol{V}(t) = \left[V_1^\alpha(t), \ldots V_{N_\alpha}^\alpha(t), V_1^\beta(t), \ldots V_{N_\alpha}^\beta(t), \right]^T , \quad (10.63a)$$

$$\boldsymbol{I}(t) = \left[I_1^\alpha(t), \ldots I_{N_\alpha}^\alpha(t), I_1^\beta(t), \ldots I_{N_\alpha}^\beta(t), \right]^T . \quad (10.63b)$$

and obtain the *general network form of Tellegen's theorem* [6–8, 10]

$$\boldsymbol{V}'^{T}(t')\,\boldsymbol{I}''(t'') = 0\,. \tag{10.64}$$

where $\boldsymbol{V}(t)$ and $\boldsymbol{I}(t)$ denote the voltage and current vectors of the connection circuit. The prime $'$ and double prime $''$ again denote different circuit elements and different times in both cases. It is only required that the topological structure of the connection circuit remains unchanged.

Taking (10.64) and

$$\boldsymbol{I}'^{T}(t')\,\boldsymbol{V}''(t'') = 0\,. \tag{10.65}$$

which follows directly from (10.64) by interchanging the primed with the double primed variables and considering the symmetry of the inner product and inserting (10.37a) and (10.37b) in both equations we obtain, after forming the sum and the difference of the resulting equations *Tellegen's Theorem in wave amplitude representation* [11, 12]

$$\boldsymbol{a}'^{T}(t')\,\boldsymbol{b}''(t'') = \boldsymbol{b}'^{T}(t')\,\boldsymbol{a}''(t'')\,, \tag{10.66a}$$

$$\boldsymbol{a}'^{T}(t')\,\boldsymbol{a}''(t'') = \boldsymbol{b}'^{T}(t')\,\boldsymbol{b}''(t'')\,. \tag{10.66b}$$

In a *network representation* of a circuit the connection circuit contains all connections but no circuit elements [5]. Expanding the tangential electric and magnetic fields on the boundaries again into basis functions allows us to give an equivalent circuit representation for the boundary surfaces. The equivalent circuit of the boundary surfaces is a connection circuit exhibiting only connections and ideal transformers.

Consistent choices of independent and dependent fields do not violate Tellegen's theorem and allow to draw canonical networks, which are based only on connections and ideal transformers. Figure 10.8 shows the canonical form of the connection network when using as independent fields the vectors $\boldsymbol{V}^{\beta}$ (dimension N_{β}) and $\boldsymbol{I}^{\alpha}$ (dimension N_{α}). In this case the dependent fields are $\underline{V}^{\alpha}$ (dimension N_{α}) and $\boldsymbol{I}^{\beta}$ (dimension N_{β}). In all cases we have $N_{\beta} + N_{\alpha}$ independent quantities and the same number of dependent quantities. Note that scattering representations are also allowed and that the connection network is frequency-independent. It is apparent from the canonical network representations that the scattering matrix is symmetric, i.e., $\boldsymbol{S}^{T} = \boldsymbol{S}$, orthogonal, i.e., $\boldsymbol{S}^{T}\boldsymbol{S} = \boldsymbol{1}$ and unitary, i.e., $\boldsymbol{S}\boldsymbol{S}^{\dagger} = \boldsymbol{1}$. The † denotes the hermitian conjugate matrix.

10.7 THE POWER PROPERTIES

The total active power P flowing into a multiport is equal to the sum of the active power flowing into every port. We obtain from (8.46)

$$P = \sum_{i=1}^{n} |\underline{a}_i|^2 - |\underline{b}_i|^2 = \underline{\boldsymbol{a}}^\dagger\, \underline{\boldsymbol{a}} - \underline{\boldsymbol{b}}^\dagger\, \underline{\boldsymbol{b}}\,. \tag{10.67}$$

The superscript † denotes the *Hermitian conjugate* of a vector or matrix respectively as defined in (C.35). The term $\underline{\boldsymbol{a}}^\dagger\, \underline{\boldsymbol{a}}$ represents the power carried by the waves incident into the multiport and $\underline{\boldsymbol{b}}^\dagger\, \underline{\boldsymbol{b}}$ is the power carried by the waves scattered from the multiport. The difference is the power P absorbed in the multiport. Inserting (10.38) and its Hermitian conjugate

$$\underline{\boldsymbol{b}}^\dagger = \underline{\boldsymbol{a}}^\dagger\, \boldsymbol{S}^\dagger \tag{10.68}$$

we can represent this equation by

$$P = \underline{\boldsymbol{a}}^\dagger\, (\boldsymbol{1} - \boldsymbol{S}^\dagger\, \boldsymbol{S})\, \underline{\boldsymbol{a}}\,. \tag{10.69}$$

For a *passive multiport* for an arbitrary choice of the incident waves $\underline{\boldsymbol{a}}$ always $P \geq 0$ must be valid, i.e., the right-hand side of (10.69) must be a *positive semidefinite Hermitian form* in $\underline{\boldsymbol{a}}$. This is fulfilled if and only if

$$\det\nolimits_k(\boldsymbol{1} - \boldsymbol{S}^\dagger\, \boldsymbol{S}) \geq 0 \qquad \text{for } k = 1 \ldots n \tag{10.70}$$

is valid, where the symbol $\det_k$ denotes the *leading principal minor* of order k of a matrix as defined in (C.47).

If the multiport is *lossless*, the total absorbed active power is $P = 0$. A lossless multiport is called a *reactance multiport* With (10.69) we obtain

$$\boldsymbol{S}^\dagger = \boldsymbol{S}^{-1} \qquad \text{for lossless multiports.} \tag{10.71}$$

A matrix fulfilling (10.71) is called *unitary*. The scattering matrix describing a *lossless multiport* is unitary. In impedance and admittance representation we obtain from (10.42) and (10.71) for *reactance multiports*

$$\underline{\boldsymbol{Z}}(\omega) + \underline{\boldsymbol{Z}}^\dagger(\omega) = \boldsymbol{0}\,, \qquad \underline{\boldsymbol{Y}}(\omega) + \underline{\boldsymbol{Y}}^\dagger(\omega) = \boldsymbol{0}\,. \tag{10.72}$$

A *lossless one-port* is called a *reactance one-port*. It has the property

$$\Re\{Z(\omega)\} = 0\,, \quad \Re\{Y(\omega)\} = 0\,, \quad |\rho(\omega)| = 1\,. \tag{10.73}$$

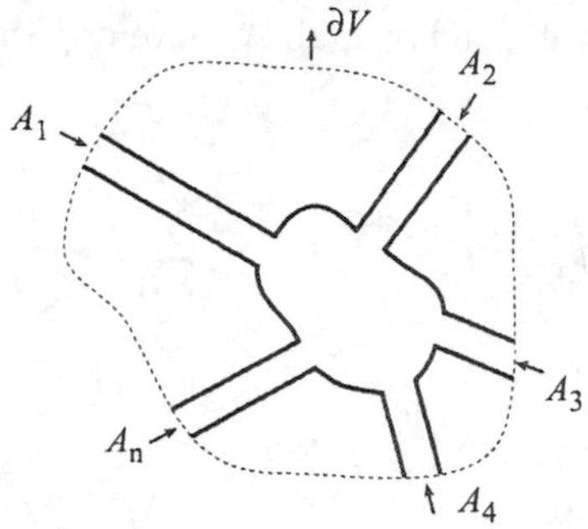

Figure 10.9: Embedding of a multiport into the domain of integration V.

10.8 RECIPROCAL MULTIPORTS

We already have treated reciprocity in Section 4.6 from a field point of view. We now derive the Lorentz reciprocity theorem for source-free multiports. To do this we consider the multiport depicted schematically in Figure 10.9 and apply the integral form of the Lorentz reciprocity theorem to this structure.

The planes of reference of the m ports are part of the boundary ∂V of the volume V. From (8.9a) and (8.9b) we obtain the transverse components of the field quantities in the plane of reference of the kth port, $(k = 1 \dots n)$:

$$\underline{\mathcal{E}}^{(i)}_{tr,k} = \underline{V}^{(i)}_k e_k \,, \tag{10.74}$$

$$\underline{\mathcal{H}}^{(i)}_{tr,k} = \underline{I}^{(i)}_k h_k \,. \tag{10.75}$$

The index i refers to the excitation of field by the ith group, $(i = 1 \dots n)$, of impressed field sources where the impressed field sources all are located outside the volume V and may be represented by sources connected to the ports of the multiport. The voltages $\underline{V}^{(i)}_k$ and the currents $\underline{I}^{(i)}_k$ are the (generalized) port voltages and (generalized) port currents of the kth port in the case of excitation of the ith group of sources. The one-forms e_k and h_k are the normalized structure forms according to (8.8a) and (8.8b).

We compute the area integral of $\underline{\mathcal{E}}_i \wedge \underline{\mathcal{H}}_j$ over the boundary ∂V. Since an electromagnetic field only occurs inside the waveguides, the integral only needs to be performed over the cross-sectional areas $A_1 \dots A_n$ of the n waveguides. Since $A_1 \dots A_n$ are transverse cross-sectional areas only the transverse field components in the waveguide cross-section give contributions to the area integral. We follow the usual conventions that the orientation of the boundary surface ∂V of the volume V is outwards whereas the cross-sectional areas of the waveguides are oriented inwards.

Therefore the integrals over ∂V and over all A_i have opposite sign. With (10.74) and (10.75) we obtain

$$\oint_{\partial V} \underline{\mathcal{E}}_i \wedge \underline{\mathcal{H}}_j = -\sum_{k=1}^{n} \int_{A_k} \underline{\mathcal{E}}^{(i)}_{tr,k} \wedge \underline{\mathcal{H}}^{(i)}_{tr,k} = -\sum_{k=1}^{n} \underline{V}^{(i)}_k \underline{I}^{(j)}_k \int_{A_k} \mathsf{e}_k \wedge \mathsf{h}_k \,. \quad (10.76)$$

From (8.24a) it follows that

$$\int_{A_k} \mathsf{e}_k \wedge \mathsf{h}_k = 1 \quad (10.77)$$

and therewith

$$\oint_{\partial V} \underline{\mathcal{E}}_i \wedge \underline{\mathcal{H}}_j = -\sum_{k=1}^{n} \underline{V}^{(i)}_k \underline{I}^{(j)}_k \,. \quad (10.78)$$

Inserting (10.78) into (4.58), we obtain

$$\sum_{k=1}^{n} \underline{V}^{(i)}_k \underline{I}^{(j)}_k = \sum_{k=1}^{n} \underline{V}^{(j)}_k \underline{I}^{(i)}_k \,. \quad (10.79)$$

We summarize the $\underline{V}^{(i)}_k$ and $\underline{I}^{(i)}_k$ into the n-dimensional column vectors $\underline{\boldsymbol{V}}^{(i)}$ and $\underline{\boldsymbol{I}}^{(i)}$. With the transposed vectors $\underline{\boldsymbol{V}}^{T(i)}$ and $\underline{\boldsymbol{I}}^{T(i)}$, we obtain according to

$$\boldsymbol{a}^T \boldsymbol{b} = a_1 b_1 + a_2 b_2 + \dots a_n b_n \quad (10.80)$$

from (10.79)

$$\underline{\boldsymbol{V}}^{T(i)} \underline{\boldsymbol{I}}^{(j)} = \underline{\boldsymbol{I}}^{T(i)} \underline{\boldsymbol{V}}^{(j)} \,. \quad (10.81)$$

With (10.5) and the transposed equation

$$\underline{\boldsymbol{V}}^T = \underline{\boldsymbol{I}}^T \boldsymbol{Z}^T \quad (10.82)$$

we obtain

$$\underline{\boldsymbol{I}}^{T(i)} \boldsymbol{Z}^T \underline{\boldsymbol{I}}^{(j)} = \underline{\boldsymbol{I}}^{T(i)} \boldsymbol{Z} \underline{\boldsymbol{I}}^{(j)} \,. \quad (10.83)$$

This equation is valid for connection of arbitrary groups of external sources i and j. This is only possible for

$$\boldsymbol{Z} = \boldsymbol{Z}^T \,. \quad (10.84)$$

For a *reciprocal multiport* the $\boldsymbol{Z}$ matrix is symmetric. In component notation this means

$$Z_{ik} = Z_{ki} \,. \quad (10.85)$$

With (10.8) it follows that

$$\boldsymbol{Y} = \boldsymbol{Y}^T \,. \quad (10.86)$$

Inserting (10.84) into (10.42) and considering that the bracketed expressions in (10.42) are commuting and that $\boldsymbol{g} = \boldsymbol{g}^T$ and

$$(\boldsymbol{A}\,\boldsymbol{B})^T = \boldsymbol{B}^T\,\boldsymbol{A}^T, \tag{10.87}$$

we obtain

$$\boldsymbol{S} = \boldsymbol{S}^T. \tag{10.88}$$

10.9 THE SYMMETRY PROPERTIES OF WAVEGUIDE JUNCTIONS

The complete analysis of distributed circuits by solving Maxwell's equations requires a considerable mathematical effort also in the case of simple structures. For waveguide circuits exhibiting geometric symmetries from the symmetry properties of the circuits symmetry properties of the scattering matrixes also follow. Also from general considerations we can get important information about the scattering matrix of a multiport. For example, we know that the scattering matrix of a reciprocal passive lossless multiport must be symmetric and unitary.

We will show in the following that from geometric symmetry considerations we can get additional information about the scattering matrix and the scattering matrix from this may be determined with the exception of only a few unknown parameters. If, for example, only a single parameter may not be determined from computation, in many cases the experimental determination of this parameter may be a good solution.

We consider multiports exhibiting certain geometric symmetries. If, for example, the geometric structure of a multiport exhibits symmetry with respect to the rotation around an axis or with respect to mirroring at a symmetry plane, this rotation and this mirroring are *symmetry operations* for that multiport. The multiport is said to be invariant with respect to these symmetry operations. In that case the field solution remains unchanged, if the symmetry operation is performed to the multiport. If the arrows indicating the direction of the field remain unchanged when the symmetry operation is performed the scattering matrix remains unchanged under this symmetry operation.

The mirroring of the multiport structure may yield a change of sign of field components. Considering the mirroring in the yz-plane, we obtain

$$x \to -x\,, \qquad y \to y\,, \qquad z \to z\,. \tag{10.89}$$

Making this substitution in Maxwell's equation, this will yield the following changes in the field components:

$$\begin{aligned} &\underline{E}_x(x,y,z) \to -\underline{E}_x(-x,y,z)\,, && \underline{E}_y(x,y,z) \to \underline{E}_y(-x,y,z)\,, \\ &\underline{E}_z(x,y,z) \to \underline{E}_z(-x,y,z)\,, && \underline{H}_x(x,y,z) \to \underline{H}_x(-x,y,z)\,, \\ &\underline{H}_y(x,y,z) \to -\underline{H}_y(-x,y,z)\,, && \underline{H}_z(x,y,z) \to -\underline{H}_z(-x,y,z)\,. \end{aligned} \tag{10.90}$$

Let $\underline{E}_{tr,i}$ be the transverse component of the electric field intensity in a certain point of the reference plane of the ith multiport. We assume that by a certain symmetry operation ports i and j are transformed into ports k and l respectively. From this it follows that

$$\begin{aligned} \underline{E}_{tr,i} &\to \delta_1\,\underline{E}_{tr,k} \\ & && \delta_1, \delta_2 = \pm 1\,. \\ \underline{E}_{tr,j} &\to \delta_2\,\underline{E}_{tr,l} \end{aligned} \tag{10.91}$$

Depending on the change of sign of the transverse electric fields in ports j and k the matrix element S_{kl} remains unchanged or changes its sign when the symmetry operation is performed, following the rule

$$S_{ij} = \delta_1 \delta_2\, S_{kl}\,. \tag{10.92}$$

10.9.1 Symmetric Three-Port Waveguide Junctions

We consider the fully symmetric three-port *parallel waveguide junction* according to Figure 10.10. The three waveguide segments are long enough that any perturbation of the field distribution occurring in the center region where all three waveguides are connected already has decayed in the port plane. We can assume that in the port plane the transverse field distribution corresponds to the TE_{10} mode. We assume the waveguide three-port parallel waveguide junction to be reciprocal and lossless. In the case of the parallel waveguide junction the broader side of the rectangular waveguide is in parallel to the plane of the drawing. The electric field is normal to the plane of the drawing. This corresponds to the case of parallel circuited transmission lines. Therefore this junction is called a parallel junction. The circuit is invariant with respect to rotation by 120° around an axis normal to the plane of the drawing. Furthermore the structure exhibits three symmetry planes normal to the plane of the drawing. We allow a lossless scattering object in the center of the waveguide junction and we make the general assumption that the symmetry properties of the waveguide junction are not disturbed by this scatterer. The rotation of the waveguide branch by 120° effects the following interchange of field quantities:

$$\underline{\mathbf{E}}_1 \to \underline{\mathbf{E}}_2\,, \qquad \underline{\mathbf{E}}_2 \to \underline{\mathbf{E}}_3\,, \qquad \underline{\mathbf{E}}_3 \to \underline{\mathbf{E}}_1\,. \tag{10.93}$$

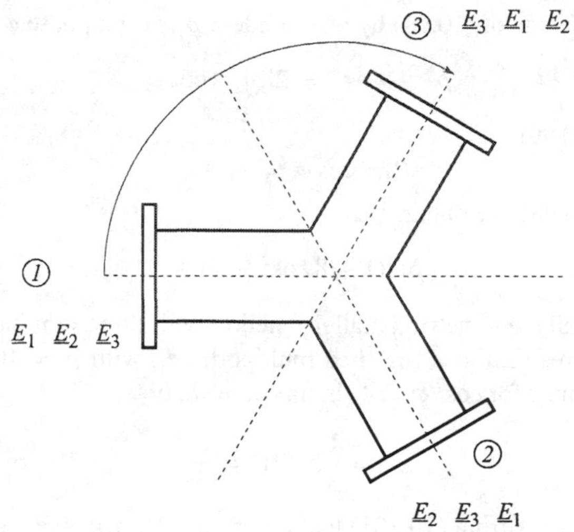

Figure 10.10: Fully symmetric three-port parallel waveguide junction.

With (10.92) we obtain from this

$$\begin{aligned} S_{11} &= S_{22} = S_{33}\,, \\ S_{12} &= S_{23} = S_{31}\,. \end{aligned} \tag{10.94}$$

The mirroring in the symmetry plane parallel to the axis of waveguide branch 1 effects the following interchanges of electric field components

$$\underline{\boldsymbol{E}}_1 \rightarrow \underline{\boldsymbol{E}}_1\,, \qquad \underline{\boldsymbol{E}}_2 \rightarrow \underline{\boldsymbol{E}}_3\,, \qquad \underline{\boldsymbol{E}}_3 \rightarrow \underline{\boldsymbol{E}}_2\,. \tag{10.95}$$

From this it follows that

$$S_{23} = S_{32}\,. \tag{10.96}$$

Due to the reciprocity, this relation also follows from (10.88). Due to (10.94) and (10.96) the scattering matrix now is determined with the exception of the two complex parameters ρ and τ:

$$\boldsymbol{S} = \begin{bmatrix} \rho & \tau & \tau \\ \tau & \rho & \tau \\ \tau & \tau & \rho \end{bmatrix}. \tag{10.97}$$

Assuming the junction to be lossless, we obtain from (10.71)

$$|\rho|^2 + 2|\tau|^2 = 1\,, \tag{10.98}$$

$$\rho^*\tau + \rho\tau^* = -|\tau|^2\,. \tag{10.99}$$

Expressing the left side of (10.99) by magnitude $2|\rho||\tau|$ and phase φ,

$$\rho^*\tau + \rho\tau^* = 2|\rho||\tau|\cos\varphi \tag{10.100}$$

we obtain with (10.99)

$$2|\rho|\cos\varphi = -|\tau|\,. \tag{10.101}$$

After inserting (10.98), we obtain

$$|\rho|^2(1 + 8\cos^2\varphi) = 1\,. \tag{10.102}$$

We see that the fully symmetric parallel junction with three terminations cannot simultaneously be realized as a matched multiport, i.e., with $\rho = 0$. The minimum value of ρ is obtained for $\cos^2\varphi = 1$. In this case we obtain

$$|\rho| = \frac{1}{3}, \quad |\tau| = \frac{2}{3}\,. \tag{10.103}$$

Furthermore it follows from (10.101) that $\cos\varphi$ must be negative. With $\cos\varphi = -1$, we obtain from (10.100)

$$\Re\{\rho\tau^*\} = -|\rho||\tau| \tag{10.104}$$

and from this with (10.103)

$$\rho = \frac{-1}{3}\,\mathrm{e}^{-\mathrm{j}\varphi_1}, \quad \tau = \frac{2}{3}\,\mathrm{e}^{-\mathrm{j}\varphi_1}\,, \tag{10.105}$$

where $\mathrm{e}^{-\mathrm{j}\varphi_1}$ is a parameter depending on the length of the waveguide arms. Using symmetry and power considerations we could determine the scattering matrix of the completely symmetric parallel junction with three arms with the exception of a remaining scalar parameter φ_1. The optimization of the parameter φ_1 may be accomplished empirically by variation of the scatterer introduced in the center of the parallel junction.

We have seen that a matched completely symmetric lossless three-port parallel junction cannot be realized. It can be shown in general that a matched lossless reciprocal three-port cannot be realized. A matched three-port must be characterized by the scattering matrix

$$\mathbf{S} = \begin{bmatrix} 0 & S_{12} & S_{13} \\ S_{12} & 0 & S_{23} \\ S_{13} & S_{23} & 0 \end{bmatrix}. \tag{10.106}$$

If the three-port is lossless, the scattering matrix must be unitary. With (10.71) we obtain from (10.106)

$$\begin{aligned} S_{13}S_{23}^* &= 0\,, \\ S_{12}S_{13}^* &= 0\,, \\ S_{12}S_{23}^* &= 0\,. \end{aligned} \tag{10.107}$$

These equations only may be fulfilled if at least one of the S_{ik} is vanishing.

10.9.2 Symmetric Four-Port Waveguide Junctions

The so-called *directional coupler* according to Figure 10.11 exhibits two symmetry planes normal to the plane of the drawing. Mirroring in symmetry plane 1 yields the following interchanges of field components

$$\underline{E}_1 \to \underline{E}_3\,, \qquad \underline{E}_2 \to \underline{E}_4\,, \qquad \underline{E}_3 \to \underline{E}_1 \qquad \underline{E}_4 \to \underline{E}_2\,. \tag{10.108}$$

From this it follows for the scattering matrix

$$\begin{aligned} S_{11} &= S_{33} & S_{22} &= S_{44}\,, \\ S_{12} &= S_{34} & (S_{13} &= S_{31})\,, \\ S_{14} &= S_{32} & (S_{24} &= S_{42})\,. \end{aligned} \tag{10.109}$$

The bracketed conditions already followed from reciprocity. From the mirroring in the second symmetry plane we obtain the following exchanges

$$\underline{E}_1 \to \underline{E}_2\,, \qquad \underline{E}_2 \to \underline{E}_1\,, \qquad \underline{E}_3 \to \underline{E}_4 \qquad \underline{E}_4 \to \underline{E}_3 \tag{10.110}$$

and from this the following additional conditions for the scattering matrix

$$S_{11} = S_{22}\,, \qquad S_{33} = S_{44}\,, \qquad S_{13} = S_{24}\,. \tag{10.111}$$

With this the scattering matrix is determined with the exception of four complex parameters ρ, τ_1, τ_2 and τ_3.

$$\boldsymbol{S} = \begin{bmatrix} \rho & \tau_1 & \tau_2 & \tau_3 \\ \tau_1 & \rho & \tau_3 & \tau_2 \\ \tau_2 & \tau_3 & \rho & \tau_1 \\ \tau_3 & \tau_2 & \tau_1 & \rho \end{bmatrix}. \tag{10.112}$$

We now assume the multiport to be matched so that $\rho = 0$ is valid. This may be achieved, for example, if both waveguides in Figure 10.11 are only weakly coupled over small holes or if the coupling region of both waveguides is long enough so that a coupling of waves in forward and backward direction will not occur. Assuming the directional coupler to be lossless, we obtain from (10.71)

$$\Re\{\tau_2^* \tau_3\} = 0\,, \quad \Re\{\tau_1^* \tau_3\} = 0\,, \quad \Re\{\tau_2^* \tau_1\} = 0\,. \tag{10.113}$$

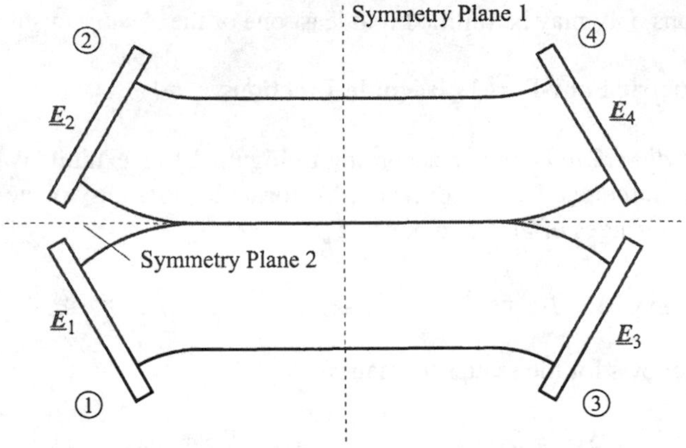

Figure 10.11: Directional coupler.

These three equations mean that the three complex quantities τ_1, τ_2 and τ_3 are mutually orthogonal in the complex number plane. Hence one of these quantities must vanish. Without restriction in generality we assume $\tau_1 = 0$ and obtain the scattering matrix

$$\boldsymbol{S} = \begin{bmatrix} \boldsymbol{0} & \boldsymbol{S}_{12} \\ \boldsymbol{S}_{12}^T & \boldsymbol{0} \end{bmatrix} \tag{10.114}$$

where the submatrix $\boldsymbol{S}_{12}$ is given by

$$\boldsymbol{S}_{12} = \boldsymbol{S}_{12}^T = \begin{bmatrix} \tau_2 & \tau_3 \\ \tau_3 & \tau_2 \end{bmatrix}. \tag{10.115}$$

Due to (10.71), $\boldsymbol{S}_{12}$ also must be unitary and we obtain

$$|\tau_2|^2 + |\tau_3|^2 = 1\,. \tag{10.116}$$

The scattering matrix of the matched lossless direction coupler only depends on a parameter τ_3 and with the exception of a phase factor it is given by

$$\begin{bmatrix} 0 & 0 & \sqrt{1-k^2} & \pm jk \\ 0 & 0 & \pm jk & \sqrt{1-k^2} \\ \sqrt{1-k^2} & \pm jk & 0 & 0 \\ \pm jk & \sqrt{1-k^2} & 0 & 0 \end{bmatrix}. \tag{10.117}$$

For the 3 dB coupler we obtain

$$k = \frac{1}{\sqrt{2}}. \tag{10.118}$$

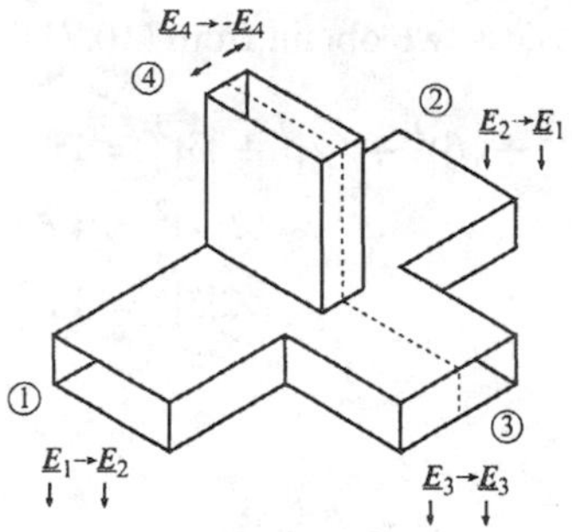

Figure 10.12: Hybrid junction.

The *hybrid junction* or *magic T* respectively according to Figure 10.12 exhibits only a symmetry plane $z = 0$. In the ports 1 to 3 the electric field lines are parallel to the symmetry plane, whereas in port 4 the electric field is normal to the symmetry plane. According to (10.90) the mirroring in the symmetry plane yields the following interchange of the electric field components

$$\underline{E}_1 \to \underline{E}_2\,, \qquad \underline{E}_2 \to \underline{E}_1\,, \qquad \underline{E}_3 \to \underline{E}_3\,, \qquad \underline{E}_4 \to -\underline{E}_4\,. \tag{10.119}$$

Ports 1 and 2 are interchanged in the mirroring, whereas ports 3 and 4 are transformed into themselves. Since the electric field in port 4 is normal to the symmetry plane, the sign of $\underline{E}_4$ is changed in the case of the mirroring. With (10.91) and (10.92) we obtain the following relations for the scattering parameters:

$$S_{11} = S_{22}, \qquad S_{13} = S_{23}\,, \tag{10.120}$$

$$S_{14} = -S_{24}, \qquad S_{34} = -S_{34} = 0\,. \tag{10.121}$$

From this it follows that the ports 3 and 4 are not directly coupled with each other. Therefore it must be possible to insert independently matching elements in arms 3 and 4 of the hybrid junction and to tune S_{33} and S_{44} independently such that

$$S_{33} = S_{44} = 0 \tag{10.122}$$

is valid.

The tuning elements must be inserted in such a way that the symmetry is maintained. If ports 3 and 4 are tuned to be free of reflection, the scattering matrix assumes the following form:

$$\begin{bmatrix} \alpha & \beta & \gamma & \delta \\ \beta & \alpha & \gamma & -\delta \\ \gamma & \gamma & 0 & 0 \\ \delta & -\delta & 0 & 0 \end{bmatrix}. \tag{10.123}$$

Assuming the circuit to be lossless, we obtain from (10.71)

$$\begin{aligned} |\alpha|^2 + |\beta|^2 + |\gamma|^2 + |\delta|^2 &= 1, \\ 2|\gamma|^2 &= 1, \\ 2|\delta|^2 &= 1. \end{aligned} \tag{10.124}$$

From this we obtain

$$\begin{aligned} \alpha = \beta &= 0, \\ |\gamma| = |\delta| &= \frac{1}{\sqrt{2}}. \end{aligned} \tag{10.125}$$

Therefore ports 1 and 2 are not coupled directly with each other. The scattering matrix according to (10.123) again describes a directional coupler. It may be shown in general that every matched lossless reciprocal four-port is a directional coupler. We can determine the lengths of arms 3 and 4 independently and therefore we may choose independently the phases of γ and δ. Without loss of generality we are choosing γ and δ to be positive and real and obtain

$$\boldsymbol{S} = \frac{1}{\sqrt{2}} \begin{bmatrix} 0 & 0 & 1 & 1 \\ 0 & 0 & 1 & -1 \\ 1 & 1 & 0 & 0 \\ 1 & -1 & 0 & 0 \end{bmatrix}. \tag{10.126}$$

REFERENCES

[1] C. Montgomery, R. Dicke, and E. Purcell, *Principles of Microwave Circuits*. New York: McGraw-Hill, 1948.

[2] H. Brand, *Schaltungslehre linearer Mikrowellennetze*. Stuttgart: Hirzel, 1970.

[3] R. E. Collin, *Foundations of Microwave Engineering*. New York: McGraw-Hill, 1992.

[4] G. Owyang, *Foundations for Microwave Circuits*. Berlin: Springer, 1989.

[5] H. Carlin and A. Giordano, *Network Theory*. Englewood Cliffs, NJ: Prentice Hall, 1964.

[6] B. Tellegen, "A general network theorem with applications," *Philips Research Reports*, vol. 7, pp. 259–269, 1952.

[7] B. Tellegen, "A general network theorem with applications," *Proc. Inst. Radio Engineers*, vol. 14, pp. 265–270, 1953.

[8] P. Penfield, R. Spence, and S. Duinker, *Tellegen's theorem and electrical networks*. Cambridge, Massachusetts: MIT Press, 1970.

[9] P. Russer, M. Mongiardo, and L. Felsen, "Electromagnetic field representations and computations in complex structures III: Network representations of the connection and subdomain circuits," *Int. J. Numer. Modeling*, vol. 15, pp. 127–145, 2002.

[10] C. Desoer and E. Kuh, *Basic Circuit Theory*. New York: McGraw-Hill, 1969.

[11] K. Gupta, R. Garg, and R. Chadha, *Computer-Aided Design of Microwave Circuits*. Boston: Artech House, 1981.

[12] J. Dobrowolski, *Introduction to Computer Methods for Microwave Circuit Analysis and Design*. Boston: Artech House, 1991.

Chapter 11

Antennas

11.1 INTRODUCTION

An antenna is any structure or device used to collect or radiate electromagnetic waves. Antennas allow the transformation of electromagnetic waves propagating along a transmission line into electromagnetic waves propagating in free space. This transformation may be performed in both directions. A transmitting antenna transforms a transmission line wave into a free-space wave, whereas a receiving antenna converts part of the free-space wave into a transmission line wave. An antenna usually is a reciprocal device and, in principle, each antenna may be used for receiving as well as for transmitting electromagnetic waves. However, the choice of the type of antenna and details in their construction depend on the special application of the antenna. An antenna is designed to achieve certain *radiation characteristics*, i.e., angular distribution of the radiated power. Furthermore volume, weight and mechanical stability play a role. In the case of transmitting antennas, the power of the transmitted signal also has to be considered. With decreasing wavelength the antenna dimensions decrease as well. Due to this circumstance the scope for antenna design at higher frequencies is larger. Figure 11.1 shows some antenna types. One of the most common antenna types is the dipole antenna shown in Figure 11.1(b). The dipole antenna usually is formed by two straight wire segments and excited by a source inserted between these segments. Using the mirror principle, we can put one wire segment in a position normal to a conducting plane. This yields the *monopole antenna* according to Figure 11.1(a). This antenna type is used frequently in the medium-wave and short-wave range. In this case the conducting plane is formed by the Earth's surface. The loop antenna depicted in Figure 11.1(c) excites a field that is dual to the field excited by the dipole antenna. The loop is formed by one or several turns of a wire. Figure 11.1(e) depicts a horn antenna. The horn antenna is formed by conically expanding a waveguide. If the aperture of the horn is large compared to the wavelength, the radiated power may be concentrated in a certain direction. We say then that this antenna has a

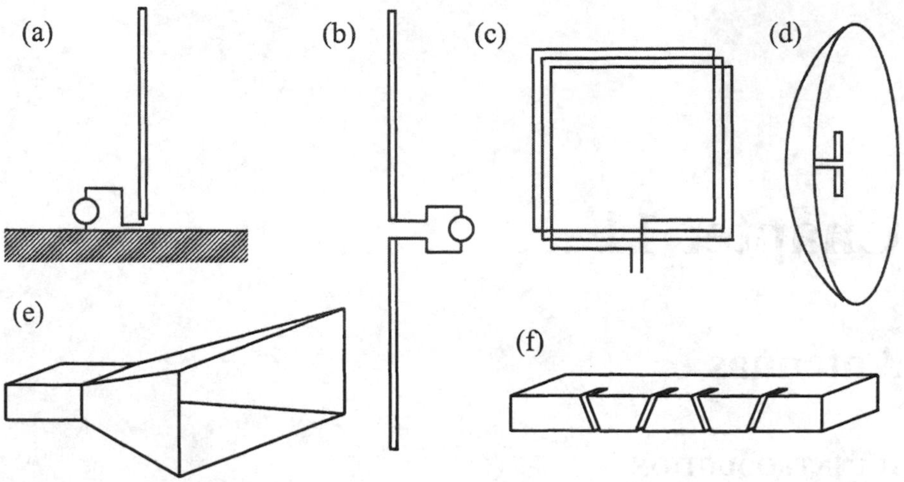

Figure 11.1: Different antenna types: (a) monopole antenna, (b) dipole antenna, (c) loop antenna, (d) parabolic reflector antenna, (e) horn antenna and (f) slot antenna.

high *directivity* compared with an imagined isotropic radiator. If in a waveguide wall currents are interrupted by slots, from these slots electromagnetic energy can also be radiated. Figure 11.1(f) shows a slot antenna. The directivity of antennas may be increased by combining several antennas into antenna arrays or by the usage of reflectors. Figure 11.1(d) shows as an example the combination of a dipole with a parabolic reflector. The dipole, that excites a primary wave, is positioned into the focal point of the parabolic reflector. By the parabolic reflector the spherical wave excited by the dipole is transformed into an almost plane wave.

A fundamental problem in computing the electromagnetic field excited by an antenna is to compute the electromagnetic field excited by the surface currents flowing in the metallic conductors forming the antenna. In general also the current distribution on the antenna surface is completely unknown. We may consider the antenna to be excited by some primary field or by currents or voltages impressed at certain points. Our task then will be first to compute the current distribution on the antenna and then the electromagnetic field generated by this current distribution. An exact treatment of this problem requires us also to consider the influence of the radiation field on the current distribution in the antenna. This usually requires solution of integral equations as discussed in Section 12.6. However, the treatment of the problem may be simplified considerably if the current distribution on the antenna is already known. For simple antenna structures the current distribution may be computed with satisfactory accuracy without taking into account the radiation. In this cases the problem of computing the field radiated by the antenna may be done in two steps: In a first step the current distribution in the antenna is computed. In a second step the field radiated by the antenna is computed by superimposing the contributions of the infinitesimally small antenna elements. For a detailed treatment of antennas the reader is referred to [1–6].

11.2 THE GREEN'S FUNCTION

A point-like excitation by an impressed electric or magnetic polarization yields a spherical electromagnetic wave. The computation of the electromagnetic wave due to a point-like excitation is of fundamental importance for the computation of electromagnetic waves excited by arbitrary source distributions since the field generated by arbitrary distributions may be computed by superimposing the fields originating from point-like sources. This follows from the application of the superposition principle and from the circumstance that every source distribution may be considered as a continuous superposition of point-like sources. The mathematical formulation of this problem yields the *Green's function.*

To compute the electromagnetic field generated by electric or magnetic polarizations impressed on surfaces we have to solve the inhomogeneous Helmholtz' equations (3.34) or (3.45), respectively, for arbitrary spatial distribution of the impressed electric polarization $\underline{\mathcal{M}}_{e0}$ or the impressed magnetic polarization $\underline{\mathcal{M}}_{m0}$, respectively. Using (3.53), we obtain the Helmholtz' equations

$$\Delta \underline{\Pi}_e + k^2 \underline{\Pi}_e = -\frac{1}{\varepsilon} \star \underline{\mathcal{M}}_{e0}\,, \tag{11.1a}$$

$$\Delta \underline{\Pi}_m + k^2 \underline{\Pi}_m = -\frac{1}{\underline{\mu}} \star \underline{\mathcal{M}}_{m0}\,. \tag{11.1b}$$

Since we can make use of the principle of duality in the following, it will be sufficient to solve (11.1a). To compute the field at a point $\boldsymbol{x}$ excited by a point-like source located at $\boldsymbol{x}'$ we use the *Green's double one-form* [7,8], already introduced in (5.58)

$$\begin{aligned} \underline{\mathcal{G}} = \underline{G}_{11}\, \mathrm{d}x\, \mathrm{d}x' + \underline{G}_{12}\, \mathrm{d}x\, \mathrm{d}y' + \underline{G}_{13}\, \mathrm{d}x\, \mathrm{d}z' \\ + \underline{G}_{21}\, \mathrm{d}y\, \mathrm{d}x' + \underline{G}_{22}\, \mathrm{d}y\, \mathrm{d}y' + \underline{G}_{23}\, \mathrm{d}y\, \mathrm{d}z' \\ + \underline{G}_{31}\, \mathrm{d}z\, \mathrm{d}x' + \underline{G}_{32}\, \mathrm{d}z\, \mathrm{d}y' + \underline{G}_{33}\, \mathrm{d}z\, \mathrm{d}z'\,. \end{aligned} \tag{11.2}$$

Using Green's double one-form we can express the Helmholtz equation (11.1a) for a point-like unit source at $\boldsymbol{x}'$ by

$$\Delta \underline{\mathcal{G}}(\boldsymbol{x}, \boldsymbol{x}') + k^2 \underline{\mathcal{G}}(\boldsymbol{x}, \boldsymbol{x}') = -\frac{1}{\varepsilon} \mathcal{I}(\boldsymbol{x}, \boldsymbol{x}')\,, \tag{11.3}$$

where $\mathcal{I}(\boldsymbol{x}, \boldsymbol{x}')$ is the identity kernel, introduced in (5.60). In the above equation the Laplace operator acts on the unprimed differentials whereas the primed differentials are treated as constants. Forming the exterior product with $\underline{\mathcal{M}}_{e0}(\boldsymbol{x}')$ integrating over the primed variables and using (5.61b) yields

$$\int' \left[\Delta \underline{\mathcal{G}}(\boldsymbol{x}, \boldsymbol{x}') \, \underline{\mathcal{M}}_{e0}(\boldsymbol{x}') + k^2 \, \underline{\mathcal{G}}(\boldsymbol{x}, \boldsymbol{x}') \, \underline{\mathcal{M}}_{e0}(\boldsymbol{x}') \right] = -\frac{1}{\varepsilon} \underline{\mathcal{M}}_{e0}(\boldsymbol{x}) \, . \tag{11.4}$$

Since the sequence of partial derivation with respect to the unprimed coordinates and integration over the primed coordinates may be interchanged under very general conditions we obtain

$$\Delta \left[\int' \underline{\mathcal{G}}(\boldsymbol{x}, \boldsymbol{x}') \, \underline{\mathcal{M}}_{e0}(\boldsymbol{x}') \right] + k^2 \left[\int' \underline{\mathcal{G}}(\boldsymbol{x}, \boldsymbol{x}') \, \underline{\mathcal{M}}_{e0}(\boldsymbol{x}') \right] = -\frac{1}{\varepsilon} \underline{\mathcal{M}}_{e0}(\boldsymbol{x}) \, . \tag{11.5}$$

Comparing this equation with (11.1a) yields

$$\underline{\Pi}_e(\boldsymbol{x}) = \int' \underline{\mathcal{G}}(\boldsymbol{x}, \boldsymbol{x}') \wedge \underline{\mathcal{M}}_{e0}(\boldsymbol{x}') \, . \tag{11.6}$$

In cartesian coordinates the Laplace operator for one-forms (A.71) is symmetric in its three components and leaves the three components uncoupled. Therefore (11.5) may be solved with

$$\underline{\mathcal{G}}(\boldsymbol{x}, \boldsymbol{x}') = \underline{G}(\boldsymbol{x}, \boldsymbol{x}')(\mathrm{d}x \, \mathrm{d}x' + \mathrm{d}y \, \mathrm{d}y' + \mathrm{d}z \, \mathrm{d}z') \, . \tag{11.7}$$

where $\underline{G}(\boldsymbol{x}, \boldsymbol{x}')$ is a *scalar Green's function*, obtained by solving the scalar Helmholtz equation

$$(\Delta + k^2) \underline{G}(\boldsymbol{x}, \boldsymbol{x}') = -\frac{1}{\varepsilon} \, \delta(\boldsymbol{x} - \boldsymbol{x}') \, . \tag{11.8}$$

In the infinitely extended homogeneous isotropic space the scalar Green's function $\underline{G}(\boldsymbol{x}, \boldsymbol{x}')$ must exhibit spherical symmetry and therefore can only depend on the magnitude $r = |\boldsymbol{r}|$ of the distance $\boldsymbol{r} = \boldsymbol{x} - \boldsymbol{x}'$ between the points x and x'. We introduce spherical coordinates at the origin $r = 0$. With

$$\Delta f = \frac{\partial^2 f}{\partial r^2} + \frac{2}{r} \frac{\partial f}{\partial r} + \frac{1}{r^2} \frac{\partial^2 f}{\partial \vartheta^2} + \frac{\cot \vartheta}{r^2} \frac{\partial f}{\partial \vartheta} + \frac{1}{r^2 \sin^2 \vartheta} \frac{\partial^2 f}{\partial \varphi^2} \tag{11.9}$$

we obtain from (11.8) considering spherical symmetry

$$\frac{1}{r} \frac{\mathrm{d}^2}{\mathrm{d}r^2} (r \underline{G}(r)) + k^2 \underline{G}(r) = -\frac{1}{\varepsilon} \delta(r) \, . \tag{11.10}$$

With the exception of the origin $r = 0$, $G(r)$ satisfies the homogeneous differential equation

$$\frac{\mathrm{d}^2}{\mathrm{d}r^2}(r\underline{G}(r)) + k^2(r\underline{G}(r)) = 0\,. \tag{11.11}$$

The solution is given by

$$r\underline{G}(r) = \underline{A}\,\mathrm{e}^{-\mathrm{j}kr} + \underline{B}\,\mathrm{e}^{\mathrm{j}kr}\,, \tag{11.12}$$

where $\underline{A}$ and $\underline{B}$ are complex coefficients not yet determined. The first term on the right side of (11.12) describes a wave propagating from the origin $r = 0$ into the space, whereas the second term describes a wave propagating into the origin. The second solution does not occur for physical reasons; therefore we set $\underline{B} = 0$ in the following. To determine $\underline{A}$ we consider the field in a small neighborhood of the point source. We assume the extension of this neighborhood to be much smaller than the wavelength λ. Therefore $|kr| \ll 1$, and k may be neglected in (11.8) and (11.11). Within this approximation we obtain

$$\underline{G}_0(r) = \lim_{k\to 0}\underline{G}(r) = \frac{\underline{A}}{r}\,. \tag{11.13}$$

From (11.8) it follows that

$$\Delta\,\underline{G}_0(\boldsymbol{x},\boldsymbol{x}') = -\frac{1}{\varepsilon}\delta(\boldsymbol{x}-\boldsymbol{x}')\,. \tag{11.14}$$

This equation corresponds to the *Poisson equation* from electrostatics

$$\mathrm{d}\star\,\mathrm{d}\underline{G}_0(\boldsymbol{x},\boldsymbol{x}') = -\frac{1}{\varepsilon}\delta(\boldsymbol{x}-\boldsymbol{x}')\,\mathrm{d}x\wedge\mathrm{d}y\wedge\mathrm{d}z\,. \tag{11.15}$$

We integrate both sides of (11.14) over a spherical volume of radius r and center point $\boldsymbol{x}'$. On the left side we can transform the volume integral into a surface integral. Considering

$$\star\,\mathrm{d}\underline{G}_0(\boldsymbol{x},\boldsymbol{x}') = r^2\sin\vartheta\frac{\partial\underline{G}_0}{\partial r}\,\mathrm{d}\vartheta\wedge\mathrm{d}\varphi + \sin\vartheta\frac{\partial\underline{G}_0}{\partial\vartheta}\,\mathrm{d}\varphi\wedge\mathrm{d}r + \frac{1}{\sin\vartheta}\frac{\partial\underline{G}_0}{\partial\varphi}\,\mathrm{d}r\wedge\mathrm{d}\vartheta\,, \tag{11.16}$$

we obtain

$$\int_{\partial V} r^2\sin\vartheta\frac{\partial\underline{G}_0}{\partial r}\,\mathrm{d}\vartheta\wedge\mathrm{d}\varphi = -\frac{1}{\varepsilon}\int_V \delta(\boldsymbol{x}-\boldsymbol{x}')\,\mathrm{d}x\wedge\mathrm{d}y\wedge\mathrm{d}z\,. \tag{11.17}$$

Since $\underline{G}_0$ is uniform over the surface of the sphere, we obtain

$$4\pi r^2\frac{\partial\underline{G}_0}{\partial r} = -\frac{1}{\varepsilon}\,. \tag{11.18}$$

From this we obtain

$$\underline{G}_0(r) = \frac{1}{4\pi\varepsilon r}\,. \tag{11.19}$$

Comparing with (11.13) yields

$$\underline{A} = \frac{1}{4\pi\varepsilon} \tag{11.20}$$

and therewith from (11.12)

$$\underline{G}(r) = \frac{\mathrm{e}^{-\mathrm{j}kr}}{4\pi\varepsilon r}\,. \tag{11.21}$$

Inserting (11.21) into (11.6) we obtain the Green's double one-form

$$\underline{\mathcal{G}}(\boldsymbol{x},\boldsymbol{x}') = \frac{\mathrm{e}^{-\mathrm{j}k|\boldsymbol{x}-\boldsymbol{x}'|}}{4\pi\varepsilon|\boldsymbol{x}-\boldsymbol{x}'|}(\mathrm{d}x\,\mathrm{d}x' + \mathrm{d}y\,\mathrm{d}y' + \mathrm{d}z\,\mathrm{d}z') \tag{11.22}$$

and with this the *retarded electric Hertz differential form*

$$\underline{\Pi}_e(\boldsymbol{x}) = \int' \frac{\mathrm{e}^{-\mathrm{j}k|\boldsymbol{x}-\boldsymbol{x}'|}}{4\pi\varepsilon|\boldsymbol{x}-\boldsymbol{x}'|}(\mathrm{d}x\,\mathrm{d}x' + \mathrm{d}y\,\mathrm{d}y' + \mathrm{d}z\,\mathrm{d}z') \wedge \underline{\mathcal{M}}_{e0}(\boldsymbol{x}')\,. \tag{11.23}$$

In the same way we obtain for the impressed magnetic polarization $\underline{\mathcal{M}}_{m0}(\boldsymbol{x}')$ from (11.1b) the *retarded magnetic Hertz differential form*

$$\underline{\Pi}_m(\boldsymbol{x}) = \int' \frac{\mathrm{e}^{-\mathrm{j}k|\boldsymbol{x}-\boldsymbol{x}'|}}{4\pi\mu|\boldsymbol{x}-\boldsymbol{x}'|}(\mathrm{d}x\,\mathrm{d}x' + \mathrm{d}y\,\mathrm{d}y' + \mathrm{d}z\,\mathrm{d}z') \wedge \underline{\mathcal{M}}_{m0}(\boldsymbol{x}')\,. \tag{11.24}$$

By inserting $\underline{\mathcal{M}}_{e0}(\boldsymbol{x}')$ and $\underline{\mathcal{M}}_{m0}(\boldsymbol{x}')$ respectively in component form into (11.23) and (11.24) we obtain

$$\underline{\Pi}_e(\boldsymbol{x}) = \int' \frac{\mathrm{e}^{-\mathrm{j}k|\boldsymbol{x}-\boldsymbol{x}'|}}{4\pi\varepsilon|\boldsymbol{x}-\boldsymbol{x}'|}(\underline{M}_{e0x}(\boldsymbol{x}')\,\mathrm{d}x + \underline{M}_{e0y}(\boldsymbol{x}')\,\mathrm{d}y + \underline{M}_{e0z}(\boldsymbol{x}')\,\mathrm{d}z)\; \mathrm{d}x' \wedge \mathrm{d}y' \wedge \mathrm{d}z'\,, \tag{11.25a}$$

$$\underline{\Pi}_m(\boldsymbol{x}) = \int' \frac{\mathrm{e}^{-\mathrm{j}k|\boldsymbol{x}-\boldsymbol{x}'|}}{4\pi\mu|\boldsymbol{x}-\boldsymbol{x}'|}(\underline{M}_{m0x}(\boldsymbol{x}')\,\mathrm{d}x + \underline{M}_{m0y}(\boldsymbol{x}')\,\mathrm{d}y + \underline{M}_{m0z}(\boldsymbol{x}')\,\mathrm{d}z)\; \mathrm{d}x' \wedge \mathrm{d}y' \wedge \mathrm{d}z'\,. \tag{11.25b}$$

The computation of the field quantities $\underline{\boldsymbol{E}}$ and $\underline{\boldsymbol{H}}$ from $\underline{\boldsymbol{M}}_{e0}$ or $\underline{\boldsymbol{M}}_{m0}$, respectively, is performed using (3.20) and (3.21) or (3.24) and (3.25). If impressed electric polarization as well as impressed magnetic polarization exist the electromagnetic field is obtained by superposition of the electromagnetic fields computed from $\underline{\boldsymbol{M}}_{e0}$ and $\underline{\boldsymbol{M}}_{m0}$, respectively.

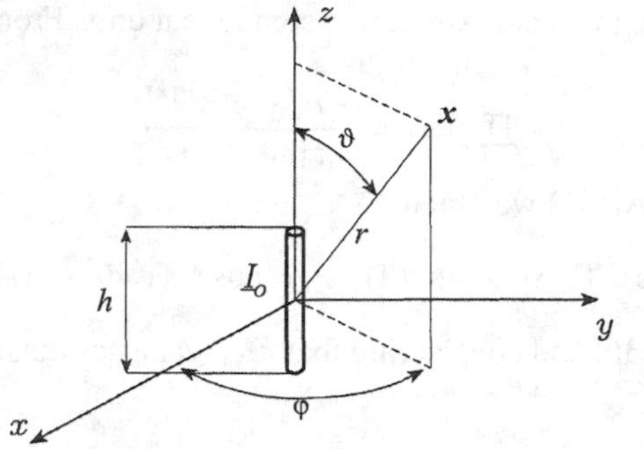

Figure 11.2: Conductor of the length h.

11.3 THE HERTZIAN DIPOLE

We compute the radiation field of a short straight wire segment with impressed harmonic current. The length h of this conductor is assumed to be small compared to the wavelength. Therefore the current $\underline{I}$ may be considered to be uniform over the length h of the wire segment. Such an arrangement is called a short electric dipole or Hertzian dipole. In his early experiments, Heinrich Hertz realized such a dipole by attaching spheres at the end of the line segment. These spheres are storing the charge accumulated at the end of the wire, if a uniformly distributed current is flowing through the wire. Complex wire antennas or other radiating wire structures may be modelled by segments of Hertzian dipoles. The current $\underline{I}_0$ may be impressed at some intersection introduced into the Hertzian dipole. For r much larger than h we may consider the polarization excited by the current $\underline{I}$ to be concentrated into the origin. From (11.25a) we obtain under this assumption the approximate solution for the Hertz form

$$\underline{\Pi}_e(\boldsymbol{x}) = \frac{e^{-jkr}}{4\pi\varepsilon_0 r}\int_V' (\underline{M}_{e0x}(\boldsymbol{x}')\,dx + \underline{M}_{e0y}(\boldsymbol{x}')\,dy + \underline{M}_{e0z}(\boldsymbol{x}')\,dz)\,dx' \wedge dy' \wedge dz' . \tag{11.26}$$

Due to (3.28) the impressed electric polarization $\underline{\mathcal{M}}_{e0}$ and the impressed current density $\underline{\mathcal{J}}_0$ are related by

$$\underline{\mathcal{M}}_{e0} = \frac{1}{j\omega}\underline{\mathcal{J}}_0 . \tag{11.27}$$

According to Figure 11.2, the current density $\underline{\boldsymbol{J}}_0(\boldsymbol{x})$ exhibits only a z component. Since the surface integral of $\underline{\boldsymbol{J}}_0$ over the cross-section of the conductor is equal to $\boldsymbol{I}_0(\boldsymbol{x})$, we obtain

$$\int_V' \underline{\mathcal{J}}_0 = h\,\underline{I}_0 . \tag{11.28}$$

The polarization $\underline{M}_{e0}(\boldsymbol{x}')$ also exhibits a z component only. From (11.26) and (11.28) it follows that

$$\underline{\Pi}_{ez}(\boldsymbol{x}) = \frac{h\underline{I}_0}{4\pi \mathrm{j}\omega\varepsilon_0} \frac{\mathrm{e}^{-\mathrm{j}kr}}{r} . \tag{11.29}$$

From (A.148) and (A.151) we obtain

$$\underline{\Pi}_e = \Pi_z \, \mathrm{d}z = \cos\vartheta \, \Pi_z \, s_1 - \sin\vartheta \, \Pi_z \, s_2 = \cos\vartheta \, \Pi_z \, \mathrm{d}r - r\sin\vartheta \, \Pi_z \, \mathrm{d}\vartheta . \tag{11.30}$$

Using (3.35) and (3.36) and considering that $\underline{M}_{e0}$ vanishes outside the conductor, we can compute $\underline{\mathcal{E}}$ and $\underline{\mathcal{H}}$.

$$\underline{\mathcal{H}} = \mathrm{j}\omega\underline{\varepsilon} \star \mathrm{d}\underline{\Pi}_e , \tag{11.31}$$

$$\underline{\mathcal{E}} = -\tilde{\mathrm{d}}\,\mathrm{d}\underline{\Pi}_e . \tag{11.32}$$

From (11.30) we obtain

$$\mathrm{d}\underline{\Pi}_e = \frac{\partial}{\partial\vartheta} (\cos\vartheta \, \Pi_z) \, \mathrm{d}\vartheta \wedge \mathrm{d}r - \frac{\partial}{\partial r} (r \sin\vartheta \, \Pi_z) \, \mathrm{d}r \wedge \mathrm{d}\vartheta . \tag{11.33}$$

Inserting (11.29) it follows that

$$\mathrm{d}\underline{\Pi}_e = \frac{h\underline{I}_0}{4\pi \mathrm{j}\omega\varepsilon_0} \left(\frac{1}{r^2} + \frac{\mathrm{j}k}{r} \right) \mathrm{e}^{-\mathrm{j}kr} \sin\vartheta \, r \, \mathrm{d}r \wedge \mathrm{d}\vartheta . \tag{11.34}$$

Applying the Hodge operator yields

$$\star \, \mathrm{d}\underline{\Pi}_e = \frac{h\underline{I}_0}{4\pi \mathrm{j}\omega\varepsilon_0} \left(\frac{1}{r^2} + \frac{\mathrm{j}k}{r} \right) \mathrm{e}^{-\mathrm{j}kr} \sin\vartheta \, r \sin\vartheta \, \mathrm{d}\varphi . \tag{11.35}$$

Using (11.31) yields

$$\underline{\mathcal{H}} = \frac{h\underline{I}_0}{4\pi} \left(\frac{1}{r^2} + \frac{\mathrm{j}k}{r} \right) \mathrm{e}^{-\mathrm{j}kr} \sin\vartheta \, r \sin\vartheta \, \mathrm{d}\varphi . \tag{11.36}$$

Consequently the magnetic field only exhibits a φ component

$$\underline{H}_\varphi = \frac{h\underline{I}_0}{4\pi} \left(\frac{\mathrm{j}k}{r} + \frac{1}{r^2} \right) \mathrm{e}^{-\mathrm{j}kr} \sin\vartheta . \tag{11.37}$$

In order to obtain $\underline{\mathcal{E}}$ from (11.32) we first compute the exterior derivative of (11.35) and obtain

$$\begin{aligned} \mathrm{d}\star \mathrm{d}\underline{\Pi}_e = \frac{h\underline{I}_0}{4\pi \mathrm{j}\omega\varepsilon_0} \Bigg[& \left(-\frac{k^2}{r} + \frac{\mathrm{j}k}{r^2} + \frac{1}{r^3} \right) \mathrm{e}^{-\mathrm{j}kr} \sin\vartheta \, r \sin\vartheta \, \mathrm{d}\varphi \wedge \mathrm{d}r + \\ & + 2 \left(\frac{\mathrm{j}k}{r^2} + \frac{1}{r^3} \right) \mathrm{e}^{-\mathrm{j}kr} \cos\vartheta \, r^2 \sin\vartheta \, \mathrm{d}\vartheta \wedge \mathrm{d}\varphi \Bigg] . \end{aligned} \tag{11.38}$$

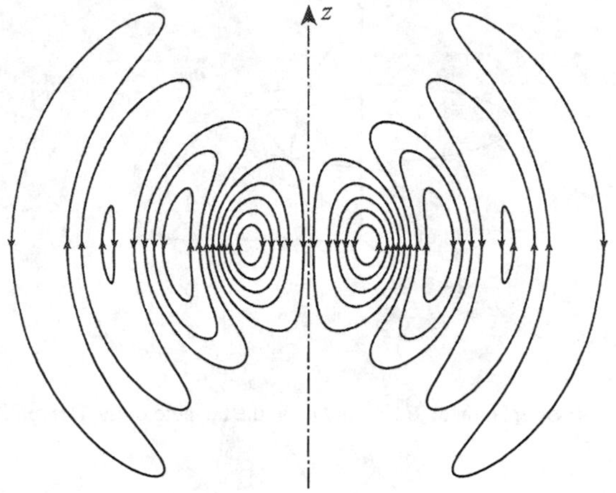

Figure 11.3: Electric flux lines in the near field of the Hertzian dipole.

Applying the Hodge operator yields

$$\underline{\mathcal{E}} = \star\,\mathrm{d}\star\mathrm{d}\,\underline{\Pi}_e = \frac{h\underline{I}_0}{4\pi \mathrm{j}\omega\varepsilon_0}\left[\left(-\frac{k^2}{r}+\frac{\mathrm{j}k}{r^2}+\frac{1}{r^3}\right)\mathrm{e}^{-\mathrm{j}kr}\sin\vartheta\, r\,\mathrm{d}\vartheta + \right.$$

$$\left. + 2\left(\frac{\mathrm{j}k}{r^2}+\frac{1}{r^3}\right)\mathrm{e}^{-\mathrm{j}kr}\cos\vartheta\,\mathrm{d}r\right]. \tag{11.39}$$

The electric field exhibits ϑ components and r components.

$$\underline{E}_\vartheta = \frac{h\underline{I}_0}{4\pi \mathrm{j}\omega\varepsilon_0}\left(-\frac{k^2}{r}+\frac{\mathrm{j}k}{r^2}+\frac{1}{r^3}\right)\mathrm{e}^{-\mathrm{j}kr}\sin\vartheta\,, \tag{11.40}$$

$$\underline{E}_r = \frac{h\underline{I}_0}{2\pi \mathrm{j}\omega\varepsilon_0}\left(\frac{\mathrm{j}k}{r^2}+\frac{1}{r^3}\right)\mathrm{e}^{-\mathrm{j}kr}\cos\vartheta\,. \tag{11.41}$$

For $kr \ll 1$ the magnetic field component $\underline{H}_\varphi$ is only determined by the term proportional to $1/r^2$ and $\underline{E}_r$ and $\underline{E}_\varphi$ are only determined by the terms proportional to $1/r^3$. This region is the *near-field* region. In the near-field the magnetic field is in phase with the current. The magnetic field there corresponds to the magnetic field distribution generated by the stationary current. Also the electric near-field corresponds to the field-distribution excited by a static electric dipole.

Figure 11.3 shows the electric field lines in the near-field. The field in the region $kr \gg 1$ is called the *far-field*. With the exception of the electric field in the directions $\vartheta = 0$ and $\vartheta = \pi$, respectively, the field quantities proportional to $1/r$ are dominant,

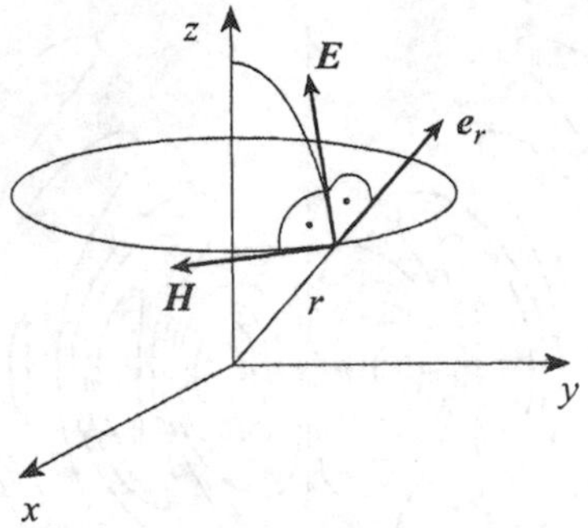

Figure 11.4: Orientation of $\boldsymbol{E}$, $\boldsymbol{H}$ and $\boldsymbol{e}_r$ in the far-field of the Hertzian dipole.

and the electric field exhibits a ϑ component only. For the far-field region we obtain the approximate differential forms

$$\underline{\mathcal{E}} = \frac{\mathrm{j}\omega\mu_0 h \underline{I}_0}{4\pi} \frac{\mathrm{e}^{-\mathrm{j}kr}}{r} \sin\vartheta \, r \, \mathrm{d}\vartheta \,, \tag{11.42}$$

$$\underline{\mathcal{H}} = \frac{\mathrm{j}kh\underline{I}_0}{4\pi} \frac{\mathrm{e}^{-\mathrm{j}kr}}{r} \sin\vartheta \, r \sin\vartheta \, \mathrm{d}\varphi \,. \tag{11.43}$$

and the corresponding field components

$$\underline{E}_\vartheta = \frac{\mathrm{j}\omega\mu_0 h \underline{I}_0}{4\pi} \frac{\mathrm{e}^{-\mathrm{j}kr}}{r} \sin\vartheta, \tag{11.44}$$

$$\underline{H}_\varphi = \frac{\mathrm{j}kh\underline{I}_0}{4\pi} \frac{\mathrm{e}^{-\mathrm{j}kr}}{r} \sin\vartheta. \tag{11.45}$$

In the far-field the electric and magnetic field components are mutually orthogonal and both are orthogonal to the direction of propagation (Figure 11.4). Furthermore the electric field and the magnetic field are proportional to each other and in phase. The *radiation diagram* of an antenna depicts the angular distribution of the radiated field. The antenna characteristics of the Hertzian dipole is given by the dependence of the magnitude of the electric field $|\underline{E}_\vartheta|$ from the angles ϑ and φ (Figure 11.5). From (11.44), (11.45) and (6.37) it follows that the ratio of electric and magnetic field in the far-field is given by

$$\frac{\underline{E}_\vartheta}{\underline{H}_\varphi} = Z_{F0} \cong 377\Omega \,. \tag{11.46}$$

From (4.21), (11.42) and (11.43) we obtain the complex Poynting form $\mathcal{T}$ for the

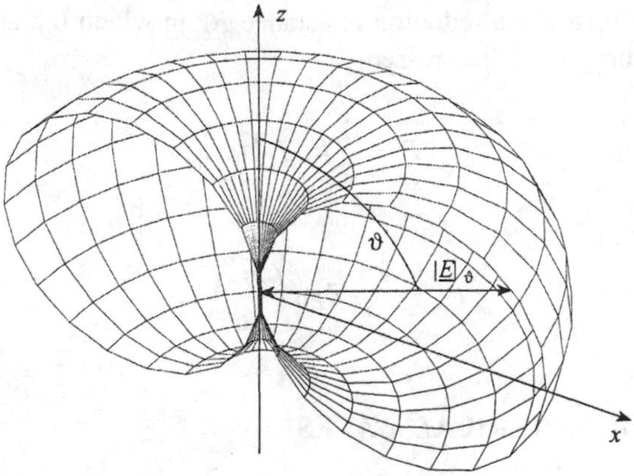

Figure 11.5: Radiation diagram of the Hertzian dipole.

far-field

$$\mathcal{T} = \frac{1}{2}\underline{\mathcal{E}} \wedge \underline{\mathcal{H}}^* = \frac{1}{2}\underline{E}_\vartheta \underline{H}_\varphi^* r^2 \sin\vartheta \, \mathrm{d}\vartheta \wedge \mathrm{d}\varphi = \frac{Z_{F0}k^2h^2|\underline{I}_0|^2}{32\pi^2}\sin^3\vartheta \, \mathrm{d}\vartheta \wedge \mathrm{d}\varphi \,. \tag{11.47}$$

In the far-field the complex Poynting vector exhibits only a radial component

$$\underline{T}_r = \frac{Z_{F0}k^2h^2|\underline{I}_0|^2}{32\pi^2 r^2}\sin^2\vartheta \,. \tag{11.48}$$

In this case $\underline{T}_r$ is real and the active power P radiated from the Hertzian dipole is obtained by integrating $\mathcal{T}$ over a closed surface surrounding the Hertzian dipole in the far-field.

$$P = \int_{\vartheta=0}^{\pi}\int_{\varphi=0}^{2\pi} \mathcal{T}. \tag{11.49}$$

With

$$\int_0^{\pi} \sin^3\vartheta \, \mathrm{d}\vartheta = \frac{4}{3} \tag{11.50}$$

we obtain from (11.48) and (11.49)

$$P = \frac{1}{12\pi} Z_{F0}k^2h^2|\underline{I}_0|^2 \,. \tag{11.51}$$

With $k = \frac{2\pi}{\lambda_0}$ we obtain from this

$$P = \frac{\pi}{3} Z_{F0}\left(\frac{h}{\lambda_0}\right)^2 |\underline{I}_0|^2 \,. \tag{11.52}$$

We can introduce formally a radiation resistance R_r in which the active power P is anticipated, if a current I_0 is impressed.

$$P = \frac{1}{2} R_r |\underline{I}_0|^2 . \tag{11.53}$$

Therewith we obtain

$$R_r = \frac{2\pi}{3} Z_{F0} \left(\frac{h}{\lambda_0} \right)^2 . \tag{11.54}$$

11.4 APERIODIC SPHERICAL WAVES

To investigate the emission of *aperiodic spherical waves* we analyze the Hertzian dipole in the time domain. In the time domain the impressed polarization $m_{e0}(t)$ and the impressed current $i_0(t)$ are related via the inverse Fourier transform of (11.27):

$$\mathcal{M}_{e0}(\boldsymbol{x}, t) = \int_0^t \mathcal{J}_0(\boldsymbol{x}, t_1)\, \mathrm{d}t_1 . \tag{11.55}$$

Integrating the electric polarization form $\mathcal{M}_{e0}$ over the volume V of the Hertzian dipole yields

$$m_{e0}(t) = \int_V \mathcal{M}_{e0}(\boldsymbol{x}, t) = h \int_0^t i_0(t_1)\, \mathrm{d}t_1 , \tag{11.56}$$

where $i_0(t)$ is the current through the Hertzian dipole, $m_{e0}(t)$ the polarization due to this current and h the dipole length. This yields

$$\Pi_{ez}(\boldsymbol{x}, t) = \frac{h}{4\pi \varepsilon_0 r}\, m_{e0}\left(t - \frac{r}{c}\right) . \tag{11.57}$$

Using (3.25) and (3.26) and considering that $\mathcal{M}_{e0}(\boldsymbol{x}, t)$ vanishes outside the conductor, we can compute $\mathcal{E}(\boldsymbol{x}, t)$ and $\mathcal{H}(\boldsymbol{x}, t)$:

$$\mathcal{H}(\boldsymbol{x}, t) = \star\, \mathrm{d}\varepsilon \frac{\partial}{\partial t}\, \Pi_e(\boldsymbol{x}, t) , \tag{11.58a}$$

$$\mathcal{E}(\boldsymbol{x}, t) = -\tilde{\mathrm{d}}\, \mathrm{d}\, \Pi_e(\boldsymbol{x}, t) . \tag{11.58b}$$

Using (11.58a) yields

$$\mathcal{H} = \frac{h}{4\pi} \left[\frac{1}{r^2} m'_{e0}\left(t - \frac{r}{c}\right) + \frac{1}{cr} m''_{e0}\left(t - \frac{r}{c}\right) \right] \sin\vartheta\, r \sin\vartheta\, \mathrm{d}\varphi . \tag{11.59}$$

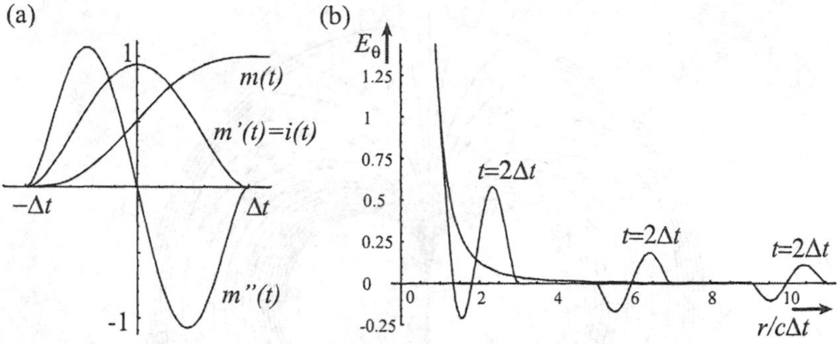

Figure 11.6: Wave pulse: (a) pulse waveforms and (b) radial dependence of the wave pulse.

The magnetic field only exhibits a φ component

$$H_\varphi = \frac{h}{4\pi}\left[\frac{1}{r^2}m'_{e0}\left(t-\frac{r}{c}\right)+\frac{1}{cr}m''_{e0}\left(t-\frac{r}{c}\right)\right]\sin\vartheta\,. \tag{11.60}$$

The electric field form is

$$\begin{aligned}\mathcal{E} = \frac{h}{4\pi\varepsilon_0}&\left\{\left[\frac{1}{r^3}m_{e0}\left(t-\frac{r}{c}\right)+\frac{1}{cr^2}m'_{e0}\left(t-\frac{r}{c}\right)\right.\right.\\&+\left.\frac{1}{c^2r}m''_{e0}\left(t-\frac{r}{c}\right)\right]\sin\vartheta\, r\,\mathrm{d}\vartheta\\&+\left.2\left[\frac{1}{r^3}m_{e0}\left(t-\frac{r}{c}\right)+\frac{1}{cr^2}m'_{e0}\left(t-\frac{r}{c}\right)\right]\cos\vartheta\,\mathrm{d}r\right\}.\end{aligned} \tag{11.61}$$

The electric field exhibits the ϑ- and r components

$$\begin{aligned}E_\vartheta = \frac{h}{4\pi\varepsilon_0}&\left[\frac{1}{r^3}m_{e0}\left(t-\frac{r}{c}\right)+\frac{1}{cr^2}m'_{e0}\left(t-\frac{r}{c}\right)\right.\\&+\left.\frac{1}{c^2r}m''_{e0}\left(t-\frac{r}{c}\right)\right]\sin\vartheta\,,\end{aligned} \tag{11.62}$$

$$E_r = \frac{h}{2\pi\varepsilon_0}\left[\frac{1}{r^3}m_{e0}\left(t-\frac{r}{c}\right)+\frac{1}{cr^2}m'_{e0}\left(t-\frac{r}{c}\right)\right]\cos\vartheta\,. \tag{11.63}$$

As an example we consider a wave pulse emitted from a Hertzian dipole excited by a current pulse. In Figure 11.6(a) the dipole current pulse $i(t) = m'(t)$ of width $2\Delta t$, its integral over time $m(t)$ and its time derivative $m''(t)$ are depicted. Figure 11.6(b) shows the time evolution of $E_\vartheta(r, 0, 0, t)$. The wave front of width $2\Delta t$

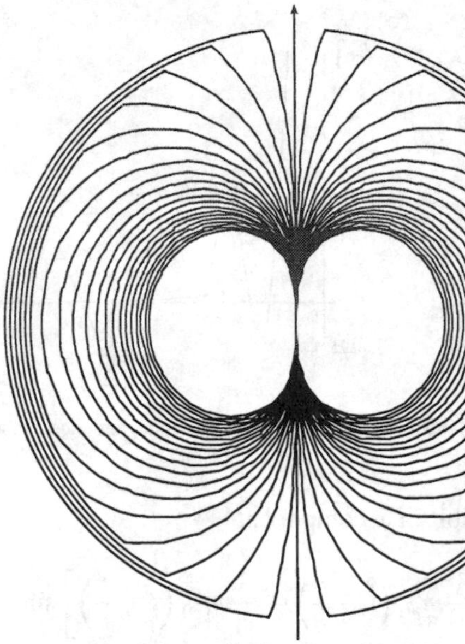

Figure 11.7: Near field of the Hertzian dipole under pulse excitation.

mainly depends on $m'(t)$ and $m''(t)$. In the far-field region, defined by $r \gg c\Delta t$ the terms proportional to $1/r$ in E_ϑ and H_φ exhibit the double pulse shape specified by $m''(t)$. The energy connected with this term is constrained within the shell of width $2c\Delta t$ at the wave front and transported into the infinity. This is the radiated part of the field. The electric and magnetic far-field time wave-forms E_ϑ and H_φ of the wave pulse are proportional to the time derivative of the driving current $i(t)$ of the dipole. The near-field parts of the electric and magnetic field proportional to $m'(t - r/c)/r^2$ also are confined to the wave front in a shell of width $2\Delta t$. This part of the wave front is carrying the electromagnetic energy for building up the near-field. It leaves behind the wave front an electric field proportional to $m(t - r/c)/r^3$. This field behind the wave front corresponds to the electrostatic field excited by a static dipole. Figure 11.7 shows the electric field in a meridional plane.

In the far-field we obtain the approximate differential forms

$$\mathcal{E}(r, \vartheta, t) = \frac{\mu_0 h}{4\pi} \frac{m''_{e0}\left(t - \frac{r}{c}\right)}{r} \sin\vartheta \, r \, \mathrm{d}\vartheta \,, \tag{11.64}$$

$$\mathcal{H}(r, \vartheta, t) = \frac{h}{4\pi c} \frac{m''_{e0}\left(t - \frac{r}{c}\right)}{r} \sin\vartheta \, r \sin\vartheta \, \mathrm{d}\varphi \,. \tag{11.65}$$

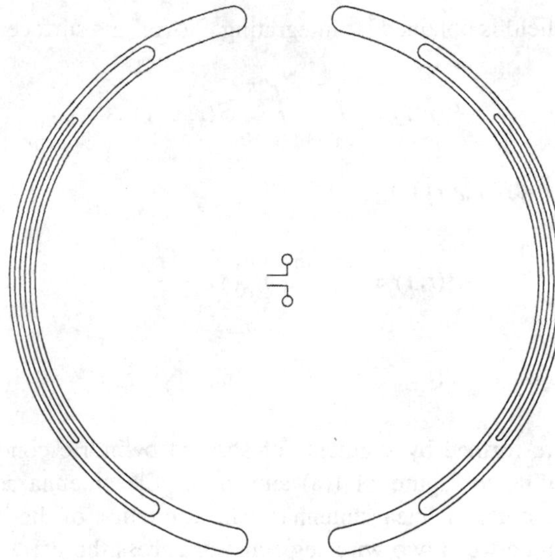

Figure 11.8: Far field of the Hertzian dipole under pulse excitation..

and the corresponding field components

$$E_\vartheta(r, \vartheta, t) = \frac{\mu_0 h}{4\pi} \frac{m''_{e0}\left(t - \frac{r}{c}\right)}{r} \sin\vartheta, \tag{11.66}$$

$$H_\varphi(r, \vartheta, t) = \frac{h}{4\pi c} \frac{m''_{e0}\left(t - \frac{r}{c}\right)}{r} \sin\vartheta. \tag{11.67}$$

The far-field is depicted in Figure 11.8. From (11.66), (11.67) and (6.37) it follows that the ratio of the electric and magnetic fields in the far-field is given by Z_{F0}. From (4.21), (11.64) and (11.65) we obtain the time-dependent Poynting form $\mathcal{S}$ for the far-field

$$\mathcal{S}(r, \vartheta, t) = \frac{1}{2}\mathcal{E} \wedge \mathcal{H} = \frac{Z_{F0} h^2}{32\pi^2 c^2} m''^2_{e0}\left(t - \frac{r}{c}\right) \sin^3\vartheta \, d\vartheta \wedge d\varphi \,. \tag{11.68}$$

In the far-field the complex Poynting vector exhibits only a radial component

$$S_r(r, \vartheta, t) = \frac{Z_{F0} h^2}{32\pi^2 r^2 c^2} m''^2_{e0}\left(t - \frac{r}{c}\right) \sin^2\vartheta \,, \tag{11.69}$$

The power $P(r, t)$ radiated from the Hertzian dipole through a spherical surface with radius r in the far-field is obtained by integrating $\mathcal{S}$ over this surface.

$$P(r,t) = \int_{\vartheta=0}^{\pi} \int_{\varphi=0}^{2\pi} \mathcal{S}(r,\vartheta,t) \,. \tag{11.70}$$

We obtain from (11.69) and (11.70)

$$P(r,t) = \frac{Z_{F0}h^2}{12\pi c^2} m_{e0}''^2 \left(t - \frac{r}{c}\right) . \tag{11.71}$$

11.5 LINEAR ANTENNAS

Linear antennas are formed by segments of straight cylindric conductors. The linear antenna according to Figure 11.1(a) and the dipole antenna according to Figure 11.1(b) are examples of linear antennas. The excitation of the linear antenna is performed in a gap between two wire segments or across the gap between the wire segment and the conducting plane in Figure 11.1(a). We assume the cross-sectional dimensions of the conductor small; furthermore we can assume that short segments of the linear conductor act as Hertzian dipoles. To compute the field produced by the linear antenna we can superimpose the field contributions of small wire segments, which may be considered as Hertzian dipoles. In general the current will be varying over the length of a linear antenna. Therefore at first we will investigate the spatial variation of a current along a linear conductor. For an exact analysis of a linear antenna the current distribution must be computed by taking into consideration the electromagnetic field radiated by the antenna. In many cases, however, we can obtain a good approximation by splitting up the problem into first computing the current distribution over the linear wire without considering the radiated field and then, in a second step of our analysis, by computing the radiated field from the given current distribution. In this section we will proceed in that way. Consider an infinitely extended straight wire with a circular cross-section. Figure 11.9 shows the magnetic field lines in a transverse plane of the infinitely extended straight wire.

For symmetry reasons the magnetic field lines are concentric circles in the transverse planes with the center in the wire axis. Assuming no longitudinal electric field, we obtain from Ampere's law

$$\underline{\mathcal{H}}(r,z) = \underline{H}_\varphi(r,z)\, r \,\mathrm{d}\varphi = \frac{1}{2\pi} \underline{I}(z) \,\mathrm{d}\varphi \,. \tag{11.72}$$

Due to (3.34) and (3.36) the magnetic field must fulfill the Helmholtz equation in free space.

$$\Delta \underline{\mathcal{H}} + k^2 \underline{\mathcal{H}} = 0 \,. \tag{11.73}$$

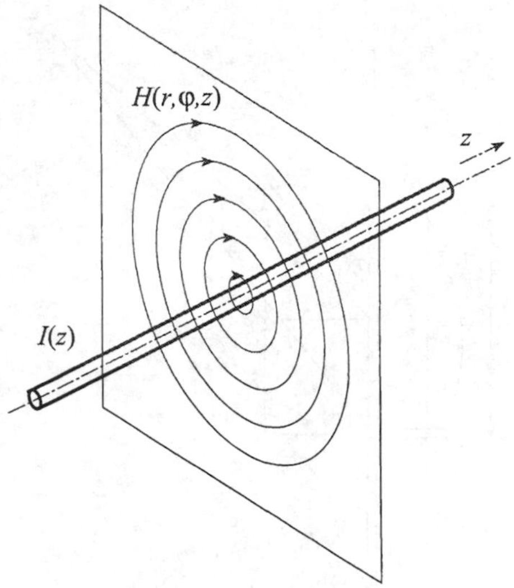

Figure 11.9: Magnetic field of the linear conductor.

From (3.16) we obtain for the cylindric coordinate system $(r, \varphi, z))$ defined by

$$x = r\cos\varphi\,, \quad y = r\sin\varphi\,, \quad z = z\,, \tag{11.74}$$

and with

$$g_1 = 1\,, \quad g_2 = r\,, \quad g_3 = 1\,, \tag{11.75}$$

the Laplace operator for one-forms,

$$\begin{aligned}
\Delta\underline{\mathcal{H}} = &\left(\frac{\partial^2 \underline{H}_r}{\partial r^2} + \frac{1}{r}\frac{\partial \underline{H}_r}{\partial r} + \frac{1}{r^2}\frac{\partial^2 \underline{H}_r}{\partial \varphi^2} + \frac{\partial^2 \underline{H}_r}{\partial z^2} - \frac{1}{r^2}\underline{H}_r - \frac{2}{r^2}\frac{\partial \underline{H}_\varphi}{\partial \varphi}\right) \mathrm{d}r \\
&+ \left(\frac{\partial^2 \underline{H}_\varphi}{\partial r^2} + \frac{1}{r}\frac{\partial \underline{H}_\varphi}{\partial r} + \frac{1}{r^2}\frac{\partial^2 \underline{H}_\varphi}{\partial \varphi^2} + \frac{\partial^2 \underline{H}_\varphi}{\partial z^2} - \frac{1}{r^2}\underline{H}_\varphi + \frac{2}{r^2}\frac{\partial \underline{H}_r}{\partial \varphi}\right) r\,\mathrm{d}\varphi + \\
&+ \left(\frac{\partial^2 \underline{H}_z}{\partial r^2} + \frac{1}{r}\frac{\partial \underline{H}_z}{\partial r} + \frac{1}{r^2}\frac{\partial^2 \underline{H}_z}{\partial \varphi^2} + \frac{\partial^2 \underline{H}_z}{\partial z^2}\right) \mathrm{d}z\,.
\end{aligned}$$

If $\underline{\mathcal{H}}$ exhibits only a φ component depending only on r and z the Laplace operator reduces to

$$\Delta\underline{\mathcal{H}} = \left(\frac{\partial^2 \underline{H}_\varphi}{\partial r^2} + \frac{1}{r}\frac{\partial \underline{H}_\varphi}{\partial r} - \frac{1}{r^2}\underline{H}_\varphi + \frac{\partial^2 \underline{H}_\varphi}{\partial z^2}\right) r\,\mathrm{d}\varphi\,. \tag{11.76}$$

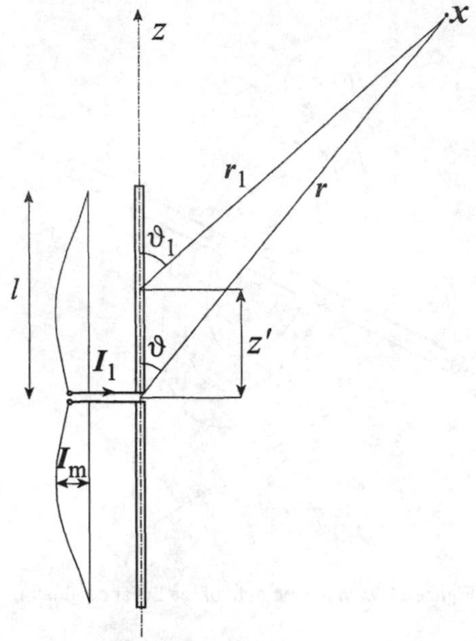

Figure 11.10: Dipole antenna.

Inserting into (11.73) yields

$$\frac{\partial^2 \underline{H}_\varphi}{\partial r^2} + \frac{1}{r}\frac{\partial \underline{H}_\varphi}{\partial r} - \frac{1}{r^2}\underline{H}_\varphi + \frac{\partial^2 \underline{H}_\varphi}{\partial z^2} + k^2 \underline{H}_\varphi = 0\,. \tag{11.77}$$

Inserting (11.72) into (11.77) the first three terms on the left side of (11.77) vanish, and we obtain

$$\frac{\mathrm{d}^2 \underline{I}}{\mathrm{d}z^2} + k^2 \underline{I} = 0\,. \tag{11.78}$$

This is the well-known transmission line equation for the TEM wave with the solution

$$\underline{I}(z) = \underline{I}^{(+)}\,\mathrm{e}^{-\mathrm{j}kz} + \underline{I}^{(-)}\,\mathrm{e}^{\mathrm{j}kz}, \tag{11.79}$$

where $\underline{I}^{(+)}$ and $\underline{I}^{(-)}$ are the amplitudes of current waves propagating in positive and negative z-direction, respectively. A current wave propagates on the straight wire with the velocity c_0 of the plane wave in free space. Let us consider the symmetric dipole antenna according to Figure 11.10. Both segments of the dipole antenna have a length l. With $\underline{I} = 0$ we obtain from (11.79) the current distribution

$$\underline{I}(z) = \begin{cases} \underline{I}_m \sin[k(l - z)] & \text{for} \quad z > 0 \\ \underline{I}_m \sin[k(l + z)] & \text{for} \quad z < 0 \end{cases} \tag{11.80}$$

on the dipole. According to (11.44) the contribution of the current flowing through a line segment of length dz to the electric far-field in point $\boldsymbol{x}$ is given by

$$\mathrm{d}\underline{E}_\vartheta = \frac{\mathrm{j}\omega\mu_0 \underline{I}(z)}{4\pi} \frac{\mathrm{e}^{-\mathrm{j}kr_1}}{r_1} \sin\vartheta_1 \mathrm{d}z\,. \tag{11.81}$$

To determine $\underline{E}$ we have to integrate $\mathrm{d}E_\vartheta$ from $z = -l$ to $z = l$. For points $\boldsymbol{x}$ in the far-field all lines drawn from a certain point of the far-field to any point of the linear antenna may be considered to be parallel. Assuming $\vartheta_1 = \vartheta$ we obtain

$$r_1 = r - z\cos\vartheta\,. \tag{11.82}$$

The exponential factor $\mathrm{e}^{-\mathrm{j}kr_1}$ is strongly varying with r_1, whereas in the denominator of (11.81) the variable r_1 may be substituted by r. By that way we obtain from (11.81) and (11.82) the approximate formula

$$\underline{E}_\vartheta \cong \frac{\mathrm{j}\omega\mu_0}{4\pi} \frac{\mathrm{e}^{-\mathrm{j}kr}}{r} \sin\vartheta \int_{-l}^{+l} \underline{I}(z)\,\mathrm{e}^{\mathrm{j}kz\cos\vartheta}\,\mathrm{d}z\,. \tag{11.83}$$

With (11.80) it follows

$$\underline{E}_\vartheta = \frac{\mathrm{j}\omega\mu_0 \underline{I}_m}{4\pi} \frac{\mathrm{e}^{-\mathrm{j}kr}}{r} \sin\vartheta \left\{ \int_{-l}^{0} \mathrm{e}^{\mathrm{j}kz\cos\vartheta} \sin[k(l+z)]\mathrm{d}z \right.$$
$$\left. + \int_{0}^{+l} \mathrm{e}^{\mathrm{j}kz\cos\vartheta} \sin[k(l-z)]\mathrm{d}z \right\}. \tag{11.84}$$

With

$$\int \mathrm{e}^{ax}\sin(bx+c)\mathrm{d}x = \frac{\mathrm{e}^{ax}}{a^2+b^2}[a\sin(bx+c) - b\cos(bx+c)] \tag{11.85}$$

and

$$\frac{\omega\mu_0}{k} = Z_{F0} \tag{11.86}$$

we obtain

$$\underline{E}_\vartheta = \frac{\mathrm{j}Z_{F0}\underline{I}_m}{2\pi} \frac{\mathrm{e}^{-\mathrm{j}kr}}{r} \frac{\cos(kl\cos\vartheta) - \cos kl}{\sin\vartheta}\,. \tag{11.87}$$

Since in the far-field (11.46) is valid for the contributions of all differential current elements, we obtain for the total far-field of the linear dipole antenna also

$$\underline{H}_\varphi = \frac{\underline{E}_\vartheta}{Z_{F0}}\,. \tag{11.88}$$

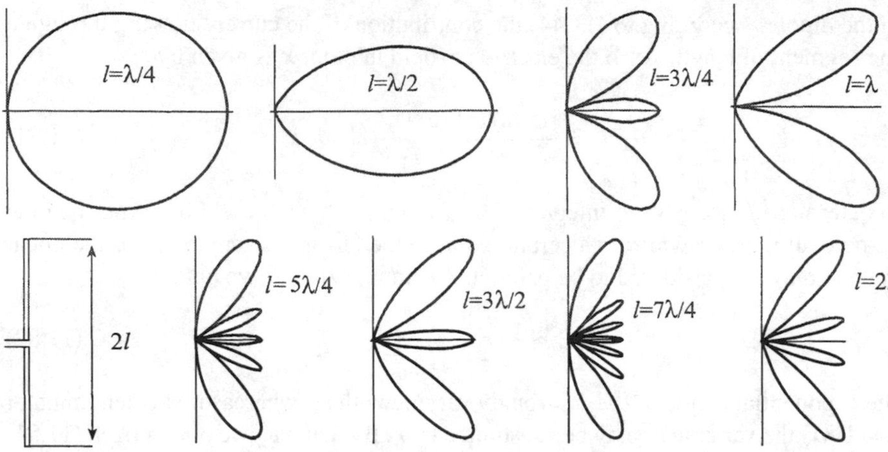

Figure 11.11: Radiation diagram of linear dipole antennas with different lengths.

Figure 11.11 shows the radiation diagram of linear dipole antennas with a half-length l, which is an integer multiple of $\lambda_0/4$. If l is an odd integer multiple of $\lambda_0/4$, the number of the maxima of the radiation diagram is equal $4l/\lambda_0$. If l is an integer multiple of $\lambda_0/2$, the number of the maxima in the radiation diagram is equal $2l/\lambda_0$. With (11.47) we obtain the radial component T_r of the complex Poynting vector

$$\underline{T}_r = \frac{Z_{F0}|\underline{I}_m|^2}{8\pi^2 r^2}\left[\frac{\cos(kl\cos\vartheta) - \cos kl}{\sin\vartheta}\right]^2 . \tag{11.89}$$

In the far-field $\underline{T}_r$ is real, i.e., there is only an active power flow. The active power radiated by the antenna is obtained by integrating $\underline{T}_r$ over a closed surface surrounding the antenna in the far-field. Integrating over the surface of a sphere, we obtain from (11.49)

$$P = \frac{Z_{F0}|\underline{I}_m|^2}{4\pi} f(kl) \tag{11.90}$$

with

$$f(kl) = \int_0^{\pi} \frac{[\cos(kl\cos\vartheta) - \cos kl]^2}{\sin\vartheta} \mathrm{d}\vartheta\, . \tag{11.91}$$

The evaluation of the integral yields

$$f(kl) = C + \ln 2kl - Ci(2kl) + \frac{1}{2}[Si(4kl) - 2Si(2kl)]\sin 2kl$$
$$+ \frac{1}{2}[C + \ln kl + Ci(4kl) - 2Ci(2kl)]\cos 2kl \tag{11.92}$$

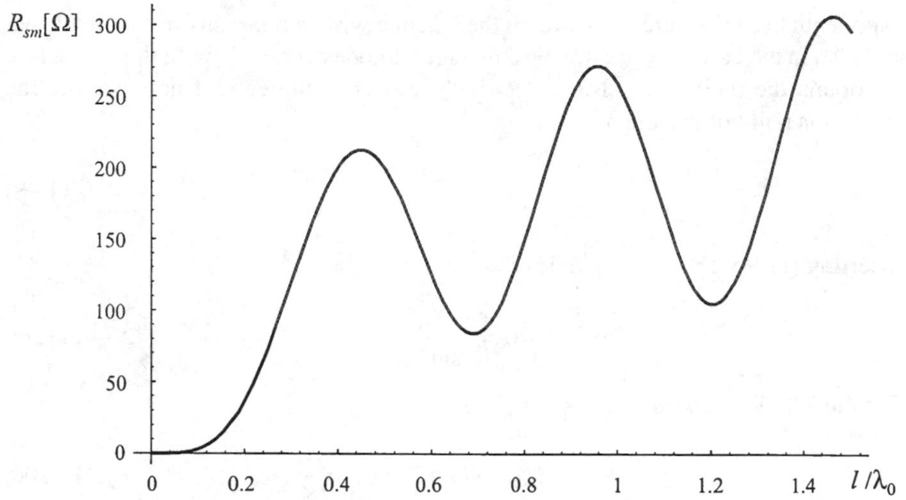

Figure 11.12: Radiation resistance R_{rm} of the dipole antenna referred to the maximum current I_m in dependence of l/λ_0.

where $C = 0.5772157\ldots$ is known as the *Euler constant* and the integral sine Si and the integral cosine Ci are given by

$$Si(x) = \int_0^x \frac{\sin x}{x} \mathrm{d}x\,, \tag{11.93}$$

$$Ci(x) = -\int_x^\infty \frac{\cos x}{x} \mathrm{d}x\,. \tag{11.94}$$

With reference to the current maximum $\underline{I}_m$ we may define the *radiation resistance* R_{rm} by

$$P = \frac{1}{2} R_{rm} |\underline{I}_m|^2\,. \tag{11.95}$$

With (11.90) we obtain

$$R_{rm} = \frac{1}{2\pi} Z_{F0} f(kl) = 60 f(kl)\Omega\,. \tag{11.96}$$

In Figure 11.12 R_{rm} is represented as a function of $kl = 2\pi l/\lambda_0$. The current $\underline{I}_1$ at the excitation point $z = 0$ of the antenna is given by

$$\underline{I}_1 = \underline{I}_m \sin kl \quad \text{for } kl \neq n\pi\,. \tag{11.97}$$

In the derivation of (11.79) we did not consider the attenuation of the current wave by the radiation of electromagnetic energy. Due to the radiation the wave is attenuated exponentially. Therefore the wave on the antenna wire is not really a standing wave, and also in the case $kl = n\pi$ we have no current nodes at $z = 0$. With this restriction we obtain the radiation resistance R_r with respect to the current impressed at the excitation point of the antenna,

$$R_r = \frac{2P}{|\underline{I}_1|^2} . \tag{11.98}$$

Inserting (11.96) and (11.97) yields

$$R_r = \frac{R_{rm}}{\sin^2 kl} . \tag{11.99}$$

For the $\lambda/2$ dipole with $l = \lambda/4$ we obtain

$$R_r = R_{rm} = 73.4\Omega \text{ for } l = \frac{\lambda}{4} . \tag{11.100}$$

11.6 THE LOOP ANTENNA

A loop antenna consists of a wire loop of one or more turns. The loop antenna primarily excites a magnetic dipole moment, which may be considered as the source of the electromagnetic wave. Figure 11.13 shows a circular loop antenna of diameter $2a$ with one turn. We consider the circumference of the loop to be small compared to the wavelength so that we can assume the current flowing through the loop to be spatially uniform. We compute the electric Hertz vector using (11.25a). The loop antenna exhibits rotational symmetry with respect to rotation around the z-axis. The necessary feeding line yields no perturbation of this symmetry since the field contributions of the two wires of the feeding line compensate each other. Due to the symmetry properties of the circular loop antenna the electromagnetic field also exhibits rotational symmetry; therefore it will be sufficient in the following to compute the Hertz vector $\underline{\Pi}_e(\boldsymbol{x})$ in the plane $y = 0$. The volume integration in (11.25a) needs to be performed over the volume filled by the conductor only. According to (11.27) we obtain for the region

$$\star\underline{\mathcal{M}}_{e0} = \frac{\underline{I}_0}{\mathrm{j}\omega A}\, r\,\mathrm{d}\varphi = \frac{\underline{I}_0}{\mathrm{j}\omega A}(-\sin\varphi\,\mathrm{d}x + \cos\varphi\,\mathrm{d}y) , \tag{11.101}$$

where A is the cross-sectional area of the conductor. The volume element is given by $Aa\,\mathrm{d}\varphi$. For $\underline{\Pi}_{e,y}$ we obtain from (11.25a)

$$\underline{\Pi}_{e,y}(\boldsymbol{x}_0) = \frac{a\underline{I}_0}{4\pi\mathrm{j}\omega\varepsilon_0}\int_0^{2\pi}\frac{\cos\varphi\,\mathrm{e}^{-\mathrm{j}kr_1}}{r_1}\mathrm{d}\varphi , \tag{11.102}$$

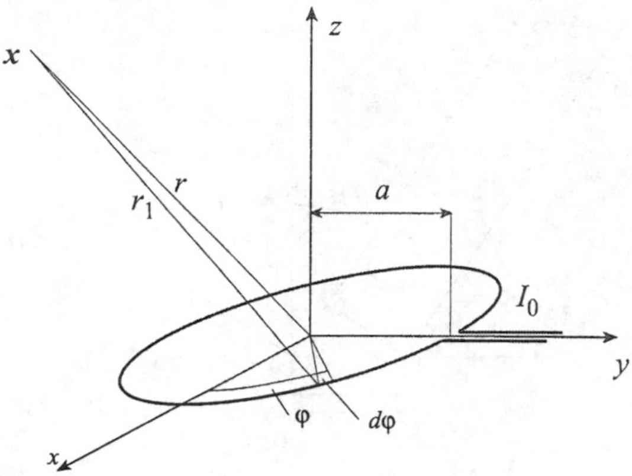

Figure 11.13: Loop antenna.

where r_1 marks the connection from the point $\boldsymbol{x}$ in the far-field to the conductor. For a point $\boldsymbol{x}$ in the far-field we can assume the lines r and r_1 to be parallel. Since $\underline{\boldsymbol{M}}_{e0}$ exhibits no z component, also $\underline{\Pi}_{e,z} = 0$ is valid. Furthermore it follows from (11.101) that in the plane $y = 0$ the x component of the Hertz vector disappears, i.e., $\underline{\Pi}_{e,x}(x, 0, z) = 0$. According to Figure 11.14 we obtain

$$r_1 = r - a \cos\varphi \sin\vartheta \,. \tag{11.103}$$

We use this expression in the exponent of the integrand, since the exponential function is strongly varying with r_1, whereas in the denominator r_1 may be replaced by r. Furthermore due to the rotational symmetry of the electromagnetic field we obtain

$$\underline{\Pi}_\varphi(r, \vartheta) = \underline{\Pi}_y(r, \vartheta)\Big|_{\varphi=0} \,. \tag{11.104}$$

This yields

$$\underline{\Pi}_\varphi = \frac{a\underline{I}_0}{4\pi \mathrm{j}\omega\varepsilon_0} \frac{\mathrm{e}^{-\mathrm{j}kr}}{r} \int_0^{2\pi} \cos\varphi \, \mathrm{e}^{\mathrm{j}ka\cos\varphi\sin\vartheta} \, \mathrm{d}\varphi \,. \tag{11.105}$$

Since the circumference of the current loop is small compared with the wavelength, $ka \ll 1$ is valid, and we can make the approximation

$$\mathrm{e}^{\mathrm{j}ka\cos\varphi\sin\vartheta} \approx 1 + \mathrm{j}ka\cos\varphi\sin\vartheta \,. \tag{11.106}$$

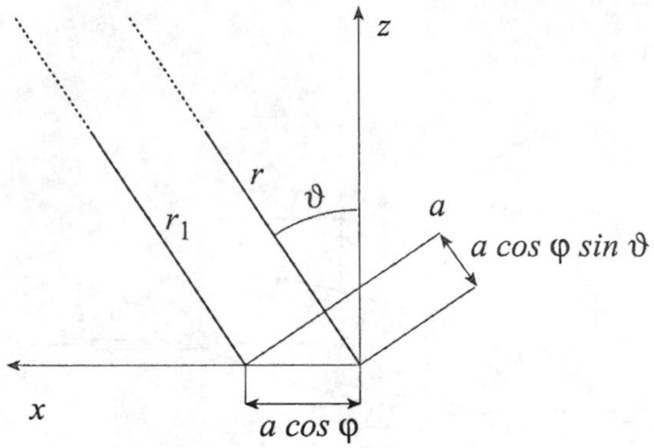

Figure 11.14: Definition of r_1.

Therewith we obtain from (11.105)

$$\underline{\Pi}_e = \frac{ka^2 \underline{I}_0}{4\omega\varepsilon_0}\, \mathrm{e}^{-\mathrm{j}kr} \sin^2\vartheta\, \mathrm{d}\varphi\,. \tag{11.107}$$

With (11.31) we obtain

$$\underline{\mathcal{H}} = \mathrm{j}\omega\varepsilon_0 \star \mathrm{d}\underline{\Pi}_e = \mathrm{j}\frac{ka^2}{4}\underline{I}_0\, \mathrm{e}^{-\mathrm{j}kr}\left(\frac{2}{r^2}\cos\vartheta\ \mathrm{d}r\ + \frac{\mathrm{j}k}{r}\sin\vartheta\ r\,\mathrm{d}\vartheta\right), \tag{11.108}$$

and therewith

$$\underline{H}_\vartheta = -\frac{1}{4}k^2 a^2 \underline{I}_0 \frac{\mathrm{e}^{-\mathrm{j}kr}}{r}\sin\vartheta\,, \tag{11.109a}$$

$$\underline{H}_r = +\frac{1}{2}\mathrm{j}ka^2 \underline{I}_0 \frac{\mathrm{e}^{-\mathrm{j}kr}}{r^2}\cos\vartheta\,. \tag{11.109b}$$

With (11.32) and $Z_{F0} = k/\omega\varepsilon_0$ we obtain

$$\underline{\mathcal{E}} = \star\ \mathrm{d}\star\ \mathrm{d}\underline{\Pi}_e = \frac{1}{4}a^2 Z_{F0}\underline{I}_0\, \mathrm{e}^{-\mathrm{j}kr}\left(\frac{k^2}{r} + \frac{2}{r^3}\right) r\sin^2\vartheta\ \mathrm{d}\varphi\,, \tag{11.110}$$

and therefrom

$$\underline{E}_\varphi = \frac{1}{4}a^2 Z_{F0}\underline{I}_0 \left(\frac{k^2}{r} + \frac{2}{r^3}\right) \mathrm{e}^{-\mathrm{j}kr}\sin\vartheta\,. \tag{11.111}$$

In the far-field $r \to \infty$ we obtain from (11.108) and (11.110)

$$\underline{\mathcal{H}} = -\frac{1}{4}k^2a^2\underline{I}_0\frac{\mathrm{e}^{-\mathrm{j}kr}}{r}\sin\vartheta\, r\,\mathrm{d}\vartheta\,, \tag{11.112a}$$

$$\underline{\mathcal{E}} = \frac{1}{4}Z_{F0}k^2a^2\underline{I}_0\frac{\mathrm{e}^{-\mathrm{j}kr}}{r}\sin\vartheta\, r\sin^2\vartheta\,\mathrm{d}\varphi\,. \tag{11.112b}$$

and the corresponding field components

$$\underline{H}_\vartheta = -\frac{1}{4}k^2a^2\underline{I}_0\frac{\mathrm{e}^{-\mathrm{j}kr}}{r}\sin\vartheta\,, \tag{11.113a}$$

$$\underline{E}_\varphi = \frac{1}{4}Z_{F0}k^2a^2\underline{I}_0\frac{\mathrm{e}^{-\mathrm{j}kr}}{r}\sin\vartheta\,. \tag{11.113b}$$

The far-field is given by (11.109a) and

$$\underline{E}_\varphi = -Z_{F0}\underline{H}_\vartheta = \frac{1}{4}Z_{F0}k^2a^2\underline{I}_0\frac{\mathrm{e}^{-\mathrm{j}kr}}{r}\sin\vartheta\,. \tag{11.114}$$

The far field of the loop antenna with a small diameter is dual to the far-field of the Hertzian dipole. From (4.21), (11.109a) and (11.114) we obtain the complex Poynting form $\mathcal{T}$ for the far-field

$$\mathcal{T} = -\frac{1}{2}\underline{E}_\varphi\underline{H}_\vartheta^* r^2\sin\vartheta\,\mathrm{d}\vartheta\wedge\mathrm{d}\varphi = \frac{1}{32}Z_{F0}k^4a^4|\underline{I}_0|^2\sin^3\vartheta\,\mathrm{d}\vartheta\wedge\mathrm{d}\varphi\,. \tag{11.115}$$

In the far-field the complex Poynting vector exhibits only a radial component

$$T_r = \frac{Z_{F0}k^4a^4|\underline{I}_0|^2\sin^2\vartheta}{32r^2}\,. \tag{11.116}$$

The power radiated from the loop antenna follows from this with (11.49) and (11.50)

$$P = \frac{\pi}{12}Z_{F0}k^4a^4|\underline{I}_0|^2\,. \tag{11.117}$$

We again introduce the radiation resistance defined in (11.54) and obtain

$$R_r = \frac{2P}{|\underline{I}_0|^2} = \frac{\pi}{6}Z_{F0}k^4a^4\,. \tag{11.118}$$

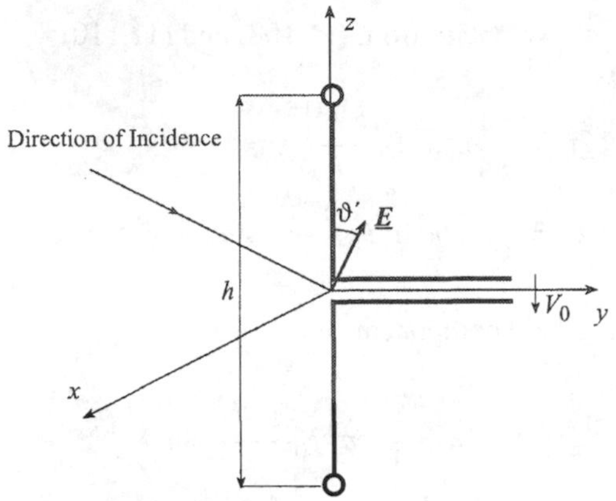

Figure 11.15: The Hertzian dipole as receiving antenna.

11.7 RECEIVING ANTENNAS

11.7.1 The Hertzian Dipole as Receiving Antenna

We consider the Hertzian dipole as depicted in Figure 11.15 in the field of a plane electromagnetic wave. The magnetic field $\underline{\boldsymbol{H}}$ may be oriented normal to the dipole axis. The direction of the electric field $\underline{\boldsymbol{E}}$ may enclose an angle ϑ' with the dipole axis. The Hertzian dipole of length l is formed by two wires with spheres attached at the end. The wires are assumed to be thin enough, so that only the spheres at the end of the wires are contributing to the capacitance of the antenna. The current flowing in the short linear conductors between the spheres may be considered to be spatially uniform. The circuit loop is closed via the displacement current between the two spheres. The potential difference $\underline{V}_0$ between both spheres is given by

$$\underline{V}_0 = -\int_{-l/2}^{+l/2} \underline{\mathcal{E}} = -l\underline{E}\cos\vartheta' \,. \tag{11.119}$$

This voltage is equal to the open circuit voltage in the feeding point of the antenna. If the antenna is oriented in parallel to the direction of the electric field the node voltage $\underline{V}_0$ assumes a maximum value $\underline{V}_{0max}$. We obtain

$$|\underline{V}_{0\,\mathrm{max}}| = h|\underline{E}| \,. \tag{11.120}$$

The length h of the Hertzian dipole determines the ratio between the open circuit voltage across the antenna port and the electric field intensity. Via (11.120) an *effective antenna length* l_{eff} may be defined for arbitrary antennas.

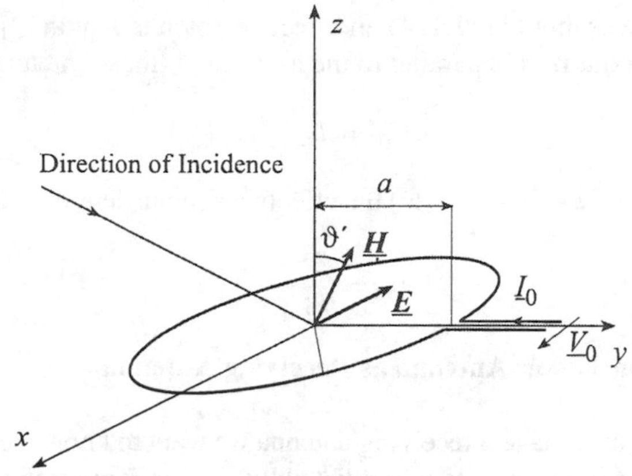

Figure 11.16: The loop antenna as receiving antenna.

11.7.2 The Loop Antenna as Receiving Antenna

A plane wave is incident on a loop antenna according to Figure 11.16. We assume that $\underline{\boldsymbol{E}}$ is parallel to the plane of the loop, whereas the direction of the magnetic field $\underline{\boldsymbol{H}}$ encloses with the normal to the loop plane an angle ϑ'. We assume $a \ll \lambda$. From (2.121) we obtain the open-circuit node voltage of the loop antenna

$$\underline{V}_0 = \mathrm{j}\omega\underline{\Phi} \tag{11.121}$$

where the magnetic flux according to (2.31) is given by

$$\underline{\Phi} = \int_A \mathcal{B} = A\mu_0\underline{H}\cos\vartheta' \tag{11.122}$$

where A is the surface of the loop antenna. For the circular loop antenna according to Figure 11.16 we obtain

$$A = a^2\pi \,. \tag{11.123}$$

We want to represent $\underline{V}_0$ as a function of the electric field intensity $\underline{\boldsymbol{E}}$ and obtain with

$$\frac{\omega\mu_0}{Z_{F0}} = \frac{2\pi}{\lambda_0} \tag{11.124}$$

from (11.88), (11.121) and (11.122)

$$\underline{V}_0 = \frac{2\pi\mathrm{j}A}{\lambda_0}\underline{E}\cos\vartheta' \,. \tag{11.125}$$

We now define according to (11.124) an effective antenna length l_{eff} for the case of an incident wave polarized in parallel to the axis of the linear antenna:

$$|\underline{V}_0| = l_{\text{eff}} |\underline{E}| \,. \tag{11.126}$$

We obtain from (11.124) and (11.126) the effective antenna length of the loop antenna

$$l_{\text{eff}} = \frac{2\pi A}{\lambda_0} \,. \tag{11.127}$$

11.7.3 The Linear Dipole Antenna as Receiving Antenna

If we are using an antenna as a receiving antenna we want to know the signal obtained at the antenna port if the antenna is irradiated by an electromagnetic field. Usually a receiving antenna is positioned in the far-field of the transmitter and the distance between the transmitter and the receiving antenna is by orders of magnitude larger than the linear dimensions of the antenna. Therefore we can assume the received electromagnetic wave within a spatial region of the size of the antenna to be a planar wave. However arranging the antenna in the received field will create a considerable perturbation. The primary field will be scattered by the antenna and the resulting electromagnetic field will no longer be a planar field. In the following we shall overcome this difficulty by applying the theorem of reciprocity. We will demonstrate a way to analyze the receiving antenna on the basis of the unperturbed primary incident plane wave field.

Let us consider the two antennas, (1) and (2), respectively, depicted in Figure 11.17(a). We assume both antennas to be coupled via the far-field – that means the distance between the antennas is by orders of magnitude larger than their linear dimensions. Let antenna (1) be a linear dipole antenna according to Figure 11.10, whereas antenna (2) may be of an arbitrary type. At the feeding nodes of antenna (1) and antenna (2), respectively, the currents $\underline{I}_{01}$ and $\underline{I}_{02}$ may be impressed. The current distribution $\underline{I}_1(z)$ on the antenna wires of antenna (1) is due to the impression of the node current $\underline{I}_{01}$. The reaction of the far-field of antenna (1) on the current $\underline{I}_{02}$ impressed into antenna (2) is given by R_{12}, whereas the reaction of the far-field of antenna (2) on the current source $\underline{I}_{01}$ is given by R_{21}. From (4.63) we obtain

$$\mathsf{R}_{12} = \mathsf{R}_{21} \tag{11.128}$$

and (4.67) yields

$$\mathsf{R}_{21} = -\underline{V}_2 \underline{I}_{01} \,, \tag{11.129}$$

where V_2 is the open circuit voltage excited from the far-field of antenna (2) in the nodes of antenna (1). We now replace the current distribution $\underline{I}_1(z)$ due to the excitation of antenna (1) with the current source $\underline{I}_{01}$ by a polarization $\underline{\boldsymbol{M}}_{e03}(\boldsymbol{x})$ impressed

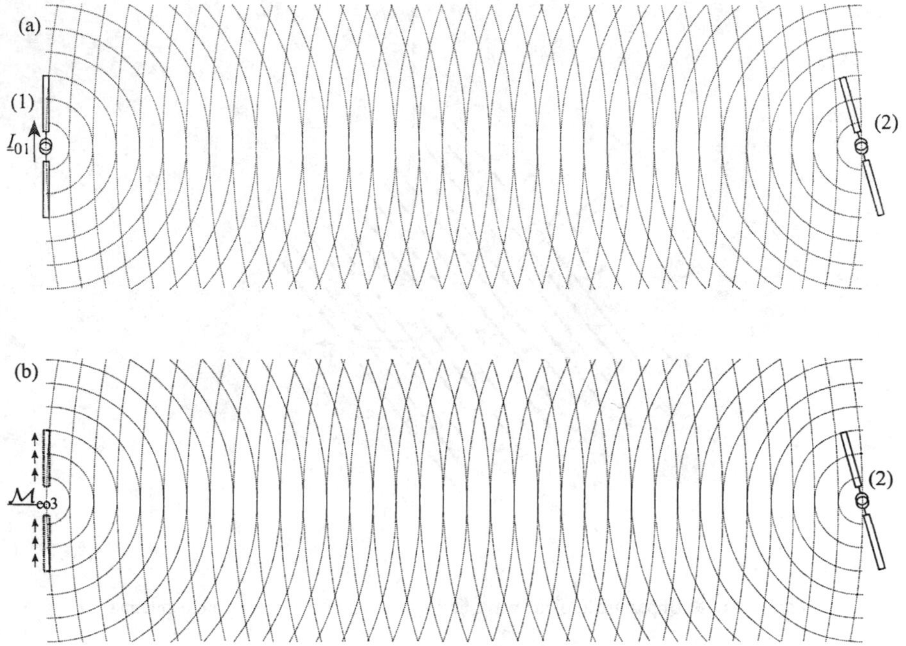

Figure 11.17: For the calculation of the h_{eff} of the dipole antenna.

into free space. Similar as in (4.63) we obtain

$$\underline{\mathcal{M}}_{e03}(\boldsymbol{x}) = \frac{1}{\mathrm{j}\omega A}\underline{I}_1(z)\,\mathrm{d}x \wedge \mathrm{d}y \tag{11.130}$$

where A is the cross-sectional area of the linear dipole antenna (1). The impressed polarization $\underline{\boldsymbol{M}}_{e03}(\boldsymbol{x})$ excites the same field as antenna (1). Therefore also the reaction R_{32} of the far-field excited by $\underline{\boldsymbol{M}}_{e03}(\boldsymbol{x})$ on the current source $\underline{I}_{02}$ must be equal to the reaction R_{12} of the far-field excited from antenna (1) on the current source $\underline{I}_{02}$.

$$\mathsf{R}_{32} = \mathsf{R}_{12}\,. \tag{11.131}$$

On the other hand due to the reciprocity of the radio link formed by the two antennas we obtain from (4.63)

$$\mathsf{R}_{32} = \mathsf{R}_{23}\,. \tag{11.132}$$

From (11.128), (11.131) and (11.132) it follows that

$$\mathsf{R}_{21} = \mathsf{R}_{23}\,. \tag{11.133}$$

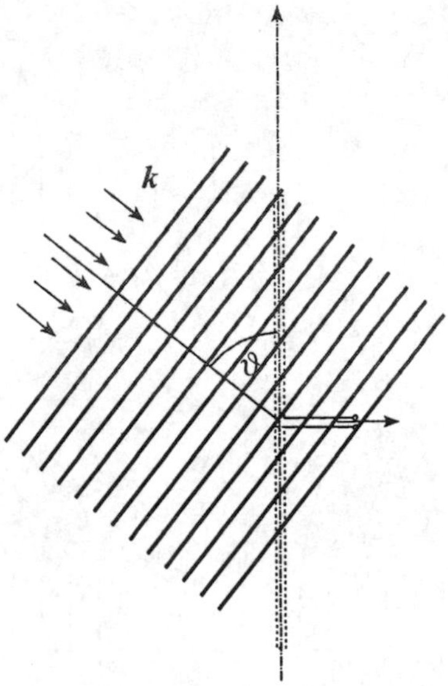

Figure 11.18: Unperturbed plane wave at the position of the linear dipole antenna.

The field excited from antenna (2) at the location of the impressed polarization $\underline{M}_{e03}(\boldsymbol{x})$ is equal to the unperturbed electric field since we have removed antenna (1). This procedure allows us to compute the voltage induced in the receiving antenna without considering the field perturbation by bringing the receiving antenna into the primary field. The reaction R_{23} of the field $\underline{E}(\boldsymbol{x})$ on $\underline{M}_{e03}(\boldsymbol{x})$ according to (4.62) is given by

$$\mathsf{R}_{23} = \mathrm{j}\omega \int_V \underline{\mathcal{E}}_2 \wedge \underline{\mathcal{M}}_{e03} \,. \tag{11.134}$$

where the integration is performed over the volume $\mathcal{V}_{03}$, where $\underline{M}_{e03}(\boldsymbol{x})$ is impressed. Since $\underline{E}(\boldsymbol{x})$ and $\underline{M}_{e03}(\boldsymbol{x})$ are uniformly distributed over the cross-sectional area, we obtain from (11.130)

$$\mathsf{R}_{23} = \int_{-l}^{+l} \underline{I}_1(z)\underline{E}_z(z)\,\mathrm{d}z \,. \tag{11.135}$$

The integration is performed over a length $2l$ of the linear dipole antenna from $z = -l$ to $z = +l$. The current distribution $\underline{I}_1(z)$ is given by (11.80).

For an electromagnetic wave incident under an angle ϑ relative to the dipole axis, as shown in Figure 11.18, with the wave number k and the electric field parallel to the plane $x = 0$ we obtain

$$k_z = -k\cos\vartheta \tag{11.136}$$

for the z component of the unperturbed electric field

$$\underline{E}_z = \underline{E}_0 \sin\vartheta \, e^{jkz\cos\vartheta} . \tag{11.137}$$

Inserting (11.85) and (11.137) into (11.135) we obtain

$$\mathsf{R}_{23} = \frac{2\underline{E}_0 \underline{I}_m}{k\sin\vartheta}\left[\cos(kl\cos\vartheta) - \cos(kl)\right] . \tag{11.138}$$

With (11.97) we relate R_{23} to the node current $\underline{I}_{01}$ and obtain

$$\mathsf{R}_{23} = 2\underline{E}_0 \underline{I}_{01} \frac{\cos(kl\cos\vartheta) - \cos kl}{k\sin\vartheta\sin kl} . \tag{11.139}$$

With (11.129) and (11.133) we obtain with $\underline{V}_0 = \underline{V}_2$

$$\underline{V}_0 = -2\underline{E}_0 \frac{\cos(kl\cos\vartheta) - \cos kl}{k\sin\vartheta\sin kl} . \tag{11.140}$$

A comparison with (11.87) shows that the antenna characteristics of the receiving antenna is the same as the antenna characteristics of the transmitting antenna. We will demonstrate the identity of the receiving antenna characteristics and transmitting antenna characteristics for arbitrary antennas in Section 11.8. This identity is a consequence of the reciprocity theorem.

For the case of an incident wave polarized in parallel to the axis of the linear antenna we obtain from (11.126) and (11.140) for $\vartheta = \pi/2$

$$l_{\text{eff}} = 2\frac{1-\cos kl}{k\sin kl} = \frac{\lambda_0}{\pi}\frac{1-\cos kl}{\sin kl} . \tag{11.141}$$

For the short linear antenna with $kl \ll 1$ we obtain

$$l_{\text{eff}} = l \quad \text{for } kl \ll 1 . \tag{11.142}$$

For the half-wave dipole with $2l = \lambda_0/2$ we obtain

$$l_{\text{eff}} = \frac{\lambda_0}{\pi} \quad \text{for } 2l = \frac{\lambda_0}{2} . \tag{11.143}$$

For arbitrary directions of incidence $\vartheta \neq \pi/2$ we obtain from (11.140) and (11.126) the *general effective antenna length*

$$l_{\text{eff}} = \left| 2\frac{\cos(kl\cos\vartheta) - \cos kl}{k\sin\vartheta\sin kl} \right| . \tag{11.144}$$

11.8 GAIN AND EFFECTIVE ANTENNA APERTURE

The power radiated by an antenna into a certain solid angle depends on the direction. We define an antenna gain G as the ratio of the active power density radiated by the antenna to the power density radiated by an antenna of reference. As the reference antenna we use the isotropic spherical radiator, which distributes the radiated power uniformly over all directions. Such a uniform radiator cannot be realized physically. The best approximation to the uniform radiator is the Hertzian dipole which, however, exhibits a non-uniform angular distribution of the radiated power. Let P_r be the active power emitted by a radiator. In this case the power density of an isotropic spherical radiator at a distance r will be $P_r/4\pi r^2$. Using the isotropic spherical radiator as a reference, the antenna gain is given by

$$G = \frac{4\pi r^2 \Re\{\underline{T}_r(r, \vartheta, \varphi)\}}{P_r} . \tag{11.145}$$

The direction in which the active power radiated by the antenna per unit of solid angle is maximum, is called the direction of maximum radiation. The antenna gain in the direction of maximum radiation is the maximum antenna gain. For a Hertzian dipole it follows from (11.51), (11.54) and (11.145) that

$$G = \frac{3}{2} \sin^2 \vartheta . \tag{11.146}$$

We obtain the same result from (11.116) and (11.117) for the small loop antenna according to Figure 11.13. The maximum antenna gain is obtained for $\vartheta = \frac{\pi}{2}$

$$G_{\max} = \frac{3}{2} . \tag{11.147}$$

For receiving antennas an effective aperture (or effective area) A_e may be defined. The effective aperture is an equivalent area through which the incident wave transports a power equal to the power received by the antenna. Multiplying the power density of the incident wave with the effective antenna aperture yields the power received by the antenna. Let $\Re\{\underline{T}\}$ be the power density of the incident wave and P_r the active power received by the antenna, we obtain the following relation

$$P_r = A_e \Re\{\underline{T}\} . \tag{11.148}$$

Since the received power depends on the orientation of the antenna as well as on the matching of the load to the antenna, the effective aperture also depends on these conditions. The magnitude of the open-circuit node voltage $\underline{V}_0$ may be obtained from

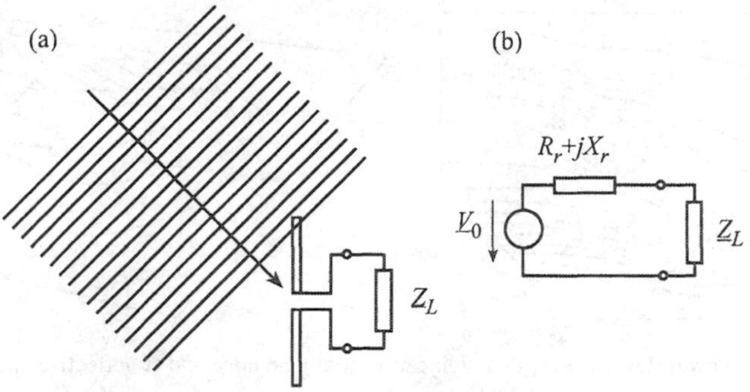

Figure 11.19: (a) Receiving antenna loaded with Z_L and (b) two-port equivalent circuit.

(11.126). If the antenna exhibits no losses, the real part of the antenna impedance is equal to the radiation resistance R_r. For power matching the load impedance is given by

$$Z_L = R_r - \mathrm{j}X_s \,. \tag{11.149}$$

In this case the antenna delivers the power

$$P_r = \frac{1}{8}\frac{|\underline{V}_0|^2}{R_r} \tag{11.150}$$

to the load. For an incident plane wave the power density is given by

$$\underline{T} = \frac{1}{2Z_{F0}}|\underline{E}|^2 \,. \tag{11.151}$$

From (11.126), (11.145), (11.150) and (11.151) we obtain the following relation between effective antenna aperture and effective antenna length

$$A_e = \frac{Z_{F0} l_{\text{eff}}^2}{4R_r} \quad \text{for power matching} \,. \tag{11.152}$$

For the Hertzian dipole oriented parallel to the electric field we obtain from (11.54)

$$A_e = \frac{3\lambda_0^2}{8\pi} \,. \tag{11.153}$$

The effective aperture of the Hertzian dipole is independent from h. Therefore the power as specified in (11.148) may be received irrespective of how short the antenna may be. This statement, however, is only valid if the conductive losses in the antenna

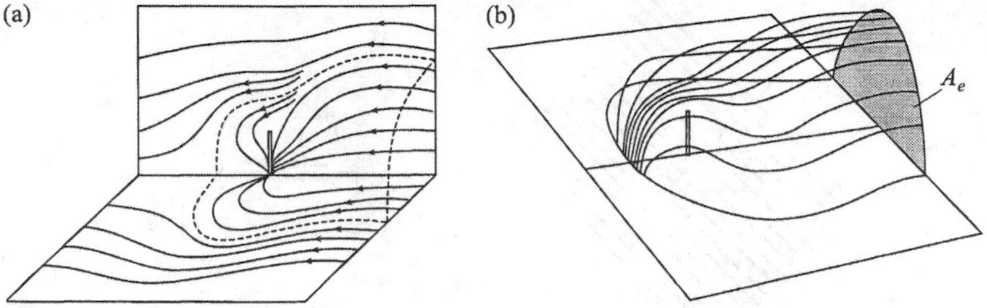

Figure 11.20: (a) Power flow lines $\Re\{\mathbf{T}\}$ at a linear receiving antenna and (b) effective aperture A_e of the linear receiving antenna [9, 10].

may be neglected and the antenna can be matched to the load. Since the radiation resistance of short antennas according to (11.54) is proportional to the square of the antenna length. For very short antennas it may not be possible to neglect the ohmic losses in comparison with the radiation resistance. Furthermore, if the radiation resistance is very small, a power matching of the antenna will only be possible in a very narrow frequency band.

Considering the Poynting vector field of power flow we can visualize the meaning of the effective aperture. Figure 11.20 depicts the Poynting vector field in the vicinity of an antenna. If a plane wave is incident on an antenna, this wave will be scattered from the antenna. The scattering of the incident wave also depends on the matching of the antenna to the load. An antenna matched to the load is equivalent to a resonant circuit. If the plane wave is incident at the resonant frequency of this resonant circuit, an oscillation of considerable amplitude may be built up in the resonant circuit. Due to this excitation the antenna will radiate a secondary wave. This scattered field of the antenna now is superimposed to the primary incident wave and the Poynting vector field now depends on the superposition of the primary field and the secondary field. As can be seen in Figure 11.20(a) the antenna is perturbing the power flow in its neighborhood considerably. A part of the power flow lines are now flowing in the feed line of the receiving antenna. In Figure 11.20(b) the surface A_1 is shown separating the region where all power flow lines are flowing into the antenna feed line from the outer region, where the power flow lines are passing the antenna. By that way we may construct a tube, which has the property that all the field energy flowing inside this tube is fed into the antenna. We can follow this tube to a region far distant enough from the antenna so that there is no distortion of the plane wave field anymore. At this point the cross-sectional area of the tube corresponds exactly with the effective aperture of the antenna. We now can understand why the effective antenna area may exhibit a much larger dimension than the antenna. When the antenna is very small the radiation resistance becomes small, too. If the antenna exhibits no

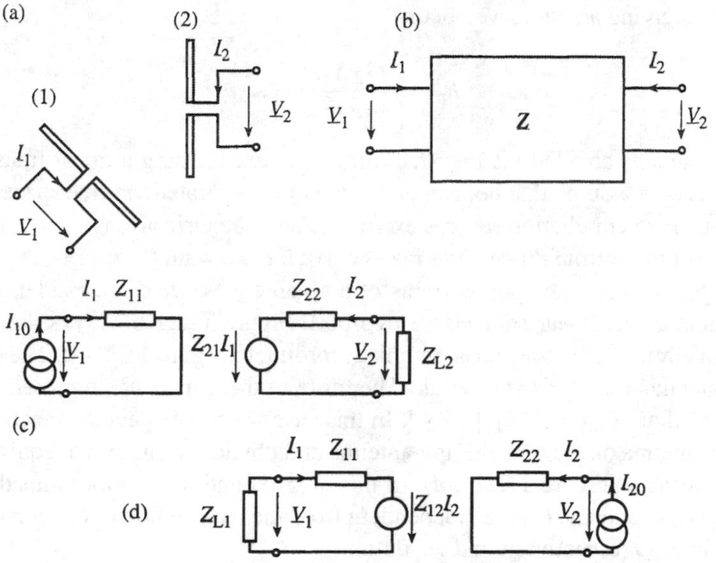

Figure 11.21: Two coupled antennas: (a) schematic presentation, (b) two-port-equivalent circuit, (c)/(d) approximated equivalent circuit for antenna (1)/(2) as transmitting antenna.

losses and is terminated by matched load, the equivalent circuit of the terminated antenna is a resonant circuit with very high Q factor. An incident wave at the resonant frequency will excite an oscillation of high amplitude and in this way large scattered field contribution. Such a small antenna will only exhibit a large effective antenna aperture within a very narrow frequency band.

Let us now consider the transmission properties of a radio-link formed by two antennas depicted schematically in Figure 11.21(a). We assume the antennas to be sufficiently coupled via their far-field only. We may consider one of the antennas to be the transmitting antenna and the other antenna to be the receiving antenna. Let us first consider antenna (1) to be the transmitting antenna and antenna (2) the receiving antenna. In this case from (11.145) and (11.148) we obtain the following relation between the power P_{t1} emitted from antenna 1 and the power P_{r2} received from antenna (2)

$$P_{r2} = \frac{G_1 A_{e2}}{4\pi r^2} P_{t1} . \tag{11.154}$$

In this equation, G_1 is the gain of antenna (1) and A_{e2} is the effective aperture of antenna (2). If we otherwise use antenna (2) as the transmitting antenna and antenna

(1) as the receiving antenna, we obtain

$$P_{r1} = \frac{G_2 A_{e1}}{4\pi r^2} P_{t2} . \tag{11.155}$$

The two antennas coupled via their radiation field are forming a linear transmission link. Let us now assume that both antennas are only separated via free space and that there are no further radiation sources existing. The geometric arrangement of both antennas and of the surrounding matter may be fixed. If we want to describe the relations between port voltages and port currents of both antennas, we can consider this transmission link as the linear source free two-port. Figure 11.21(b) shows the two-port equivalent circuit of the transmission link according to Figure 11.21(a). The coupling of two antennas usually is very weak. Therefore in the impedance representation we can assume that $|Z_{12}| \ll |Z_{11}|, |Z_{22}|$. In this case we can neglect the reaction of the receiving antenna on the transmitting antenna and obtain the simplified equivalent circuits according to Figure 11.21(c/d). In this approximation the input impedances of antenna (1) and antenna (2) are independent from the termination of the other antenna by Z_{11} and Z_{22}, respectively, and we obtain

$$\Re\{Z_{ii}\} = R_{ri}, \quad i = 1, 2 , \tag{11.156}$$

where R_{ri} is the radiation resistance of antenna (i). We now assume in both cases of operation according to Figure11.21(c) and 11.21(d), respectively, power-matching to be fulfilled, in other words

$$Z_{Li} = Z_{ii}^{\star} . \tag{11.157}$$

For antenna (1) as the transmitting antenna and antenna (2) as the receiving antenna, we obtain

$$P_{t1} = \frac{1}{2} R_{r1} |\underline{I}_{10}|^2 , \tag{11.158}$$

$$P_{r2} = \frac{1}{2} \Re\{Z_{L2}\} |\underline{I}_2|^2 . \tag{11.159}$$

For the equivalent circuit according to Figure 11.21(c), we obtain from (11.156) to (11.159)

$$\frac{P_{r2}}{P_{t1}} = \frac{|Z_{21}|^2}{4 R_{r1} R_{r2}} . \tag{11.160}$$

If antenna (2) is the transmitting antenna and antenna (1) the receiving antenna it follows from Figure 11.21(d) that

$$\frac{P_{r1}}{P_{t2}} = \frac{|Z_{12}|^2}{4 R_{r1} R_{r2}} . \tag{11.161}$$

Due to the reciprocity (4.63) is fulfilled and therefore

$$Z_{12} = Z_{21} \,. \tag{11.162}$$

From this it follows that

$$\frac{P_{r2}}{P_{t1}} = \frac{P_{r1}}{P_{t2}} \tag{11.163}$$

for the case of power matching of the load to the receiving antenna. The ratio of the available active power at the port of the receiving antenna to the input power of the transmitting antenna is called the transmission factor. According to (11.163) the transmission factor of a radio link corresponding to Figure 11.21(a) is of equal size in both directions. From (11.154), (11.155) and (11.163) we obtain

$$\frac{G_1}{A_{e1}} = \frac{G_2}{A_{e2}} \,. \tag{11.164}$$

We see that the ratio of gain to effective aperture is the same for both antennas. The quantities G_1, A_{e1} only depend on antenna (1) and the quantities G_2, A_{e2} only depend on antenna 2, and we also have considered the possibility of using different antenna types (1) and (2). Equation (11.164) only may be satisfied if the ratio of the gain and effective aperture is the same for all types of antennas. For an optimally oriented Hertzian dipole ($\vartheta = \pi/2$) we can compute from (11.147) and (11.153) the following relation between gain and effective antenna aperture

$$A_e = \frac{\lambda_0^2}{4\pi} G \,. \tag{11.165}$$

This relation also holds for any type of antenna and for any orientation of the antenna. Therefore the transmission ratio may be expressed by the gain of both antennas or by the effective aperture of both antennas or by the gain of one antenna and the effective aperture of the other antenna. The relation

$$\frac{P_{r2}}{P_{t1}} = \frac{P_{r1}}{P_{t2}} = \frac{G_1 A_{e2}}{4\pi r^2} = \frac{G_2 A_{e1}}{4\pi r^2} = \frac{\lambda_0^2 G_1 G_2}{16\pi^2 r^2} = \frac{A_{e1} A_{e2}}{\lambda_0^2 r^2} \tag{11.166}$$

is known as the *Friis transmission formula.*

11.9 ANTENNA ARRAYS

An *antenna array* is an arrangement of antenna elements, distributed in space. The waves radiated by these antenna elements have defined amplitudes and phases and are

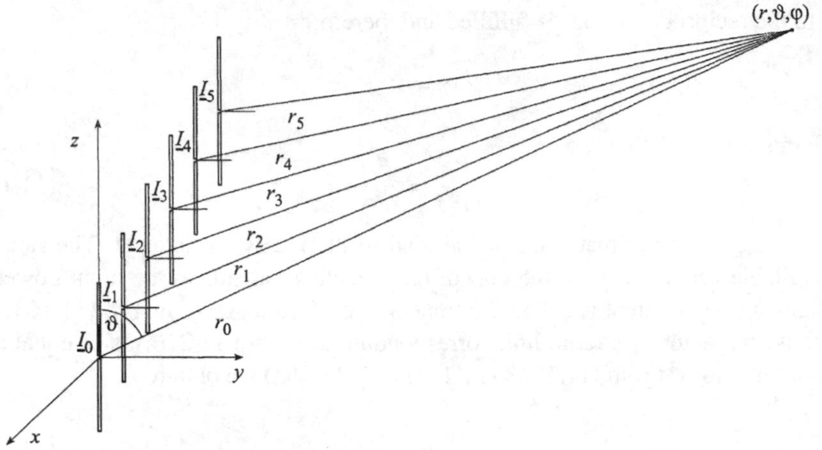

Figure 11.22: Linear antenna array formed by linear dipole antennas.

superimposed in space. Antenna arrays allow the formation of antenna characteristics of high gain and strong directivity. The characteristics of an antenna array depend on the characteristics of the antenna elements, the spatial arrangement of the antenna elements and the amplitude and phase of the feeding signals of each antenna element. Figure 11.22 shows a linear antenna array, i.e., a linear array of dipole antennas. The computation of the characteristics of the linear antenna array is simplified if all antenna elements exhibit identical shapes and if their arrangement obeys some spatial symmetry.

Let us consider a single linear dipole antenna element radiating a vertically polarized wave. In polar coordinates the far-field of the antenna is given by

$$\underline{E}_\vartheta = Z_{F0}\underline{H}_\varphi = \underline{A}\frac{\mathrm{e}^{-\mathrm{j}kr}}{2\pi r}F(\vartheta, \varphi) \tag{11.167}$$

where $\underline{A}$ is the complex amplitude and $F(\vartheta, \varphi)$ is the *element characteristics*. For the linear dipole antenna according to Figure 11.10 we obtain from (11.87)

$$A = \mathrm{j}Z_{F0}I_m, \tag{11.168}$$

$$F(\vartheta, \varphi) = \frac{\cos(kl\cos\vartheta) - \cos kl}{\sin\vartheta}. \tag{11.169}$$

Let us now consider an arrangement of n parallel antenna elements with identical element characteristics. The contribution of the νth antenna element ($\nu = 1\ldots n$) to the far-field is given by

$$\underline{E}_{\vartheta\nu} = Z_{F0}\underline{H}_{\varphi\nu} = \underline{A}_\nu \frac{e^{-jkr_\nu}}{2\pi r_\nu} F(\vartheta, \varphi), \tag{11.170}$$

where r_ν is the distance from the center of the νth dipole element to the point of observation in the far-field. The complex amplitudes $\underline{A}_\nu$ of the antenna elements are put into relation to the amplitude $\underline{A}_0$ by

$$\underline{A}_\nu = p_\nu \underline{A}_0 \, e^{-j\delta_\nu} \tag{11.171}$$

where p_ν and δ_ν are the corresponding amplitude and phase ratios. We obtain the total far-field of the antenna array by superimposing the contributions of all antenna elements:

$$\underline{E}_\vartheta = Z_{F0}\underline{H}_\varphi = \underline{A}_0 F(\vartheta, \varphi) \sum_{\nu=1}^{n} p_\nu \frac{e^{-j(kr_\nu + \delta_\nu)}}{2\pi r_\nu}. \tag{11.172}$$

We choose the origin of our coordinate system near to the antenna array. Let r_0 be the distance from the point of observation to the antenna elements and r_ν the distance from the far-field point of observation to the center of the νth antenna element we obtain for the far-field the following approximation:

$$r_\nu = r_0 - x_\nu \sin\vartheta \cos\varphi - y_\nu \sin\vartheta \sin\varphi - z_\nu \cos\vartheta. \tag{11.173}$$

In (11.172) we can approximate r_ν in the denominator by r_0 and obtain

$$\underline{E}_\vartheta = \underline{A}_0 \frac{e^{-jkr_0}}{2\pi r_0} F(\vartheta, \varphi) M(\vartheta, \varphi), \tag{11.174}$$

where the so-called *array factor* $M(\vartheta, \varphi)$ is given by

$$M(\vartheta, \varphi) = \sum_{\nu=1}^{n} p_\nu \exp\{-j[k(r_\nu - r_0) + \delta_\nu]\}. \tag{11.175}$$

We obtain the following simple result: *The characteristics of an antenna array with equal, equally oriented and equidistant antenna elements is the product of characteristics of a single antenna element and the array factor.* This law is called the *multiplicative law.*

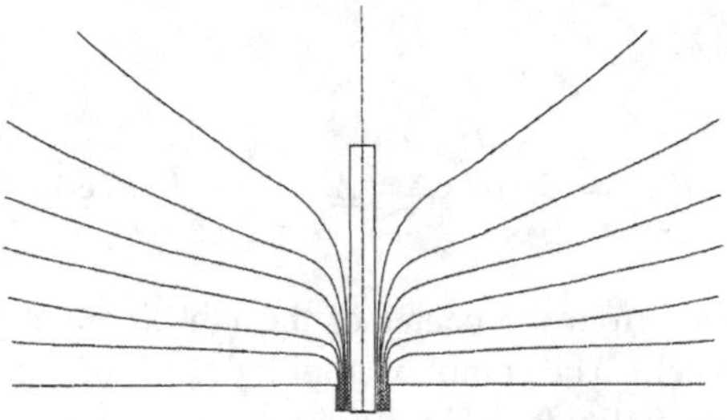

Figure 11.23: Poynting vector field of a vertical antenna.

11.10 APERTURE ANTENNAS

11.10.1 Radiating Apertures

The antennas treated so far have been composed of linear conductors. To design antennas we first have computed the current distribution on the linear conductor. Then we have treated this current distribution as an impressed current distribution. In a further step we have computed the radiated field from the impressed current distribution using (11.25a) and (11.27). The subdivision of the design procedure into computation of the current distribution without considering radiation, and subsequently computing the radiation on the basis of a given current distribution allows a straightforward computation of the radiation characteristics of the antenna; however, this method neglects the reaction of the radiation field on the current distribution in the antenna. For example, the attenuation of the current wave on the antenna conductors due to the radiation has not been considered.

Another consequence of this model has been that it has supported a picture in which the current flowing in the conductor at some point is the source of the field. However, at this point we only have changed a model from the network and transmission line model to the field model. A consequent application of the field picture shows that the electromagnetic energy is carried in the feedline of the antenna as well as in the free space by the electromagnetic field. Figure 11.23 shows a vertical linear antenna fed by a coaxial line and it illustrates the continuous transition of the field-lines of the Poynting vector field from the feedline region into the free-space region. In the coaxial feed line the Poynting vector field lines are oriented in parallel to the axis of the coaxial line, and these Poynting field lines pass through the aperture and then spread into free space surrounding the antenna. In this field picture the opening from the feed line into free space, i.e., the aperture is the source of the radiation. In principle also the open end of a coaxial line or any other line will radiate electromagnetic energy into space. As long as the transverse dimensions of a line are small

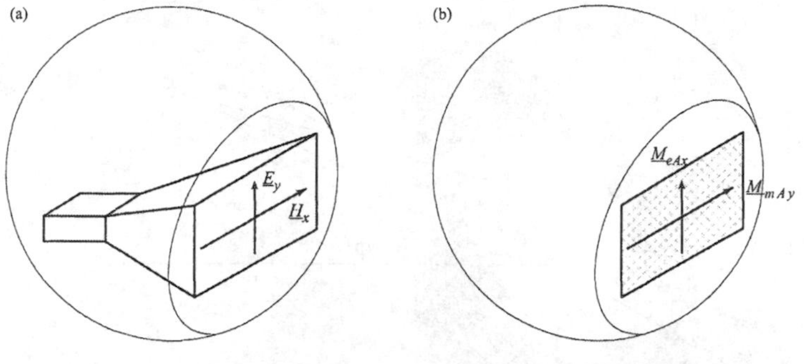

Figure 11.24: (a) Horn antenna and (b) impressed equivalent area polarizations.

compared with the wavelength the main part of an electromagnetic wave propagating in the line towards the open end will be reflected and only a very small part will be radiated. This situation changes considerably, if the transverse dimensions of a waveguide are in the order of magnitude of the wavelength λ_0 or exceed the wavelength of the electromagnetic wave. In this case the main part of the wave may be radiated into space.

The horn antenna shown in Figure 11.1(e) was obtained by continuously widening a waveguide with a rectangular cross-section such that the transverse dimensions of the aperture are larger than the wavelength λ_0. If the widening of the waveguide is smooth the transverse field distribution of the waveguide mode will be scaled up according to the widening of the waveguide. With an increasing widening of the cross-section the cut-off wavelength increases, and according to (7.106) the wave impedance of the waveguide approaches the field impedance of free space. In this case the waveguide wave is well matched to the free space and is radiated with only low reflection.

If we know the transverse field distribution in the aperture of the horn antenna or if we know the tangential electric or magnetic field distribution in an arbitrary surface enclosing the antenna we can compute the radiation field of the antenna using Huygens' principle. Figure 11.24 illustrates this procedure for the horn antenna. Knowing the tangential components of $\underline{\mathcal{E}}$ and $\underline{\mathcal{H}}$ we can compute the equivalent magnetic area polarization $\underline{\mathcal{M}}_{mA}$ and the equivalent electric area polarization $\underline{\mathcal{M}}_{eA}$ in the aperture plane.

We consider an arbitrarily shaped surface emitter aperture in the xy-plane at $z = 0$ as depicted in Figure 11.25. The electric and magnetic field components tangential to the aperture are described by the differential forms

$$\underline{\mathcal{E}}_t(\boldsymbol{x}') = \underline{E}_x(\boldsymbol{x}')\,\mathrm{d}x' + \underline{E}_y(\boldsymbol{x}')\,\mathrm{d}y' , \tag{11.176}$$

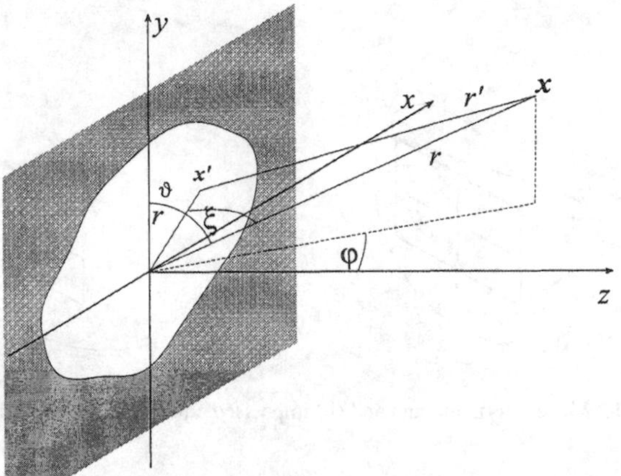

Figure 11.25: Aperture of a surface emitter.

$$\underline{\mathcal{H}}_t(\boldsymbol{x}') = \underline{E}_x(\boldsymbol{x}')\,\mathrm{d}x' + \underline{E}_y(\boldsymbol{x}')\,\mathrm{d}y', \tag{11.177}$$

where $\boldsymbol{x}'$ denotes a point in the aperture plane. The area polarizations $\underline{\mathcal{M}}_{eA}(\boldsymbol{x}')$ and $\underline{\mathcal{M}}_{mA}(\boldsymbol{x}')$ are represented by the twisted one-forms

$$\underline{\mathcal{M}}_{eA}(\boldsymbol{x}') = \underline{M}_{eAy}(\boldsymbol{x}')\,\mathrm{d}x' - \underline{M}_{eAx}(\boldsymbol{x}')\,\mathrm{d}y', \tag{11.178}$$

$$\underline{\mathcal{M}}_{mA}(\boldsymbol{x}') = \underline{M}_{mAy}(\boldsymbol{x}')\,\mathrm{d}x' - \underline{M}_{mAx}(\boldsymbol{x}')\,\mathrm{d}y'. \tag{11.179}$$

In the aperture plane we obtain from (4.51) and (4.52):

$$\underline{\mathcal{M}}_{mA}(\boldsymbol{x}') = -\frac{1}{\mathrm{j}\omega}\,\mathrm{d}z' \lrcorner \left(\mathrm{d}z' \wedge \underline{\mathcal{E}}(x')\right), \tag{11.180}$$

$$\underline{\mathcal{M}}_{eA}(\boldsymbol{x}') = \frac{1}{\mathrm{j}\omega}\,\mathrm{d}z' \lrcorner \left(\mathrm{d}z' \wedge \underline{\mathcal{H}}(x')\right). \tag{11.181}$$

From this we can compute the electric Hertz form $\underline{\Pi}_e$ and the magnetic Hertz form $\underline{\Pi}_m$ using (11.25a) and (11.27). Since $\underline{\mathcal{M}}_{eA}$ and $\underline{\mathcal{M}}_{mA}$ describe area polarizations the integrals have to be performed over surfaces, and we obtain

$$\underline{\Pi}_m(\boldsymbol{x}) = \frac{1}{4\pi\mu_0}\int_A' (\mathrm{d}x\,\mathrm{d}x' + \mathrm{d}y\,\mathrm{d}y') \wedge \frac{\underline{\mathcal{M}}_{mA}(\boldsymbol{x}')\,\mathrm{e}^{-\mathrm{j}k|\boldsymbol{x}-\boldsymbol{x}'|}}{|\boldsymbol{x}-\boldsymbol{x}'|}, \tag{11.182}$$

$$\underline{\Pi}_e(\boldsymbol{x}) = \frac{1}{4\pi\varepsilon_0}\int_A' (\mathrm{d}x\,\mathrm{d}x' + \mathrm{d}y\,\mathrm{d}y') \wedge \frac{\underline{\mathcal{M}}_{eA}(\boldsymbol{x}')\,\mathrm{e}^{-\mathrm{j}k|\boldsymbol{x}-\boldsymbol{x}'|}}{|\boldsymbol{x}-\boldsymbol{x}'|}. \tag{11.183}$$

If we are interested only in the far-field of the surface emitter, the integrals (11.182) and (11.183) may be simplified in a similar way as we have done it for the linear antenna. Let us consider the aperture in a plane surface emitter located in $z = 0$, depicted in Figure 11.25. For a far-field point $\boldsymbol{x}$ the lines r and r' can be assumed to be parallel so that

$$|\boldsymbol{x} - \boldsymbol{x}'| = r - r_0 \cos\xi \tag{11.184}$$

is valid. Applying this approximation in the exponent of the integrand of (11.182) and (11.183) and setting $|\boldsymbol{x} - \boldsymbol{x}'| = r$ we obtain

$$\underline{\Pi}_m(\boldsymbol{x}) = \frac{\mathrm{e}^{-\mathrm{j}kr}}{4\pi\mu_0 r}\int_A' (\mathrm{d}x\,\mathrm{d}x' + \mathrm{d}y\,\mathrm{d}y') \wedge \underline{\mathcal{M}}_{mA}(\boldsymbol{x}')\,\mathrm{e}^{\mathrm{j}kr_0\cos\xi}\,, \tag{11.185}$$

$$\underline{\Pi}_e(\boldsymbol{x}) = \frac{\mathrm{e}^{-\mathrm{j}kr}}{4\pi\varepsilon_0 r}\int_A' (\mathrm{d}x\,\mathrm{d}x' + \mathrm{d}y\,\mathrm{d}y') \wedge \underline{\mathcal{M}}_{eA}(\boldsymbol{x}')\,\mathrm{e}^{\mathrm{j}kr_0\cos\xi}\,. \tag{11.186}$$

The forms $\underline{\mathcal{E}}$ and $\underline{\mathcal{H}}$ are computed from $\underline{\Pi}_m$ and $\underline{\Pi}_e$ using (3.35), (3.36), (3.46) and (3.47). The partial fields computed from $\underline{\Pi}_e$ and $\underline{\Pi}_m$ have to be superimposed. For the computation of the far-field it is useful to represent $\underline{\Pi}_e$ and $\underline{\Pi}_m$ in spherical coordinates. The computations can be simplified by neglecting all terms going to zero stronger than by r^{-1} for $r \to \infty$. Considering (11.185) and (11.186), we see that the r dependence of $\underline{\Pi}_e$ and $\underline{\Pi}_m$ is due to the factor $\mathrm{e}^{-\mathrm{j}kr}/r$. Computing the $\underline{\Pi}_e$ and $\underline{\Pi}_m$ we only consider these terms where the negative power of r is not increased. For the far-field computation of a one-form

$$\mathcal{U} = U_r\,\mathrm{d}r + U_\vartheta r\,\mathrm{d}\vartheta + U_\varphi r\sin\vartheta\,\mathrm{d}\varphi\,, \tag{11.187}$$

we have to consider that

$$U_i \sim \frac{\mathrm{e}^{-\mathrm{j}kr}}{r} \quad \text{for} \quad kr \gg 1, \quad i = r, \varphi, \theta\,, \tag{11.188}$$

therefore may replace $\partial/\partial r$ with $-\mathrm{j}k$ and obtain the following approximation for the far field computation:

$$\mathrm{d}\mathcal{U} \cong -\mathrm{j}k\,\mathrm{d}r \wedge \mathcal{U} \quad \text{for} \quad kr \gg 1\,, \tag{11.189a}$$

$$\star\,\mathrm{d}\star\,\mathrm{d}\mathcal{U} \cong k^2\mathcal{U}\,. \tag{11.189b}$$

Using this approximation we obtain from (3.24), (3.25), (3.46) and (3.47) the far-field components

$$\underline{E}_\vartheta = Z_{F0}\underline{H}_\varphi = k^2\left(\underline{\Pi}_{e\vartheta} + Z_{F0}\underline{\Pi}_{m\varphi}\right)\,, \tag{11.190a}$$

$$\underline{E}_\varphi = -Z_{F0}\underline{H}_\vartheta = k^2\left(\underline{\Pi}_{e\varphi} - Z_{F0}\underline{\Pi}_{m\vartheta}\right)\,. \tag{11.190b}$$

11.10.2 Horn Antennas

Let us now compute the far field of the horn antenna excited in the TE_{10} mode according to Figure 11.26. From (7.125), (7.130a) and (7.130b) we obtain the transverse field components of the rectangular waveguide excited in the TE_{10} mode.

$$\underline{E}_y = -Z_{\mathrm{TE}}\underline{H}_x = \sqrt{\frac{2}{ab}}\underline{V}(z)\cos\frac{\pi x}{a}\,. \tag{11.191}$$

If the waveguide cross-section is gradually increased in the z-direction, this transverse field distribution will be scaled up correspondingly. According to (7.106) a field impedance of the waveguide $Z_{\mathrm{TE}10}$ approaches the free-space field impendence Z_{F0}. If the wave impedance of the waveguide is smoothly varying with Z only low reflections will occur. We assume that the transverse widening of the aperture plane is sufficiently large that $Z_{\mathrm{TE}10}$ may be approximated by Z_{F0} in the aperture plane. Using (11.178) and (11.179) we now can determine the equivalent area polarizations in the aperture plane A:

$$\underline{M}_{mAx} = \frac{1}{\mathrm{j}\omega}\underline{E}_y\Big|_{z=0} = \frac{1}{\mathrm{j}\omega}\sqrt{\frac{2}{ab}}\underline{V}\cos\frac{\pi x}{a}\,, \tag{11.192a}$$

$$\underline{M}_{eAy} = \frac{1}{\mathrm{j}\omega}\underline{H}_x\Big|_{z=0} = -\frac{1}{Z_{F0}}\underline{M}_{mAx}\,. \tag{11.192b}$$

We now have approximated $Z_{\mathrm{TE}01}$ by Z_{F0}. Inserting (11.192a) into (11.185) we obtain

$$\underline{\Pi}_{mx}(\boldsymbol{x}) = \frac{\mathrm{e}^{-\mathrm{j}kr}}{4\pi\mu_0 r}\int_A \underline{M}_{mAx}(\boldsymbol{x}')\,\mathrm{e}^{\mathrm{j}kr_0\cos\xi}\,\mathrm{d}A\,. \tag{11.193}$$

Let us now introduce spherical coordinates with respect to the y axis (Figure 11.26). With

$$r_0\cos\xi = x'\sin\varphi\sin\vartheta + y'\cos\vartheta \tag{11.194}$$

we obtain from (11.192a), (11.193) and (11.194):

$$\underline{\Pi}_{mx}(\boldsymbol{x}) = \frac{\mathrm{e}^{-\mathrm{j}kr}}{4\pi\mathrm{j}\omega\mu_0 r}\sqrt{\frac{2}{ab}}XY\underline{V} \tag{11.195a}$$

with

$$X = \int_{-\frac{a}{2}}^{+\frac{a}{2}}\cos\frac{\pi x'}{a}\,\mathrm{e}^{\mathrm{j}kx'\sin\varphi\sin\vartheta}\,\mathrm{d}x'\,, \tag{11.195b}$$

$$Y = \int_{-\frac{b}{2}}^{+\frac{b}{2}}\mathrm{e}^{\mathrm{j}ky'\cos\vartheta}\,\mathrm{d}y'\,. \tag{11.195c}$$

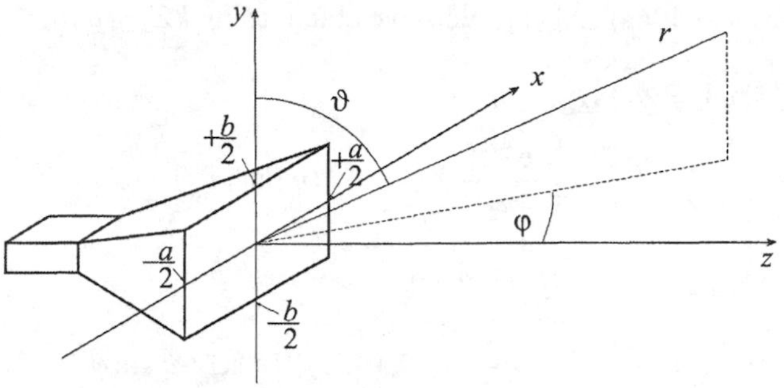

Figure 11.26: Horn antenna.

After evaluation of the integrals we obtain

$$X = \frac{2\pi a \cos\left(\frac{1}{2}ka \sin\varphi \sin\vartheta\right)}{\pi^2 - k^2a^2 \sin^2\varphi \sin^2\vartheta}\,, \tag{11.196a}$$

$$Y = \frac{2\sin\left(\frac{1}{2}kb\cos\vartheta\right)}{k\cos\vartheta}\,. \tag{11.196b}$$

From (11.185), (11.186) and (11.192b) we obtain

$$\underline{\Pi}_y = -Z_{F0}\underline{\Pi}_{mx}\,. \tag{11.197}$$

The Hertz vector has the following spherical components

$$\underline{\Pi}_{m\vartheta} = \underline{\Pi}_{mx}\cos\vartheta\sin\varphi\,, \tag{11.198a}$$

$$\underline{\Pi}_{m\varphi} = \underline{\Pi}_{mx}\cos\varphi\,, \tag{11.198b}$$

$$\underline{\Pi}_{e\vartheta} = -\underline{\Pi}_{ey}\sin\vartheta\,. \tag{11.198c}$$

From (11.190a), (11.190b), (11.197) and (11.198a) – (11.198c) we obtain the far-field of the horn antenna.

$$\underline{E}_\vartheta = Z_{F0}\underline{H}_\varphi = k^2 Z_{F0}\underline{\Pi}_{mx}(\sin\vartheta + \cos\varphi)\,, \tag{11.199a}$$

$$\begin{aligned}\underline{E}_\varphi &= -Z_{F0}\underline{H}_\vartheta \\ &= -k^2 Z_{F0}\underline{\Pi}_{mx}\cos\vartheta\sin\varphi\,.\end{aligned} \tag{11.199b}$$

With (11.195a), (11.196a) and (11.196b) we obtain using $kZ_{F0}/\omega\mu_0 = 1$:

$$\begin{aligned} \underline{E}_\vartheta(r, \vartheta, \varphi) &= Z_{F0}\underline{H}_\varphi(r, \vartheta, \varphi) \\ &= \sqrt{\frac{2}{ab}}\, k\underline{V}\frac{\mathrm{e}^{-\mathrm{j}kr}}{4\pi \mathrm{j} r} X(\vartheta, \varphi)Y(\vartheta)(\sin\vartheta + \cos\varphi)\,, \end{aligned} \tag{11.200a}$$

$$\begin{aligned} \underline{E}_\varphi(r, \vartheta, \varphi) &= -Z_{F0}\underline{H}_\vartheta(r, \vartheta, \varphi) \\ &= -\sqrt{\frac{2}{ab}}\, k\underline{V}\frac{\mathrm{e}^{-\mathrm{j}kr}}{4\pi \mathrm{j} r} X(\vartheta, \varphi)Y(\vartheta)\cos\vartheta \sin\varphi\,. \end{aligned} \tag{11.200b}$$

In the main beam direction ($\vartheta = \frac{\pi}{2}, \varphi = 0$) we obtain from (11.196a) and (11.196b)

$$X\left(\frac{\pi}{2}, 0\right) Y\left(\frac{\pi}{2}\right) = \frac{2}{\pi}ab\,. \tag{11.201}$$

The field intensity in the main beam direction is

$$\underline{E}_\vartheta = Z_{F0}\underline{H}_\varphi = \frac{-\mathrm{j}}{\pi^2}\sqrt{2ab}\, k\underline{V}\frac{\mathrm{e}^{-\mathrm{j}kr}}{r}\,. \tag{11.202}$$

The power density in the main beam direction is

$$T\left(r, \frac{\pi}{2}, 0\right) = \frac{1}{2}Z_{F0}^{-1}|\underline{E}_\vartheta|^2 = \frac{ab}{\pi^4 r^2}k^2 Z_{F0}^{-1}|\underline{V}|^2\,. \tag{11.203}$$

Since we have assumed that the wave travelling in the waveguide towards the aperture is completely radiated, the complete active power radiated from the horn antenna according to (4.139) and (4.141) is given by

$$P_r = \frac{1}{2}Z_{F0}^{-1}|\underline{V}|^2\,. \tag{11.204}$$

With (11.145) we obtain from this the maximum antenna gain

$$G_{\max} = \frac{8abk^2}{\pi^3} = \frac{32ab}{\pi\lambda_0^2}\,. \tag{11.205}$$

The maximum antenna gain is proportional to the aperture area $a \cdot b$ of the horn antenna. From (11.165) and (11.205) we obtain the effective antenna area

$$A_e = \frac{8}{\pi^2}ab \cong 0.81ab\,. \tag{11.206}$$

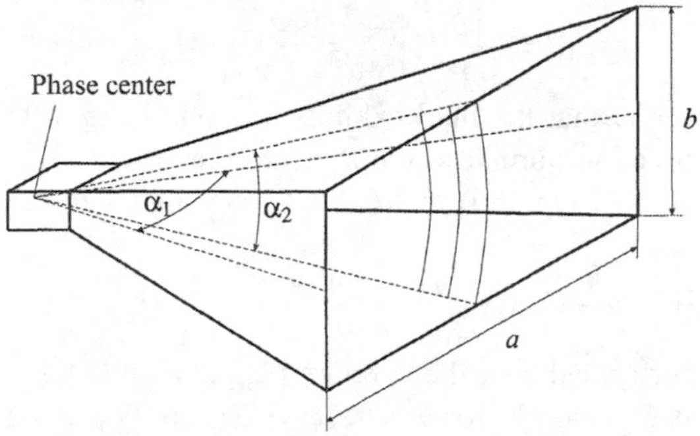

Figure 11.27: Curvature of the phase front.

The effective antenna area is only 81% of the geometric area. The high gain in the case of large apertures means a high directivity. For $ka \gg 1$ and $kb \gg 1$ the main radiation lobe in the $\varphi = 0$ plane exhibits the angular width

$$\Delta\vartheta = 2\frac{\lambda_0}{b}. \tag{11.207}$$

In the $\vartheta = \frac{\pi}{2}$ plane and the angular width of the main radiation lobe is

$$\Delta\varphi = 3\frac{\lambda_0}{a}. \tag{11.208}$$

In our considerations up to now we have neglected the curvature of the phase front in the aperture. This assumption is only justified if the beam angles α_1 and α_2 in Figure 11.27 fulfill the conditions $a\alpha_1 \ll \lambda_0$, $b\alpha_2 \ll \lambda_0$. The real or virtual origin of a spherical wave is called the *phase center*.

11.10.3 Gain and Effective Area of Aperture Antennas

Let us now treat the gain and effective antenna area of the plane surface emitter more generally. According to Figure 11.25 we make no special assumptions about the aperture of the antenna. To achieve a high gain of the surface emitter we assume a plane phase front in the aperture and the electrical field to be vertically polarized in the aperture. In the aperture the electric and magnetic field in (11.192a) and (11.192b) may be replaced by equivalent area polarizations

$$\underline{M}_{mAx} = \frac{1}{j\omega}\underline{E}_y\Big|_{z=0}, \tag{11.209}$$

$$\underline{M}_{eAy} = \frac{1}{j\omega}\underline{H}_x\Big|_{z=0} = -\frac{1}{Z_{F0}}\underline{M}_{mAx}\,. \tag{11.210}$$

Furthermore we assume that the field exhibits no node lines in the aperture plane and therefore the direction normal to the aperture plane will be the main radiation direction. From (11.185), (11.186) and (11.210) we obtain for the main radiation direction

$$\underline{\Pi}_{ey} = -Z_{F0}\underline{\Pi}_{mx} = -\frac{c_0}{4\pi}\frac{e^{-jkr}}{r}\int_A \underline{M}_{eAx}\mathrm{d}A\,. \tag{11.211}$$

The integration is performed over the aperture plane. With (11.190a) we obtain the far field components $\underline{E}_\vartheta$ and $\underline{H}_\varphi$ for $\vartheta = \frac{\pi}{2}, \varphi = 0$ (with $\underline{\Pi}_\vartheta = -\underline{\Pi}_y$, $\underline{\Pi}_\varphi = \underline{\Pi}_x$)

$$\underline{E}_\vartheta = Z_{F0}\underline{H}_\varphi = -k^2(\underline{\Pi}_{ey} - Z_{F0}\underline{\Pi}_{mx}) = -2k^2\underline{\Pi}_y \quad \text{for} \quad \vartheta = \frac{\pi}{2}, \varphi = 0\,. \tag{11.212}$$

From this we obtain

$$\underline{E}_\vartheta = -jk\frac{e^{-jkr}}{2\pi r}\int_A \underline{E}_y\Big|_{z=0}\mathrm{d}A\,. \tag{11.213}$$

With (11.44) we obtain the radiation density in the main beam direction

$$T = \frac{k^2}{8\pi^2 r^2 Z_{F0}}\left|\int_A \underline{E}_y\Big|_{z=0}\mathrm{d}A\right|^2\,. \tag{11.214}$$

The total active power radiated from the aperture is obtained by integrating T over the aperture,

$$P_r = \frac{1}{2Z_{F0}}\int_A |\underline{E}_y|^2\Big|_{z=0}\mathrm{d}A\,. \tag{11.215}$$

The gain G_{max} in the main beam direction follows from (11.127), (11.214) and (11.215)

$$G_{\max} = \frac{4\pi r^2 T}{P_r} = \frac{4\pi}{\lambda_0^2}\frac{\left|\int_A \underline{E}_y\Big|_{z=0}\mathrm{d}A\right|^2}{\int_A |\underline{E}_y|^2\Big|_{z=0}\mathrm{d}A}\,. \tag{11.216}$$

With (11.145) we obtain from this the effective area in the main beam direction

$$A_e = \frac{|\int_A \underline{E}_y|_{z=0}\mathrm{d}A|^2}{\int_A |\underline{E}_y|^2|_{z=0}\mathrm{d}A}\,. \tag{11.217}$$

If the aperture is uniformly illuminated, i.e., the electromagnetic field is uniform over the aperture, we obtain the maximum value of the effective aperture, given by $A_e = A$, and it follows: *For a plane surface emitter the aperture dimensions of which are large compared with the wavelength λ_0, the aperture of which is uniformly illuminated and the phase front is plane and parallel to the aperture plane, the effective antenna area is equal to the geometric aperture area.*

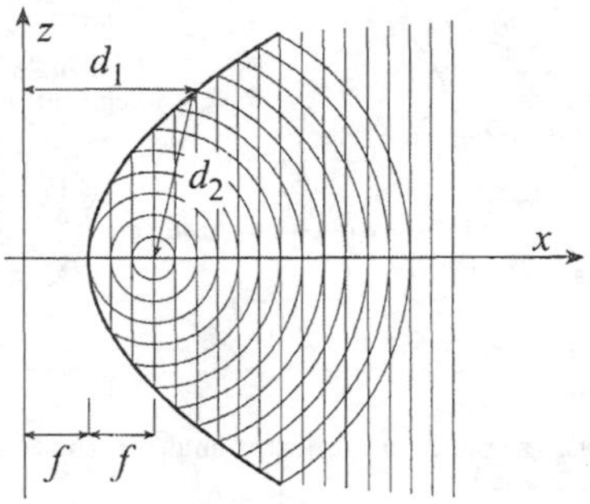

Figure 11.28: Phase surfaces in a parabolic reflector antenna.

11.10.4 Mirror and Lens Antennas

To realize large apertures with plane phase fronts, methods from optics are used to transform a spherical wave radiated by a primary radiator into a plane wave. This can be achieved using lenses or mirrors. Figure 11.28 gives a schematic illustration of a parabolic reflector antenna. A primary radiator in the focal point of a paraboloid emits a spherical wave. The spherical wave is reflected by the conducting paraboloid and in this way transformed into a plane wave. Transformation occurs due to the circumstance that the distances d_1 and d_2 are equal. Removing the paraboloid reflector and bringing the spherical wave originating from the focal point to interfere with a plane wave propagating in the x-direction from the left produces a node surface of paraboloid shape. Inserting a conducting paraboloid this node surface meets with the boundary conditions. Now the plane wave incident from the left is converted into a spherical wave and the spherical wave originating from a source in the focal point of the paraboloid is converted into a plane wave when incident on the reflector. In the next step of our consideration we may omit the plane wave incident from the left.

Figure 11.29 shows various types of reflector antennas. The reflector antenna in Figure 11.29(a) is excited by a small horn antenna or a small dipole antenna positioned in the focal point of the paraboloid. To achieve an effective antenna area coming as close as possible to the geometric antenna area a uniform irradiation of the paraboloid is required. The primary radiator should only irradiate the paraboloid reflector and not the regions beside the reflector. Therefore, the primary radiator should exhibit an appropriate directivity. These requirements will yield a higher focal length of the parabolic antenna and consequently large antenna dimensions in the longitudinal direction.

The antenna dimension in the longitudinal direction can be reduced by introducing a second mirror as shown in Figure 11.29(b). This antenna type is called a

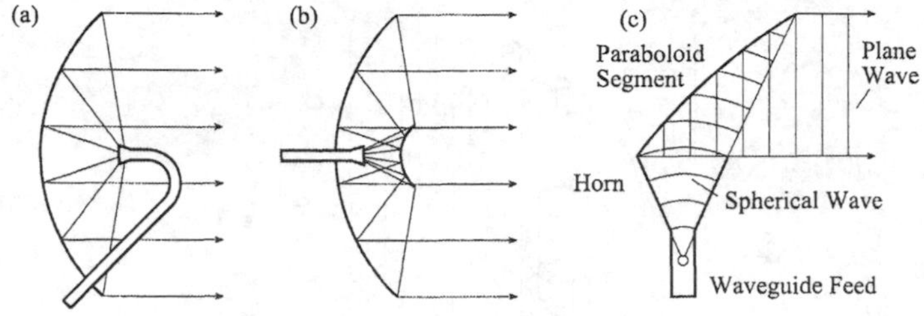

Figure 11.29: Reflector antennas: (a) parabolic reflector antenna, (b) cassegrain antenna, (c) horn parabolic antenna.

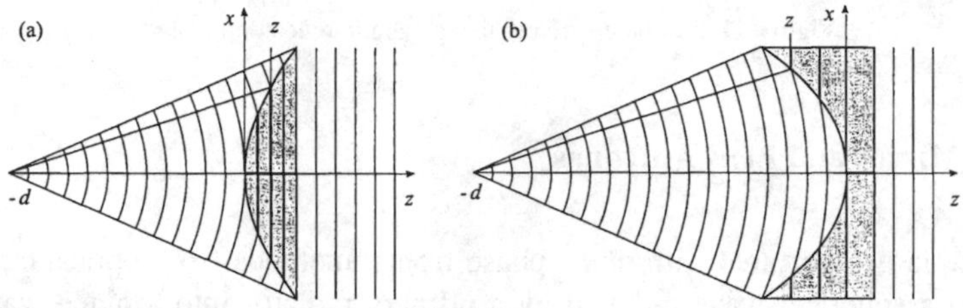

Figure 11.30: Lenses for transformation of spherical waves into plane waves: (a) $n > 1$ and (b) $n < 1$.

cassegrain antenna. Both antenna types shown in Figure 11.29(a) and Figure 11.29(b) suffer from the disadvantage that the primary radiator is in the path of the secondary beam. The scattering of the secondary beam by the primary radiator not only degrades the antenna characteristics, but also yields a frequency-dependent mismatch of the antenna to the feed line. Both disadvantages can be avoided with the horn paraboloid antenna, shown in Figure 11.29(c). In this antenna for the reflection only a sector of the paraboloid is used; it is crossed by the axis of the paraboloid. Therefore the reflected secondary beam does not pass the focal point and will not be scattered by the primary radiator.

Figure 11.30 illustrates the application of lenses for the transformation of a spherical wave into a wave with plane phase front. The dimensions of the lens are assumed to be large compared with the wavelength. Therefore we can apply geometric optic methods to compute the shape of the lens. The phase velocity c of a ray corresponds to the phase velocity of a plane wave. In the lens the phase velocity c of the electromagnetic wave deviates from the free-space phase velocity c_0. The refractive index n is given by

$$n = \frac{c_0}{c} . \tag{11.218}$$

The refractive index n may be either larger or smaller than 1. A refractive index $n > 1$, given by (2.76) may be realized using dielectric materials. To transform a spherical

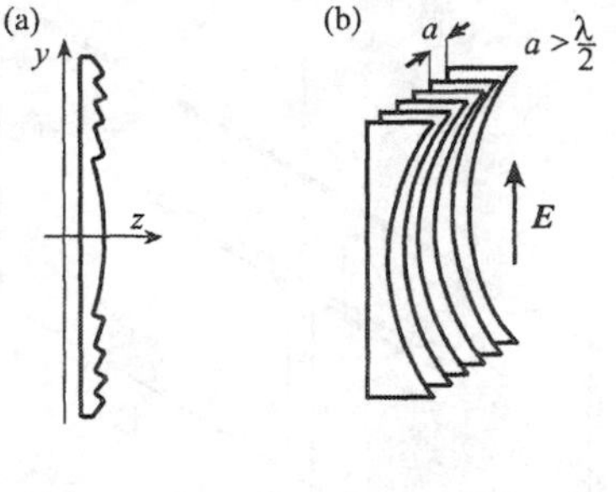

Figure 11.31: (a) Fresnel lens and (b) Waveguide lens

wave into a plane wave, the optical length of the path, i.e., the geometrical length weighted with the reflective index, must be equal for the central path C_1 and any path C_2 in Figure 11.30(a) yields

$$d + nz = \sqrt{(d+z)^2 + x^2}\,. \tag{11.219}$$

From this it follows that the lens surface is given by a hyperboloid according to

$$z^2(n^2 - 1) + 2dz(n - 1) - x^2 = 0\,. \tag{11.220}$$

In order to save material and weight a dielectric lens may be designed as a step-lens or Fresnel lens as in Figure 11.31(a). A lens with $n < 1$ can be realized using a stack of equidistant parallel conducting plates as in Figure 11.31(b). If an incident plane wave is polarized parallel to the conducting plates TE_{10} waveguide modes are excited between every two plates. According to (7.101) the phase velocity of the TE_{10} waves is given by

$$c = \frac{c_0}{\sqrt{1 - \left(\frac{\lambda_0}{2a}\right)^2}}\,. \tag{11.221}$$

From this we obtain an ellipsoid lens surface given by

$$z^2(1 - n^2) - 2dz(1 - n) + x^2 = 0\,. \tag{11.222}$$

11.10.5 Slot Antennas

A slot antenna is an aperture antenna with a narrow aperture, extended only in one dimension. The slot width is small compared with the wavelength λ_0. Figure 11.32 shows a slot antenna excited by a coaxial line, and Figure 11.1(f) shows an array of a slot antenna excited by rectangular waveguide. The slot in Figure 11.32 behaves as

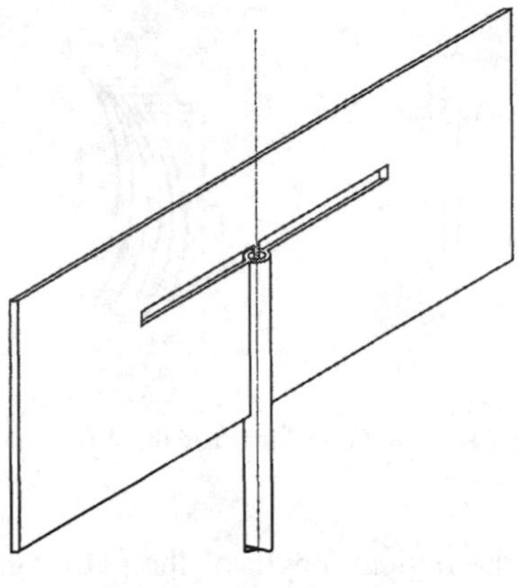

Figure 11.32: Slot antenna fed via coaxial line.

a line short-circuited on both ends. By the coaxial line a standing wave with voltage nodes at both ends of the slot is excited. The voltage occurring along the slot $\underline{V}(z)$ and the magnetic area polarization equivalent to the electric aperture field are depicted in Figure 11.33. The magnetic area polarization is given by

$$\underline{M}_{mAz} = \frac{1}{\mathrm{j}\omega} \frac{1}{s} \underline{V}(z) . \tag{11.223}$$

The slot width is s. Usually a slot antenna is shielded on the rear side so that the slot is radiating only in one half-space. Electric and magnetic area polarization give the same far-field contribution as shown in (11.212). We therefore can take twice the far-field contribution obtained from the magnetic area polarization instead of also computing a far-field contribution from the electric area polarization. From (11.193), (11.199b), (11.223), (11.77) and (11.190b) we obtain

$$\underline{H}_\vartheta = -\frac{\underline{E}_\varphi}{Z_{F0}} = +\frac{\mathrm{j}\omega\varepsilon_0}{2\pi} \frac{\mathrm{e}^{-\mathrm{j}kr}}{r} \sin\vartheta \int_{-l}^{+l} \underline{V}(z)\, \mathrm{e}^{\mathrm{j}kz\cos\vartheta} \mathrm{d}z . \tag{11.224}$$

This equation is dual to the equation derived for the linear dipole (11.83), if one considers that the radiator is radiating only into a half-space, therefore creating twice the field intensity in this half-space. The duality with respect to the linear antenna becomes obvious, if we are performing the following substitution:

$$\underline{I}(z) \to \frac{2}{Z_{F0}} \underline{V}(z) , \tag{11.225a}$$

$$\underline{E}_\vartheta \to Z_{F0} \underline{H}_\vartheta , \tag{11.225b}$$

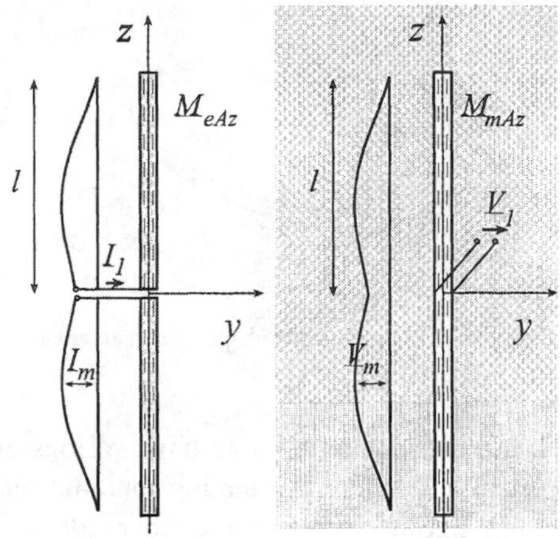

Figure 11.33: Slot antenna.

$$\underline{H}_\varphi \to -\frac{1}{Z_{F0}}\underline{E}_\varphi\,. \tag{11.225c}$$

Since the power P radiated from the slot antenna is proportional to $|\underline{V}_m|^2$, we can define a radiation conductance G_{rm} related to the voltage maximum $|\underline{V}_m|$ via

$$P = \frac{1}{2}G_{rm}|\underline{V}_m|^2\,. \tag{11.226}$$

In this way we obtain from (11.90) and (11.225a) the following relation between the radiation resistance related to the voltage maximum $|\underline{V}_m|$ of the linear antenna and the radiation conductance G_{rm} of the slot antenna

$$G_{rm} = \frac{4R_{rm}}{Z_{F0}^2}\,. \tag{11.227}$$

11.11 MICROSTRIP ANTENNAS

The microstrip antenna is a metallic patch printed on a thin, grounded dielectric substrate [11–15]. Microstrip antennas can be realized using printed circuit technology. The advantages of printed antennas are small dimensions, light weight, easy manufacturing and easy integration into arrays. Microstrip antennas can also be printed on curved surfaces to make conformal antennas. Disadvantages of printed antennas are usually a narrow bandwidth and comparatively high losses.

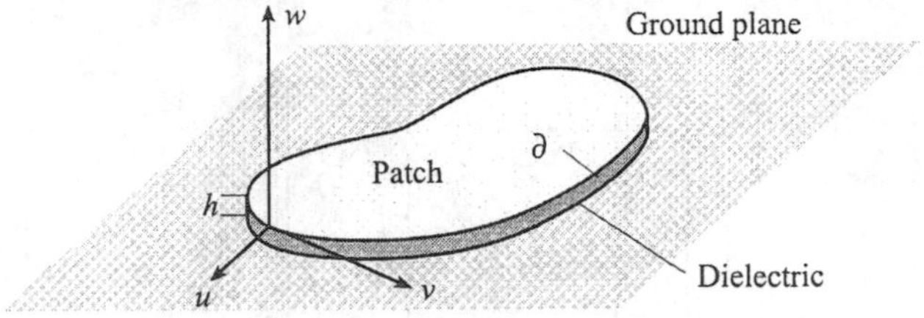

Figure 11.34: Arbitrarily shaped patch antenna.

The patches of a microstrip antenna may have various geometries. The most simple type of a microstrip antenna is a radiating metallic patch on one side of a dielectric substrate that has a ground plate on the other side. In Figure 11.34 an arbitrarily shaped patch antenna is depicted. The substrate thickness is h. If the transverse size of the patch is large compared to h, then the electromagnetic field between the patch and the ground plane is uniform in w direction, the electric field being normal to the patch, and the magnetic field being tangential to the patch. Therefore

$$\underline{\mathcal{E}}(u, v) = \underline{E}_w(u, v)s_3 \,, \tag{11.228a}$$

$$\underline{\mathcal{H}}(u, v) = \underline{H}_u(u, v)s_1 + \underline{H}_v(u, v)s_2 \,. \tag{11.228b}$$

The electromagnetic field in the interior region can be described by TM modes, for w being the direction of propagation. It can be computed by solving the Helmholtz equation (3.34) with the ansatz

$$\underline{\Pi}_e(u, v) = \underline{\Pi}_{ew}(u, v)s_3 \,. \tag{11.229}$$

The structure formed by the patch on the grounded dielectric can be understood as a resonator, too. In the cavity model of the patch antenna the field computation is performed in two steps: In a first step the interior field of this region under the patch is modelled as a cavity bounded by electric walls on the top and bottom and a magnetic wall along the periphery ∂V_s. In this first step the radiation from the open periphery of the resonator is neglected. In a second step Huygen's principle is applied to compute from (4.51) the equivalent magnetic area polarization on the periphery ∂V_s

$$\underline{\mathcal{M}}_{mA}(u, v) = -\frac{1}{\mathrm{j}\omega} s_1 \lrcorner (s_1 \wedge \underline{\mathcal{E}}(u, v)) = -\frac{1}{\mathrm{j}\omega} \underline{E}_w(u, v)s_3 \,, \tag{11.230}$$

and from this the field radiated from the patch antenna.

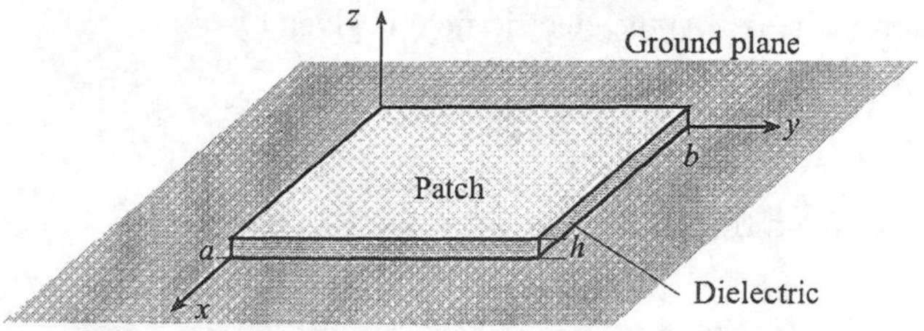

Figure 11.35: Rectangular patch antenna.

From (11.185) we obtain the magnetic Hertz form describing the radiated field. In the far-field the electric and magnetic field components are

$$\underline{\Pi}_m(\boldsymbol{x}) = \frac{\mathrm{e}^{-\mathrm{j}kr}}{4\pi\mu_0 r} \int'_{\partial V_s} (s_2 s_2' + s_3 s_3') \wedge \underline{\mathcal{M}}_{mA}(u', v')\, \mathrm{e}^{\mathrm{j}kr_0 \cos\xi}\,, \tag{11.231}$$

where the angle ξ is specified in Figure 11.25. In the far-field the electric and magnetic field components may be computed from (11.190a), (11.190b):

$$\underline{E}_\vartheta = Z_{F0}\underline{H}_\varphi = k^2 Z_{F0}\underline{\Pi}_{m\varphi}\,, \tag{11.232a}$$

$$\underline{E}_\varphi = -Z_{F0}\underline{H}_\vartheta = -k^2 Z_{F0}\underline{\Pi}_{m\vartheta}\,. \tag{11.232b}$$

The patch antenna either may be fed by a coaxial feed through the ground plane or by a planar feed line [11–15]. Exciting the microstrip antenna by a microstrip line on the same substrate allows to fabricate antenna and feed line simultaneously.

A microstrip antenna fabricated on a plane substrate is called a planar antenna. As in the case of conventional antennas, also microstrip antennas may be combined to form arrays [11–14]. The elements of an array may be spatially distributed to form a linear or a two-dimensional array.

Conformal microstrip antennas and antenna arrays are fabricated on curved substrates [12, 13, 16, 17]. The shape of a conformal antenna is usually defined by the contours of the object on which it is mounted, e.g., a vehicle or an aircraft. This means that the shape of the antenna is not at the disposal of the antenna designer.

11.11.1 Planar Rectangular Patch Antenna

We apply the cavity model to the rectangular patch antenna shown in Figure 11.35 [14]. In the resonator the electromagnetic field is represented by

$$\underline{\mathcal{E}}(\boldsymbol{x}) = \underline{E}_z(\boldsymbol{x})\,\mathrm{d}z\,, \tag{11.233a}$$

$$\underline{\mathcal{H}}(\boldsymbol{x}) = \underline{H}_x(\boldsymbol{x})\,\mathrm{d}x + \underline{H}_y(\boldsymbol{x})\,\mathrm{d}y\,. \tag{11.233b}$$

For the resonant mode (m, n) the electric field is given by

$$\underline{E}_z(\boldsymbol{x}) = \underline{E}_0 \cos\frac{m\pi x}{a} \cos\frac{n\pi y}{b} . \tag{11.234}$$

With (2.145) we obtain from this

$$\underline{H}_x(\boldsymbol{x}) = -\frac{\mathrm{j}}{\omega\mu}\frac{n\pi}{b}\underline{E}_0 \cos\frac{m\pi x}{a} \sin\frac{n\pi y}{b} , \tag{11.235a}$$

$$\underline{H}_y(\boldsymbol{x}) = -\frac{\mathrm{j}}{\omega\mu}\frac{m\pi}{b}\underline{E}_0 \sin\frac{m\pi x}{a} \cos\frac{n\pi y}{b} . \tag{11.235b}$$

From (11.185) we obtain

$$\underline{\Pi}_m(\boldsymbol{x}) = \frac{\mathrm{j}\,\mathrm{e}^{-\mathrm{j}kr}}{4\pi\omega\mu_0 r}\int'_{\partial V_s}(\mathrm{d}x\,\mathrm{d}x' + \mathrm{d}y\,\mathrm{d}y') \wedge \mathrm{d}z'\,\underline{E}_z(\boldsymbol{x})\,\mathrm{e}^{\mathrm{j}kr_0\cos\xi} , \tag{11.236}$$

We choose $a > b$. In this case the mode $(1, 0)$ is the fundamental mode. In this case the field in the resonator is given by

$$\underline{E}_z(\boldsymbol{x}) = \underline{E}_0 \cos\frac{\pi x}{a} , \tag{11.237a}$$

$$\underline{H}_x(\boldsymbol{x}) = 0 , \tag{11.237b}$$

$$\underline{H}_y(\boldsymbol{x}) = \frac{\mathrm{j}}{\omega\mu}\frac{\pi}{a}\underline{E}_0 \sin\frac{\pi x}{a} . \tag{11.237c}$$

Using (11.236) we can compute $\underline{\Pi}_m(\boldsymbol{x})$ for the far-field and from this with (11.232*a*) and (11.232*b*) the far-field components for the electric field. We give the solution for $\varphi = 0$ and $\varphi = \pi/2$. The plane $\varphi = 0$ is parallel to the electric field and is called the E-plane whereas the plane $\varphi = \pi/2$ is parallel to the magnetic field and is called the H-plane. For the E-plane we obtain

$$\underline{E}_\vartheta(r, \vartheta, 0) = \mathrm{j}\frac{bk\,\mathrm{e}^{-\mathrm{j}kr}}{r\pi}h\underline{E}_0 \exp\left(\mathrm{j}\frac{ka}{2}\sin\vartheta\right)\cos\left(\frac{ka}{2}\sin\vartheta\right) , \tag{11.238a}$$

$$\underline{E}_\varphi(r, \vartheta, 0) = 0 , \tag{11.238b}$$

and for the H-plane

$$\underline{E}_\vartheta(r, \vartheta, \frac{\pi}{2}) = 0 , \tag{11.239a}$$

$$\underline{E}_\varphi(r, \vartheta, \frac{\pi}{2}) = -\mathrm{j}\frac{bk\,\mathrm{e}^{-\mathrm{j}kr}}{r\pi}h\underline{E}_0 \exp\left(\mathrm{j}\frac{kb}{2}\sin\vartheta\right)\frac{\sin\left(\frac{kb}{2}\sin\vartheta\right)}{\frac{kb}{2}\sin\vartheta} . \tag{11.239b}$$

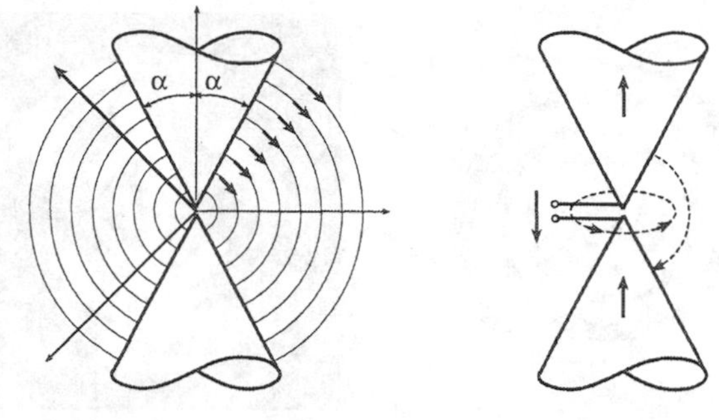

Figure 11.36: Biconical antenna: (a) electric and magnetic fields, (b) voltages and currents.

11.12 BROADBAND ANTENNAS

For applications that require coverage of a wide range of frequencies broadband antennas are required. An antenna may be scaled in wavelength by changing its linear dimensions. Therefore an antenna structure which is invariant under linear scaling of its dimensions by an arbitrary scale factor must be frequency-independent. A simple antenna structure with broadband characteristics is the biconical antenna [3]. Strictly speaking, a structure invariant under scaling of its dimensions must be of infinite extension. Truncating such a structure to finite extension, however, will yield a broadband antenna structure [18–20].

We analyze the biconical antenna geometry depicted in Figure 11.36. The cone surface is defined by $\vartheta = \alpha/2$ where α is the angle enclosed by the cone. For infinite extension of the cones in the r-direction the problem may be easily treated analytically. Furthermore in the case of infinite extension it is obvious that the electromagnetic properties of the structure are frequency independent since the structure is invariant under a scaling transformation in radial direction.

For symmetry reasons we expect a field solution with only an $\underline{E}_\vartheta$ and an $\underline{E}_\vartheta$ component. We make the ansatz

$$\underline{\mathcal{E}}(r, \vartheta, \varphi) = \underline{E}_\vartheta(r, \vartheta, \varphi) r \, \mathrm{d}\vartheta \,, \tag{11.240a}$$

$$\underline{\mathcal{H}}(r, \vartheta, \varphi) = \underline{H}_\varphi(r, \vartheta, \varphi) r \sin\vartheta \, \mathrm{d}\varphi \,. \tag{11.240b}$$

With (2.137), (2.138) and (A.158) we obtain from this the solution

$$\underline{\mathcal{E}}(r, \vartheta) = Z_{F0} \frac{\underline{H}_0}{\sin\vartheta} \frac{\mathrm{e}^{-\mathrm{j}kr}}{r} r \, \mathrm{d}\vartheta \,, \tag{11.241a}$$

$$\underline{\mathcal{H}}(r, \vartheta) = \underline{H}_0 \frac{\mathrm{e}^{-\mathrm{j}kr}}{r \sin\vartheta} r \sin\vartheta \, \mathrm{d}\varphi \,. \tag{11.241b}$$

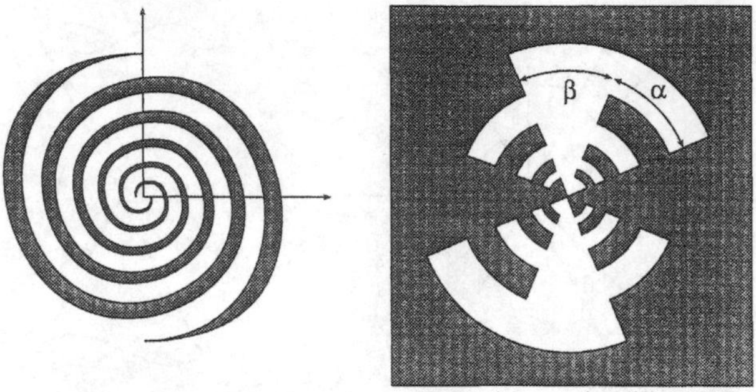

Figure 11.37: Logarithmic-periodic antenna: (a) log-spiral antenna, (b) log-periodic slot antenna.

A voltage at $V(r)$ and a current $I(r)$ may be defined by

$$V(r) = -\int_{C_1} \underline{\mathcal{E}}(r,\vartheta) = 2Z_{F0}\underline{H_0}\,\mathrm{e}^{-\mathrm{j}kr}\ln\left[\cot\left(\frac{\alpha}{4}\right)\right], \tag{11.242a}$$

$$I(r) = \int_{C_2} \underline{\mathcal{H}}(r,\vartheta) = 2\pi\underline{H}_0\,\mathrm{e}^{-\mathrm{j}kr} \tag{11.242b}$$

The power P_r radiated by the antenna is

$$P_r = \frac{1}{2}\Re\{\underline{V}(0)\underline{I}^*(0)\} = 2\pi Z_{F0}|\underline{H_0}^2|\ln\left[\cot\left(\frac{\alpha}{4}\right)\right]. \tag{11.243}$$

The radiation resistance R_r is given by

$$R_r = \frac{\underline{V}(0)}{\underline{I}(0)} = \frac{Z_{F0}}{\pi}\ln\left[\cot\left(\frac{\alpha}{4}\right)\right]. \tag{11.244}$$

The logarithmic-periodic geometry allows to realize frequency-independent antennas. A logarithmic-periodic antenna is a set of adjoining cells, each cell being scaled in dimensions relative to the adjacent cells by a factor that remains fixed throughout [18–20]. Figure 11.37 shows two examples of planar logarithmic-periodic antennas. Apart from the truncation of the structures the spiral antenna structure (see Figure 11.37(a)) is invariant under arbitrary linear scaling, whereas the structure depicted in Figure 11.37(b) is invariant under scaling by a given factor and its powers, respectively.

Due to Babinet's principle (11.227) holds [3, 21] for a pair of dual planar antenna structures complementary in the metallized and nonmetallized regions. The structure in Figure 11.37(b) is congruent with its dual structure and therefore the radiation impedance is real and frequency-independent and given by $R_r = Z_{F0}/2$.

REFERENCES

[1] S. Silver, *Microwave Antenna Theory and Design*. New York: McGraw-Hill, 1949.

[2] W. Stutzman and G. Thiele, *Antenna Theory and Design*. New York: John Wiley & Sons, 1981.

[3] C. A. Balanis, *Antenna Theory*. New York: John Wiley & Sons, 1982.

[4] E. Wolff, *Antenna Analysis*. Boston: Artech House, 1988.

[5] E. Roubine and J. Bolomey, *Antennas, Volume 1, General Principles*. London: North Oxford Academic, 1987.

[6] S. Drabowitch and C. Ancona, *Antennas, Volume 1, General Principles*. London: North Oxford Academic, 1987.

[7] G. de Rham, *Differentiable Manifolds*. New York: Springer, 1984.

[8] K. F. Warnick and D. Arnold, "Electromagnetic Green functions using differential forms," *J. Electromagn. Waves and Appl.*, vol. 10, no. 3, pp. 427–438, 1996.

[9] F. Landstorfer, H. Liska, H. Meinke, and B. Müller, "Energieströmung in elektromagnetischen Wellenfeldern," *Nachrichtentechn. Z.*, pp. 225–231, May 1972.

[10] F. Landstorfer and R. Sacher, *Optimisation of Wire Antennas*. New York: Letchworth, Research Studies Press Ltd, Wiley, 1990.

[11] R. Garg, P. Bhartia, I. Bahl, and A. Ittipiboon, *Microstrip Antenna Design Handbook*. Boston: Artech House, 2000.

[12] J. James and P. Hall, *Handbook of Microstrip Antennas, Volume 1*. London: Peter Peregrinus, 1989.

[13] J. James and P. Hall, *Handbook of Microstrip Antennas, Volume 2*. London: Peter Peregrinus, 1989.

[14] S. Drabowitch, A. Papiernik, H. Griffith, J. Encinas, and B. Smith, *Modern Antennas*. London: Chapman and Hall, 1998.

[15] D. Pozar, "Microstrip antennas," *Proceedings of the IEEE*, vol. 80, pp. 79–91, Jan. 1992.

[16] R. Munson, "Conformal microstrip antennas and microstrip phased arrays," *IEEE Trans. Antennas Propagat.*, vol. 22, pp. 74– 78, Jan. 1974.

[17] V. Sohtell, "Microstrip antennas on a cylindrical surface," in *Handbook of Microstrip Antennas, Volume 2* (J. James and P. Hall, eds.), pp. 496–591, London: Peter Peregrinus, 1989.

[18] V. Rumsey, *Frequency-Independent Antennas*. New York: Academic Press, 1985.

[19] E. Jordan, G. Deschamps, J. Dyson, and P. Mayes, "Developments in broadband antennas," *IEEE Spectrum*, vol. 1, pp. 58–71, Apr. 1964.

[20] P. Mayes, "Frequency-independent antennas and broad-band derivates thereof," *Proceedings of the IEEE*, vol. 80, pp. 103–112, Jan. 1992.

[21] R. S. Elliott, *Antenna Theory and Design*. New York: IEEE Press, 2003.

Chapter 12

Numerical Electromagnetics

12.1 INTRODUCTION

For many electromagnetic structures, exact analytical solutions cannot be found. It is therefore necessary to consider numerical methods to obtain approximate solutions of field problems. A great variety of methods for electromagnetic field modelling has been developed [1–3]. In order to obtain results of the required accuracy with a minimum of computational effort a method matched to the problem has to be chosen. The optimum design of radio-frequency devices, circuits and systems strongly depends on the availability of advanced *computer-aided design* (CAD) tools for modelling and optimization.

The *method of moments* (MoM) plays a crucial role in numerical electromagnetics [4–6]. In the MoM the field functions are expanded into series of *basis functions*. The problem of solving partial differential equations or integral equations for the field functions is converted into the problem of solving linear systems of equations for determining the coefficients of the series expansions of the field functions. Within the methods for field computation the MoM holds a special position since most of the methods of field computation – for example the integral equation method, the spectral domain method, the partial wave synthesis, the transmission line matrix method and the finite difference method – may be considered in connection with the MoM. The MoM is a very general scheme for the discretization of the field problem, whereas the other methods specify in detail how to the discretization is performed.

Table 12.1 lists some of the most widely used methods for electromagnetic field computation [1]. The *integral equation* (IE) method introduces the far-field interaction via Green's functions and may reduce the dimension of the field problem [6,7]. Since the interaction with infinite space is included in the Green's function integral equation methods are especially powerful in the case of radiating structures. In the *spectral domain method* the integral equations are transformed into algebraic equations by Fourier transformation with respect to the space coordinates [8]. Integral equation

Table 12.1: Methods for Numerical Electromagnetic Field Computation

Method	*Memory Requirement*	*CPU Time*	*Preprocessing*	*Generality*
Integral equation method	*SM*	*SM*	*M*	+
Spectral domain method	*S*	*S*	*L*	−
Partial wave synthesis	*M*	*S/M*	*M*	+
Method of lines	*M*	*S*	*L*	+
Finite difference method	*L*	*L*	*S*	++
Transmission line matrix method	*L*	*L*	*S*	++
Finite element method	*L*	*M/L*	*S*	++

Notes: *L* ... large, *M* .... moderate, *S* ... small, ++ ... very good, + ... good, − ... marginal,

methods as well as spectral domain methods are computationally efficient but require considerable analytic preprocessing for the specific class of structures to be modelled.

In the *partial wave synthesis* method or *mode-matching* method the space is subdivided into subdomains and within each subdomain the electromagnetic field is expanded into eigensolutions of the field equation. The partial wave synthesis method is efficient if boundary surfaces separating the subdomains are coordinate surfaces belonging to coordinate systems for which the analytic field solutions are known [9]. The partial wave synthesis method has been applied successfully to coplanar transmission line structures [10] and to coplanar transmission line discontinuities [11] with a small transverse dimension where electromagnetic full-wave modelling also inside the conductor is required.

In the *method of lines* (MoL) [12,13] all but one of the independent variables of the field equations are discretized to obtain a system of ordinary differential equations. With respect to the continuous variable analytic solutions are sought. Concerning the treatment of the continuous variables the MoL may be compared with the mode-matching method, whereas the treatment of the discretized variables corresponds to the finite difference method.

The *finite difference* (FD) method [14, 15], the *finite integration* (FI) method [16–18], the *finite element* (FE) method [19] and *the transmission line matrix* (TLM) method [20–23] are suited for modelling general three-dimensional structures without analytic preprocessing of the problem. However, all these space-discretizing methods require a high computational effort and usually need a long time for computation. In principle, the time-discretizing methods may be applied either in the frequency domain or time domain. The FE methods usually are applied in the frequency domain, the FD methods are applied in the frequency domain (FDFD) as well as in the time domain (FDTD), whereas the TLM method mainly is used in time domain. Frequency domain modelling is more appropriate for narrow-band simulation of high-Q structures. Time domain modelling allows for the characterization of the electromagnetic properties of the structures under consideration in a broad frequency band by computing the response to a single impulsive excitation [24]. The space-discretizing methods are less suitable to model radiating structures and structures including large free-space regions. For such applications hybrid methods, combining the TLM method with the integral equation method allow for accurate and efficient modelling [25].

In the following we give a short overview over the fundamentals of numerical field modelling. The discussion of the Sturm-Liouville equation yields the concept of function spaces and shows the way to approximate field functions by series expansion into basis functions. The MoM allows one to convert problems formulated by differential equations or integral equations into a representation by a system of algebraic equations. This corresponds to the transformation of a field problem into a network problem. Finally we will discuss, as two specific examples for numerical field computation, the integral equation method and the TLM method.

12.2 THE STURM-LIOUVILLE EQUATION

The treatment of many boundary value problems in electromagnetics yields to the *Sturm-Liouville differential equation* [26,27]. The homogeneous Sturm-Liouville differential equation has the general form

$$\frac{d}{dx}\left[p(x)\frac{d\psi(x)}{dx}\right] + \left[q(x) + \lambda\sigma(x)\right]\psi(x) = 0, \tag{12.1}$$

where $\psi(x)$ is the unknown function whereas $p(x)$, $q(x)$ and $\sigma(x)$ are real and continuous functions within the considered domain of x. Furthermore in general $p(x)$ and $\sigma(x)$ are considered to be positive. The constant λ in general may also be complex. In the following we investigate the solutions of the Sturm-Liouville differential equation in the interval $[a, b]$ for the boundary conditions

$$\psi_i + c_a\frac{d\psi_i}{dx} = 0 \quad \text{for} \quad x = a, \tag{12.2}$$

$$\psi_i + c_b\frac{d\psi_i}{dx} = 0 \quad \text{for} \quad x = b, \tag{12.3}$$

where c_a and c_b are real. The solutions of the differential equation (12.1) exist for certain *eigenvalues* λ_n of λ. For chosen boundary conditions at $x = a$ and $x = b$, we obtain an infinite number of solutions ψ_n of the differential equation (12.1) with the corresponding eigenvalues λ_n.

If the complex function $f(x)$ and the real positive function $\sigma(x)$ are defined in the interval $[a, b]$ and the integral

$$\langle f | f \rangle = \int_a^b \sigma(x) |f(x)|^2 \, \mathrm{d}x < \infty \tag{12.4}$$

exists, the function $f(x)$ is called *square integrable* with the *kernel function* $\sigma(x)$. The collection of all functions, square integrable in a given interval $[a, b]$, form an infinite-dimensional linear vector space. An infinite-dimensional complex linear vector space is called a *Hilbert space* [28, 29]. In the Hilbert space inner products and a metric upon these are defined. The expression

$$\langle f(x) | g(x) \rangle = \int_a^b \sigma(x) f^*(x) g(x) \, \mathrm{d}x \tag{12.5}$$

is the *inner product* of the functions $f(x)$ and $g(x)$. The expression

$$||f|| = \sqrt{\langle f | f \rangle} \tag{12.6}$$

is called the *norm* of $f(x)$.

We show that in the case of fulfilled boundary conditions (12.2) and (12.3) the eigenvalues λ_n are real and that the eigenfunctions $\psi_m(x)$ and $\psi_n(x)$ belonging to different eigenvalues λ_m and λ_n are orthogonal with respect to the weighting function $\sigma(x)$, in other words

$$\langle \psi_n | \psi_m \rangle = 0 \quad \text{for} \quad \lambda_m \neq \lambda_n \, . \tag{12.7}$$

To prove the above assumption we insert ψ_m and λ_m into (12.1) and multiply the equation from the left with ψ_n^*. Then we form the same expression with interchanged indices m and n. The difference of these expressions is integrated over the interval $[a, b]$ and we obtain

$$\int_a^b \left[\psi_n^* \frac{\mathrm{d}}{\mathrm{d}x} \left(p \frac{\mathrm{d}\psi_m}{\mathrm{d}x} \right) - \psi_m \frac{\mathrm{d}}{\mathrm{d}x} \left(p \frac{\mathrm{d}\psi_n^*}{\mathrm{d}x} \right) \right] \mathrm{d}x = \int_a^b (\lambda_n^* - \lambda_m) \, \sigma \psi_n^* \psi_m \, \mathrm{d}x \, . \tag{12.8}$$

By partial integration of the left side we obtain

$$\begin{aligned} p \left[\psi_n^* \frac{\mathrm{d}\psi_m}{\mathrm{d}x} - \psi_m \frac{\mathrm{d}\psi_n^*}{\mathrm{d}x} \right] \Bigg|_a^b - \int_a^b p \left[\frac{\mathrm{d}\psi_n^*}{\mathrm{d}x} \frac{\mathrm{d}\psi_m}{\mathrm{d}x} - \frac{\mathrm{d}\psi_n^*}{\mathrm{d}x} \frac{\mathrm{d}\psi_m}{\mathrm{d}x} \right] \mathrm{d}x = \\ = p \left[\psi_n^* \frac{\mathrm{d}\psi_m}{\mathrm{d}x} - \psi_m \frac{\mathrm{d}\psi_n^*}{\mathrm{d}x} \right] \Bigg|_a^b = (\lambda_n^* - \lambda_m) \int_a^b \sigma \psi_n^* \psi_m \, \mathrm{d}x \, . \end{aligned} \tag{12.9}$$

If $\psi_m(x)$ and $\psi_n(x)$ fulfill the same boundary conditions (12.2) and (12.3) it follows that

$$\psi_n^* \frac{d\psi_m}{dx} - \psi_m \frac{d\psi_n^*}{dx} = 0 \quad \text{for} \quad x = a = b\,. \tag{12.10}$$

From (12.9) and (12.10) it follows for $n \neq m$ that

$$\left(\lambda_n^* - \lambda_m\right) \langle \psi_n | \psi_m \rangle = 0\,. \tag{12.11}$$

For $m = n$ due to

$$\langle \psi_n | \psi_n \rangle = \int_a^b \sigma |\psi_n|^2 \, dx \tag{12.12}$$

(12.11) may only be fulfilled for $(\lambda_n^* - \lambda_n) = 0$, i.e., the eigenvalues λ_n must be real. With this it follows from (12.11) that

$$(\lambda_n - \lambda_m) \langle \psi_n | \psi_m \rangle = 0\,. \tag{12.13}$$

and from this we obtain (12.7).

The functions ψ_n are orthogonal in the interval $[a, b]$ with respect to the integration kernel $\sigma(x)$. For *degenerate eigenvalues*, $\lambda_i = \lambda_j$ the corresponding eigenfunctions $\psi_i(x)$ and $\psi_j(x)$ need not to be orthogonal. In this case we may introduce the new eigenfunctions ψ_i^1 and ψ_j^1 by a linear transformation

$$\psi_i^1 = \psi_i\,, \qquad \psi_j^1 = \psi_j + \alpha\psi_i \tag{12.14}$$

where the coefficient α is given by

$$\alpha = -\frac{\langle \psi_i, \psi_j \rangle}{\langle \psi_i, \psi_i \rangle}\,. \tag{12.15}$$

This method can be extended to an arbitrary number of degenerate eigenvalues and is called the *Gram-Schmidt orthogonalization method.*

In the following we assume an *orthonormal system of eigenfunctions*, i.e., that all eigenfunctions are normalized and mutually orthogonal,

$$\langle \psi_m | \psi_n \rangle = \delta_{mn}\,. \tag{12.16}$$

The eigenfunctions ψ_n form *complete sets of functions* and allow the expansion of an arbitrary piecewise continuous function $f(x)$ into a series

$$f(x) = \sum_{n=1}^{\infty} a_n \psi_n(x)\,. \tag{12.17}$$

To determine the coefficients a_n we multiply both sides of the equation with $\sigma(x)\psi_m^*(x)$ and integrate from a to b.

$$\int_a^b \sigma(x)\psi_m^*(x) f(x)\,\mathrm{d}x = \sum_{n=1}^{\infty} a_n \int_a^b \sigma(x)\psi_m^*(x)\psi_n(x)\,\mathrm{d}x = \sum_{n=1}^{\infty} a_n \delta_{mn}\,. \tag{12.18}$$

From this it follows that

$$a_m = \int_a^b \sigma(x)\psi_m^*(x) f(x)\,\mathrm{d}x\,. \tag{12.19}$$

The function $f(x)$ is assumed to be square integrable with the kernel function $\sigma(x) > 0$ for $x \in [a, b]$. We investigate the convergence of the series expansion of $f(x)$ into $\psi_n(x)$ and consider the series truncated after the Nth term,

$$f_N(x) = \sum_{n=1}^{N} a_n \psi_n(x)\,, \tag{12.20}$$

where the coefficients a_n are given by (12.19). The $\psi_n(x)$ forms a complete set of functions if for every function $f(x)$, square integrable in the interval $[a, b]$, the relation

$$\lim_{N\to\infty} \int_a^b |f(x) - f_N(x)|^2 \sigma(x)\,\mathrm{d}x = 0 \tag{12.21}$$

is valid. After inserting of (12.20), we obtain from this

$$\lim_{N\to\infty} \int_a^b \left| f(x) - \sum_{n=1}^{N} a_n \psi_n(x) \right|^2 \sigma(x)\,\mathrm{d}x = 0\,. \tag{12.22}$$

For $N \to \infty$ we obtain the *Parseval's Theorem*

$$\int_a^b |f(x)|^2 \sigma(x)\,\mathrm{d}x = \sum_{n=1}^{\infty} |a_n|^2\,. \tag{12.23}$$

If (12.21) is fulfilled in the interval (a, b) for any piecewise continuous function $f(x)$ the set of functions $\psi_n(x)$ is complete.

12.3 SPECTRAL REPRESENTATION OF GREEN'S FUNCTIONS

The *inhomogeneous Sturm-Liouville differential equation* for a perturbation $g(x)$ is given by

$$\frac{\mathrm{d}}{\mathrm{d}x} p(x) \frac{\mathrm{d}\psi}{\mathrm{d}x} + \left[q(x) + \lambda\sigma(x)\right]\psi(x) = g(x)\,. \tag{12.24}$$

The inhomogeneous Sturm-Liouville differential equation for a perturbation Dirac point-like perturbation at x' described by the delta distribution $\delta(x - x')$ is

$$\frac{\mathrm{d}}{\mathrm{d}x} p(x) \frac{\mathrm{d}G(x,x')}{\mathrm{d}x} + \left[q(x) + \lambda\sigma(x)\right] G(x,x') = \delta(x - x') . \tag{12.25}$$

The Green's function $G(x, x')$ is the solution at x for the point-like perturbation at x'. We expand the Green's function $G(x, x')$ into eigenfunctions $\psi_n(x)$ with the eigenvalues λ_n. If the same boundary conditions (12.2) and (12.3) are assumed in (12.24) and (12.25), the solution of (12.24) is given by

$$\psi(x) = \int_a^b G(x,x')g(x')\,\mathrm{d}x' . \tag{12.26}$$

The Green's function $G(x, x')$ may be considered as a function of the variable x with the parameter x' denoting the position of the point-like source. Since the solutions $\psi_n(x)$ of the homogeneous Sturm-Liouville differential equation (12.1) form a *complete set of basis functions*, we can expand $G(x, x')$ into a series

$$G(x,x') = \sum_{-\infty}^{\infty} a_n(x')\psi_n(x) , \tag{12.27}$$

where the expansion coefficients $a_n(x')$ depend on the location x' of the source. After inserting into (12.25), it follows with (12.1) that

$$\sum_{n=-\infty}^{\infty} a_n(x')(\lambda - \lambda_n)\sigma(x)\psi_n(x) = \delta(x - x') . \tag{12.28}$$

Multiplying both sides with $\psi_m^*(x)$ and integrating over the interval $[a, b]$, we obtain

$$\sum_{n=-\infty}^{\infty} a_n(x')(\lambda - \lambda_n) \int_a^b \sigma(x)\psi_m^*(x)\psi_n(x)\,\mathrm{d}x = a_m(x')(\lambda - \lambda_m)$$

$$= \int_a^b \psi_m^*(x)\delta(x - x')\,\mathrm{d}x = \psi_m^*(x') . \tag{12.29}$$

With this it follows from (12.27) that

$$G(x,x') = \sum_{n=-\infty}^{\infty} \frac{\psi_n^*(x')\psi_n(x)}{\lambda - \lambda_n} . \tag{12.30}$$

The Green's function $G(x, x')$ exhibits poles at $\lambda = \lambda_n$. We note that the Green's function is symmetric in x and x'. This expresses the property of reciprocity since we may interchange the point of source with the point of observation.

12.4 THE METHOD OF MOMENTS

The *method of moments* is a general concept allowing the expansion of the field solutions into a set of *basis functions*. In this way the electromagnetic field problem, which is primarily described by partial differential equations or integral equations, is transformed into a problem described by a linear system of algebraic equations [4–6]. The linear system of equations relates the coefficients of the series expansion of the electric field with the coefficients of the series expansions of the magnetic field. Formally the expansion coefficients of electric and magnetic fields can be interpreted as generalized voltages and currents respectively, and the coefficient matrix of the linear system of equations as a generalized impedance matrix or admittance matrix. We therefore may interpret the application of the MoM as the transformation of a field problem into a network problem. If the electric and the magnetic fields are expanded into complete sets of basis functions the series expansions usually will exhibit an infinite number of elements resulting in a system of equations of infinite dimension. Truncating the series expansions after a finite number of elements reduces the system of equations to a finite dimension and allows an approximate numerical solution of the problem. By this reduction the field problem becomes equivalent to a network problem with a finite number of network elements.

We distinguish between basis functions defined on the entire domain under consideration (*entire domain basis functions*) and basis functions, which are only defined on subdomains of the considered domain (*subsectional basis functions*). If the domain of definition of the solution function is within the domain of definition of the basis functions and if the system of basis functions is *complete* within this domain, the representation of the solution by an infinite series of the basis functions is possible under quite general conditions. Representing the solution with a series of basis functions, we obtain from the system of differential or integral equations to be solved a linear system of algebraic equations for the expansion coefficients of the series representing the solution. Truncation of the series expansion after finite number of series elements yields a finite-dimensional system of algebraic equations, which may be solved numerically.

In the geometrical picture of the Hilbert space the solution function is represented by a vector. We are seeking linear combinations of a finite number of basis functions or basis vectors respectively, which approximate the exact solution function or solution vector respectively. The approximation of the solution by a series expansion is found if the components of the corresponding vector in Hilbert space are known. Linear differential equations and linear integral equations are both linear operator equations. In the following we will restrict our considerations to functions of only one variable. We consider the operator equation

$$L_{op} f(x) = g(x) , \tag{12.31}$$

where the linear differential operator $\boldsymbol{L}_{op}$ is applied to the unknown function $f(x)$. The function $g(x)$ is known. Examples for linear differential operators are $\boldsymbol{L}_{op} = -\,\mathrm{d}^2/\,\mathrm{d}x^2$ or $\boldsymbol{L}_{op} = \mathrm{d}^2/\,\mathrm{d}x^2 + k^2$. The function $g(x)$ is given and $f(x)$ has to be determined. The unknown function $f(x)$ may be represented by a linear combination of a set of functions $\varphi_n(x)$, as

$$f(x) = \sum_n \alpha_n \varphi_n(x) \tag{12.32}$$

with the unknown expansion coefficients α_n. Inserting (12.32) into (12.31) yields

$$\sum_n \alpha_n \boldsymbol{L}_{op} \varphi_n(x) = g(x)\,. \tag{12.33}$$

We choose a test function ψ_m, multiply it with (12.33) and compute the following integral

$$\sum_n \alpha_n \int_a^b \psi_m^* \boldsymbol{L}_{op} \varphi_n \,\mathrm{d}x = \int_a^b \psi_m^*(x) g(x) \,\mathrm{d}x\,. \tag{12.34}$$

The functions $f(x)$, $g(x)$, $\varphi_n(x)$ are defined in the interval $[a, b]$. Introducing the matrix elements

$$L_{mn} = \int_a^b \psi_m^*(x) \boldsymbol{L}_{op} \varphi_n(x) \,\mathrm{d}x \tag{12.35}$$

of the linear operator $\boldsymbol{L}_{op}$ and the expansion coefficients

$$\beta_m = \int_a^b \psi_m^*(x) g(x) \,\mathrm{d}x \tag{12.36}$$

of the function $g(x)$ yields the linear system of equations

$$\sum_n L_{mn} \alpha_n = \beta_m \tag{12.37}$$

for the determination of the unknown expansion coefficients α_n of the function $f(x)$. Truncating the series expansions with $m = 1 \ldots M$ and $n = 1 \ldots N$ yields a finite-dimensional linear system of equations. With the vectors

$$\boldsymbol{\beta} = [\beta_1 \ldots \beta_M]^T\,, \qquad \boldsymbol{\alpha} = [\alpha_1 \ldots \alpha_N]^T \tag{12.38}$$

and the matrix

$$\boldsymbol{L} = \begin{bmatrix} L_{11} & \cdots & L_{1N} \\ \cdots & \ddots & \cdots \\ L_{M1} & \cdots & L_{MN} \end{bmatrix} \tag{12.39}$$

we obtain the linear system of equations in matrix notation

$$\boldsymbol{L}\boldsymbol{\alpha} = \boldsymbol{\beta}\,. \tag{12.40}$$

For $M = N$ we obtain a quadratic matrix $\boldsymbol{L}$, which may be inverted for non singular $\boldsymbol{L}$. The solution of the linear system of equations (12.40) yields

$$\boldsymbol{\alpha} = \boldsymbol{L}^{-1}\boldsymbol{\beta}\,. \tag{12.41}$$

As an example we investigate the following boundary value problem. The differential equation

$$\frac{\mathrm{d}^2 f}{\mathrm{d}x^2} = -1 + 3x - 12x^2 \qquad x \,\varepsilon\, [0, 1] \tag{12.42}$$

has to be solved for the Dirichlet boundary conditions $f(0) = f(1) = 0$. The exact solution of (12.42) is

$$f(x) = x - \frac{x^2}{2} + \frac{x^3}{2} - x^4\,. \tag{12.43}$$

To demonstrate the solution using the MoM we use the expansion function

$$\varphi_n(x) = x^n(1-x)\,. \tag{12.44}$$

The special case of the MoM in which the test function is equal to the expansion function,

$$\varphi_n = \psi_n\,, \tag{12.45}$$

is called *Galerkin's method.* Choosing

$$f = \sum_{n=1}^{N} \alpha_n \varphi_n(x) \tag{12.46}$$

we obtain with $\boldsymbol{L}_{op} = \mathrm{d}^2/\,\mathrm{d}x^2$

$$L_{mn} = \frac{-2\,m\,n}{\left[(m+n)^2 - 1\right]\,(m+n)} \tag{12.47}$$

and

$$\beta_m = \frac{-2\,\left(12 + 14m + 5m^2\right)}{(1+m)\,(2+m)\,(3+m)\,(4+m)}\,. \tag{12.48}$$

For $N = 3$ we obtain the linear system of equations

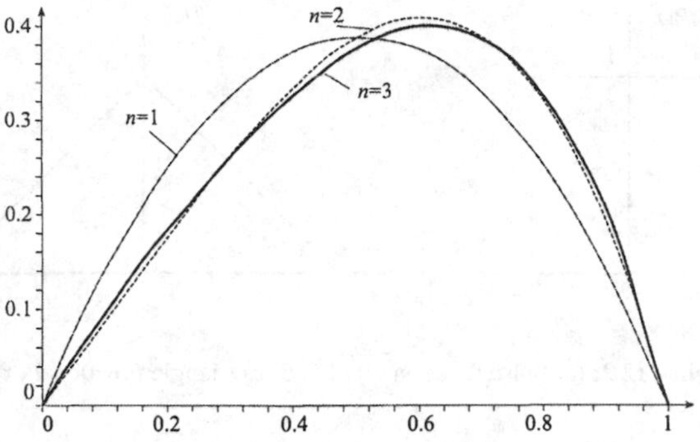

Figure 12.1: The approximations $f_1(x)$, $f_2(x)$ and $f_3(x)$.

$$\begin{bmatrix} -\frac{1}{3} & -\frac{1}{6} & -\frac{1}{10} \\ -\frac{1}{6} & -\frac{2}{15} & -\frac{1}{10} \\ -\frac{1}{10} & -\frac{1}{10} & -\frac{3}{35} \end{bmatrix} \begin{bmatrix} \alpha_1 \\ \alpha_2 \\ \alpha_3 \end{bmatrix} = \begin{bmatrix} -\frac{31}{60} \\ -\frac{1}{3} \\ -\frac{33}{140} \end{bmatrix} . \qquad (12.49)$$

The solution is given by $[\alpha_1, \alpha_2, \alpha_3] = \left[1, \frac{1}{2}, 1\right]$. From this we obtain

$$f_3(x) = \varphi_1(x) + \frac{1}{2}\varphi_2(x) + \varphi_3(x) = x - \frac{x^2}{2} + \frac{x^3}{2} - x^4 = f(x) . \qquad (12.50)$$

This approximation already agrees with the exact solution.

In many cases for the approximation of functions $f(x)$ *subdomain basis functions* or *subsectional basis functions* respectively are useful. For N sampling points in the interval $[a, b]$ we obtain

$$x_n = a + \frac{n(b-a)}{N+1} . \qquad (12.51)$$

The pulse function according to Figure 12.2(a) is defined by

$$P(x) = \begin{cases} 1 & \text{for } |x| < \frac{1}{2} \\ 0 & \text{for } |x| > \frac{1}{2} . \end{cases} \qquad (12.52)$$

Using pulse functions as basis, a function may be approximated by a staircase function. Triangle functions according to Figure 12.2(b) yield a continuous linear approximation by a polygon curve.

$$T(x) = \begin{cases} 1 - |x| & \text{for } |x| < 1 \\ 0 & \text{for } |x| > 1 \end{cases} . \qquad (12.53)$$

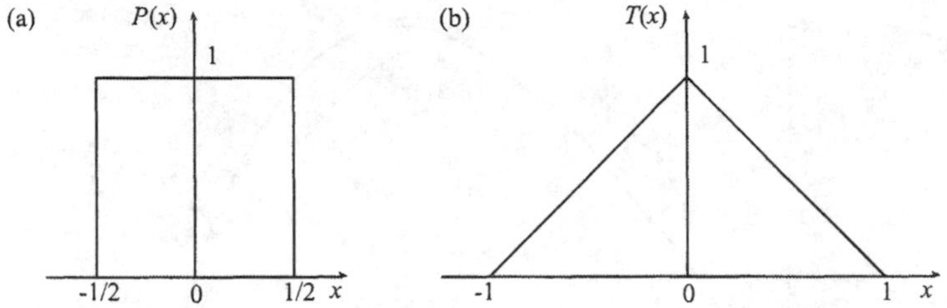

Figure 12.2: (a) Pulse function $P(x)$ and (b) triangle function $T(x)$.

As an example we investigate the boundary value problem according to (12.42) for the Dirichlet boundary conditions $f(0) = f(1) = 0$. We subdivide the interval [0, 1] by N sampling points into $N + 1$ intervals and apply triangular functions according to (12.53) as expansion functions. We make the series expansion

$$f_N(x) = \sum_{n=1}^{N} \alpha_n \varphi_n(x) \tag{12.54}$$

with the basis functions

$$\varphi_{N,n}(x) = T((N+1)x - n)\,. \tag{12.55}$$

As test functions we apply the pulse functions

$$\psi_{N,n}(x) = P((N+1)x - n)\,. \tag{12.56}$$

We compute the matrix elements of the operator $\boldsymbol{L}_{op} = \mathrm{d}^2/\,\mathrm{d}x^2$:

$$\mathrm{d}^2/\,\mathrm{d}x^2 \varphi_{N,n}(x) = (N+1) \left[\delta\left(x - \frac{n-1}{N+1}\right) - 2\delta\left(x - \frac{n}{N+1}\right) + \delta\left(x - \frac{n+1}{N+1}\right)\right]. \tag{12.57}$$

From (12.35) we obtain

$$L_{N,mn} = \int_a^b P((N+1)x - m) \frac{d^2\varphi_{N,n}(x)}{\mathrm{d}x^2}\,\mathrm{d}x = (N+1)\left(\delta_{m,n-1} - 2\delta_{m,n} + \delta_{m,n+1}\right). \tag{12.58}$$

The components of the vector $\boldsymbol{\beta}_N$ are

$$\beta_{N,m} = \int_a^b \psi^*_{N,m}(x) g(x)\,\mathrm{d}x = \int_{\frac{m-1/2}{N+1}}^{\frac{m+1/2}{N+1}} g(x)\,\mathrm{d}x\,. \tag{12.59}$$

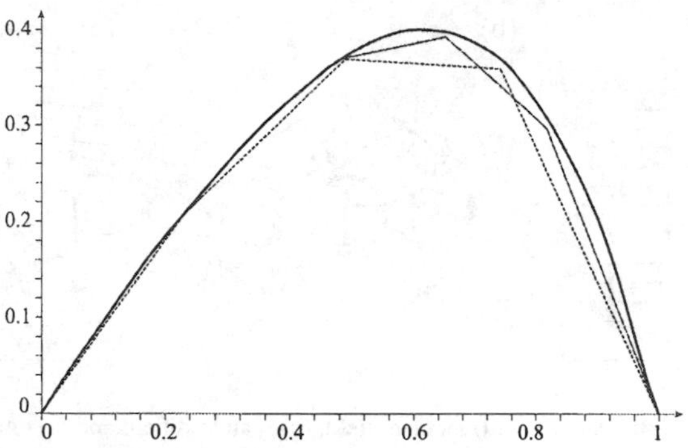

Figure 12.3: The approximation by triangle functions for $N = 3$, $N = 5$ and $N = 10$ in comparison with the exact solution.

For $N = 5$ we obtain the linear system of equations

$$\begin{bmatrix} -12 & 6 & 0 & 0 & 0 \\ 6 & -12 & 6 & 0 & 0 \\ 0 & 6 & -12 & 6 & 0 \\ 0 & 0 & 6 & -12 & 6 \\ 0 & 0 & 0 & 6 & -12 \end{bmatrix} \begin{bmatrix} \alpha_1 \\ \alpha_2 \\ \alpha_3 \\ \alpha_4 \\ \alpha_5 \end{bmatrix} = \begin{bmatrix} -\frac{31}{216} \\ -\frac{49}{216} \\ -\frac{91}{216} \\ -\frac{157}{216} \\ -\frac{247}{216} \end{bmatrix} . \tag{12.60}$$

The solution of (12.60) is

$$\boldsymbol{\alpha}^T_{<5>} = \begin{bmatrix} \frac{395}{2592} & \frac{91}{324} & \frac{107}{288} & \frac{127}{324} & \frac{755}{2592} \end{bmatrix} . \tag{12.61}$$

For $N = 3$ we obtain the solution vector

$$\boldsymbol{\alpha}^T_{<3>} = \begin{bmatrix} \frac{111}{512} & \frac{47}{128} & \frac{183}{512} \end{bmatrix} \tag{12.62}$$

and for $N = 10$ the solution vector

$$\boldsymbol{\alpha}^T_{<10>} = [0.087, 0.167, 0.239, 0.303, 0.355, 0.388, 0.398, 0.375, 0.309, 0.189] . \tag{12.63}$$

12.5 THE INTEGRAL EQUATION METHOD

The electromagnetic field of a three-dimensional structure can be computed from the unknown quantities over certain boundary surfaces that are solved by the integral

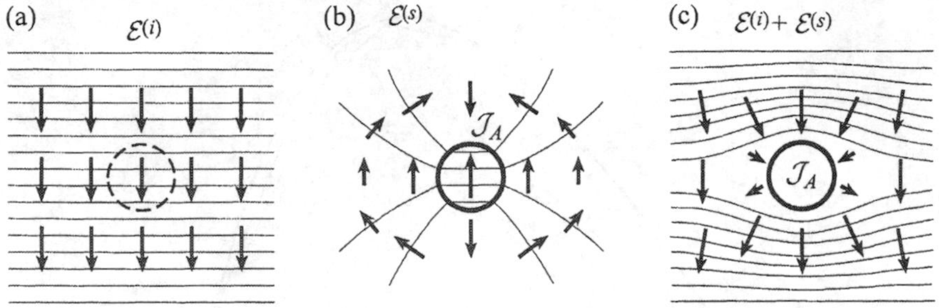

Figure 12.4: Scattering: (a) incident field, (b) scattered field and (c) total field.

equation method [6, 7, 26, 30, 31]. If the distribution of the sources of the electromagnetic field is known, the field may be computed in a straightforward manner by computing the Hertz vector fields via (11.25a) and (11.25b) and from these the electric and magnetic fields. However, in electromagnetic field computation the source current or polarization distribution is not known initially; its determination is part of the problem. To show this we consider the scattering of the electromagnetic field by a conducting sphere, illustrated in Figure 12.4. Let $\underline{\mathcal{E}}^{(i)}$ be the incident or primary field, existing without the conducting sphere (Figure 12.4(a)). At the surface of conducting sphere primary field $\underline{\mathcal{E}}^{(i)}$ will not satisfy the boundary conditions. Therefore in the conducting sphere currents will be induced, creating a secondary or scattered field $\underline{\mathcal{E}}^{(s)}$, so that the total field has no tangential electric field component on the surface of the conducting scattering body. The complete field $\underline{\mathcal{E}}$ on the surface S of the conductive body in Figure 12.4 results from the superposition of the incident field $\underline{\mathcal{E}}^{(i)}$ and the scattered field $\underline{\mathcal{E}}^{(s)}$,

$$\underline{\mathcal{E}} = \underline{\mathcal{E}}^{(i)} + \underline{\mathcal{E}}^{(s)} . \tag{12.64}$$

The source of the scattered field is the electric surface current, described by the surface current density form $\underline{\mathcal{J}}_A(\boldsymbol{x})$. Inserting (3.28) into (11.6)

$$\underline{\Pi}_e(\boldsymbol{x}) = \frac{1}{\mathrm{j}\omega} \int \underline{\mathcal{G}}_{e0}(\boldsymbol{x}, \boldsymbol{x}') \wedge \underline{\mathcal{J}}(\boldsymbol{x}') , \tag{12.65}$$

where the integration is performed over the surface S of the sphere. The free-space Green's dyadic form $\underline{\mathcal{G}}_{e0}$ is given by (11.22),

$$\underline{\mathcal{G}}_{e0}(\boldsymbol{x}, \boldsymbol{x}') = \underline{G}_{e0}(\boldsymbol{x}, \boldsymbol{x}')(\mathrm{d}x\,\mathrm{d}x' + \mathrm{d}y\,\mathrm{d}y' + \mathrm{d}z\,\mathrm{d}z') \tag{12.66}$$

with

$$\underline{G}_{e0}(\boldsymbol{x}, \boldsymbol{x}') = \frac{\mathrm{e}^{-jk|\boldsymbol{x}-\boldsymbol{x}'|}}{4\pi\varepsilon|\boldsymbol{x} - \boldsymbol{x}'|} . \tag{12.67}$$

With (3.35) we obtain the scattered electric field form

$$\underline{\mathcal{E}}^{(s)} = \mathrm{d}\tilde{\mathrm{d}}\,\underline{\Pi}_e + k^2\,\underline{\Pi}_e \tag{12.68}$$

with $k^2 = \omega^2\varepsilon\mu$. We introduce the dyadic Green's function

$$\underline{\mathcal{G}}_{e1}(\boldsymbol{x}, \boldsymbol{x}') = \frac{1}{\mathrm{j}\omega}\,(\mathrm{d}\tilde{\mathrm{d}} + k^2)\left(\underline{G}_{e0}(\boldsymbol{x}, \boldsymbol{x}')(\,\mathrm{d}x\,\mathrm{d}x' + \mathrm{d}y\,\mathrm{d}y' + \mathrm{d}z\,\mathrm{d}z')\right), \tag{12.69}$$

where the exterior differential operators d and $\tilde{\mathrm{d}}$ act on the variable x only and not on x' and obtain

$$\underline{\mathcal{E}}^{(s)} = \int' \underline{\mathcal{G}}_{e1}(\boldsymbol{x}, \boldsymbol{x}') \wedge \underline{\mathcal{J}}(\boldsymbol{x}')\,. \tag{12.70}$$

The integration is performed over x'. In a more explicit notation (12.69) may be written as

$$\begin{aligned}\underline{\mathcal{G}}_{e1}(\boldsymbol{x}, \boldsymbol{x}') = \frac{1}{\mathrm{j}\omega}\Bigg[&\left(\frac{\partial^2 \underline{G}_{e0}}{\partial x^2} + k^2 \underline{G}_{e0}\right)\mathrm{d}x\,\mathrm{d}x' + \frac{\partial^2 \underline{G}_{e0}}{\partial x \partial y}\,\mathrm{d}x\,\mathrm{d}y' + \frac{\partial^2 \underline{G}_{e0}}{\partial z \partial x}\,\mathrm{d}x\,\mathrm{d}z'\\ &+ \frac{\partial^2 \underline{G}_{e0}}{\partial x \partial y}\,\mathrm{d}y\,\mathrm{d}x' + \left(\frac{\partial^2 \underline{G}_{e0}}{\partial y^2} + k^2 \underline{G}_{e0}\right)\mathrm{d}y\,\mathrm{d}y' + \frac{\partial^2 \underline{G}_{e0}}{\partial y \partial z}\,\mathrm{d}y\,\mathrm{d}z'\\ &+ \frac{\partial^2 \underline{G}_{e0}}{\partial z \partial x}\,\mathrm{d}z\,\mathrm{d}x' + \frac{\partial^2 \underline{G}_{e0}}{\partial y \partial z}\,\mathrm{d}z\,\mathrm{d}y' + \left(\frac{\partial^2 \underline{G}_{e0}}{\partial z^2} + k^2 \underline{G}_{e0}\right)\mathrm{d}z\,\mathrm{d}z'\Bigg].\end{aligned} \tag{12.71}$$

Using (2.169) to express the current on the sphere by the surface current density we obtain from (12.70) after integration over the direction normal to the surface the relation between the surface current density on the scatterer $\underline{\mathcal{J}}_A(\boldsymbol{x}')$ and the scattered electric field

$$\underline{\mathcal{E}}^{(s)} = \int_S' \underline{\mathcal{G}}_{e1}(\boldsymbol{x}, \boldsymbol{x}') \wedge \underline{\mathcal{J}}_A(\boldsymbol{x}')\,. \tag{12.72}$$

The integration is performed over the surface S with x' as the variable of integration.

On the surface of the scatterer the tangential electric field component must vanish. With (2.191) this yields

$$\underline{\mathcal{E}}_t = \underline{\mathcal{E}}_t^{(i)} + \underline{\mathcal{E}}_t^{(s)} = n \lrcorner\, n \wedge (\underline{\mathcal{E}}^{(i)} + \underline{\mathcal{E}}^{(s)}) = 0 \quad \text{for } \boldsymbol{x} \in S\,. \tag{12.73}$$

From (12.72) and (12.73) we obtain the *integral equation*

$$\underline{\mathcal{E}}_t^{(i)} = -n \lrcorner\, n \wedge \int_S' \underline{\mathcal{G}}_{e1}(\boldsymbol{x}, \boldsymbol{x}') \wedge \underline{\mathcal{J}}_A(\boldsymbol{x}') \quad \text{for } \boldsymbol{x}, \boldsymbol{x}' \in S\,. \tag{12.74}$$

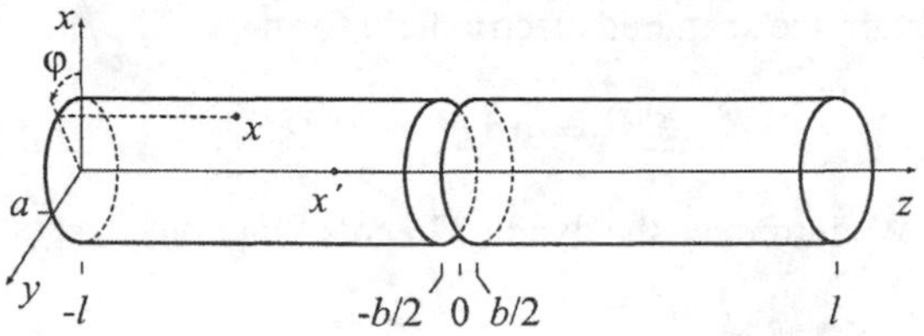

Figure 12.5: Dipole antenna.

This integral equation can be written as a functional equation

$$L_{op}(\underline{\mathcal{J}}_s) = \underline{\mathcal{E}}^{(i)}\Big|_{\tan} . \tag{12.75}$$

This functional equation may be solved numerically using the MoM.

12.6 THE INTEGRAL EQUATION FOR THE LINEAR ANTENNA

Up to now we have investigated linear antennas on the basis of a given current distribution on the antenna. This allowed us to determine the radiation characteristics of the antenna with satisfactory accuracy; however, we could not determine the antenna impedance in this way. For a more accurate computation of the radiation characteristic and for the computation of the antenna impedance an accurate determination of the current distribution under consideration of the radiation is necessary. In the following we apply the integral equation method in connection with the MoM to determine the current distribution on a linear antenna [32–35].

We investigate the linear dipole antenna depicted in Figure 12.5. The linear antenna consists of a straight circular cylindric conductor of length $2l$ with diameter $2a$ – much smaller than the wavelength. If a primary wave is incident on an ideally conducting surface, on this surface currents are induced, causing the tangential component of the total electric field to vanish on the surface of the ideal conductor. Assuming an ideal conductor, the surface current density $\underline{\mathcal{J}}_{\mathcal{A}}$ is bounded to an infinitesimally thin surface layer of the antenna. The linear dipole antenna exhibits a current flow in the z-direction only. From (12.70) we obtain

$$\underline{\mathcal{E}}^{(s)} = \int_{V_{ant}}' \underline{\mathcal{G}}_{e1}(\boldsymbol{x}, \boldsymbol{x}') \wedge \underline{\mathcal{J}}(\boldsymbol{x}') . \tag{12.76}$$

The integration is performed over the coordinates $\boldsymbol{x}'$ over the volume of the antenna wires V_{ant}. The points of observation $\boldsymbol{x}$ are located on the surface of the antenna wires. If the antenna exhibits infinite conductivity the current is confined to the wire surface. We can simplify the computation by replacing the current on the wire surface by a current line source in the axis of the cylinder. For the field excited by the current on the wire surface and outside the conductor this will be a good approximation. The virtual field excited by the equivalent line source inside the wire will be ignored, since the field vanishes inside an ideal conductor. In this case we can set $\boldsymbol{x}' = (0, 0, z')$ in $\underline{\mathcal{G}}_{e1}(\boldsymbol{x}, \boldsymbol{x}')$. Integrating the current density over the wire cross-section $A_{ant}(z)$ yields the antenna current

$$\underline{I}(z) = \int_{A_{ant}(z)} \underline{\mathcal{J}}(\boldsymbol{x}) \,. \tag{12.77}$$

Integrating (12.76) over the wire cross-section $A_{ant}(z)$ we obtain with (12.66) and (12.71) *Pocklington's integral equation* [36]

$$\underline{E}_z^{(s)}(z)\Big|_{r=a} = \frac{1}{4\pi \mathrm{j}\omega\varepsilon_0} \int_{z'=-l}^{z'=l} \left(\frac{\partial^2}{\partial z^2} + k^2 \right) \frac{\exp(-\mathrm{j}kr)}{r} \underline{I}(z')dz' \tag{12.78}$$

with

$$r = |\boldsymbol{x} - \boldsymbol{x}'| = \sqrt{a^2 + (z - z')^2} \,. \tag{12.79}$$

The kernel of Pocklington's integral equation (12.78) may be written in the more convenient *Richmond form* [37] as

$$\underline{E}_z^{(s)}(z)\Big|_{r=a} = \frac{1}{4\pi \mathrm{j}\omega\varepsilon_0} \int_{z'=-l}^{z'=l} \frac{e^{-\mathrm{j}kr}}{r^5} \left[(1 + \mathrm{j}kr)(2r^2 - 3a^2) + k^2 a^2 r^2 \right] \underline{I}(z')dz' \,. \tag{12.80}$$

Exciting the dipole in the gap $[-b/2, b/2]$ with a voltage $\underline{V}_0$ yields

$$\underline{E}_{iz}(z)\Big|_{\boldsymbol{x} \in \mathcal{A}} = \begin{cases} \underline{V}_0/b & \text{for } |z| \le b/2 \\ 0 & \text{for } b/2 < |z| < l \end{cases} \tag{12.81}$$

The solution of this integral equation (12.80) of type

$$\underline{E}_z^{(s)}(z)\Big|_{r=a} = \int_{z'=-l}^{z'=l} K(z, z')\underline{I}(z')dz' \tag{12.82}$$

is performed using the method of moment. We make the following staircase approximation of the current distribution

$$I(z) = \sum_{n=1}^{N} I_n P(\frac{z - z_n}{\Delta z}) \tag{12.83}$$

with

$$\Delta z = \frac{2l}{N+1}, \tag{12.84a}$$

$$z_n = -l + n\Delta z. \tag{12.84b}$$

As the test function we use the delta distributions

$$\psi_n = \delta(z - z_n) \tag{12.85}$$

and obtain

$$\sum_n L_{mn}\alpha_n = \beta_m \tag{12.86}$$

with

$$\beta_m = \Delta z E_{iz}(z_m) \tag{12.87}$$

and

$$L_{mn} = \Delta z \int_{z_n - \Delta z/2}^{z_n + \Delta z/2} K(z_m, z') dz'. \tag{12.88}$$

12.7 THE TRANSMISSION LINE MATRIX METHOD

The *transmission line matrix* (TLM) method introduced by Johns [38] is a space- and time discretizing method of electromagnetic field computation. Originally TLM is based on the analogy between the electromagnetic field and a mesh of transmission lines [39]. The TLM method allows one to model complex electromagnetic structures. Detailed descriptions are given in [20–23, 40, 41]. The TLM scheme has been derived from Maxwell's equation using the finite difference approximation [42], the finite integration approximation [43] and the MoM [44]. In the following we introduce the TLM scheme via finite integration [45, 46].

Figure 12.6 illustrates the principle of TLM. In a first step the space is discretized in TLM cells. Although other discretization geometries are possible, we assume in the following cells with a cube shape. As shown in Figure 12.6(a) on every surface of the TLM cell samples of tangential electric and magnetic fields are taken. This may be done either by taking a probe of the tangential field components in the centers of the surface or by averaging over a certain volume. The first approach is called the finite difference approach, whereas the second one is called the finite integration approach. In both cases we obtain 12 electric field samples and 12 magnetic field samples per TLM cell. The sampled electric and magnetic field components are summarized in

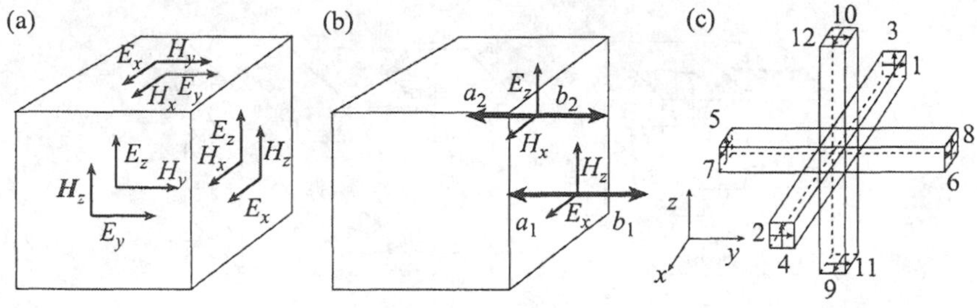

Figure 12.6: (a) The TLM cell, (b) the wave amplitudes and (c) the condensed symmetric TLM node.

12-dimensional vectors

$$
\begin{aligned}
{}_k\boldsymbol{E}_{l,m,n} &= {}_k[E_1, E_2, E_3, \dots E_{10}, E_{11}, E_{12}]^T_{l,m,n} \,, \\
{}_k\boldsymbol{H}_{l,m,n} &= {}_k[H_1, H_2, H_3, \dots H_{10}, H_{11}, H_{12}]^T_{l,m,n} \,. \qquad (12.89a)
\end{aligned}
$$

The orientation of electric and magnetic field samples is chosen in such a way that the power flow is directed into the TLM cell if the electric and magnetic field components have the same sign. Assuming a spatial discretization with Δl and a time discretization with Δt and introducing the discrete space coordinates l, m, n and the discrete time coordinate k the relation between the continuous coordinates x, y, z, t and the discrete coordinates are given by

$$
x = l\Delta l \,, \quad y = m\Delta l \,, \quad z = n\Delta l \,, \quad t = k\Delta t \,. \qquad (12.90)
$$

We now introduce the wave amplitude vectors

$$
\begin{aligned}
{}_k\boldsymbol{a}_{l,m,n} &= {}_k[a_1, a_2, a_3, \dots a_{10}, a_{11}, a_{12}]^T_{l,m,n} \,, \\
{}_k\boldsymbol{b}_{l,m,n} &= {}_k[b_1, b_2, b_3, \dots b_{10}, b_{11}, b_{12}]^T_{l,m,n} \,, \qquad (12.91a)
\end{aligned}
$$

where ${}_k\boldsymbol{a}_{l,m,n}$ summarizes the waves incident in the TLM cell and ${}_k\boldsymbol{b}_{l,m,n}$ contains the amplitudes of the waves scattered by the TLM cell. The incident and the scattered waves propagate normal to the tangential planes. The wave amplitude and the field components are related via

$$
{}_k\boldsymbol{a}_{l,m,n} = \frac{1}{2\sqrt{Z_F}}\, {}_k\boldsymbol{E}_{l,m,n} + \frac{\sqrt{Z_F}}{2}\, {}_k\boldsymbol{H}_{l,m,n} \,, \qquad (12.92a)
$$

$$
{}_k\boldsymbol{b}_{l,m,n} = \frac{1}{2\sqrt{Z_F}}\, {}_k\boldsymbol{E}_{l,m,n} - \frac{\sqrt{Z_F}}{2}\, {}_k\boldsymbol{H}_{l,m,n} \qquad (12.92b)
$$

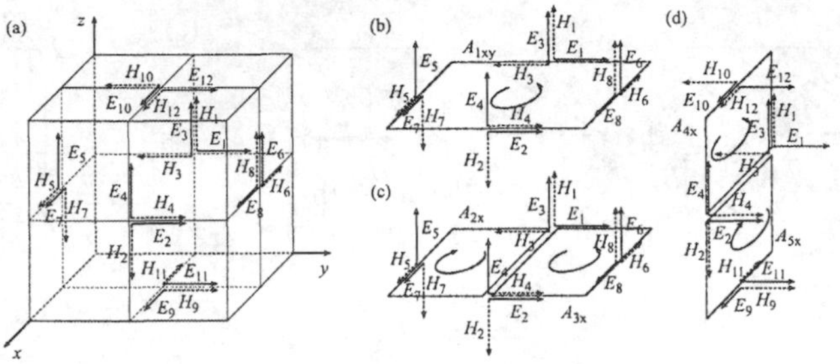

Figure 12.7: (a) The TLM cell, (b) integration path ∂A_{1xy}, (c) integration paths ∂A_{2x} and ∂A_{3x} and (d) integration paths ∂A_{4x} and ∂A_{5x}.

and

$$ {}_k\boldsymbol{E}_{l,m,n} = \sqrt{Z_F}\left({}_k\boldsymbol{a}_{l,m,n} + {}_k\boldsymbol{b}_{l,m,n}\right), \tag{12.93a} $$

$$ {}_k\boldsymbol{H}_{l,m,n} = \frac{1}{\sqrt{Z_F}}\left({}_k\boldsymbol{a}_{l,m,n} - {}_k\boldsymbol{b}_{l,m,n}\right), \tag{12.93b} $$

where $Z_F = \sqrt{\frac{\mu}{\varepsilon}}$ is the field impedance. In the network model of TLM, in each sampling point, one port is assigned to each polarization. Now we can replace the geometric model by a network model, i.e., the TLM *node* depicted in Figure 12.6(c). In the following, we use the term TLM *cell* for the geometrical object we have defined in the continuous space, whereas the term TLM *node* is used for the abstract network model.

To derive relations between the incident waves ${}_k\boldsymbol{a}_{l,m,n}$ and the scattered waves ${}_k\boldsymbol{b}_{l,m,n}$ of the cell l, m, n we apply finite integration to the TLM cell shown in Figure 12.7(a). Ampère's law (2.50) and Faraday's law (2.51) yield

$$ \oint_{\partial A_{iuv}} \mathcal{H}(\boldsymbol{x}, t) = \frac{\mathrm{d}}{\mathrm{d}t}\int_{A_{iuv}} \mathcal{D}(\boldsymbol{x}, t), \tag{12.94a} $$

$$ \oint_{\partial A_{iuv}} \mathcal{E}(\boldsymbol{x}, t) = -\frac{\mathrm{d}}{\mathrm{d}t}\int_{A_{iuv}} \mathcal{B}(\boldsymbol{x}, t). \tag{12.94b} $$

In order to obtain a system of 12 linear equations relating the scattered wave amplitudes $\boldsymbol{b}$ to the incident wave amplitudes $\boldsymbol{b}$ we proceed as follows. First we apply both laws to three surfaces of integration A_{1xy}, A_{1yz} and A_{1zx}. The surface A_{1xy}, shown in Figure 12.7(b) is parallel to the xy-plane and goes through the center of the TLM cell. The surfaces A_{1yz} and A_{1zx} respectively are parallel to the yz-plane and the zx-plane. This yields six independent equations. For the time discretization we apply a *Crank-Nicolson scheme* [47]. That means, we replace the time-derivative by forward differences and time-dependent quantities by the arithmetic mean of the two time steps involved. From (12.94a) and (12.94b) we obtain in this way

$$\frac{1}{2}\oint_{\partial A_{iuv}} ({}_k\mathcal{H} + {}_{k-1}\mathcal{H}) = \frac{\varepsilon}{\Delta t}\int_{A_{iuv}} \star({}_k\mathcal{E} - {}_{k-1}\mathcal{E}) \,, \tag{12.95a}$$

$$\frac{1}{2}\oint_{\partial A_{iuv}} ({}_k\mathcal{E} + {}_{k-1}\mathcal{E}) = -\frac{\mu}{\Delta t}\int_{A_{iuv}} \star({}_k\mathcal{H} - {}_{k-1}\mathcal{H}) \,. \tag{12.95b}$$

We compute these integrals for the surface A_{1xy} depicted in Figure 12.7(b). Sampling the electric and magnetic fields in the center points of the TLM cell surfaces we obtain

$$\oint_{\partial A_{1xy}} \mathcal{H} \cong \Delta l(H_5 + H_4 + H_6 + H_3) \,, \tag{12.96a}$$

$$\int_{A_{1xy}} \star\mathcal{E} \cong \frac{\Delta l^2}{4}(E_5 + E_4 + E_6 + E_3) \,, \tag{12.96b}$$

$$\oint_{\partial A_{1xy}} \mathcal{E} \cong \Delta l(E_7 + E_2 - E_8 - E_1) \,, \tag{12.96c}$$

$$\int_{A_{1xy}} \star\mathcal{H} \cong \frac{\Delta l^2}{4}(-H_7 - H_2 + H_8 + H_1) \,. \tag{12.96d}$$

Inserting (12.96a) and (12.96b) into (12.95a) and considering $\varepsilon = 1/Z_F c$ we obtain

$$\begin{aligned}&({}_kH_5 + {}_{k-1}H_5 + {}_kH_4 + {}_{k-1}H_4 + {}_kH_6 + {}_{k-1}H_6 + {}_kH_3 + {}_{k-1}H_3) = \\ &\frac{\Delta l}{2Z_F c\Delta t}({}_kE_5 - {}_{k-1}E_5 + {}_kE_4 - {}_{k-1}E_4 + {}_kE_6 - {}_{k-1}E_6 + {}_kE_3 - {}_{k-1}E_3)\end{aligned} \tag{12.97}$$

Choosing the ratio of the space interval Δl and the time interval Δt as $\frac{\Delta l}{\Delta t} = 2c$ we obtain with (12.93a) and (12.93b)

$${}_k(b_5 + b_4 + b_6 + b_3) = {}_{k-1}(a_5 + a_4 + a_6 + a_3) \,. \tag{12.98}$$

Inserting (12.96c), (12.96d) and $\frac{\Delta l}{\Delta t} = 2c$ into (12.95b) and considering $\mu = Z_F/c$ we obtain

$$\begin{aligned}&({}_kE_7 + {}_{k-1}E_7 + {}_kE_2 + {}_{k-1}E_2 - {}_kE_8 - {}_{k-1}E_8 - {}_kE_1 - {}_{k-1}E_1) = \\ &\frac{Z_F\Delta l}{2c\Delta t}({}_kH_7 - {}_{k-1}H_7 + {}_kH_2 - {}_{k-1}H_2 - {}_kH_8 + {}_{k-1}H_8 - {}_kH_1 + {}_{k-1}H_1) \,.\end{aligned} \tag{12.99}$$

We obtain with (12.93a) and (12.93b)

$${}_k(b_7 + b_2 - b_8 - b_1) = -{}_{k-1}(a_7 + a_2 - a_8 - a_1) \,. \tag{12.100}$$

Performing these integrations also over the surfaces A_{1yz} and A_{1zx} yields a total of the following six equations

$$_k(b_5 + b_4 + b_6 + b_3) = {}_{k-1}(a_5 + a_4 + a_6 + a_3)\,, \tag{12.101a}$$
$$_k(b_9 + b_8 + b_{10} + b_7) = {}_{k-1}(a_9 + a_8 + a_{10} + a_7)\,, \tag{12.101b}$$
$$_k(b_1 + b_{12} + b_2 + b_{11}) = {}_{k-1}(a_1 + a_{12} + a_2 + a_{11})\,, \tag{12.101c}$$
$$_k(b_7 + b_2 - b_8 - b_1) = -{}_{k-1}(a_7 + a_2 - a_8 - a_1)\,, \tag{12.101d}$$
$$_k(b_{11} + b_6 - b_{12} - b_5) = -{}_{k-1}(a_{11} + a_6 - a_{12} - a_5)\,, \tag{12.101e}$$
$$_k(b_3 + b_{10} - b_4 - b_9) = -{}_{k-1}(a_3 + a_{10} - a_4 - a_9)\,. \tag{12.101f}$$

A first order finite difference scheme based on Ampère's law and Faraday's law may only exhibit six equations. However, to relate the 12 scattered waves to the 12 incident waves of a TLM cell we need 12 equations. To obtain six additional independent equations, we perform the integrations of (12.95a) and (12.95b) over the area A_{2x} and A_{3x} depicted in Figure 12.7(c) and over the areas A_{4x} and A_{5x} in Figure 12.7(d). The contributions to the line integrals through the center of the cells in Figures 12.7(c) and 12.7(d) compensate each other. By that way we obtain two further equations, and by performing this procedure for all three spatial orientations we obtain in total the following additional six equations:

$$_k(b_7 - b_2 - b_8 + b_1) = {}_{k-1}(a_7 - a_2 - a_8 + a_1)\,, \tag{12.102a}$$
$$_k(b_{11} - b_6 - b_{12} + b_5) = {}_{k-1}(a_{11} - a_6 - a_{12} + a_5)\,, \tag{12.102b}$$
$$_k(b_3 - b_{10} - b_4 + b_9) = {}_{k-1}(a_3 - a_{10} - a_4 + a_9)\,, \tag{12.102c}$$
$$_k(b_5 - b_4 + b_6 - b_3) = -{}_{k-1}(a_5 - a_4 + a_6 - a_3)\,, \tag{12.102d}$$
$$_k(b_9 - b_8 + b_{10} - b_7) = -{}_{k-1}(a_9 - a_8 + a_{10} - a_7)\,, \tag{12.102e}$$
$$_k(b_1 - b_{12} + b_2 - b_{11}) = -{}_{k-1}(a_1 - a_{12} + a_2 - a_{11})\,. \tag{12.102f}$$

We put (12.101a) to (12.101f) and (12.102a) to (12.102f) in the form

$$\boldsymbol{M}\ {}_k\boldsymbol{b} = \boldsymbol{L}\boldsymbol{M}\ {}_{k-1}\boldsymbol{a} \tag{12.103}$$

with

$$\boldsymbol{L} = \mathrm{diag}\,[1, 1, 1, -1, -1, -1, 1, 1, 1, -1, -1, -1]\,, \tag{12.104}$$

where we have used the notation introduced in (C.18) for the diagonal matrix. $\boldsymbol{L}$ is a diagonal matrix with the diagonal elements ± 1 and turns out to be the scattering matrix of the *symmetric condensed* TLM *node* in its eigensystem. The scattering matrix $\boldsymbol{S}$ of the symmetric condensed TLM node is given by

$$\boldsymbol{S} = \boldsymbol{M}^{-1}\boldsymbol{L}\boldsymbol{M}\,. \tag{12.105}$$

The scattering matrix S is given by

$$S = \begin{bmatrix} 0 & \mathbf{S_0} & \mathbf{S_0}^T \\ \mathbf{S_0}^T & 0 & \mathbf{S_0} \\ \mathbf{S_0} & \mathbf{S_0}^T & 0 \end{bmatrix} \quad \text{with } \mathbf{S_0} = \begin{bmatrix} 0 & 0 & \frac{1}{2} & -\frac{1}{2} \\ 0 & 0 & -\frac{1}{2} & \frac{1}{2} \\ \frac{1}{2} & \frac{1}{2} & 0 & 0 \\ \frac{1}{2} & \frac{1}{2} & 0 & 0 \end{bmatrix} . \tag{12.106}$$

The scattering matrix S has the property $S = S^T = S^\dagger = S^{-1}$, i.e. it is real, symmetric, hermitian and unitary. Consequently the TLM scheme fulfills energy conservation, reciprocity and invariance with respect to time reversal exactly. We note that the scattering matrix S may also be determined completely by considering only symmetry and energy conservation.

We consider the TLM mesh to be composed of condensed symmetric TLM nodes as shown in Figure 12.6(c), where each of the six arms is of length $\Delta l/2$. We assume a homogeneous lossless space with no sources. All incident and scattered wave amplitudes at the node (l, m, n) can be summarized in the vectors ${}_k\boldsymbol{a}_{l,m,n}$ and ${}_k\boldsymbol{b}_{l,m,n}$.

In order to describe the complete discretized mesh state, we introduce the field state space. To the node with the discrete space coordinate (l, m, n) at the discrete time coordinate k a base vector $|k; l, m, n\rangle$ is assigned. The set of basis vectors $|k_2; l_2, m_2, n_2\rangle$ is orthonormal. The orthogonality relations are given by

$$\langle k_1; l_1, m_1, n_1 | k_2; l_2, m_2, n_2\rangle = \delta_{k_1,k_2}\, \delta_{l_1,l_2}\, \delta_{m_1,m_2}\, \delta_{n_1,n_2} . \tag{12.107}$$

The electric field vector $|F_E\rangle$ and the magnetic field vector $|F_M\rangle$ combine all tangential field samples of the TLM mesh

$$|F_E\rangle = \frac{1}{\sqrt{Z}} \sum_{k,l,m,n} \boldsymbol{E}^k_{l,m,n}\, |k; l, m, n\rangle \ , \quad |F_M\rangle = \sqrt{Z} \sum_{k,l,m,n} \boldsymbol{H}^k_{l,m,n}\, |k; l, m, n\rangle \ . \tag{12.108}$$

All incident and scattered wave amplitudes of the TLM mesh can be combined in two vectors $|\boldsymbol{a}\rangle$ and $|\boldsymbol{b}\rangle$ respectively

$$|\boldsymbol{a}\rangle = \sum_{k,l,m,n} \boldsymbol{a}^k_{l,m,n}\, |k; l, m, n\rangle \ , \qquad |\boldsymbol{b}\rangle = \sum_{k,l,m,n} \boldsymbol{b}^k_{l,m,n}\, |k; l, m, n\rangle \ . \tag{12.109}$$

The cell boundary mapping is a bijective one-to-one mapping between the 24 electric and magnetic field components and the 24 incident and scattered wave amplitudes at one condensed symmetric TLM node. Since all tangential electric and magnetic field components in each cell boundary surface are also specified in the neighboring cell boundary surfaces, only 12 field components per TLM cell are linearly independent. Specifying, for example, all 12 incident wave amplitudes per TLM cell yields a complete description of the field state.

The time shift operator T_s and its hermitian conjugate $T_s^\dagger$ increment or decrement k by 1, i.e., it shifts the field state by Δt,

$$T_s \; |k; l, m, n\rangle = |k+1; l, m, n\rangle \; , \quad T_s^\dagger \; |k; l, m, n\rangle = |k-1; l, m, n\rangle \; . \quad (12.110)$$

If we connect a time delay Δt with the scattering, the simultaneous scattering at all TLM mesh nodes can be described by the operator equation

$$|\boldsymbol{b}\rangle = \boldsymbol{T}_s \; \boldsymbol{S} \; |\boldsymbol{a}\rangle \; . \quad (12.111)$$

In order to describe the propagation of the wave amplitudes in the TLM mesh, we define the spatial shift operators $\boldsymbol{X}_s$, $\boldsymbol{Y}_s$, $\boldsymbol{Z}_s$ and their hermitian conjugates $\boldsymbol{X}_s^\dagger$, $\boldsymbol{Y}_s^\dagger$ and $\boldsymbol{Z}_s^\dagger$, which increment and decrement the three discrete spatial coordinates l, m and n in the same way as the operators $\boldsymbol{T}_s$ and $\boldsymbol{T}_s^\dagger$ do this with the discrete time coordinate k. We introduce the connection operator

$$\begin{aligned} \boldsymbol{\Gamma} = {} & \boldsymbol{X}_s(\boldsymbol{\Delta}_{1,2} + \boldsymbol{\Delta}_{3,4}) + \boldsymbol{X}_s^\dagger(\boldsymbol{\Delta}_{2,1} + \boldsymbol{\Delta}_{4,3}) + \boldsymbol{Y}_s(\boldsymbol{\Delta}_{5,6} + \boldsymbol{\Delta}_{7,8}) \\ & + \boldsymbol{Y}_s^\dagger(\boldsymbol{\Delta}_{6,5} + \boldsymbol{\Delta}_{8,7}) + \boldsymbol{Z}_s(\boldsymbol{\Delta}_{9,10} + \boldsymbol{\Delta}_{11,12}) + \boldsymbol{Z}_s^\dagger(\boldsymbol{\Delta}_{10,9} + \boldsymbol{\Delta}_{12,11}) \, . \end{aligned} \quad (12.112)$$

with the 12×12 (m,n)-matrix $\left(\boldsymbol{\Delta}_{i,j}\right)_{m,n} = \delta_{i,m} \; \delta_{j,n}$. The scattered wave amplitudes are incident into the neighboring TLM cells. If we assume an instantaneous propagation, we may describe the propagation of all wave amplitudes in the TLM mesh by

$$|\boldsymbol{a}\rangle = \boldsymbol{\Gamma} \; |\boldsymbol{b}\rangle \; . \quad (12.113)$$

The connection operator has the properties $\boldsymbol{\Gamma} = \boldsymbol{\Gamma}^\dagger = \boldsymbol{\Gamma}^{-1}$, i.e., it is hermitian and unitary. The two equations (12.111) and (12.113) constitute the complete TLM scheme.

REFERENCES

[1] T. Itoh, *Numerical Techniques for Microwave and Millimeter-Wave Passive Structures.* New York: John Wiley & Sons, 1989.

[2] R. Sorrentino, *Numerical Methods for Passive Microwave and Millimeter Wave Structures.* IEEE Press, 1989.

[3] E. Yamashita, *Analysis Methods for Electromagnetic Wave Problems.* Boston, London: Artech House, 1990.

[4] R. F. Harrington, "Matrix methods for field problems," *Proceedings of the IEEE,* vol. 55, pp. 136–149, Feb. 1967.

[5] R. F. Harrington, *Field Computation by Moment Methods.* San Francisco: IEEE Press, 1968.

[6] J. J. Wang, *Generalized Moment Methods in Electromagnetics*. New York: John Wiley & Sons, 1991.

[7] J. Mosig, "Integral equation technique," in *Numerical Techniques for Microwave and Millimeter Wave Passive Structures* (T. Itoh, ed.), pp. 133–213, New York: John Wiley & Sons, 1989.

[8] D. Mirshekar-Syahkal, *Spectral Domain Methods for Microwave Integrated Circuits*. Taunton, Somerset, England: Research Studies Press, 1990.

[9] M. Mongiardo, P. Russer, C. Tomassoni, and L. Felsen, "Analysis of N-furcation in elliptical waveguides via the generalized network formulation," *IEEE Trans. Microwave Theory Techn.*, vol. 47, pp. 2473–2478, Dec. 1999.

[10] J. Kessler, R. Dill, and P. Russer, "Field theory investigation of high-t_c superconducting coplanar wave guide transmission lines and resonators," *IEEE Trans. Microwave Theory Techn.*, vol. 39, pp. 1566–1574, Sept. 1991.

[11] R. Schmidt and P. Russer, "Modelling of cascaded coplanar waveguide discontinuities by the mode-matching approach," *1995 Int. Microwave Symposium Digest, Orlando*, pp. 281–284, May 1995.

[12] U. Schulz and R. Pregla, "A new technique for the analysis of the dispersion characteristics of planar waveguides and its application to microstrips with tunig septums," *Radio Sci*, vol. 16, pp. 1173–1178, Nov.–Dec. 1981.

[13] R. Pregla and W. Pascher, "The method of lines," in *Numerical Techniques for Microwave and Millimeter Wave Passive Structures* (T. Itoh, ed.), pp. 381–446, New York: John Wiley & Sons, 1989.

[14] K. Yee, "Numerical solution of initial boundary value problems involving Maxwell's equations in isotropic media," *IEEE Trans. Antennas Propagat.*, vol. 14, pp. 302–307, May 1966.

[15] A. Taflove, *Computational Electrodynamics - The Finite-Difference Time-Domain Method*. Boston, London: Artech House, 1995.

[16] T. Weiland, "A discretization method for the solution of Maxwell's equations for six-component fields," *Electronics and Communications (AEU)*, vol. 31, pp. 116–120, 1977.

[17] T. Weiland, "On the unique numerical solution of maxwellian eigenvalue problems in three dimensions," *Particle Accelerators*, vol. 17, pp. 227–242, 1985.

[18] T. Weiland, "Time domain electromagnetic field computation with finite difference methods," *Int. J. Numer. Modeling*, vol. 9, pp. 295–319, 1996.

[19] J. Jin, *The Finite Element Method in Electromagnetics*. New York: John Wiley & Sons, 1993.

[20] W. Hoefer, "The transmission line matrix method-theory and applications," *IEEE Trans. Microwave Theory Techn.*, vol. 33, pp. 882–893, Oct. 1985.

[21] W. Hoefer, "The transmission line matrix (TLM) method," in *Numerical Techniques for Microwave and Millimeter Wave Passive Structures* (T. Itoh, ed.), pp. 496–591, New York: John Wiley & Sons, 1989.

[22] C. Christopoulos, *The Transmission-Line Modeling Method TLM*. New York: IEEE Press, 1995.

[23] P. Russer, "The transmission line matrix method," in *Applied Computational Electromagnetics*, NATO ASI Series, pp. 243–269, Cambridge, Massachusetts, London, England: Springer, 2000.

[24] T. Mangold and P. Russer, "Full-wave modeling and automatic equivalent-circuit generation of millimeter-wave planar and multilayer structures," *IEEE Trans. Microwave Theory Techn.*, vol. 47, pp. 851–858, June 1999.

[25] S. Lindenmeier, L. Pierantoni, and P. Russer, "Hybrid space discretizing-integral equation methods for numerical modeling of transient interference," *IEEE Trans. Electromagn. Compatibility*, vol. 41, pp. 425–430, Nov. 1999.

[26] R. E. Collin, *Field Theory of Guided Waves*. New York: IEEE Press, 1991.

[27] D. G. Dudley, *Mathematic Foundations for Electromagnetic Theory*. New York: IEEE Press, 1994.

[28] B. Friedman, *Lectures on Applications-Oriented Mathematics*. New York: John Wiley & Sons, 1969.

[29] S. Hassani, *Mathematical Physics*. Berlin: Springer, 2002.

[30] C. A. Balanis, *Advanced Engineering Electromagnetics*. New York: John Wiley & Sons, 1989.

[31] Morita, *Integral Equation Methods for Electromagnetics*. Artech House, 1990.

[32] J. D. Kraus, *Antennas*. New York: McGraw-Hill, 1988.

[33] W. Stutzman and G. Thiele, *Antenna Theory and Design*. New York: John Wiley & Sons, 1981.

[34] C. A. Balanis, *Antenna Theory*. New York: John Wiley & Sons, 1982.

[35] F. Landstorfer and R. Sacher, *Optimisation of Wire Antennas*. New York: Letchworth, Research Studies Press Ltd, Wiley, 1990.

[36] H. Pocklington, "Electrical oscillations in wire," *Cambridge Philos. Soc. Proc.*, vol. 9, pp. 324–332, 1897.

[37] J. Richmond, "Digital computer solutions of the rigorous equations for scattering problems," *Proceedings of the IEEE*, vol. 53, pp. 796–804, Aug. 1965.

[38] P. Johns and R. Beurle, "Numerical solution of 2-dimensional scattering problems using a transmission-line matrix," *Proc. IEE*, vol. 118, pp. 1203–1208, Sept. 1971.

[39] G. Kron, "Equivalent circuit of the field equations of Maxwell I," *Proc. IRE*, vol. 32, pp. 289–299, May 1944.

[40] C. Christopoulos and P. Russer, "Application of TLM to microwave circuits," in *Applied Computational Electromagnetics*, NATO ASI Series, pp. 300–323, Cambridge, Massachusetts, London, England: Springer, 2000.

[41] C. Christopoulos and P. Russer, "Application of TLM to EMC problems," in *Applied Computational Electromagnetics*, NATO ASI Series, pp. 324–350, Cambridge, Massachusetts, London, England: Springer, 2000.

[42] S. Hein, "Consistent finite difference modelling of Maxwell's equations with lossy symmetrical condensed TLM node," *Int. J. Numer. Modeling*, vol. 6, pp. 207–220, 1993.

[43] M. Aidam and P. Russer, "Derivation of the TLM method by finite integration," *AEÜ Int. J. Electron. Commun.*, vol. 51, pp. 35–39, 1997.

[44] M. Krumpholz and P. Russer, "A field theoretical derivation TLM," *IEEE Trans. Microwave Theory Techn.*, vol. 42, pp. 1660–1668, Sept. 1994.

[45] N. Peña and M. Ney, "A general and complete two-dimensional TLM hybrid node formulation based on Maxwell's integral equations," in *Proc. 12th Annual Review of Progress in Applied Computational Electromagnetics, Monterey*, (Monterey, CA), pp. 254–261, Mar. 1996.

[46] N. Peña and M. Ney, "A general formulation of a three-dimensional TLM condensed node with the modeling of electric and magnetic losses and current sources," in *Proc. 12th Annual Review of Progress in Applied Computational Electromagnetics, Monterey*, (Monterey, CA), pp. 262–269, Mar. 1996.

[47] J. Thomas, *Numerical partial differential equations*. New York: Springer, 1995.

Appendix A

Vectors and Differential Forms

A.1 VECTORS

This section gives a compact summary of the vector algebra for the linear three-dimensional vector space. For a more detailed treatment see for example [1, 2]. A *vector* is a variable quantity, such as force, electric or magnetic field that has magnitude and direction and can be resolved into components. In the *Cartesian coordinate system* (x, y, z) a vector $\boldsymbol{a}$ is represented by its Cartesian coordinates (a_x, a_y, a_z). The vector may be visualized by an arrow starting fom the origin and terminating at the point (a_x, a_y, a_z) as depicted in Figure A.1(a). However, we could start from any point in our Cartesian reference frame. The origin only is chosen for simplicity.

Let $\boldsymbol{e}_x$ be a vector of unit magnitude pointing in the positive x-direction and $\boldsymbol{e}_y$ and $\boldsymbol{e}_z$ vectors of unit magnitude pointing in the positive y- and z-directions respectively. By vector addition we obtain

$$\boldsymbol{a} = a_x\boldsymbol{e}_x + a_y\boldsymbol{e}_y + a_z\boldsymbol{e}_z = \begin{bmatrix} a_x \\ a_y \\ a_z \end{bmatrix} . \tag{A.1}$$

The vectors $\boldsymbol{e}_x$, $\boldsymbol{e}_y$ and $\boldsymbol{e}_z$ form a *basis* of our three-dimensional linear vector space. Figure A.1(a) shows the vector $\boldsymbol{a}$ in a right-handed Cartesian coordinate system (x, y, z), whereas in Figure A.1(b) the same vector $\boldsymbol{a}$ is represented in a left-handed coordinate system (x', y', z') = (-x,-y,-z). In the new coordinate system (x', y', z') the vector a exhibits the coordinates $(a'_x, a'_y, a'_z) = (-a_x, -a_y, -a_z)$. The vector $\boldsymbol{a}$ itself remains unchanged under this coordinate transformation. We have

$$\boldsymbol{a} = a_x\boldsymbol{e}_x + a_y\boldsymbol{e}_y + a_z\boldsymbol{e}_z = a'_x\boldsymbol{e}_{x'} + a'_y\boldsymbol{e}_{y'} + a'_z\boldsymbol{e}_{z'} . \tag{A.2}$$

The components of the vector are odd functions of the coordinates. A vector with this

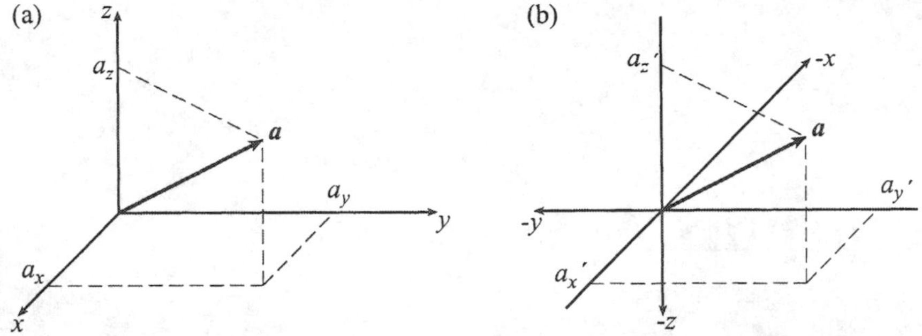

Figure A.1: Cartesian components of a vector (a) in a right-handed coordinate system and (b) in a left-handed coordinate system.

property is called a *polar vector*. Vectors occurring in curve integrals like the vectors representing the electric or magnetic field are represented by polar vectors.

The sum of two vectors $\boldsymbol{a}$ and $\boldsymbol{b}$ is given by

$$\boldsymbol{a} + \boldsymbol{b} = \begin{bmatrix} a_x + b_x \\ a_y + b_y \\ a_z + b_z \end{bmatrix}. \tag{A.3}$$

The *magnitude* $|\boldsymbol{a}|$ of the vector $\boldsymbol{a}$ is given by

$$|\boldsymbol{a}| = \sqrt{|a_x|^2 + |a_y|^2 + |a_z|^2}. \tag{A.4}$$

The *inner product* or *scalar product* of two vectors $\boldsymbol{a}$ and $\boldsymbol{b}$ is defined as

$$\boldsymbol{a} \cdot \boldsymbol{b} = a_x b_x + a_y b_y + a_z b_z \,. \tag{A.5}$$

For real vectors $\boldsymbol{a}$ and $\boldsymbol{b}$ the inner product is given by

$$\boldsymbol{a} \cdot \boldsymbol{b} = |\boldsymbol{a}||\boldsymbol{b}| \cos\varphi \,, \tag{A.6}$$

where φ is the angle enclosed by $\boldsymbol{a}$ and $\boldsymbol{b}$. The vectors $\boldsymbol{e}_x$, $\boldsymbol{e}_y$ and $\boldsymbol{e}_z$ fulfill the relation

$$\boldsymbol{e}_i \cdot \boldsymbol{e}_j = \delta_{ij} \quad \text{for} \quad i, j = x, y, z \,, \tag{A.7}$$

where the *Kronecker delta symbol* δ_{mn} is given by

$$\delta_{mn} = \begin{cases} 0 & \text{for } m \neq n \\ 1 & \text{for } m = n \end{cases}. \tag{A.8}$$

The vectors $\boldsymbol{e}_x$, $\boldsymbol{e}_y$ and $\boldsymbol{e}_z$ form a complete set of *orthonormal basis vectors* of our three-dimensional vector space.

The *cross-product* or *vector product*

$$\boldsymbol{c} = \boldsymbol{a} \times \boldsymbol{b} \tag{A.9}$$

of two vectors $\boldsymbol{a}$ and $\boldsymbol{b}$ is a vector normal to both vectors $\boldsymbol{a}$ and $\boldsymbol{b}$ and with the magnitude

$$|\boldsymbol{a} \times \boldsymbol{b}| = |\boldsymbol{a}||\boldsymbol{b}| \sin\varphi. \tag{A.10}$$

If the two vectors $\boldsymbol{a}$ and $\boldsymbol{b}$ define a parallelogram, then the magnitude $|\boldsymbol{a} \times \boldsymbol{b}|$ of the cross-product is the area of the parallelogram. The vectors $\boldsymbol{a}$, $\boldsymbol{b}$ and $\boldsymbol{a} \times \boldsymbol{b}$ form a right-handed system. We assume the orthonormal basis vectors $\boldsymbol{e}_x$, $\boldsymbol{e}_y$ and $\boldsymbol{e}_z$ to form a right-handed system according to Figure A.1(a). In this case the relations

$$\begin{aligned} \boldsymbol{e}_1 \times \boldsymbol{e}_2 &= -\boldsymbol{e}_2 \times \boldsymbol{e}_1 &= \boldsymbol{e}_3\,, \\ \boldsymbol{e}_2 \times \boldsymbol{e}_3 &= -\boldsymbol{e}_3 \times \boldsymbol{e}_2 &= \boldsymbol{e}_1\,, \\ \boldsymbol{e}_3 \times \boldsymbol{e}_1 &= -\boldsymbol{e}_1 \times \boldsymbol{e}_3 &= \boldsymbol{e}_2 \end{aligned} \tag{A.11}$$

are fulfilled. The *Kronecker tensor* is defined as

$$\delta_{ijk} = \begin{cases} +1 & \text{if } (ijk) \text{ is an even permutation of } (xyz) \\ -1 & \text{if } (ijk) \text{ is an odd permutation of } (xyz) \\ 0 & \text{if } (ijk) \text{ is no permutation of } (xyz) \end{cases} . \tag{A.12}$$

Permutations from an original ordered sequence such as ijk can be achieved by successive interchanges of elements. A permutation is said to be *even* or *odd* if an even or odd number of transpositions is required for the rearrangement.

We can express (A.11) in the following form:

$$\boldsymbol{e}_i \times \boldsymbol{e}_j = \sum_k \delta_{ijk} \boldsymbol{e}_k\,. \tag{A.13}$$

For two vectors $\boldsymbol{a}$ and $\boldsymbol{b}$ we obtain

$$\boldsymbol{a} \times \boldsymbol{b} = \sum_{i,j,k} a_i b_j \delta_{ijk} \boldsymbol{e}_k\,. \tag{A.14}$$

From this we obtain (A.9) in component representation

$$\begin{bmatrix} c_x \\ c_y \\ c_z \end{bmatrix} = \begin{bmatrix} a_y b_z - a_z b_y \\ a_z b_x - a_x b_z \\ a_x b_y - a_y b_x \end{bmatrix} . \tag{A.15}$$

The vector $\boldsymbol{C}$ defined by the cross-product (A.9) behaves fundamentally different from the polar vectors. If the coordinate axes are inverted we obtain $a_i \rightarrow -a'_i$, $b_i \rightarrow -b'_i$. However, due to (A.15) this yields $c_i \rightarrow c'_i$. Under coordinate inversion the cross-product vector $\boldsymbol{c}$ does not behave like a polar vector. A vector with the transformation properties is called an *axial vector* or *pseudovector*. Vectors occuring in surface integrals like the vectors representing current or flux densities are represented by axial vectors.

With (C.20) the cross-product may also be represented as determinant:

$$\boldsymbol{a} \times \boldsymbol{b} = \begin{vmatrix} \boldsymbol{e}_x & \boldsymbol{e}_y & \boldsymbol{e}_z \\ a_1 & a_2 & a_3 \\ b_1 & b_2 & b_3 \end{vmatrix} . \tag{A.16}$$

The cross-product is anticommutative, in other words,

$$\boldsymbol{b} \times \boldsymbol{a} = -\boldsymbol{a} \times \boldsymbol{b} . \tag{A.17}$$

The *triple scalar product* of three vectors $\boldsymbol{a}$, $\boldsymbol{b}$ and $\boldsymbol{c}$, defined by

$$[\boldsymbol{abc}] = \boldsymbol{a} \cdot (\boldsymbol{b} \times \boldsymbol{c}) = (\boldsymbol{a} \times \boldsymbol{b}) \cdot \boldsymbol{c} \tag{A.18}$$

yields the volume of the parallelepiped defined by the vectors $\boldsymbol{a}$, ($\boldsymbol{b}$ and $\boldsymbol{c}$). The triple scalar product is invariant under cyclic permutation of the factors; however it changes its sign, if two factors are interchanged. From (A.5), (A.14) and (A.18) we obtain

$$[\boldsymbol{abc}] = \sum_{i,j,k} a_i b_j c_k \delta_{ijk} = \begin{vmatrix} a_1 & a_2 & a_3 \\ b_1 & b_2 & b_3 \\ c_1 & c_2 & c_3 \end{vmatrix} . \tag{A.19}$$

In the case of the inversion of all coordinate axes the triple scalar product changes its sign. A scalar with the transformation properties of a triple scalar product is called a *pseudoscalar*. Scalar quantities occurring in volume integrals, as for example charge densities or energy densities are represented by pseudoscalars. Pseudoscalars are distinguished from *scalars* describing quantities that are not integrated over space coordinates. Examples for scalars are scalar potential functions.
For the twofold cross-product of three vectors we obtain

$$(\boldsymbol{a} \times \boldsymbol{b}) \times \boldsymbol{c} = (\boldsymbol{a} \cdot \boldsymbol{c})\boldsymbol{b} - (\boldsymbol{b} \cdot \boldsymbol{c})\boldsymbol{a} , \tag{A.20}$$

$$\boldsymbol{a} \times (\boldsymbol{b} \times \boldsymbol{c}) = (\boldsymbol{a} \cdot \boldsymbol{c})\boldsymbol{b} - (\boldsymbol{a} \cdot \boldsymbol{b})\boldsymbol{c} . \tag{A.21}$$

For the threefold products of four vectors the following relations hold:

$$(\boldsymbol{a} \times \boldsymbol{b}) \cdot (\boldsymbol{c} \times \boldsymbol{d}) = \begin{vmatrix} (\boldsymbol{a} \cdot \boldsymbol{c}) & (\boldsymbol{a} \cdot \boldsymbol{d}) \\ (\boldsymbol{b} \cdot \boldsymbol{c}) & (\boldsymbol{b} \cdot \boldsymbol{d}) \end{vmatrix} , \tag{A.22}$$

$$(\boldsymbol{a} \times \boldsymbol{b}) \times (\boldsymbol{c} \times \boldsymbol{d}) = [\boldsymbol{abd}]\boldsymbol{c} - [\boldsymbol{abc}]\boldsymbol{d} = [\boldsymbol{acd}]\boldsymbol{b} - [\boldsymbol{bcd}]\boldsymbol{a} . \tag{A.23}$$

A.2 DIFFERENTIAL FORMS

In this section the framework of differential forms used in this book is presented. For further reading see [3–7]. Scalar and vector fields may be represented by *exterior differential forms*. Differential forms are essentially the expressions under an integration symbol. An exterior differential form of *order p* is called a *p-form*. In n-dimensional space the order of a differential form may assume values $1 \dots n$. In this book we only consider the three-dimensional case. The p-forms in three-dimensional space are

$$\text{zero-form:} \qquad f(\boldsymbol{x}), \tag{A.24}$$

$$\text{one-form:} \qquad \mathcal{U}(\boldsymbol{x}) = U_x \, \mathrm{d}x + U_y \, \mathrm{d}y + U_z \, \mathrm{d}z, \tag{A.25}$$

$$\text{two-form:} \qquad \mathcal{V}(\boldsymbol{x}) = V_x \, \mathrm{d}y \wedge \mathrm{d}z + V_y \, \mathrm{d}z \wedge \mathrm{d}x + V_z \, \mathrm{d}x \wedge \mathrm{d}y, \tag{A.26}$$

$$\text{three-form:} \qquad \mathcal{Q}(\boldsymbol{x}) = \rho \, \mathrm{d}x \wedge \mathrm{d}y \wedge \mathrm{d}z. \tag{A.27}$$

For the exterior differential form $\mathrm{d}x \wedge \mathrm{d}y$ the following *commutation relation* holds:

$$\mathrm{d}x \wedge \mathrm{d}y = -\,\mathrm{d}y \wedge \mathrm{d}x. \tag{A.28}$$

Deciding

$$\mathrm{d}x \wedge \mathrm{d}y = \mathrm{d}x \, \mathrm{d}y \tag{A.29}$$

assigns to $\mathrm{d}x \wedge \mathrm{d}y$ the positive orientation and to $\mathrm{d}y \wedge \mathrm{d}x$ the negative orientation. For a p-form $\mathcal{U}$ and a q-form $\mathcal{V}$ the commutation relation is

$$\mathcal{U} \wedge \mathcal{V} = (-1)^{p+q+1} \, \mathcal{V} \wedge \mathcal{U}. \tag{A.30}$$

We introduce the bracket symbol $[\ ]^{(p)}$ to express the relation of a scalar field ($p = 0, 3$) or a vector field ($p = 1, 2$) to a p-form. The p-forms are related to scalar and vector fields via

$$f(\boldsymbol{x}) = [f(\boldsymbol{x})]^{(0)}, \tag{A.31}$$

$$\mathcal{U}(\boldsymbol{x}) = [\boldsymbol{U}(\boldsymbol{x})]^{(1)} = U_x \, \mathrm{d}x + U_y \, \mathrm{d}y + U_z \, \mathrm{d}z, \tag{A.32}$$

$$\mathcal{V}(\boldsymbol{x}) = [\boldsymbol{V}(\boldsymbol{x})]^{(2)} = V_x \, \mathrm{d}y \wedge \mathrm{d}z + V_y \, \mathrm{d}z \wedge \mathrm{d}x + V_z \, \mathrm{d}x \wedge \mathrm{d}y, \tag{A.33}$$

$$\mathcal{Q}(\boldsymbol{x}) = [\rho(\boldsymbol{x})]^{(3)} = \rho \, \mathrm{d}x \wedge \mathrm{d}y \wedge \mathrm{d}z. \tag{A.34}$$

In these equations f is a scalar, $\boldsymbol{U}$ is a vector, $\boldsymbol{V}$ is a pseudovector, and ρ is a pseudoscalar. In differential form notation a clear distinction between scalars, pseudoscalars, vectors and pseudovectors is made. Scalars are represented by zero-forms, pseudoscalars by three-forms, vectors by one-forms and pseudovectors by two-forms. The domains of integration are a curve C for a one-form, an area A for a two-form and a volume V for a three-form (Figure A.2). The corresponding integrals are

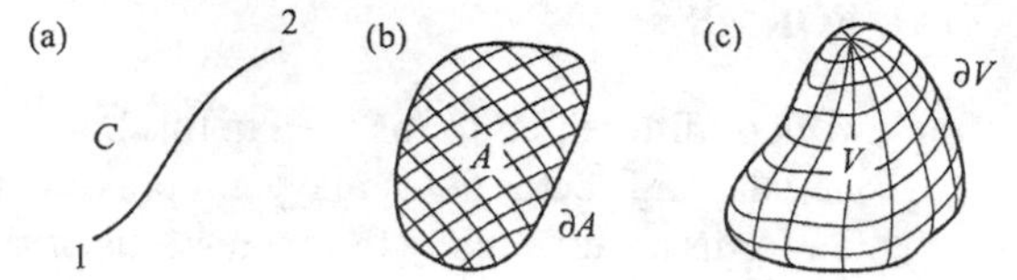

Figure A.2: Domains of integration: (a) curve, (b) surface and (c) volume.

$$\text{one-form:} \quad \int_1^2 \mathcal{U} = \int_1^2 U_x \,\mathrm{d}x + U_y \,\mathrm{d}y + U_z \,\mathrm{d}z \,, \tag{A.35}$$

$$\text{two-form:} \quad \int_A \mathcal{V} = \int_A V_x \,\mathrm{d}y \wedge \mathrm{d}z + V_y \,\mathrm{d}z \wedge \mathrm{d}x + V_z \,\mathrm{d}x \wedge \mathrm{d}y \,, \tag{A.36}$$

$$\text{three-form:} \quad \int_V \mathcal{Q} = \int_V \rho \,\mathrm{d}x \wedge \mathrm{d}y \wedge \mathrm{d}z \,. \tag{A.37}$$

We introduce the *star operator* or *Hodge operator* $\star$ defined via

$$\begin{aligned}
\star f &= f \,\mathrm{d}x \wedge \mathrm{d}y \wedge \mathrm{d}z \,, \\
\star (A_x \,\mathrm{d}x + A_y \,\mathrm{d}y + A_z \,\mathrm{d}z) &= A_x \,\mathrm{d}y \wedge \mathrm{d}z + A_y \,\mathrm{d}z \wedge \mathrm{d}x + A_z \,\mathrm{d}x \wedge \mathrm{d}y \,, \\
\star (A_x \,\mathrm{d}y \wedge \mathrm{d}z + A_y \,\mathrm{d}z \wedge \mathrm{d}x + A_z \,\mathrm{d}x \wedge \mathrm{d}y)n &= A_x \,\mathrm{d}x + A_y \,\mathrm{d}y + A_z \,\mathrm{d}z \,, \\
\star f \,\mathrm{d}x \wedge \mathrm{d}y \wedge \mathrm{d}z &= f \,.
\end{aligned} \tag{A.38}$$

The star operator has the property

$$\star\star = 1 \,. \tag{A.39}$$

A.2.1 Products of Exterior Differential Forms

The product of a p-form and a q-form is a $(p+q)$-form. The product of two one-forms $\mathcal{A}$ and $\mathcal{B}$ is

$$\begin{aligned}
\mathcal{A} \wedge \mathcal{B} &= (A_x \,\mathrm{d}x + A_y \,\mathrm{d}y + A_z \,\mathrm{d}z) \wedge (B_x \,\mathrm{d}x + B_y \,\mathrm{d}y + B_z \,\mathrm{d}z) \\
&= (A_y B_z - A_z B_y) \,\mathrm{d}y \wedge \mathrm{d}z + (A_z B_x - A_x B_z) \,\mathrm{d}z \wedge \mathrm{d}x + \\
&\quad (A_x B_y - A_y B_x) \,\mathrm{d}x \wedge \mathrm{d}y \,.
\end{aligned} \tag{A.40}$$

The relation to vector notation is given by

$$\mathcal{A} = [\boldsymbol{A}]^{(1)} \,, \quad \mathcal{B} = [\boldsymbol{B}]^{(1)} \quad \rightarrow \quad \mathcal{A} \wedge \mathcal{B} = [\boldsymbol{A} \times \boldsymbol{B}]^{(2)} \,. \tag{A.41}$$

The exterior product of two one-forms corresponds to the cross-product (A.9) of two vectors. The exterior product of a one-form $\mathcal{A}$ and a two-form $\mathcal{D}$ is

$$\begin{aligned}\mathcal{A} \wedge \mathcal{D} &= (A_x\,\mathrm{d}x + A_y\,\mathrm{d}y + A_z\,\mathrm{d}z) \wedge (C_x\,\mathrm{d}y \wedge \mathrm{d}z + C_y\,\mathrm{d}z \wedge \mathrm{d}x + C_z\,\mathrm{d}x \wedge \mathrm{d}y) \\ &= (A_x D_x + A_y D_y + A_z D_z)\,\mathrm{d}x \wedge \mathrm{d}y \wedge \mathrm{d}z\,. \end{aligned} \tag{A.42}$$

The relation to vector notation is given by

$$\mathcal{A} = [\boldsymbol{A}]^{(1)}\,, \quad \mathcal{D} = [\boldsymbol{D}]^{(2)} \quad \rightarrow \quad \mathcal{A} \wedge \mathcal{D} = [\boldsymbol{A} \cdot \boldsymbol{D}]^{(3)}\,. \tag{A.43}$$

The exterior product of a one-form with a two-form corresponds to the scalar product (A.5) of two vectors. The exterior product of a three one-forms $\mathcal{A}$, $\mathcal{B}$ and $\mathcal{C}$ is

$$\begin{aligned}\mathcal{A} \wedge \mathcal{B} \wedge \mathcal{C} &= (A_x\,\mathrm{d}x + A_y\,\mathrm{d}y + A_z\,\mathrm{d}z) \wedge (B_x\,\mathrm{d}x + B_y\,\mathrm{d}y + B_z\,\mathrm{d}z) \\ &\wedge\ (C_x\,\mathrm{d}x + C_y\,\mathrm{d}y + C_z\,\mathrm{d}z) \\ &= (A_x B_y C_z + A_y B_z C_x + A_z B_x C_y \\ &- A_z B_y C_x - A_x B_z C_y - A_y B_x C_z)\,\mathrm{d}x \wedge \mathrm{d}y \wedge \mathrm{d}y\,. \end{aligned} \tag{A.44}$$

The relation to vector notation is given by

$$\mathcal{A} = [\boldsymbol{A}]^{(1)}\,,\ \mathcal{B} = [\boldsymbol{B}]^{(1)}\,,\ \mathcal{C} = [\boldsymbol{C}]^{(1)} \ \rightarrow\ \mathcal{A} \wedge \mathcal{B} \wedge \mathcal{C} = [[\boldsymbol{abc}]]^{(3)}\,. \tag{A.45}$$

The exterior product of three one-forms corresponds to the triple scalar product (A.18) of three vectors.

The *contraction* $\mathcal{A} \lrcorner \mathcal{B}$ of two differential forms $\mathcal{A}$ and $\mathcal{B}$ is defined via

$$\mathrm{d}x_i \lrcorner \mathrm{d}x_j = \delta_{ij}\,, \tag{A.46}$$

$$\mathcal{A} \lrcorner (\mathcal{B} \wedge \mathcal{C}) = (\mathcal{A} \lrcorner \mathcal{B}) \wedge \mathcal{C} + (-1)^{deg(\mathcal{A})} \mathcal{B} \wedge (\mathcal{A} \lrcorner \mathcal{C})\,. \tag{A.47}$$

The symbol $\lrcorner$ is named "angle" and the contraction also is called *angle product*. The angle product has been introduced by Burke [4]. We use the modified form given by Warnick [8].

The contraction operator is useful to obtain the tangential component of a one-form or the normal component of a two-form with respect to a surface. For a one-form $\mathcal{E}$ and a two-form $\mathcal{D}$ we obtain

$$\mathrm{d}z \lrcorner (\mathrm{d}z \wedge \mathcal{E}) = E_x\,\mathrm{d}x + E_y\,\mathrm{d}y\,, \tag{A.48}$$

$$\mathrm{d}z \lrcorner (\mathrm{d}z \wedge \mathcal{D}) = D_z\,\mathrm{d}x \wedge \mathrm{d}y\,. \tag{A.49}$$

A.2.2 The Exterior Derivative

The *exterior derivative* $\mathrm{d}\mathcal{U}$ of an exterior differential form $\mathcal{U}$ is defined as

$$\mathrm{d}\mathcal{U} = \sum_i \mathrm{d}x_i \wedge \frac{\partial \mathcal{U}}{\partial x_i} . \tag{A.50}$$

For the exterior differentiation the following product rules are valid:

$$\mathrm{d}\,(\mathcal{U}+\mathcal{V}) = \mathrm{d}\mathcal{U} + \mathrm{d}\,\mathcal{V}\,, \tag{A.51}$$

$$\mathrm{d}\,(\mathcal{U}\wedge\mathcal{V}) = \mathrm{d}\mathcal{U}\wedge\mathcal{V} + (-1)^{(\deg\mathcal{U})}\mathcal{U}\wedge\mathrm{d}\,\mathcal{V}\,. \tag{A.52}$$

The exterior derivatives of p-forms are

zero-form: $$\mathrm{d}f(\boldsymbol{x}) = \frac{\partial f}{\partial x}\,\mathrm{d}x + \frac{\partial f}{\partial y}\,\mathrm{d}y + \frac{\partial f}{\partial z}\,\mathrm{d}z\,, \tag{A.53}$$

one-form: $$\mathrm{d}\mathcal{U}(\boldsymbol{x}) = \left(\frac{\partial U_z}{\partial y} - \frac{\partial U_y}{\partial z}\right)\mathrm{d}y\wedge\mathrm{d}z + \left(\frac{\partial U_x}{\partial z} - \frac{\partial U_z}{\partial x}\right)\mathrm{d}z\wedge\mathrm{d}x + \left(\frac{\partial U_y}{\partial x} - \frac{\partial U_x}{\partial y}\right)\mathrm{d}x\wedge\mathrm{d}y\,, \tag{A.54}$$

two-form: $$\mathrm{d}\,\mathcal{V}(\boldsymbol{x}) = \left(\frac{\partial V_x}{\partial x} + \frac{\partial V_y}{\partial y} + \frac{\partial V_z}{\partial z}\right)\mathrm{d}x\wedge\mathrm{d}y\wedge\mathrm{d}z\,, \tag{A.55}$$

three-form: $$\mathrm{d}\,\mathcal{Q}(\boldsymbol{x}) = 0\,. \tag{A.56}$$

The exterior derivative of a product of a p-form $\mathcal{A}$ with q-form $\mathcal{B}$ is given by

$$\mathrm{d}\,(\mathcal{A}\wedge\mathcal{B}) = \mathrm{d}\mathcal{A}\wedge\mathcal{B} + (-1)^p\mathcal{A}\wedge\mathrm{d}\mathcal{B}\,. \tag{A.57}$$

A form $\mathcal{V}$ for which $\mathrm{d}\mathcal{V} = 0$ is said to be *closed*, and a form $\mathcal{V}$ for which $\mathcal{V} = \mathrm{d}\mathcal{U}$ is said to be *exact*. The relation

$$\mathrm{d}\,\mathrm{d}\mathcal{U} = 0 \tag{A.58}$$

may be verified easily. For differential forms the statement $\mathcal{V} = \mathrm{d}\mathcal{U}$ implies $\mathrm{d}\mathcal{V} = 0$. In conventional vector notation this corresponds to curl grad $= \mathbf{0}$ and div curl $= 0$. All exact forms are closed. However it may also be shown, that all closed forms are exact. *Poincaré's lemma* [3,4] states

$$\mathrm{d}\mathcal{V} = 0 \quad \leftrightarrow \quad \mathcal{V} = \mathrm{d}\mathcal{U} \tag{A.59}$$

A.2.3 The Laplace Operator

We define the *covariant derivative*, expressed by the operator $\tilde{\mathrm{d}}$, by

$$\tilde{\mathrm{d}}\mathcal{U} = (-1)^{\deg \mathcal{U}+1} \star \mathrm{d} \star \mathcal{U} \,. \tag{A.60}$$

The *Laplace operator* Δ is defined by

$$\Delta = \tilde{\mathrm{d}}\,\mathrm{d} + \mathrm{d}\,\tilde{\mathrm{d}} \,. \tag{A.61}$$

From this definition it follows that for any p-form $\mathcal{U}$, the Laplace operator can be interchanged with the Hodge operator:

$$\Delta \star \mathcal{U} = \star \Delta \mathcal{U} \,. \tag{A.62}$$

Applying the Laplace operator to a zero-form Φ, a one-form $\mathcal{A}$, a two-form $\mathcal{B}$ and a three-form $\mathcal{Q}$ yields

$$\text{zero-form:} \quad \Delta\Phi = \star\, \mathrm{d} \star \mathrm{d}\Phi \,, \tag{A.63}$$

$$\text{one-form:} \quad \Delta\mathcal{A} = (\mathrm{d}\star \mathrm{d}\star - \star\, \mathrm{d} \star d)\, \mathcal{A} \,, \tag{A.64}$$

$$\text{two-form:} \quad \Delta\mathcal{B} = (\star\, \mathrm{d} \star d \; - \; \mathrm{d}\star \mathrm{d}\star)\, \mathcal{B} \,, \tag{A.65}$$

$$\text{three-form:} \quad \Delta\mathcal{Q} = \mathrm{d}\star \mathrm{d} \star \mathcal{Q} \,. \tag{A.66}$$

In Cartesian coordinates the application of the Laplace operator to a zero-form f yields

$$\Delta f = \frac{\partial^2 f}{\partial x^2} + \frac{\partial^2 f}{\partial y^2} + \frac{\partial^2 f}{\partial z^2} \,. \tag{A.67}$$

To apply the Laplace operator to a one-form $\mathcal{A}$ in Cartesian coordinates we compute

$$\begin{aligned}
\star\, \mathrm{d} \star \mathrm{d}\mathcal{A} = & \left(\frac{\partial^2 A_y}{\partial x \partial y} + \frac{\partial^2 A_z}{\partial z \partial x} - \frac{\partial^2 A_x}{\partial y^2} - \frac{\partial^2 A_x}{\partial z^2} \right) \mathrm{d}x + \\
& + \left(\frac{\partial^2 A_z}{\partial y \partial z} + \frac{\partial^2 A_x}{\partial x \partial y} - \frac{\partial^2 A_y}{\partial z^2} - \frac{\partial^2 A_y}{\partial x^2} \right) \mathrm{d}y + \\
& + \left(\frac{\partial^2 A_x}{\partial z \partial x} + \frac{\partial^2 A_y}{\partial y \partial z} - \frac{\partial^2 A_z}{\partial x^2} - \frac{\partial^2 A_z}{\partial y^2} \right) \mathrm{d}z \,,
\end{aligned} \tag{A.68}$$

$$\begin{aligned}\mathrm{d}\star\,\mathrm{d}\star\mathcal{A} = &\left(\frac{\partial^2 A_x}{\partial x^2} + \frac{\partial^2 A_x}{\partial x \partial y} + \frac{\partial^2 A_x}{\partial x \partial z}\right)\mathrm{d}x + \\ &+ \left(\frac{\partial^2 A_y}{\partial x \partial y} + \frac{\partial^2 A_y}{\partial y^2} + \frac{\partial^2 A_y}{\partial z \partial y}\right)\mathrm{d}y + \\ &+ \left(\frac{\partial^2 A_z}{\partial z \partial x} + \frac{\partial^2 A_z}{\partial y \partial z} + \frac{\partial^2 A_z}{\partial z^2}\right)\mathrm{d}z\end{aligned} \tag{A.69}$$

and from this we obtain

$$\begin{aligned}\Delta\,\mathcal{A} = &\left(\frac{\partial^2 A_x}{\partial x^2} + \frac{\partial^2 A_x}{\partial y^2} + \frac{\partial^2 A_x}{\partial z^2}\right)\mathrm{d}x + \\ &+ \left(\frac{\partial^2 A_y}{\partial x^2} + \frac{\partial^2 A_y}{\partial y^2} + \frac{\partial^2 A_y}{\partial z^2}\right)\mathrm{d}y + \\ &+ \left(\frac{\partial^2 A_z}{\partial x^2} + \frac{\partial^2 A_z}{\partial y^2} + \frac{\partial^2 A_z}{\partial z^2}\right)\mathrm{d}z\,.\end{aligned} \tag{A.70}$$

Therefore, in Cartesian coordinates the Laplace operator for one-forms is given by

$$\Delta\,\mathcal{A} = \frac{\partial^2 \mathcal{A}}{\partial x^2} + \frac{\partial^2 \mathcal{A}}{\partial y^2} + \frac{\partial^2 \mathcal{A}}{\partial z^2} = \Delta A_x\,\mathrm{d}x + \Delta A_y\,\mathrm{d}y + \Delta A_z\,\mathrm{d}z\,. \tag{A.71}$$

A.3 THE STOKES' THEOREM

The *Stokes' theorem* relates the integration of a p-form $\mathcal{U}$ over the closed p-dimensional boundary ∂V of a $p+1$-dimensional volume V to the volume integral of the $p+1$-form $\mathrm{d}\mathcal{U}$ over V via

$$\oint_{\partial V} \mathcal{U} = \oint_V \mathrm{d}\mathcal{U}\,. \tag{A.72}$$

In the following we give the proof for the two-dimensional area integral and the three-dimensional volume integral. For a more general discussion of the Stokes' theorem the reader is referred to the literature [3–7]. To give the proof of Stokes' theorem for the area integral we consider the one-form

$$\mathcal{U} = U_x\,\mathrm{d}x + U_y\,\mathrm{d}y + U_z\,\mathrm{d}z \tag{A.73}$$

and its exterior derivative

$$\begin{aligned}\mathrm{d}\mathcal{U} = &\left(\frac{\partial U_z}{\partial y} - \frac{\partial U_y}{\partial z}\right)\mathrm{d}y \wedge \mathrm{d}z + \left(\frac{\partial U_x}{\partial z} - \frac{\partial U_z}{\partial x}\right)\mathrm{d}z \wedge \mathrm{d}x \\ &+ \left(\frac{\partial U_y}{\partial x} - \frac{\partial U_x}{\partial y}\right)\mathrm{d}x \wedge \mathrm{d}y\,.\end{aligned} \tag{A.74}$$

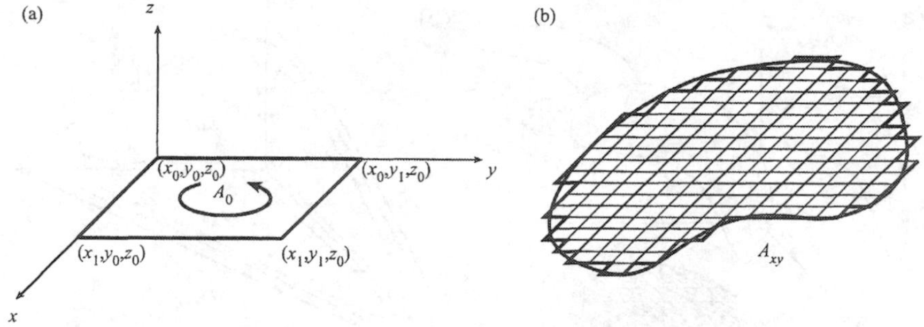

Figure A.3: (a) Surface element and (b) surface.

Integrating $d\mathcal{U}$ over the rectangular area element A_0 shown in Figure A.3(a) yields

$$\begin{aligned}
\int_{A_0} d\mathcal{U} &= \int_{A_0} \left(\frac{\partial U_y}{\partial x} - \frac{\partial U_x}{\partial y} \right) dx \wedge dy \\
&= \int_{y_0}^{y_1} \left(\int_{x_0}^{x_1} \frac{\partial U_y}{\partial x} dx \right) dy - \int_{x_0}^{x_1} \left(\int_{y_0}^{y_1} \frac{\partial U_x}{\partial y} dx \right) dx \\
&= \int_{y_0}^{y_1} \left[U_y(x_1, y) - U_y(x_0, y) \right] dy - \int_{x_0}^{x_1} \left[U_x(x, y_1) - U_x(x, y_0) \right] dx \\
&= \int_{x_0}^{x_1} U_x(x, y_0)\, dx + \int_{y_0}^{y_1} U_y(x_1, y)\, dy + \\
&\quad + \int_{x_1}^{x_0} U_x(x, y_1)\, dx + \int_{y_1}^{y_0} U_y(x_0, y)\, dy = \oint_{\partial A_0} \mathcal{U}\,. \qquad \text{(A.75)}
\end{aligned}$$

Since the above relation is valid for the rectangular area element in Figure A.3(a), we can approximate an arbitrarily shaped area A_{xy} in the xy-plane by small area elements as shown in Figure A.3(b). The sum of the surface integrals of $d\mathcal{U}$ over the areas of all surface elements is equal to the circulation integrals of $\mathcal{U}$ over the boundaries of all surface elements. We consider that the contributions of the adjacent edges of two neighboring surface elements cancel each other out. Therefore only the path elements through the outer edges contribute to the circulation integral. If we let the side lengths of the area elements go to zero, the approximated area converges to the given area A_{xy} with the smooth boundary ∂A_{xy} and we obtain

$$\int_{A_{xy}} d\mathcal{U} = \oint_{\partial A_{xy}} \mathcal{U}\,. \qquad \text{(A.76)}$$

So far we have only considered a plane surface. To derive the Stokes' theorem for

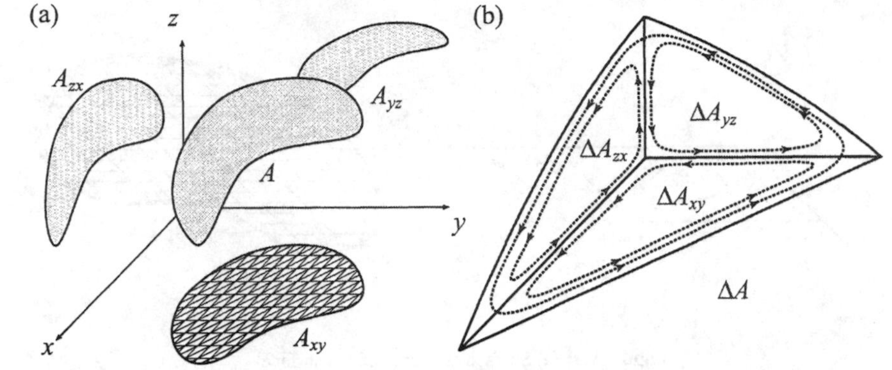

Figure A.4: (a) Curved surface and its projections on the coordinate planes and (b) triangular surface element.

curved surfaces in three-dimensional space we decompose the surface integral into three parts corresponding to the projection of the area A on the three coordinate planes as shown in Figure A.4(a). We introduce the parametric representation

$$U_i^x(y,z) = U_i\,(x(y,z)y,z)\ , \tag{A.77}$$
$$U_i^y(z,x) = U_i\,(x,y(z,x),z)\ , \tag{A.78}$$
$$U_i^z(x,y) = U_i\,(x,y,z(x,y))\ . \tag{A.79}$$

From (A.75) we obtain

$$\int_{A_{xy}} \left(\frac{\partial U_z^x}{\partial y} - \frac{\partial U_y^x}{\partial z}\right) \mathrm{d}y \wedge \mathrm{d}z = \oint_{\partial A_{xy}} U_x^x\,\mathrm{d}x + U_y^x\,\mathrm{d}y\ , \tag{A.80}$$
$$\int_{A_{yz}} \left(\frac{\partial U_x^y}{\partial z} - \frac{\partial U_z^y}{\partial x}\right) \mathrm{d}z \wedge \mathrm{d}x = \oint_{\partial A_{yz}} U_y^y\,\mathrm{d}y + U_z^y\,\mathrm{d}z\ , \tag{A.81}$$
$$\int_{A_{zx}} \left(\frac{\partial U_y^z}{\partial x} - \frac{\partial U_x^z}{\partial y}\right) \mathrm{d}x \wedge \mathrm{d}y = \oint_{\partial A_{zx}} U_z^z\,\mathrm{d}z + U_x^z\,\mathrm{d}x\ . \tag{A.82}$$

We subdivide the area A_{xy} in triangular infinitesimal surface elements as depicted in Figure A.4(a). This generates also a triangular subdivision of the curved surface A. Figure A.4(b) shows one infinitesimal triangular surface element ΔA and its projections on the coordinate planes ΔA_{yz}, ΔA_{zx} and ΔA_{xy}. The curve integral over the boundary $\partial \Delta A$ is the sum of the curve integrals over the boundaries $\partial \Delta A_{yz}$, $\partial \Delta A_{zx}$ and $\partial \Delta A_{xy}$. In the three coordinate planes the sum over all integrals over the area

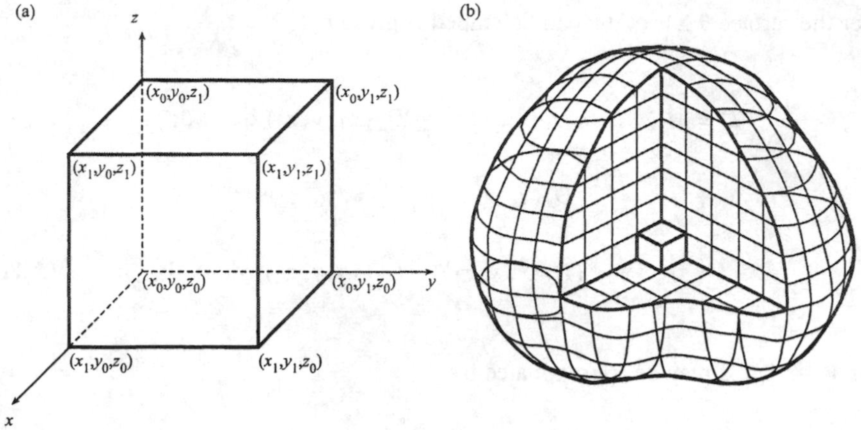

Figure A.5: (a) Volume element ΔV and (b) volume V.

elements $\partial \Delta A_{ij}$ yields the integral over the boundary ∂A_{ij} and we obtain

$$\oint_{\partial A_{xy}} U_x^x \, \mathrm{d}x + U_y^x \, \mathrm{d}y + \oint_{\partial A_{yz}} U_y^y \, \mathrm{d}y + U_z^y \, \mathrm{d}z + \oint_{\partial A_{zx}} U_z^z \, \mathrm{d}z + U_x^z \, \mathrm{d}x = \oint_{\partial A} U_x \, \mathrm{d}x + U_y \, \mathrm{d}z + U_z \, \mathrm{d}z \, . \quad \text{(A.83)}$$

Inserting (A.80), (A.81) and (A.82) yields Stokes' theorem for curved surfaces

$$\int_A \left[\left(\frac{\partial U_z}{\partial y} - \frac{\partial U_y}{\partial z} \right) \mathrm{d}y \wedge \mathrm{d}z + \left(\frac{\partial U_x}{\partial z} - \frac{\partial U_z}{\partial x} \right) \mathrm{d}z \wedge \mathrm{d}x + \left(\frac{\partial U_y}{\partial x} - \frac{\partial U_x}{\partial y} \right) \mathrm{d}x \wedge \mathrm{d}y \right] = \oint_{\partial A} U_x \, \mathrm{d}x + U_y \, \mathrm{d}y + U_z \, \mathrm{d}z \quad \text{(A.84)}$$

In order to derive Stokes' theorem for three-dimensional volumes we consider an infinitesimal rectangular parallelepiped volume element ΔV as depicted in Figure A.5(a). The integral of the two-form

$$\mathcal{D} = D_x \, \mathrm{d}y \wedge \mathrm{d}z + D_y \, \mathrm{d}z \wedge \mathrm{d}x + D_z \, \mathrm{d}x \wedge \mathrm{d}y \quad \text{(A.85)}$$

over the surface $\partial \Delta V$ of the parallelepiped is given by

$$
\begin{aligned}
\oint_{\partial \Delta V} \mathcal{D} = & \int_{y_0}^{y_1} \int_{z_0}^{z_1} (D_x(x_1, y, z) - D_x(x_0, y, z)) \, \mathrm{d}y \wedge \mathrm{d}z \\
& + \int_{z_0}^{z_1} \int_{x_0}^{x_1} (D_x(x, y_1, z) - D_x(x, y_0, z)) \, \mathrm{d}z \wedge \mathrm{d}x \\
& + \int_{x_0}^{x_1} \int_{y_0}^{y_1} (D_x(x, y, z_1) - D_x(x_0, y, z_0)) \, \mathrm{d}x \wedge \mathrm{d}y \, . \qquad \text{(A.86)}
\end{aligned}
$$

For $V \rightarrow 0$ this may be approximated by

$$
\int_{\partial \Delta V} \mathcal{D} = \int_{\Delta V} \left(\frac{\partial D_x}{\partial x} + \frac{\partial D_y}{\partial y} + \frac{\partial D_y}{\partial y} \right) \mathrm{d}x \wedge \mathrm{d}y \wedge \mathrm{d}z = \int_{\Delta V} \mathrm{d}\mathcal{D} \, . \qquad \text{(A.87)}
$$

A volume V with boundary ∂V as depicted in Figure A.5(b) may be considered to consist of infinitesimally small rectangular parallelepiped volume elements. The sum of the volume integrals over all volume elements is equal to the volume integral over V. Computing the sum of all surface integrals over the boundaries of the volume cells, we consider that the contributions from adjacent surfaces of neighboring cells cancel each other and only the outer surfaces contribute. The set of the outer surfaces of the volume cells converges to ∂V and we obtain *Stokes' theorem*

$$
\oint_{\partial V} \mathcal{D} = \int_{V} \mathrm{d}\mathcal{D} \, . \qquad \text{(A.88)}
$$

A.4 CURVILINEAR COORDINATES

A three-dimensional *curvilinear coordinate system* with coordinates u, v, w may be defined via embedding in the three-dimensional Cartesian coordinate system x, y, z and specifying the functions

$$
\begin{aligned}
u &= u(x, y, z) \, , \\
v &= v(x, y, z) \, , \qquad \text{(A.89)} \\
w &= w(x, y, z) \, .
\end{aligned}
$$

We can express the differentials dx, dy, dz by the differentials du, dv, dw via

$$dx = \frac{\partial x}{\partial u} du + \frac{\partial x}{\partial v} dv + \frac{\partial x}{\partial w} dw, \tag{A.90}$$

$$dy = \frac{\partial y}{\partial u} du + \frac{\partial y}{\partial v} dv + \frac{\partial y}{\partial w} dw, \tag{A.91}$$

$$dz = \frac{\partial z}{\partial u} du + \frac{\partial z}{\partial v} dv + \frac{\partial z}{\partial w} dw. \tag{A.92}$$

For the exterior product of two differentials we obtain the transformation rule

$$\begin{aligned} dx \wedge dy = & \left(\frac{\partial x}{\partial u}\frac{\partial y}{\partial v} - \frac{\partial x}{\partial v}\frac{\partial y}{\partial u}\right) du \wedge dv \\ & + \left(\frac{\partial x}{\partial v}\frac{\partial y}{\partial w} - \frac{\partial x}{\partial w}\frac{\partial y}{\partial v}\right) dv \wedge dw \\ & + \left(\frac{\partial x}{\partial w}\frac{\partial y}{\partial u} - \frac{\partial x}{\partial u}\frac{\partial y}{\partial w}\right) dw \wedge du. \end{aligned} \tag{A.93}$$

This defines the *Jacobian determinant*

$$\frac{\partial(x, y)}{\partial(u, v)} \equiv \begin{vmatrix} \frac{\partial x}{\partial u} & \frac{\partial x}{\partial v} \\ \frac{\partial y}{\partial u} & \frac{\partial y}{\partial v} \end{vmatrix} = \frac{\partial x}{\partial u}\frac{\partial y}{\partial v} - \frac{\partial x}{\partial v}\frac{\partial y}{\partial u}. \tag{A.94}$$

We obtain

$$dx \wedge dy = \frac{\partial(x, y)}{\partial(u, v)} du \wedge dv + \frac{\partial(x, y)}{\partial(v, w)} dv \wedge dw + \frac{\partial(x, y)}{\partial(w, u)} dw \wedge du. \tag{A.95}$$

Analogous equations are obtained for $dy \wedge dz$ and $dz \wedge dx$. To express $du \wedge dv$, $dv \wedge dw$ and $dw \wedge du$ by $dx \wedge dy$, $dy \wedge dz$ and $dz \wedge dx$ insert (A.90) – (A.92) and reorder the terms.

For three-forms we obtain

$$dx \wedge dy \wedge dz = \frac{\partial(x, y, z)}{\partial(u, v, w)} du \wedge dv \wedge dw \tag{A.96}$$

with the Jacobian determinant

$$\frac{\partial(x, y, z)}{\partial(u, v, w)} \equiv \begin{vmatrix} \frac{\partial x}{\partial u} & \frac{\partial x}{\partial v} & \frac{\partial x}{\partial w} \\ \frac{\partial y}{\partial u} & \frac{\partial y}{\partial v} & \frac{\partial y}{\partial w} \\ \frac{\partial z}{\partial u} & \frac{\partial z}{\partial v} & \frac{\partial z}{\partial w} \end{vmatrix}. \tag{A.97}$$

Expanding the Jacobian determinant yields

$$\frac{\partial(x,y,z)}{\partial(u,v,w)} = \frac{\partial x}{\partial u}\frac{\partial y}{\partial v}\frac{\partial z}{\partial w} + \frac{\partial x}{\partial v}\frac{\partial y}{\partial w}\frac{\partial z}{\partial u} + \frac{\partial x}{\partial w}\frac{\partial y}{\partial u}\frac{\partial z}{\partial v} - \frac{\partial x}{\partial u}\frac{\partial y}{\partial w}\frac{\partial z}{\partial v} - \frac{\partial x}{\partial v}\frac{\partial y}{\partial u}\frac{\partial z}{\partial w} - \frac{\partial x}{\partial w}\frac{\partial y}{\partial v}\frac{\partial z}{\partial u} . \tag{A.98}$$

Consider the vector $\boldsymbol{x} = (x, y, z)^T$. The length of a path element is given by

$$\mathrm{d}s = \sqrt{\mathrm{d}\boldsymbol{x} \cdot \mathrm{d}\boldsymbol{x}} . \tag{A.99}$$

Using (A.90) – (A.92) we obtain

$$\begin{aligned} \mathrm{d}\boldsymbol{x} \cdot \mathrm{d}\boldsymbol{x} &= \mathrm{d}x^2 + \mathrm{d}y^2 + \mathrm{d}z^2 \\ &= \frac{\partial \boldsymbol{x}}{\partial u} \cdot \frac{\partial \boldsymbol{x}}{\partial u}\,\mathrm{d}u^2 + \frac{\partial \boldsymbol{x}}{\partial v} \cdot \frac{\partial \boldsymbol{x}}{\partial v}\,\mathrm{d}v^2 + \frac{\partial \boldsymbol{x}}{\partial w} \cdot \frac{\partial \boldsymbol{x}}{\partial w}\,\mathrm{d}w^2 \\ &\quad + 2\frac{\partial \boldsymbol{x}}{\partial u} \cdot \frac{\partial \boldsymbol{x}}{\partial v}\,\mathrm{d}u\,\mathrm{d}v + 2\frac{\partial \boldsymbol{x}}{\partial v} \cdot \frac{\partial \boldsymbol{x}}{\partial w}\,\mathrm{d}v\,\mathrm{d}w + 2\frac{\partial \boldsymbol{x}}{\partial w} \cdot \frac{\partial \boldsymbol{x}}{\partial u}\,\mathrm{d}w\,\mathrm{d}u . \end{aligned} \tag{A.100}$$

A coordinate system $u(\boldsymbol{x})$, $v(\boldsymbol{x})$, $w(\boldsymbol{x})$ in the three-dimensional Euclidean space with Cartesian coordinates x, y, z is called an *orthogonal coordinate system* if the *tangent vectors* $\partial\boldsymbol{x}/\partial u$, $\partial\boldsymbol{x}/\partial v$, $\partial\boldsymbol{x}/\partial w$ are mutually perpendicular. For orthogonal coordinate systems (u, v, w) the relations

$$\frac{\partial \boldsymbol{x}}{\partial u} \cdot \frac{\partial \boldsymbol{x}}{\partial v} = 0 , \quad \frac{\partial \boldsymbol{x}}{\partial v} \cdot \frac{\partial \boldsymbol{x}}{\partial w} = 0 , \quad \frac{\partial \boldsymbol{x}}{\partial w} \cdot \frac{\partial \boldsymbol{x}}{\partial u} = 0 \tag{A.101}$$

must be fulfilled. The *metric coefficients* g_1,g_2, and g_3 defined by

$$g_1^2 = \frac{\partial \boldsymbol{x}}{\partial u} \cdot \frac{\partial \boldsymbol{x}}{\partial u} , \quad g_2^2 = \frac{\partial \boldsymbol{x}}{\partial v} \cdot \frac{\partial \boldsymbol{x}}{\partial v} , \quad g_3^2 = \frac{\partial \boldsymbol{x}}{\partial w} \cdot \frac{\partial \boldsymbol{x}}{\partial w} \tag{A.102}$$

give the lengths of the tangent vectors $\partial\boldsymbol{x}/\partial u$, $\partial\boldsymbol{x}/\partial v$ and $\partial\boldsymbol{x}/\partial w$ [9]. For suitable functions $g_1(\boldsymbol{x})$, $g_2(\boldsymbol{x})$, $g_3(\boldsymbol{x})$ we obtain the *orthonormal basis vectors*

$$\boldsymbol{e}_1 = \frac{1}{g_1}\frac{\partial \boldsymbol{x}}{\partial u} , \quad \boldsymbol{e}_2 = \frac{1}{g_2}\frac{\partial \boldsymbol{x}}{\partial v} , \quad \boldsymbol{e}_3 = \frac{1}{g_3}\frac{\partial \boldsymbol{x}}{\partial w} . \tag{A.103}$$

We attach to each point $\boldsymbol{x}$ of the three-dimensional Euclidean space a right-handed orthonormal frame $\boldsymbol{e}_1, \boldsymbol{e}_2, \boldsymbol{e}_3$ with

$$\boldsymbol{e}_i \cdot \boldsymbol{e}_j = \delta_{ij} , \tag{A.104}$$

where δ_{ij} is the Kronecker delta symbol defined in (A.8).

The orthonormal basis vectors constitute a *moving frame* [3,4]. Moving frames represent geometric objects in a basis that is tied to intrinsic geometric features of the problem. We presuppose the moving frame $\boldsymbol{e}_1, \boldsymbol{e}_2, \boldsymbol{e}_3$ is right-handed. Otherwise we change the order of the three vectors. We now express $\mathrm{d}\boldsymbol{x}$ by

$$\mathrm{d}\boldsymbol{x} = \frac{\partial \boldsymbol{x}}{\partial u}\,\mathrm{d}u + \frac{\partial \boldsymbol{x}}{\partial v}\,\mathrm{d}v + \frac{\partial \boldsymbol{x}}{\partial w}\,\mathrm{d}w\,. \tag{A.105}$$

Inserting (A.103) yields

$$\mathrm{d}\boldsymbol{x} = g_1\boldsymbol{e}_1\,\mathrm{d}u + g_2\boldsymbol{e}_2\,\mathrm{d}v + g_3\boldsymbol{e}_3\,\mathrm{d}w\,. \tag{A.106}$$

Introducing the *unit one-forms*

$$\mathrm{s}_1 = g_1\,\mathrm{d}u\,, \quad \mathrm{s}_2 = g_2\,\mathrm{d}v\,, \quad \mathrm{s}_3 = g_3\,\mathrm{d}w \tag{A.107}$$

yields

$$\mathrm{d}\boldsymbol{x} = \mathrm{s}_1\,\boldsymbol{e}_1 + \mathrm{s}_2\,\boldsymbol{e}_2 + \mathrm{s}_3\,\boldsymbol{e}_3\,. \tag{A.108}$$

The *star operator* or *Hodge operator* $\star$ as defined in (A.2) is given by

$$\begin{aligned}
\star f &= f\,\mathrm{s}_1 \wedge \mathrm{s}_2 \wedge \mathrm{s}_3\,,\\
\star(A_u\mathrm{s}_1 + A_v\mathrm{s}_2 + A_w\,\mathrm{s}_3) &= A_u\mathrm{s}_2 \wedge \mathrm{s}_3 + A_v\mathrm{s}_3 \wedge \mathrm{s}_1 + A_w\mathrm{s}_1 \wedge \mathrm{s}_2\,,\\
\star(A_u\mathrm{s}_2 \wedge \mathrm{s}_3 + A_v\mathrm{s}_3 \wedge \mathrm{s}_1 + A_w\mathrm{s}_1 \wedge \mathrm{s}_2) &= A_u\mathrm{s}_1 + A_v\mathrm{s}_2 + A_w\,\mathrm{s}_3\,,\\
\star(f\,\mathrm{s}_1 \wedge \mathrm{s}_2 \wedge \mathrm{s}_3) &= f\,.
\end{aligned} \tag{A.109}$$

Let us now consider the exterior derivative in curvilinear coordinates. The exterior derivative of a zero-form f given by

$$\mathrm{d}f = \frac{\partial f}{\partial u}\,\mathrm{d}u + \frac{\partial f}{\partial v}\,\mathrm{d}v + \frac{\partial f}{\partial w}\,\mathrm{d}w\,. \tag{A.110}$$

Inserting (A.107) yields

$$\mathrm{d}f = \frac{1}{g_1}\frac{\partial f}{\partial u}\mathrm{s}_1 + \frac{1}{g_2}\frac{\partial f}{\partial v}\mathrm{s}_2 + \frac{1}{g_3}\frac{\partial f}{\partial w}\mathrm{s}_3\,. \tag{A.111}$$

In conventional vector analysis the exterior derivative of a zero-form corresponds to the application of the gradient operator:

$$\operatorname{grad} f = \frac{1}{g_1}\frac{\partial f}{\partial u}\,\boldsymbol{e}_1 + \frac{1}{g_2}\frac{\partial f}{\partial v}\,\boldsymbol{e}_2 + \frac{1}{g_3}\frac{\partial f}{\partial w}\,\boldsymbol{e}_3\,. \tag{A.112}$$

We compute the exterior derivative of a one-form $\mathcal{A}$, given by

$$\mathcal{A} = A_u s_1 + A_v s_2 + A_w s_3 \,. \tag{A.113}$$

With (A.107) we obtain

$$\mathcal{A} = g_1 A_u \, du + g_2 A_v \, dv + g_3 A_w \, dw \,. \tag{A.114}$$

Forming the exterior derivative yields

$$\begin{aligned} \mathrm{d}\mathcal{A} = & \left(\frac{\partial(g_2 A_v)}{\partial u} - \frac{\partial(g_1 A_u)}{\partial v} \right) du \wedge dv \\ & + \left(\frac{\partial(g_3 A_w)}{\partial v} - \frac{\partial(g_2 A_v)}{\partial w} \right) dv \wedge dw \\ & + \left(\frac{\partial(g_1 A_u)}{\partial w} - \frac{\partial(g_3 A_w)}{\partial u} \right) dw \wedge du \end{aligned} \tag{A.115}$$

and after again inserting (A.107) we obtain

$$\begin{aligned} \mathrm{d}\mathcal{A} = & \frac{1}{g_2 g_3} \left(\frac{\partial(g_3 A_w)}{\partial v} - \frac{\partial(g_2 A_v)}{\partial w} \right) s_2 \wedge s_3 \\ & + \frac{1}{g_3 g_1} \left(\frac{\partial(g_1 A_u)}{\partial w} - \frac{\partial(g_3 A_w)}{\partial u} \right) s_3 \wedge s_1 \\ & + \frac{1}{g_1 g_2} \left(\frac{\partial(g_2 A_v)}{\partial u} - \frac{\partial(g_1 A_u)}{\partial v} \right) s_1 \wedge s_2 \,. \end{aligned} \tag{A.116}$$

In conventional vector analysis the exterior derivative of a one-form corresponds to the curl operator:

$$\begin{aligned} \operatorname{curl} A = & \frac{1}{g_2 g_3} \left(\frac{\partial(g_3 A_w)}{\partial v} - \frac{\partial(g_2 A_v)}{\partial w} \right) \boldsymbol{e}_1 \\ & + \frac{1}{g_3 g_1} \left(\frac{\partial(g_1 A_u)}{\partial w} - \frac{\partial(g_3 A_w)}{\partial u} \right) \boldsymbol{e}_2 \\ & + \frac{1}{g_1 g_2} \left(\frac{\partial(g_2 A_v)}{\partial u} - \frac{\partial(g_1 A_u)}{\partial v} \right) \boldsymbol{e}_3 \,. \end{aligned} \tag{A.117}$$

To compute the exterior derivative of a two-form $\mathcal{B}$, given by

$$\mathcal{B} = B_u s_2 \wedge s_3 + B_v s_3 \wedge s_1 + B_w s_1 \wedge s_2 \tag{A.118}$$

first we insert (A.107) and obtain

$$\mathcal{B} = g_2 g_3\, B_u\, \mathrm{d}v \wedge \mathrm{d}w + g_3 g_1\, B_v\, \mathrm{d}w \wedge \mathrm{d}u + g_1 g_2\, B_w\, \mathrm{d}u \wedge \mathrm{d}v\,. \tag{A.119}$$

After applying the exterior derivative and again inserting (A.107) we obtain

$$\mathrm{d}\,\mathcal{B} = \frac{1}{g_1 g_2 g_3}\left[\frac{\partial(g_2 g_3 B_u)}{\partial u} + \frac{\partial(g_3 g_1 B_v)}{\partial v} + \frac{\partial(g_1 g_2 B_w)}{\partial w}\right] s_1 \wedge s_2 \wedge s_3\,. \tag{A.120}$$

In conventional vector analysis the exterior derivative of a two-form corresponds to the divergence operator:

$$\operatorname{div} \boldsymbol{B} = \frac{1}{g_1 g_2 g_3}\left[\frac{\partial(g_2 g_3 B_u)}{\partial u} + \frac{\partial(g_3 g_1 B_v)}{\partial v} + \frac{\partial(g_1 g_2 B_w)}{\partial w}\right]. \tag{A.121}$$

Let us now apply the Laplace operator (3.14) to a scalar field f. From (3.15) we obtain

$$\Delta f = \star\, \mathrm{d} \star \mathrm{d}f\,. \tag{A.122}$$

We start with

$$\mathrm{d}f = \frac{\partial f}{\partial u}\,\mathrm{d}u + \frac{\partial f}{\partial v}\,\mathrm{d}v + \frac{\partial f}{\partial w}\,\mathrm{d}w\,. \tag{A.123}$$

Inserting (A.107) yields

$$\mathrm{d}f = \frac{1}{g_1}\frac{\partial f}{\partial u}\, s_1 + \frac{1}{g_2}\frac{\partial f}{\partial v}\, s_2 + \frac{1}{g_3}\frac{\partial f}{\partial w}\, s_3\,. \tag{A.124}$$

Applying the star operator (A.109) gives

$$\star\, \mathrm{d}f = \frac{1}{g_1}\frac{\partial f}{\partial u}\, s_2 \wedge s_3 + \frac{1}{g_2}\frac{\partial f}{\partial v}\, s_3 \wedge s_1 + \frac{1}{g_3}\frac{\partial f}{\partial w}\, s_1 \wedge s_2\,, \tag{A.125}$$

and with (A.107) we obtain

$$\star\, \mathrm{d}f = \frac{g_2 g_3}{g_1}\frac{\partial f}{\partial u}\,\mathrm{d}v \wedge \mathrm{d}w + \frac{g_3 g_1}{g_2}\frac{\partial f}{\partial v}\,\mathrm{d}w \wedge \mathrm{d}u + \frac{g_1 g_2}{g_3}\frac{\partial f}{\partial w}\,\mathrm{d}u \wedge \mathrm{d}v\,. \tag{A.126}$$

We now compute again the exterior derivative

$$\begin{aligned}\mathrm{d}\star \mathrm{d}f = \Bigg[&\frac{\partial}{\partial u}\left(\frac{g_2 g_3}{g_1}\frac{\partial f}{\partial u}\right) + \frac{\partial}{\partial v}\left(\frac{g_3 g_1}{g_2}\frac{\partial f}{\partial v}\right) + \\ &+ \frac{\partial}{\partial w}\left(\frac{g_1 g_2}{g_3}\frac{\partial f}{\partial w}\right)\Bigg]\,\mathrm{d}u \wedge \mathrm{d}v \wedge \mathrm{d}w\,,\end{aligned} \tag{A.127}$$

and obtain with (A.107)

$$\mathrm{d}\star\mathrm{d}f = \frac{1}{g_1 g_2 g_3}\left[\frac{\partial}{\partial u}\left(\frac{g_2 g_3}{g_1}\frac{\partial f}{\partial u}\right) + \frac{\partial}{\partial v}\left(\frac{g_3 g_1}{g_2}\frac{\partial f}{\partial v}\right) + \right. \\ \left. + \frac{\partial}{\partial w}\left(\frac{g_1 g_2}{g_3}\frac{\partial f}{\partial w}\right)\right] s_1 \wedge s_2 \wedge s_3 \,. \tag{A.128}$$

Using (A.122) and (A.109) yields

$$\Delta f = \frac{1}{g_1 g_2 g_3}\left[\frac{\partial}{\partial u}\left(\frac{g_2 g_3}{g_1}\frac{\partial f}{\partial u}\right) + \frac{\partial}{\partial v}\left(\frac{g_3 g_1}{g_2}\frac{\partial f}{\partial v}\right) + \right. \\ \left. + \frac{\partial}{\partial w}\left(\frac{g_1 g_2}{g_3}\frac{\partial f}{\partial w}\right)\right]. \tag{A.129}$$

A.4.1 Circular Cylindrical Coordinates

To deal with circular cylindrical electromagnetic structures we introduce *circular cylindrical coordinates*. Figure A.6 shows the circular cylindrical coordinates. The coordinates r, φ, z are defined in the following intervals:

$$\begin{aligned} u &= r, & 0 &\leq r < \infty, \\ v &= \varphi, & 0 &\leq \varphi < 2\pi, \\ w &= z, & -\infty &< z < +\infty. \end{aligned} \tag{A.130}$$

The circular cylindric coordinates r, φ, z are related to the Cartesian coordinates x, y, z via

$$\begin{aligned} x &= r\cos\varphi\,, \\ y &= r\sin\varphi\,, \\ z &= z \end{aligned} \tag{A.131}$$

and

$$\begin{aligned} r &= \sqrt{x^2 + y^2}\,, \\ \varphi &= \arctan\frac{y}{x}\,, \\ z &= z\,. \end{aligned} \tag{A.132}$$

Inserting (A.131) into (A.102) yields the metric coefficients of the circular cylindrical coordinate system:

$$g_1 = 1\,, \quad g_2 = r\,, \quad g_3 = 1 \tag{A.133}$$

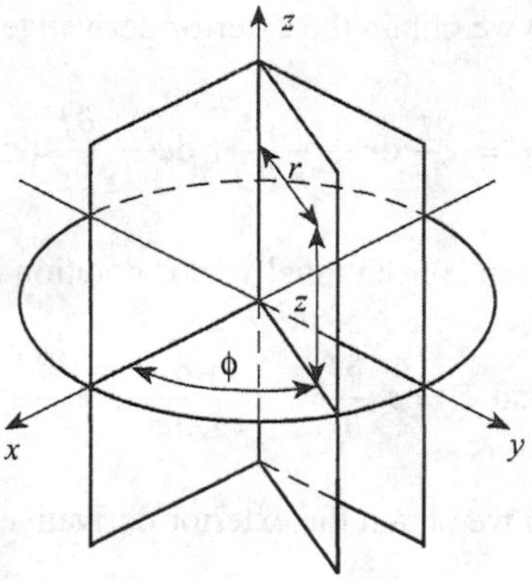

Figure A.6: Circular cylindric coordinates.

and the unit forms

$$\begin{aligned} s_1 &= \mathrm{d}r\,, \\ s_2 &= r\,\mathrm{d}\varphi\,, \\ s_3 &= \mathrm{d}z\,. \end{aligned} \tag{A.134}$$

The components of the p-forms are referred to the unit forms and given by

$$\begin{aligned} &\text{zero-form:} && f(\boldsymbol{x})\,, \\ &\text{one-form:} && \mathcal{E}(\boldsymbol{x}) = E_r\,\mathrm{d}r + E_\varphi\,r\,\mathrm{d}\varphi + E_z\,\mathrm{d}z\,, \\ &\text{two-form:} && \mathcal{D}(\boldsymbol{x}) = D_r\,r\,\mathrm{d}\varphi \wedge \mathrm{d}z + D_\varphi\,\mathrm{d}z \wedge \mathrm{d}r + D_z\,r\,\mathrm{d}r \wedge \mathrm{d}\varphi\,, \\ &\text{three-form:} && \mathcal{Q}(\boldsymbol{x}) = \rho\,r\,\mathrm{d}r \wedge \mathrm{d}\varphi \wedge \mathrm{d}z\,. \end{aligned} \tag{A.135}$$

From (A.109) and (A.134) we obtain

$$\star f = f\,r\,\mathrm{d}r \wedge \mathrm{d}\varphi \wedge \mathrm{d}z\,, \tag{A.136}$$
$$\star(A_r\,\mathrm{d}r + A_\varphi\,r\,\mathrm{d}\varphi + A_z\,\mathrm{d}z) = A_r\,r\,\mathrm{d}\varphi \wedge \mathrm{d}z + A_\varphi\,\mathrm{d}z \wedge \mathrm{d}r + A_z\,r\,\mathrm{d}r \wedge \mathrm{d}\varphi\,.$$

The length $\mathrm{d}s$ of a path element is given by

$$\mathrm{d}s = \sqrt{(\mathrm{d}r)^2 + r^2(\mathrm{d}\varphi)^2 + \mathrm{d}z^2}\,. \tag{A.137}$$

In circular cylindrical coordinates the exterior differential operator is

$$\mathrm{d}\mathcal{A} = \mathrm{d}r \wedge \frac{\partial \mathcal{A}}{\partial r} + \mathrm{d}\varphi \wedge \frac{\partial \mathcal{A}}{\partial \varphi} + \mathrm{d}z \wedge \frac{\partial \mathcal{A}}{\partial z}\,. \tag{A.138}$$

From (A.111) and (A.133) we obtain the exterior derivative of a zero-form f

$$\mathrm{d}f = \frac{\partial f}{\partial r}\,\mathrm{d}r + \frac{1}{r}\frac{\partial f}{\partial \varphi}\,r\,\mathrm{d}\varphi + \frac{\partial f}{\partial z}\,\mathrm{d}z\,. \tag{A.139}$$

The corresponding relation in conventional vector notation is

$$\operatorname{grad} f = \boldsymbol{e}_r \frac{\partial f}{\partial r} + \boldsymbol{e}_\varphi \frac{1}{r}\frac{\partial f}{\partial \varphi} + \boldsymbol{e}_z \frac{\partial f}{\partial z}\,. \tag{A.140}$$

From (A.116) and (A.133) we obtain the exterior derivative of a one-form

$$\begin{aligned}\mathrm{d}\mathcal{A} = {} & \frac{1}{r}\left(\frac{\partial A_z}{\partial \varphi} - \frac{\partial (rA_\varphi)}{\partial z}\right) r\,\mathrm{d}\varphi \wedge \mathrm{d}z + \left(\frac{\partial A_r}{\partial z} - \frac{\partial A_z}{\partial r}\right)\mathrm{d}z \wedge \mathrm{d}r \\ & + \frac{1}{r}\left(\frac{\partial (rA_\varphi)}{\partial r} - \frac{\partial A_r}{\partial \varphi}\right) r\,\mathrm{d}r \wedge \mathrm{d}\varphi\,. \end{aligned} \tag{A.141}$$

In conventional vector analysis the exterior derivative of a one-form corresponds to the curl operator:

$$\begin{aligned}\operatorname{curl} \boldsymbol{A} = {} & \frac{1}{r}\left(\frac{\partial A_z}{\partial \varphi} - \frac{\partial (rA_\varphi)}{\partial z}\right)\boldsymbol{e}_r + \left(\frac{\partial A_r}{\partial z} - \frac{\partial A_z}{\partial r}\right)\boldsymbol{e}_\varphi \\ & + \frac{1}{r}\left(\frac{\partial (rA_\varphi)}{\partial r} - \frac{\partial A_r}{\partial \varphi}\right)\boldsymbol{e}_z\,. \end{aligned} \tag{A.142}$$

From (A.120) and (A.133) we obtain the derivative of a two-form $\mathcal{D}$

$$\mathrm{d}\mathcal{D} = \frac{1}{r}\left[\frac{\partial (rD_r)}{\partial r} + \frac{\partial D_\varphi}{\partial \varphi} + \frac{\partial (rD_z)}{\partial z}\right] r\,\mathrm{d}r \wedge \mathrm{d}\varphi \wedge \mathrm{d}z\,. \tag{A.143}$$

In conventional vector analysis the exterior derivative of a two-form corresponds to the divergence:

$$\operatorname{div} \boldsymbol{D} = \frac{1}{r}\left[\frac{\partial (rD_r)}{\partial r} + \frac{\partial D_\varphi}{\partial \varphi} + \frac{\partial (rD_z)}{\partial z}\right]. \tag{A.144}$$

From (A.129) and (A.133) we determine that the scalar Laplace operator in circular cylindric coordinates

$$\Delta f = \frac{\partial^2 f}{\partial r^2} + \frac{1}{r}\frac{\partial f}{\partial r} + \frac{1}{r^2}\frac{\partial^2 f}{\partial \varphi^2} + \frac{\partial^2 f}{\partial z^2}\,. \tag{A.145}$$

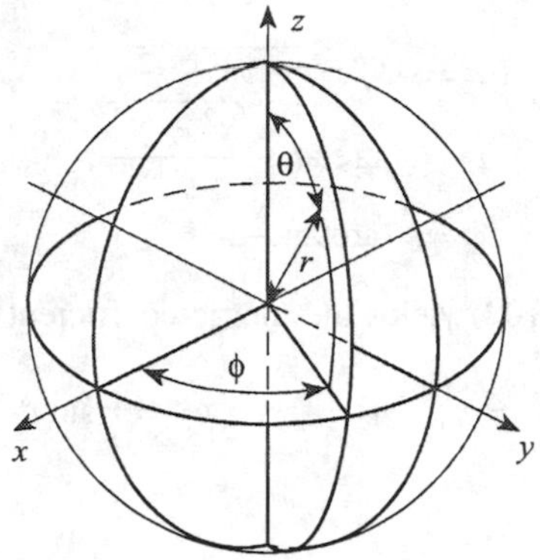

Figure A.7: Spherical coordinates.

In circular cylindric coordinates the Laplace operator applied to a one-form $\mathcal{A}$ yields

$$\begin{aligned} \Delta\mathcal{A} = \mathrm{d}r & \left(\Delta A_r - \frac{1}{r^2}A_r - \frac{2}{r^2}\frac{\partial A_\varphi}{\partial\varphi}\right) \\ & + r\,\mathrm{d}\varphi\left(\Delta A_\varphi - \frac{1}{r^2}A_\varphi + \frac{2}{r^2}\frac{\partial A_r}{\partial\varphi}\right) + \mathrm{d}z\,\Delta A_z\,. \end{aligned} \tag{A.146}$$

The Laplace operator on the right-hand side of this differential equation is the scalar Laplace operator as defined in (A.145). If the one-form $\mathcal{A}$ exhibits only a z component, the explicit Laplace operator for the one-form reduces to the explicit Laplace operator for the zero-form.

A.4.2 Spherical Coordinates

The analysis of electromagnetic structures with spherical symmetry usually is performed in *spherical coordinates*. Figure A.7 shows the spherical coordinates. The coordinates r, φ, ϑ are defined in the following intervals:

$$\begin{aligned} u &= r, & 0 &\leq r < \infty, \\ v &= \vartheta, & 0 &\leq \vartheta \leq \pi, \\ w &= \varphi, & 0 &\leq \varphi < 2\pi, \end{aligned} \tag{A.147}$$

The spherical coordinates r, ϑ, φ are related to the Cartesian coordinates x, y, z via

$$\begin{aligned} x &= r\sin\vartheta\cos\varphi\,, \\ y &= r\sin\vartheta\sin\varphi\,, \\ z &= r\cos\vartheta \end{aligned} \tag{A.148}$$

and

$$
\begin{aligned}
r &= \sqrt{x^2 + y^2 + z^2}\,, \\
\vartheta &= \arctan \frac{\sqrt{x^2 + y^2}}{z}\,, \\
\varphi &= \arctan \frac{y}{x}\,.
\end{aligned}
\tag{A.149}
$$

Inserting (A.148) into (A.102) yields the metric coefficients of the spherical coordinate system:

$$
g_1 = 1\,, \quad g_2 = r\,, \quad g_3 = r \sin \vartheta \tag{A.150}
$$

and the unit differential forms

$$
\begin{aligned}
s_1 &= \mathrm{d}r\,, \\
s_2 &= r\,\mathrm{d}\vartheta\,, \\
s_3 &= r \sin \vartheta\,\mathrm{d}\varphi\,.
\end{aligned}
\tag{A.151}
$$

The components of the p-forms are referred to the unit forms and given by

$$
\begin{aligned}
\text{zero-form:} \quad & f(\boldsymbol{x})\,, \\
\text{one-form:} \quad & \mathcal{E}(\boldsymbol{x}) = E_r\,\mathrm{d}r + E_\vartheta\, r\,\mathrm{d}\vartheta + E_\varphi\, r \sin \vartheta\,\mathrm{d}\varphi\,, \\
\text{two-form:} \quad & \mathcal{D}(\boldsymbol{x}) = D_r\, r^2 \sin \vartheta\,\mathrm{d}\vartheta \wedge \mathrm{d}\varphi + D_\vartheta\, r \sin \vartheta\,\mathrm{d}\varphi \wedge \mathrm{d}r + D_\varphi\, r\,\mathrm{d}r \wedge \mathrm{d}\vartheta\,, \\
\text{three-form:} \quad & \mathcal{Q}(\boldsymbol{x}) = \rho\, r^2 \sin \vartheta\,\mathrm{d}r \wedge \mathrm{d}\vartheta \wedge \mathrm{d}\varphi\,.
\end{aligned}
\tag{A.152}
$$

From (A.109) and (A.151) we obtain

$$
\begin{aligned}
\star f &= f\, r^2 \sin \vartheta\,\mathrm{d}r \wedge \mathrm{d}\vartheta \wedge \mathrm{d}\varphi\,, \\
\star\,(A_r\,\mathrm{d}r + A_\vartheta\, r\,\mathrm{d}\vartheta + A_\varphi\, r \sin \vartheta\,\mathrm{d}\varphi) &= A_r\, r^2 \sin \vartheta\,\mathrm{d}\vartheta \wedge \mathrm{d}\varphi + A_\vartheta\, r \sin \vartheta\,\mathrm{d}\varphi \wedge \mathrm{d}r \\
&\quad + A_\varphi\, r\,\mathrm{d}r \wedge \mathrm{d}\vartheta\,.
\end{aligned}
\tag{A.153}
$$

The length $\mathrm{d}s$ of a path element is given by

$$
\mathrm{d}s = \sqrt{(\mathrm{d}r)^2 + r^2(\mathrm{d}\vartheta)^2 + r^2 \sin^2 \vartheta (\mathrm{d}\varphi)^2}\,. \tag{A.154}
$$

In spherical coordinates the exterior differential operator is

$$
\mathrm{d}\mathcal{A} = \mathrm{d}r \wedge \frac{\partial \mathcal{A}}{\partial r} + \mathrm{d}\vartheta \wedge \frac{\partial \mathcal{A}}{\partial \vartheta} + \mathrm{d}\varphi \wedge \frac{\partial \mathcal{A}}{\partial \varphi}\,. \tag{A.155}
$$

From (A.111) and (A.151) we obtain the exterior derivative of a zero-form f is

$$
\mathrm{d}f = \frac{\partial f}{\partial r}\,\mathrm{d}r + \frac{1}{r}\frac{\partial f}{\partial z}\,\mathrm{d}\vartheta + \frac{1}{r \sin \vartheta}\frac{\partial f}{\partial \varphi}\, r \sin \vartheta\,\mathrm{d}\varphi\,. \tag{A.156}
$$

The corresponding relation in conventional vector notation is

$$\operatorname{grad} f = \boldsymbol{e}_r \frac{\partial f}{\partial r} + \boldsymbol{e}_\vartheta \frac{1}{r}\frac{\partial f}{\partial \vartheta} + \boldsymbol{e}_\varphi \frac{1}{r \sin\vartheta}\frac{\partial f}{\partial \varphi}, \tag{A.157}$$

From (A.115) and (A.151) we obtain the exterior derivative of a one-form

$$\begin{aligned} \mathrm{d}\mathcal{A} = & \frac{1}{r^2 \sin\vartheta}\left(\frac{\partial(r\sin\vartheta A_\varphi)}{\partial\vartheta} - \frac{\partial(r A_\vartheta)}{\partial\varphi}\right) r^2 \sin\vartheta \,\mathrm{d}\vartheta \wedge \mathrm{d}\varphi \\ & + \frac{1}{r\sin\vartheta}\left(\frac{\partial A_r}{\partial\varphi} - \frac{\partial(r\sin\vartheta A_\varphi)}{\partial r}\right) r\sin\vartheta \,\mathrm{d}\varphi \wedge \mathrm{d}r \\ & + \frac{1}{r}\left(\frac{\partial(r A_\vartheta)}{\partial r} - \frac{\partial A_r}{\partial\vartheta}\right) r\,\mathrm{d}r \wedge \mathrm{d}\vartheta \,. \end{aligned} \tag{A.158}$$

In conventional vector analysis the exterior derivative of a one-form corresponds to the curl operator:

$$\begin{aligned} \operatorname{curl} \boldsymbol{A} = & \frac{1}{r^2 \sin\vartheta}\left(\frac{\partial(r\sin\vartheta A_\varphi)}{\partial\vartheta} - \frac{\partial(r A_\vartheta)}{\partial\varphi}\right) \boldsymbol{e}_r \\ & + \frac{1}{r\sin\vartheta}\left(\frac{\partial A_r}{\partial\varphi} - \frac{\partial(r\sin\vartheta A_\varphi)}{\partial r}\right) \boldsymbol{e}_\vartheta \\ & + \frac{1}{r}\left(\frac{\partial(r A_\vartheta)}{\partial r} - \frac{\partial A_r}{\partial\vartheta}\right) \boldsymbol{e}_\varphi \,. \end{aligned} \tag{A.159}$$

From (A.116) and (A.151) we obtain the derivative of a two-form $\mathcal{B}$

$$\mathrm{d}\mathcal{B} = \frac{1}{r^2\sin\vartheta}\left[\frac{\partial(r^2\sin\vartheta B_r)}{\partial r} + \frac{\partial(r\sin\vartheta B_\vartheta)}{\partial\vartheta} + \frac{\partial(r B_\varphi)}{\partial\varphi}\right] r^2\sin\vartheta\,\mathrm{d}r\wedge\mathrm{d}\vartheta\wedge\mathrm{d}\varphi \,. \tag{A.160}$$

In conventional vector analysis the exterior derivative of a two-form corresponds to the divergence operator:

$$\operatorname{div} \boldsymbol{B} = \frac{1}{r^2\sin\vartheta}\left[\frac{\partial(r^2\sin\vartheta B_r)}{\partial r} + \frac{\partial(r\sin\vartheta B_\vartheta)}{\partial\vartheta} + \frac{\partial(r B_\varphi)}{\partial\varphi}\right] . \tag{A.161}$$

From (A.129) and (A.151) we obtain the scalar Laplace operator for a zero-form f

$$\Delta f = \frac{1}{r^2}\frac{\partial}{\partial r}\left(r^2 \frac{\partial f}{\partial r}\right) + \frac{1}{r^2\sin\vartheta}\frac{\partial}{\partial\vartheta}\left(\sin\vartheta\frac{\partial f}{\partial\vartheta}\right) + \frac{1}{r^2\sin^2\vartheta}\frac{\partial^2 f}{\partial\varphi^2} \,. \tag{A.162}$$

The Laplace operator for a one-form $\mathcal{A}$ is given by

$$\begin{aligned}\Delta\mathcal{A} = \mathrm{d}r &\left(\Delta A_r - \frac{2}{r^2}A_r - \frac{2}{r^2}\frac{\partial A_\vartheta}{\partial\vartheta} - \frac{2\cot\vartheta}{r^2}A_\vartheta - \frac{2}{r^2\sin\vartheta}\frac{\partial A_\varphi}{\partial\varphi}\right)\\ + r\,\mathrm{d}\vartheta &\left(\frac{2}{r^2}\frac{\partial A_r}{\partial\vartheta} + \Delta A_\vartheta - \frac{1}{r^2\sin^2\vartheta}A_\vartheta - \frac{2}{r^2}\frac{\cot\vartheta}{\sin\vartheta}\frac{\partial A_\varphi}{\partial\varphi}\right)\\ + r\sin\vartheta\,\mathrm{d}\varphi &\left(\frac{2}{r^2\sin\vartheta}\frac{\partial A_r}{\partial\varphi} + \frac{2}{r^2}\frac{\cot\vartheta}{\sin\vartheta}\frac{\partial A_\vartheta}{\partial\varphi} + \Delta A_\varphi - \frac{1}{r^2\sin^2\vartheta}A_\varphi\right).\end{aligned} \tag{A.163}$$

The Laplace operator on the right-hand side of this differential equation is the scalar Laplace operator as defined in (A.162).

In spherical coordinates an important class of solutions of the *vector Helmholtz equation*

$$\Delta\mathcal{A} + k^2\,\mathcal{A} = 0 \tag{A.164}$$

can be obtained via the solution of the *scalar Helmholtz equation*

$$\Delta f + k^2 f = 0 \tag{A.165}$$

by a method given in [10]. If f is a solution of the scalar Helmholtz equation (A.164) then

$$\mathcal{A} = \star\ \mathrm{d}r \wedge \mathrm{d}f \tag{A.166}$$

is a solution of the vector Helmholtz equation (A.165). The proof is given by inserting

$$\mathcal{A} = \star\ \mathrm{d}r \wedge \mathrm{d}f = \frac{1}{r}\frac{\partial f}{\partial\vartheta}r\sin\vartheta\,\mathrm{d}\varphi - \frac{1}{r\sin\vartheta}\frac{\partial f}{\partial\varphi}r\,\mathrm{d}\vartheta \tag{A.167}$$

into (A.163) and the resulting expression into (A.164).

A.4.3 Twisted Forms

To describe a surface and its neighborhood we introduce a right-handed curvilinear orthogonal coordinate system u, v, n, where u, v are coordinates tangential to the surface and the coordinate n is normal to the surface. The tangential unit one-forms are s_1 and s_2 and the normal unit one-form is n are defined as in (A.107).

To transform an ordinary one-form into a twisted one-form and vice versa we introduce the *twist operator* by

$$\perp_n \mathcal{U} = \star(n \wedge \mathcal{U})\ . \tag{A.168}$$

The index n of the twist operator $\perp_n$ denotes the axis of rotation. If u, v are the coordinates tangential to the surface and n is the coordinate normal to the surface, the application of the twist operator $\perp_n$ to a one-form tangential to the surface rotates the one-form around n by 90° in the positive direction,

$$\perp_n (U_u s_1 + U_v s_2) = -U_v s_1 + U_u s_2 \,, \tag{A.169}$$

$$\perp_n (U_v s_1 - U_u s_2) = U_u s_1 + U_v s_2 \,. \tag{A.170}$$

A.4.4 Integration of Differential Forms by Pullback

In contrast to the integration of vector fields the integration of differential forms requires no metric. The integration of differentials can be done simply by the *method of pullback* [11]. If a path integral

$$\int_P \mathcal{A} \tag{A.171}$$

over the path P has to be evaluated, and the path P is parametrized by

$$x = p_1(t) \,, \quad y = p_2(t) \,, \quad z = p_3(t) \tag{A.172}$$

for $t_1 \leq t \leq t_2$, we introduce the *pullback* of $\mathcal{A}$ to the path P, defined by

$$\begin{aligned} P^{\triangleleft}\mathcal{A} &= P^{\triangleleft}(A_x\,\mathrm{d}x + A_y\,\mathrm{d}y + A_z\,\mathrm{d}z) \\ &= A_x(p_1, p_2, p_3)\,\mathrm{d}p_1 + A_y(p_1, p_2, p_3)\,\mathrm{d}p_2 + A_z(p_1, p_2, p_3)\,\mathrm{d}p_1 \\ &= \left(A_x(p_1, p_2, p_3)\frac{\partial p_1}{\partial t} + A_y(p_1, p_2, p_3)\frac{\partial p_2}{\partial t} \right. \\ &\left. + A_z(p_1, p_2, p_3)\frac{\partial p_3}{\partial t} \right) \mathrm{d}t \,. \end{aligned} \tag{A.173}$$

With the pullback of $\mathcal{A}$ we convert the integral over the path P to an integral in t over the interval $t_1 \leq t \leq t_2$

$$\int_P \mathcal{A} = \int_{t_1}^{t_2} P^{\triangleleft}\mathcal{A} \,. \tag{A.174}$$

A.5 DOUBLE DIFFERENTIAL FORMS

In vector analysis a *dyadic* defines a linear mapping of vectors [12–14]. In vector calculus a dyadic is a the formal sum of a finite number of *dyads*, a dyad being a pair of vectors [15]. In differential form calculus dyadics may be represented by *double forms* [11, 16]. A *double one-form* $\mathcal{G}$ is defined by

$$\begin{aligned} \mathcal{G} &= G_{11}\,\mathrm{d}x\,\mathrm{d}x' + G_{12}\,\mathrm{d}x\,\mathrm{d}y' + G_{13}\,\mathrm{d}x\,\mathrm{d}z' \\ &+ G_{21}\,\mathrm{d}y\,\mathrm{d}x' + G_{22}\,\mathrm{d}y\,\mathrm{d}y' + G_{23}\,\mathrm{d}y\,\mathrm{d}z' \\ &+ G_{31}\,\mathrm{d}z\,\mathrm{d}x' + G_{32}\,\mathrm{d}z\,\mathrm{d}y' + G_{33}\,\mathrm{d}z\,\mathrm{d}z' \,. \end{aligned} \tag{A.175}$$

The Green's double form relates the observation space $\boldsymbol{x}$ to the source space $\boldsymbol{x}'$. Primed and unprimed differentials $\mathrm{d}x_i'$ and $\mathrm{d}x_j$ commute, i.e., in products they may be interchanged without changing the sign. The rules are

$$\mathrm{d}x_i\,\mathrm{d}x_j' = \mathrm{d}x_i'\,\mathrm{d}x_j \quad \text{with} \quad \mathrm{d}x_i = \mathrm{d}x,\ \mathrm{d}y,\ \mathrm{d}z\ . \tag{A.176}$$

A one-form $\mathcal{A}$ is mapped into a one-form $\mathcal{B}$ and a two-form $\mathcal{C}$ is mapped into a two-form $\mathcal{D}$ via

$$\mathcal{B}(\boldsymbol{x}) = \int' \mathcal{G}(\boldsymbol{x}, \boldsymbol{x}') \wedge \star\, \mathcal{A}(\boldsymbol{x}')\ , \tag{A.177a}$$

$$\mathcal{D}(\boldsymbol{x}) = \star \int' \mathcal{G}(\boldsymbol{x}, \boldsymbol{x}') \wedge \mathcal{C}(\boldsymbol{x}')\ . \tag{A.177b}$$

In these equations $\mathcal{G}(\boldsymbol{x}, \boldsymbol{x}')$ is called the *kernel* of integration. The primed integration symbol denotes that the integration is performed over the primed variables. For the integration the unprimed differentials are treated as constants.

With the *identity kernel*, defined by

$$\mathcal{I}(\boldsymbol{x}, \boldsymbol{x}') = \delta(\boldsymbol{x} - \boldsymbol{x}')\,(\,\mathrm{d}x\,\mathrm{d}x' + \mathrm{d}y\,\mathrm{d}y' + \mathrm{d}z\,\mathrm{d}z') \tag{A.178}$$

any one-form $\mathcal{A}$ and any two-form $\mathcal{C}$ respectively is mapped in itself. We obtain

$$\int' \mathcal{I}(\boldsymbol{x}, \boldsymbol{x}') \wedge \star\, \mathcal{A}(\boldsymbol{x}') = \mathcal{A}(\boldsymbol{x})\ , \tag{A.179a}$$

$$\star \int' \mathcal{I}(\boldsymbol{x}, \boldsymbol{x}') \wedge \mathcal{C}(\boldsymbol{x}') = \mathcal{C}(\boldsymbol{x})\ . \tag{A.179b}$$

A.6 RELATIONS BETWEEN EXTERIOR CALCULUS AND CONVENTIONAL VECTOR NOTATION

A.6.1 Differential Operators

In conventional vector notation three differential operators are defined. The *gradient operator* applied to a scalar yields a vector

$$\operatorname{grad} f = \begin{bmatrix} \frac{\partial f}{\partial x} \\ \frac{\partial f}{\partial y} \\ \frac{\partial f}{\partial z} \end{bmatrix}\ . \tag{A.180}$$

Table A.1: Differential Operators

Vector Differential Operator	*Exterior Differential Operator*
$\text{grad } f$	$\mathrm{d}f$
$\text{curl } \boldsymbol{A}$	$\mathrm{d}\mathcal{A}$
$\text{div } \boldsymbol{B}$	$\mathrm{d}\mathcal{B}$
$\text{curl grad } f = \boldsymbol{0}$	$\mathrm{d}\,\mathrm{d}f = 0$
$\text{div curl } \boldsymbol{A} = 0$	$\mathrm{d}\,\mathrm{d}\mathcal{A} = 0$
$\text{grad div } f$	$\mathrm{d}\star\mathrm{d}f \quad \text{or} \quad \star\,\mathrm{d}\star\mathrm{d}f$
$\text{curl curl } \boldsymbol{A}$	$\mathrm{d}\star\mathrm{d}\mathcal{A} \quad \text{or} \quad \star\,\mathrm{d}\star\mathrm{d}\mathcal{A}$
$\Delta f = \text{div grad } f$	$\Delta f = \star\,\mathrm{d}\star\mathrm{d}f$
$\Delta \boldsymbol{A} = \text{grad div } \boldsymbol{A} - \text{curl curl } \boldsymbol{A}$	$\Delta\mathcal{A} = \mathrm{d}\star\mathrm{d}\star\mathcal{A} - \star\,\mathrm{d}\star\mathrm{d}\mathcal{A}$
$\Delta \boldsymbol{B} = \text{grad div } \boldsymbol{B} - \text{curl curl } \boldsymbol{B}$	$\Delta\mathcal{B} = \star\,\mathrm{d}\star\mathrm{d}\mathcal{B} - \mathrm{d}\star\mathrm{d}\star\mathcal{B}$
$\text{grad}(f\,g) = g \text{ grad } f + f \text{ grad } g$	$g\ \mathrm{d}(gf) = \mathrm{d}f + f\,\mathrm{d}g$
$\text{curl}\,(f\boldsymbol{A}) = \text{grad } f \times \boldsymbol{A} + f \text{ curl } \boldsymbol{A}$	$\mathrm{d}\,(f\,\mathcal{A}) = \mathrm{d}f \wedge \mathcal{A} + f\ \mathrm{d}\mathcal{A}$
$\text{div}\,(\boldsymbol{A} \times \boldsymbol{B}) = \text{curl } \boldsymbol{A} \cdot \boldsymbol{B} - \boldsymbol{A} \cdot \text{curl } \boldsymbol{B}$	$\mathrm{d}\,(\mathcal{A} \wedge \mathcal{B}) = \mathrm{d}\mathcal{A} \wedge \mathcal{B} - \mathcal{A} \wedge \mathrm{d}\mathcal{B}$
$\text{div}\,(f\,\boldsymbol{C}) = \text{grad } f \cdot \boldsymbol{C} + f \cdot \text{div } \boldsymbol{C}$	$\mathrm{d}\,(f \wedge \mathcal{C}) = \mathrm{d}f \wedge \mathcal{C} + f\,\mathrm{d}\mathcal{C}$

Notes: $f = [f]^{(0)}$, $g = [g]^{(0)}$, $\mathcal{A} = [\boldsymbol{A}]^{(1)}$, $\mathcal{B} = [\boldsymbol{B}]^{(1)}$, $\mathcal{C} = [\boldsymbol{C}]^{(2)}$

Table A.2: Maxwell's Equations

Equation	*Vector Notation*	*Differential Form Notation*
Ampère's law	$\text{curl } \boldsymbol{H} = \frac{\partial \boldsymbol{D}}{\partial t} + \boldsymbol{J}$	$\mathrm{d}\mathcal{H} = \frac{\partial \mathcal{D}}{\partial t} + \mathcal{J}$
Faraday's law	$\text{curl } \boldsymbol{E} = -\frac{\partial \boldsymbol{B}}{\partial t}$	$\mathrm{d}\mathcal{E} = -\frac{\partial \mathcal{B}}{\partial t}$
Gauss's law	$\text{div } \boldsymbol{D} = \rho$	$\mathrm{d}\mathcal{D} = \mathcal{Q}$
Magnetic flux continuity	$\text{div } \boldsymbol{B} = 0$	$\mathrm{d}\mathcal{B} = 0$
Electric constitutive equation	$\boldsymbol{D} = \varepsilon\,\boldsymbol{E}$	$\mathcal{D} = \star\,\varepsilon\,\mathcal{E}$
Magnetic constitutive equation	$\boldsymbol{B} = \mu\,\boldsymbol{H}$	$\mathcal{B} = \star\,\mu\,\mathcal{H}$
Ampère's law	$\text{curl } \underline{\boldsymbol{H}} = \mathrm{j}\omega\underline{\varepsilon}\underline{\boldsymbol{E}} + \underline{\boldsymbol{J_0}}$	$\mathrm{d}\mathcal{H} = \star\mathrm{j}\omega\underline{\varepsilon}\underline{\mathcal{E}} + \underline{\mathcal{J}}$
Faraday's law	$\text{curl } \underline{\boldsymbol{E}} = -\mathrm{j}\omega\underline{\mu}\underline{\boldsymbol{H}}$	$\mathrm{d}\mathcal{E} = -\star\mathrm{j}\omega\underline{\mu}\underline{\mathcal{H}}$

The *curl operator* applied to a vector $\boldsymbol{U}$ yields the pseudovector

$$\operatorname{curl}\boldsymbol{U} = \begin{bmatrix} \frac{\partial U_z}{\partial y} - \frac{\partial U_y}{\partial z} \\ \frac{\partial U_x}{\partial z} - \frac{\partial U_z}{\partial x} \\ \frac{\partial U_y}{\partial x} - \frac{\partial U_x}{\partial y} \end{bmatrix} . \tag{A.181}$$

The *divergence operator* applied to a pseudovector $\boldsymbol{V}$ yields the pseudoscalar

$$\operatorname{div}\boldsymbol{U} = \frac{\partial U_x}{\partial x} + \frac{\partial U_y}{\partial y} + \frac{\partial U_z}{\partial z} . \tag{A.182}$$

Table A.1 shows the correspondences between the differential operators in conventional vector notation and exterior differential form notation.

A.6.2 Maxwell's Equations

Table A.2 shows the correspondences between Maxwell's equations in vector notation and differential form notation.

REFERENCES

[1] G. Arfken, *Mathematical Methods for Physicists.* New York: Academic Press, 1985.

[2] C. Wong, *Introduction to Mathematical Physics.* Oxford: Oxford University Press, 1991.

[3] H. Flanders, *Differential Forms.* New York: Academic Press, 1963.

[4] W. L. Burke, *Applied Differential Geometry.* Cambridge: Cambridge University Press, 1985.

[5] P. Bamberg and S. Sternberg, *A Course in Mathematics for Students in Physics 2.* Cambridge: Cambridge University Press, 1990.

[6] T. Frankel, *The Geometry of Physics.* Cambridge: Cambridge University Press, 1997.

[7] S. Weintraub, *Differential Forms - A Complement to Vector Calculus.* New York: Academic Press, 1997.

[8] K. F. Warnick, R. Selfridge, and D. Arnold, "Electromagnetic boundary conditions and differential forms," *IEE Proc. Microwaves, Antennas, Propagat.*, vol. 142, pp. 326–332, Aug. 1995.

[9] P. Moon and D. Spencer, *Field Theory Handbook.* Berlin: Springer, 1991.

[10] R. S. Elliott, *Electromagnetics - History, Theory, and Applications.* New York: IEEE Press, 1991.

[11] K. F. Warnick and D. Arnold, "Electromagnetic Green functions using differential forms," *J. Electromagn. Waves and Appl.*, vol. 10, no. 3, pp. 427–438, 1996.

[12] R. E. Collin, *Field Theory of Guided Waves.* New York: IEEE Press, 1991.

[13] C.-T. Tai, *Generalized Vector and Dyadic Analysis.* New York: IEEE Press, 1992.

[14] C.-T. Tai, *Dyadic Green Functions in Electromagnetic Theory.* New York: IEEE Press, 1993.

[15] R. Abraham, J. E. Marsden, and T. Ratiu, *Manifolds, Tensor Analysis and Applications.* London: Addison-Wesley, 1983.

[16] G. de Rham, *Differentiable Manifolds.* New York: Springer, 1984.

Appendix B

Special Functions

For a number of coordinate systems the Laplace equation, the Helmholtz equation and the wave equation may be solved exactly by separation into ordinary differential equations. As the solutions of these ordinary differential equations special functions occur. In this chapter special functions for circular cylindrical and spherical coordinate systems and some important formulae are summarized. For a detailed presentation of the mathematical background see, for example, [1,2]. A comprehensive presentation of the coordinate systems for which the partial differential equations mentioned above may be solved exactly and the methods of solutions are given in [3]. Comprehensive collections of formulae and theorems for the special functions of mathematical physics are provided in [4,5].

B.1 THE ORDINARY BESSEL FUNCTIONS

The separation of the Helmholtz or wave equation in circular cylindrical coordinates leads to *Bessel's differential equation*

$$z\frac{\mathrm{d}}{\mathrm{d}z}\left(z\frac{\mathrm{d}f}{\mathrm{d}z}\right) + (z^2 - n^2)f = 0\,. \tag{B.1}$$

The variable z and the parameter n can be arbitrarily complex. However, in the following n will be assumed as real and integer or half-integer. The solutions of Bessel's differential equation are the *Bessel function* of the first kind $J_n(z)$, the *Neumann function* or Bessel function of the second kind $Y_n(z)$ and the *Hankel functions* of the first kind $H_n^{(1)}(z)$ and of the second kind $H_n^{(2)}(z)$. The index n denotes the order of the function. The Bessel functions of the first kind $J_n(z)$ are defined by

$$J_n(z) = \sum_{k=0}^{\infty} \frac{(-1)^k\,(z/2)^{2k+n}}{k!\,\Gamma(n+k+1)} \tag{B.2}$$

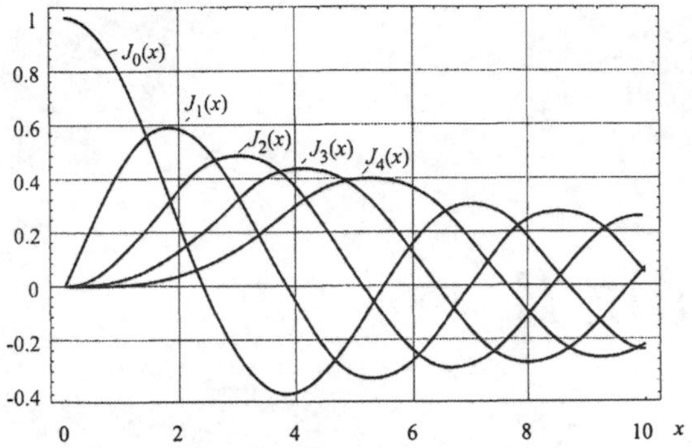

Figure B.1: The Bessel functions of the first kind of order 0,1,2,3 and 4.

with the *factorial function* $n!$ and the *gamma function* $\Gamma(n)$ defined by

$$n! = n(n-1)\dots 3\cdot 2\cdot 1 \quad \text{for integer } n \geq 1\,, \tag{B.3}$$

$$0! = 1\,, \tag{B.4}$$

$$\Gamma(n) = (n-1)! \quad \text{for integer } n \geq 1\,, \tag{B.5}$$

$$\Gamma(n+\tfrac{1}{2}) = 2^{2n}\sqrt{\pi}\,\frac{n!}{(2n)!} \quad \text{for integer } n \geq 0\,. \tag{B.6}$$

Figure B.1 shows the Bessel functions of the first kind of order 0,1,2,3 and 4. The Neumann function $Y_n(z)$ is defined by

$$Y_n(z) = \frac{J_n(z)\cos(n\pi) - J_{-n}(z)}{\sin(n\pi)}\,. \tag{B.7}$$

Figure B.2 shows the Neumann functions of order 0,1,2,3 and 4. The Hankel functions $H_n^{(1)}(z)$ and $H_n^{(2)}(z)$ are defined by

$$H_n^{(1)}(z) = \mathrm{j}\,\frac{J_{-n}(z) - J_n(z)\,\mathrm{e}^{-\mathrm{j}n\pi}}{\sin(n\pi)}\,, \tag{B.8}$$

$$H_n^{(2)}(z) = \mathrm{j}\,\frac{J_n(z)\;\mathrm{e}^{\mathrm{j}n\pi} - J_{-n}(z)}{\sin(n\pi)}\,. \tag{B.9}$$

The Hankel functions are related to the Bessel functions of the first and second kind via

$$H_n^{(1)}(z) = J_n(z) + \mathrm{j}\,Y_n(z)\,, \tag{B.10}$$

$$H_n^{(2)}(z) = J_n(z) - \mathrm{j}\,Y_n(z)\,. \tag{B.11}$$

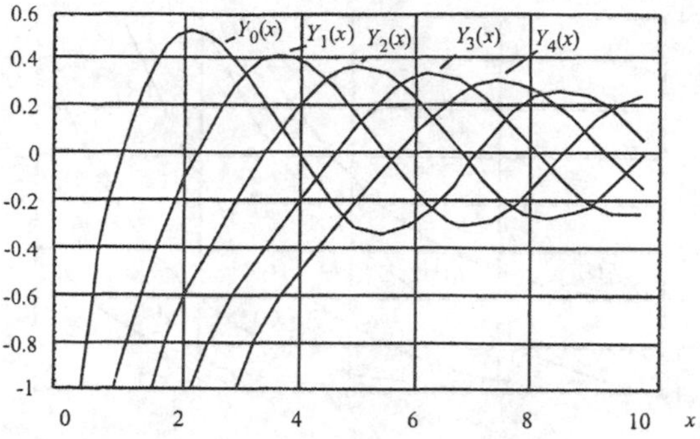

Figure B.2: The Neumann functions of order 0,1,2,3 and 4.

Denoting by $f_n(z)$ any of the functions $J_n(z)$, $Y_n(z)$, $H_n^{(1)}(z)$, $H_n^{(2)}(z)$ the following recurrence relations are valid:

$$f_{n-1}(z) + f_{n+1}(z) = \frac{2n}{z} f_n(z) , \tag{B.12}$$

$$f_{n-1}(z) - f_{n+1}(z) = 2 f_n'(z) . \tag{B.13}$$

The expression $f_n'(x)$ denotes the derivation of f_n with respect to x. The functions of positive and negative order are related via

$$f_{-n}(z) = (-1)^n f_n(z) . \tag{B.14}$$

From (B.13) and (B.14) it follows that

$$f_0'(z) = -f_1(z) . \tag{B.15}$$

The ordinary Bessel functions of the first kind are the Fourier series expansion coefficients of the *generating function*

$$\mathrm{e}^{\mathrm{j}z\sin\varphi} = \sum_{n=-\infty}^{+\infty} J_n(z)\,\mathrm{e}^{\mathrm{j}n\varphi} . \tag{B.16}$$

If ξ_{ni} and ξ_{nk} are the ith and kth zero of $J_n(x)$, in other words

$$J_n(\xi_{ni}) = 0 , \quad J_n(\xi_{nk}) = 0 \tag{B.17}$$

the following *orthogonality relation* is valid:

$$\int_0^a r J_n\left(\frac{\xi_{ni} r}{a}\right) J_n\left(\frac{\xi_{nk} r}{a}\right) \mathrm{d}r = \begin{cases} 0 & \text{for} \quad i \neq k \\ \frac{a^2}{2} J_n'^2(\xi_i) & \text{for} \quad i = k \end{cases} . \tag{B.18}$$

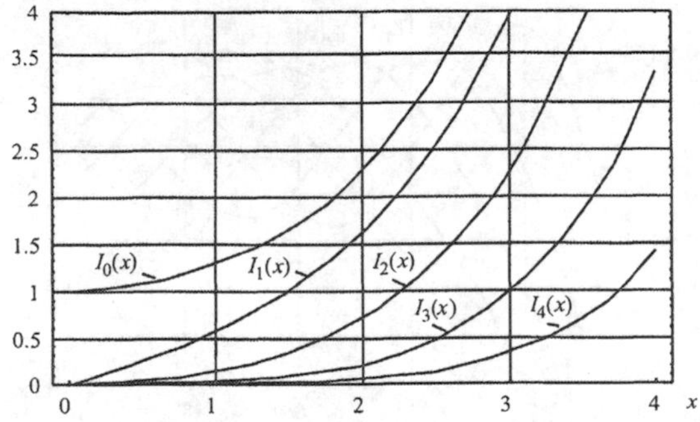

Figure B.3: The modified Bessel functions of the first kind of order 0,1,2,3 and 4.

B.2 THE MODIFIED BESSEL FUNCTIONS

The *modified Bessel functions* are the solutions of the *modified Bessel differential equation*

$$z\frac{\mathrm{d}}{\mathrm{d}z}\left(z\frac{\mathrm{d}f}{\mathrm{d}z}\right) - (z^2 + n^2)f = 0 \,. \tag{B.19}$$

The modified Bessel differential equation is obtained by replacing z by $\mathrm{j}z$ in the Bessel differential equation (B.1). Solutions are the *modified Bessel function of the first kind* $I_n(x)$ and the *modified Bessel function of the second kind* $K_n(x)$. The index n denominates the order of the modified Bessel functions. The modified Bessel function of the first kind is defined by the following series expansion:

$$I_n(z) = \sum_{k=0}^{\infty} \frac{(z/2)^{2k+n}}{k!(n+k)!} \,. \tag{B.20}$$

Figure B.3 shows the modified Bessel functions of the first kind. The modified Bessel function of the second kind is related to the modified Bessel function of the first kind via

$$K_n(z) = \frac{\pi}{2}\frac{I_{-n}(z) - I_n(z)}{\sin(n\pi)} \,. \tag{B.21}$$

The modified Bessel functions are related to the ordinary Bessel functions via

$$I_n(x) = (-\mathrm{j})^n J_n(\mathrm{j}x) \,, \tag{B.22}$$

$$K_n(x) = \frac{\pi}{2}(\mathrm{j})^{n+1}[J_n(\mathrm{j}x) + \mathrm{j}Y_n(\mathrm{j}x)] \,. \tag{B.23}$$

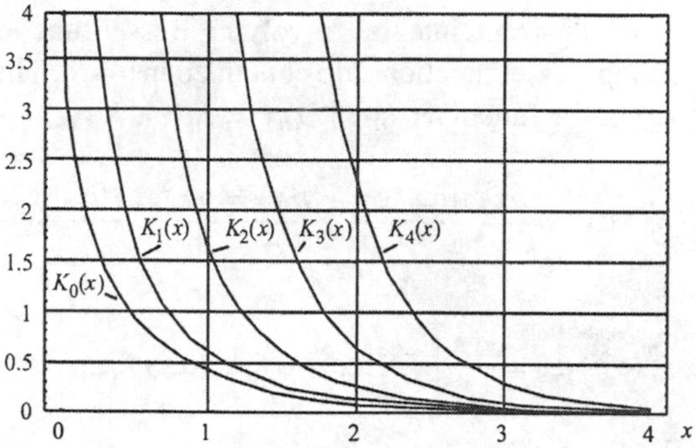

Figure B.4: The modified Bessel functions of the second kind of order 0,1,2,3 and 4.

Figure B.4 shows the modified Bessel functions of the second kind. For the modified Bessel functions the following recurrence relations are valid:

$$I_{n-1}(z) - I_{n+1}(z) = \frac{2n}{z} I_n(z) , \tag{B.24}$$

$$I_{n-1}(z) + I_{n+1}(z) = 2I_n'(z) , \tag{B.25}$$

$$K_{n-1}(z) - K_{n+1}(z) = -\frac{2n}{z} K_n(z) , \tag{B.26}$$

$$K_{n-1}(z) + K_{n+1}(z) = -2K_n'(z) . \tag{B.27}$$

The modified Bessel functions of positive and negative order are related via

$$I_{-n}(z) = I_n(z) , \tag{B.28}$$

$$K_{-n}(z) = K_n(z) . \tag{B.29}$$

From (B.25), (B.28), (B.27) and (B.29) we obtain

$$I_0'(z) = I_1(z) , \tag{B.30}$$

$$K_0'(z) = -K_1(z) . \tag{B.31}$$

The modified Bessel functions of the first kind are the Fourier series expansion coefficients of the *generating function*

$$\mathrm{e}^{z \cos\varphi} = I_0(z) + 2 \sum_{n=1}^{\infty} I_n(z) \cos(n\varphi) . \tag{B.32}$$

Now we list some important integrals involving Bessel functions. A large number of integrals involving Bessel functions are contained in [4–6]. The functions $f_n(z)$, $g_n(z)$ may denote any of the functions $J_n(z)$, $Y_n(z)$, $I_n(z)$, $K_n(z)$.

$$\int x f_n(\alpha x) g_n(\beta x)\,\mathrm{d}x = \frac{\beta x f_n(\alpha x) g_{n-1}(\beta x) - \alpha x f_{n-1}(\alpha x) g_n(\beta x)}{\alpha^2 - \beta^2}, \tag{B.33}$$

$$\int x f_n^2(\alpha x)\,\mathrm{d}x = \frac{x^2}{2}\left[f_n^2(\alpha x) - f_{n-1}(\alpha x) f_{n+1}(\alpha x)\right]. \tag{B.34}$$

From (B.15) and (B.34) we obtain

$$\int x J_0'^2(\alpha x)\,\mathrm{d}x = \frac{x^2}{2}\left[J_1^2(\alpha x) - J_0(\alpha x) J_2(\alpha x)\right]. \tag{B.35}$$

From (B.15) and (B.34) it follows that

$$\int_0^{x_0} x J_1^2(\alpha x)\,\mathrm{d}x = \int_0^{x_0} x J_0'^2(\alpha x)\,\mathrm{d}x = \begin{cases} \frac{x_0^2}{2} J_1^2(\alpha x_0) & \text{for} \quad J_0(\alpha x_0) = 0 \\ \frac{x_0^2}{2} J_0^2(\alpha x_0) & \text{for} \quad J_0'(\alpha x_0) = 0 \\ & \text{for} \quad J_1(\alpha x_0) = 0\,. \end{cases} \tag{B.36}$$

From (B.34) we obtain

$$\int_0^{x_0} x J_0^2(\alpha x)\,\mathrm{d}x = \frac{x_0^2}{2}\left[J_0^2(\alpha x_0) + J_1^2(\alpha x_0)\right]. \tag{B.37}$$

B.3 SPHERICAL BESSEL FUNCTIONS

The *spherical Bessel functions* $j_n(x)$, $y_n(x)$ and the *spherical Hankel functions* $h_n^{(1)}(x)$, $h_n^{(2)}(x)$ are solutions of the differential equation

$$x^2 \frac{\mathrm{d}^2 z_n(x)}{\mathrm{d}x^2} + 2x \frac{\mathrm{d}z_n(x)}{\mathrm{d}r} + (x^2 - n^2) z_n(x) = 0\,. \tag{B.38}$$

This differential equation is the differential equation (3.82a) normalized to $x = kr$, and describes the radial component of spherical wave functions. With the substitution

$$z_n(x) = \frac{1}{\sqrt{x}} f_{n+\frac{1}{2}} \tag{B.39}$$

we transform (B.38) into

$$x^2 \frac{\mathrm{d}^2 f_{n+\frac{1}{2}}(x)}{\mathrm{d}x^2} + r \frac{\mathrm{d}f_{n+\frac{1}{2}}(x)}{\mathrm{d}x} + \left[x^2 - \left(n + \frac{1}{2} \right)^2 \right] f_{n+\frac{1}{2}}(x) = 0 \,. \tag{B.40}$$

This is the Bessel's differential equation (B.1) for half odd integer order $n + \frac{1}{2}$. The spherical Bessel functions $j_n(x)$, $y_n(x)$ and the spherical Hankel functions $h_n^{(1)}(x)$, $h_n^{(2)}(x)$are related to the cylindrical Bessel and Hankel functions of order $n + 1/2$ by

$$j_n(x) = \sqrt{\frac{\pi}{2x}} J_{n+\frac{1}{2}}(x) \,, \tag{B.41}$$

$$y_n(x) = \sqrt{\frac{\pi}{2x}} Y_{n+\frac{1}{2}}(x) \,, \tag{B.42}$$

$$h_n^{(1)}(x) = \sqrt{\frac{\pi}{2x}} H_{n+\frac{1}{2}}^{(1)}(x) \,, \tag{B.43}$$

$$h_n^{(2)}(x) = \sqrt{\frac{\pi}{2x}} H_{n+\frac{1}{2}}^{(2)}(x) \,. \tag{B.44}$$

The spherical Bessel functions of order 0 to 2 are given by

$$j_0(x) = \frac{\sin x}{x} \,, \tag{B.45}$$

$$j_1(x) = \frac{\sin x}{x^2} - \frac{\cos x}{x} \,, \tag{B.46}$$

$$j_2(x) = \left(\frac{3}{x^3} - \frac{1}{x} \right) \sin x - \frac{3}{x^2} \cos x \tag{B.47}$$

and

$$y_0(x) = -\frac{\cos x}{x} \,, \tag{B.48}$$

$$y_1(x) = -\frac{\sin x}{x} - \frac{\cos x}{x^2} \,, \tag{B.49}$$

$$y_2(x) = \left(-\frac{3}{x^3} + \frac{1}{x} \right) \cos x - \frac{3}{x^2} \sin x \,. \tag{B.50}$$

and the spherical Hankel functions of order 0 to 2 are given by

$$h_0^{(1)}(x) = h_0^{(2)*}(x) = -\frac{\mathrm{j}}{x}\,\mathrm{e}^{\mathrm{j}x}\,, \tag{B.51}$$

$$h_1^{(1)}(x) = h_1^{(2)*}(x) = \left(-\frac{\mathrm{j}}{x} - \frac{1}{x^2}\right)\mathrm{e}^{\mathrm{j}x}\,, \tag{B.52}$$

$$h_2^{(1)}(x) = h_2^{(2)*}(x) = \left(\frac{\mathrm{j}}{x} - \frac{3}{x^2} - \frac{3\mathrm{j}}{x^3}\right)\mathrm{e}^{\mathrm{j}x}\,. \tag{B.53}$$

B.4 LEGENDRE POLYNOMIALS

The separation of the Helmholtz or wave equation in spherical coordinates leads to the following differential equation for the function of the ϑ coordinate

$$\frac{1}{\sin\vartheta}\frac{\mathrm{d}}{\mathrm{d}\vartheta}\left(\sin\vartheta\,\frac{\mathrm{d}f}{\mathrm{d}\vartheta}\right) + \left[n(n+1) - \frac{m^2}{\sin^2\vartheta}\right]f = 0\,. \tag{B.54}$$

With the substitution

$$x = \cos\vartheta \tag{B.55}$$

we can bring (B.54) in the form

$$(1-x^2)\frac{\mathrm{d}^2 f}{\mathrm{d}x^2} + 2x\frac{\mathrm{d}f}{\mathrm{d}x} + \left[n(n+1) - \frac{m^2}{1-x^2}\right]f = 0\,. \tag{B.56}$$

This is the *Legendre differential equation.* Since the variable z and the parameters n and m can be arbitrarily complex. However, in the following m and n will be assumed as real and integer. Since in the spherical coordinate system we consider the interval $0 \le \vartheta \le \pi$ this yields $-1 \le x \le 1$. The solutions of the Legendre differential equation are the *associated Legendre function* of the first kind $P_n^m(z)$, and the second kind $Q_n^m(z)$. The subscript n denotes the degree of the Legendre function.

For $m = 0$ we obtain the *ordinary Legendre differential equation*

$$(1-x^2)\frac{\mathrm{d}^2 f}{\mathrm{d}x^2} + 2x\frac{\mathrm{d}f}{\mathrm{d}x} + n(n+1)f = 0\,. \tag{B.57}$$

The solutions of the ordinary Legendre differential equation are the *ordinary Legendre functions of the first kind* $P_n(x)$, and the second kind $Q_n(x)$. For integer n the ordinary Legendre functions of the first kind are polynomials of degree n and therefore are also called *Legendre polynomials* of degree n. The explicit expression for the

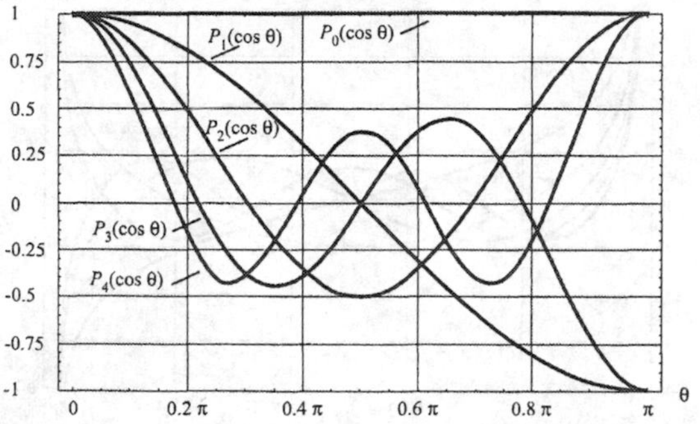

Figure B.5: The Legendre polynomials of the first kind of order 0,1,2,3 and 4.

Legendre polynomial is

$$P_n(x) = \sum_{m=0}^{N} (-1)^m \frac{(2n-2m)!x^{n-2m}}{2^n\, m!(n-m)!\,(n-2m)!}\,, \tag{B.58}$$

where $M = n/2$ for even n and $M = (n-1)/2$ for odd n. In terms of x and $\cos\vartheta$ the five Legendre polynomials of lowest order are

$$P_0(x) = 1\,, \tag{B.59}$$

$$P_1(x) = x = \cos\vartheta\,, \tag{B.60}$$

$$P_2(x) = \tfrac{1}{2}(3x^2 - 1) = \tfrac{1}{4}(1 + 3\cos 2\vartheta)\,, \tag{B.61}$$

$$P_3(x) = \tfrac{1}{2}(5x^3 - 3x) = \tfrac{1}{8}(3\cos\vartheta + 5\cos 3\vartheta)\,, \tag{B.62}$$

$$\begin{aligned} P_4(x) &= \tfrac{1}{8}(35x^4 - 30x^2 + 3) \\ &= \tfrac{1}{64}(9 + 20\cos 2\vartheta + 35\cos 4\vartheta)\,. \end{aligned} \tag{B.63}$$

Figure B.5 shows the Legendre polynomials of the first kind of order 0,1,2,3 and 4. The Legendre functions of the second kind $Q_n(x)$ are infinite at $x = \pm 1$, or at $\vartheta = 0$ and $\vartheta = \pi$. The lowest order Legendre functions of the second kind are given by

$$Q_0(x) = \frac{1}{2}\log\frac{1+x}{1-x} = \log\cot\frac{\vartheta}{2}\,, \tag{B.64}$$

$$Q_1(x) = \frac{x}{2}\log\frac{1+x}{1-x} - 1 = \cos\vartheta\cot\log\frac{\vartheta}{2} - 1\,, \tag{B.65}$$

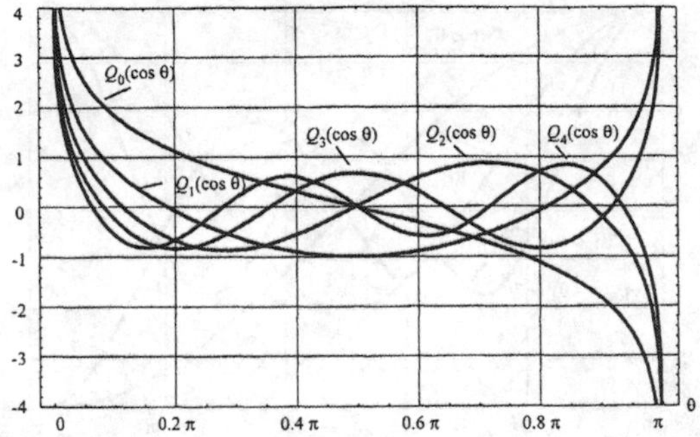

Figure B.6: The Legendre polynomials of the second kind of order 0,1,2,3 and 4.

$$\begin{aligned} Q_2(x) &= \frac{3x^2-1}{4}\log\frac{1+x}{1-x} - \frac{3u}{2} \\ &= \frac{1}{4}(3\cos^2\vartheta - 1)\log\cot\frac{\vartheta}{2} - \frac{3}{2}\cos\vartheta\,. \end{aligned} \tag{B.66}$$

The nth order Legendre function of the second kind is

$$Q_n(x) = Q_0(x)P_n(x) - \sum_{m=0}^{M} \frac{2n-4m+3}{(2m-1)(n-m+1)} P_{n-2m+1}(x)\,, \tag{B.67}$$

where $M = n/2$ for even n and $M = (n+1)/2$ for odd n. The Legendre functions with positive and negative arguments are related via

$$P_n(-x) = (-1)^n P_n(x)\,, \tag{B.68}$$

$$Q_n(-x) = (-1)^{n+1} Q_n(x)\,. \tag{B.69}$$

Figure B.6 shows the Legendre polynomials of the first kind of order 0,1,2,3 and 4. The associated Legendre functions, i.e., the solutions of the associated Legendre differential equation (B.56) are

$$P_n^m(x) = (-1)^m (1-x^2)^{\frac{m}{2}} \frac{\mathrm{d}^m P_n(x)}{\mathrm{d}x^m}\,, \tag{B.70}$$

$$Q_n^m(x) = (-1)^m (1-x^2)^{\frac{m}{2}} \frac{\mathrm{d}^m Q_n(x)}{\mathrm{d}x^m}\,. \tag{B.71}$$

Some lower order associated Legendre functions of the first kind are

$$P_1^1(x) = -(1-x^2)^{\frac{1}{2}} \,, \tag{B.72}$$

$$P_2^1(x) = -3(1-x^2)^{\frac{1}{2}} x \,, \tag{B.73}$$

$$P_2^2(x) = 3(1-x^2)^{\frac{1}{2}} \,, \tag{B.74}$$

$$P_3^1(x) = \frac{3}{2}(1-x^2)^{\frac{1}{2}}(1-5x^2) \,, \tag{B.75}$$

$$P_3^2(x) = 15(1-x^2)x \,, \tag{B.76}$$

$$P_3^3(x) = 1 - 5(1-x^2)^{\frac{3}{2}}(1-5x^2) \,. \tag{B.77}$$

Some lower order associated Legendre functions of the second kind are

$$Q_1^1(x) = -(1-x^2)^{\frac{1}{2}} \left(\frac{1}{2} \log \frac{1+x}{1-x} + \frac{x}{1-x^2} \right) , \tag{B.78}$$

$$Q_2^1(x) = -(1-x^2)^{\frac{1}{2}} \left(\frac{3}{2} \log \frac{1+x}{1-x} + \frac{3x^2-2}{1-x^2} \right) , \tag{B.79}$$

$$Q_2^2(x) = (1-x^2)^{\frac{1}{2}} \left(\frac{3}{2} \log \frac{1+x}{1-x} + \frac{5x-3x^2}{(1-x^2)^2} \right) . \tag{B.80}$$

The *orthogonality relations* for Legendre functions are

$$\int_{-1}^{1} P_{n_1}^m(x) P_{n_2}^m(x) = \frac{2}{2n_1+1} \frac{(n_1+m)!}{(n_1-m)!} \delta_{n_1,n_2} \,. \tag{B.81}$$

B.5 SPHERICAL HARMONICS

Spherical harmonics are solutions of the partial differential equation

$$\frac{1}{\sin\vartheta} \frac{\mathrm{d}}{\mathrm{d}\vartheta} \left(\sin\vartheta \, \frac{\mathrm{d}f(\vartheta,\varphi)}{\mathrm{d}\vartheta} \right) + \frac{1}{\sin^2\vartheta} \frac{\mathrm{d}^2 f(\vartheta,\varphi)}{\mathrm{d}\varphi^2} + n(n+1)f = 0 \,. \tag{B.82}$$

Setting

$$f(\vartheta,\varphi) = \Theta(\vartheta)\Phi(\varphi) \tag{B.83}$$

with the integer separation parameter m with $|m| \leq n$ the partial differential equation (B.82) may be separated into (B.54) for $\Theta(\vartheta)$ and the second order differential equation for $\Phi(\varphi)$.

$$\frac{1}{\sin\vartheta}\frac{\mathrm{d}}{\mathrm{d}\vartheta}\left(\sin\vartheta\,\frac{\mathrm{d}f}{\mathrm{d}\vartheta}\right)+\left[n(n+1)-\frac{m^2}{\sin^2\vartheta}\right]f=0\,, \tag{B.84}$$

$$\frac{\mathrm{d}^2\Phi(\varphi)}{\mathrm{d}\varphi^2}+m^2\Phi(\varphi)=0\,. \tag{B.85}$$

The normalized solution of (B.82) is the spherical harmonic

$$Y_n^m(\vartheta,\varphi)=\sqrt{\frac{2n+1}{2}\,\frac{(n-m)!}{(n+m)!}}\,P_n^m(\cos\vartheta)\,\mathrm{e}^{\mathrm{j}m\varphi}\,. \tag{B.86}$$

Some lower order spherical harmonics are

$$Y_0^0(\vartheta,\varphi)=\frac{1}{4\pi}\,, \tag{B.87}$$

$$Y_1^1(\vartheta,\varphi)=-\sqrt{\frac{1}{8\pi}}\,\sin\vartheta\,\mathrm{e}^{\mathrm{j}\varphi}\,, \tag{B.88}$$

$$Y_1^0(\vartheta,\varphi)=\sqrt{\frac{1}{4\pi}}\,\cos\vartheta\,, \tag{B.89}$$

$$Y_1^{-1}(\vartheta,\varphi)=\sqrt{\frac{1}{8\pi}}\,\sin^2\vartheta\,\mathrm{e}^{2\mathrm{j}\varphi}\,, \tag{B.90}$$

$$Y_2^2(\vartheta,\varphi)=\sqrt{\frac{5}{96\pi}}\,3\sin^2\vartheta\,\mathrm{e}^{2\mathrm{j}\varphi}\,, \tag{B.91}$$

$$Y_2^1(\vartheta,\varphi)=-\sqrt{\frac{5}{24\pi}}\,3\sin\vartheta\cos\vartheta\,\mathrm{e}^{\mathrm{j}\varphi}\,, \tag{B.92}$$

$$Y_2^0(\vartheta,\varphi)=\sqrt{\frac{5}{4\pi}}\left(\frac{3}{2}\cos^2\vartheta-\frac{1}{2}\right)\,, \tag{B.93}$$

$$Y_2^{-1}(\vartheta,\varphi)=\sqrt{\frac{5}{24\pi}}\,3\sin\vartheta\cos\vartheta\,\mathrm{e}^{-\mathrm{j}\varphi}\,, \tag{B.94}$$

$$Y_2^{-2}(\vartheta,\varphi)=\sqrt{\frac{5}{96\pi}}\,3\sin^2\vartheta\,\mathrm{e}^{-2\mathrm{j}\varphi}\,. \tag{B.95}$$

Figure B.7 shows some of the lowest order spherical harmonics.

The *orthogonality relation* for spherical harmonics is

$$\int_{\varphi=0}^{2\pi}\int_{\vartheta=0}^{\pi}Y_{n_1}^{m_1\,*}(\vartheta,\varphi)\,Y_{n_2}^{m_2}(\vartheta,\varphi)\;\mathrm{d}\vartheta\;\mathrm{d}\varphi=\delta_{n_1,n_2}\delta_{m_1,m_2}\,. \tag{B.96}$$

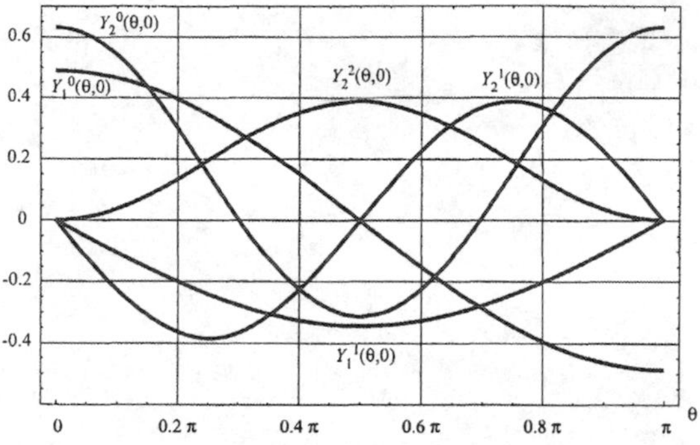

Figure B.7: The spherical harmonics $Y_1^0, Y_1^1, Y_2^0, Y_2^1, Y_2^2$.

The spherical harmonics constitute a *complete system of functions* within the interval $0 \leq \vartheta \leq \pi,\ 0 \leq \varphi \leq 2\pi$. Any function $f(\vartheta, \varphi)$ with sufficient continuity properties may be expanded in this interval in a series

$$f(\vartheta, \varphi) = \sum_{m,n} a_{m,n} Y_m(\vartheta, \varphi) . \tag{B.97}$$

The expansion coefficients a_{mn} are

$$a_{m,n} = \int_{\varphi=0}^{2\pi} \int_{\vartheta=0}^{\pi} Y_n^{m*}(\vartheta, \varphi) f(\vartheta, \varphi) \, d\vartheta \, d\varphi . \tag{B.98}$$

REFERENCES

[1] P. Morse and H. Feshbach, *Methods of Theoretical Physics, Part 1.* New York: McGraw-Hill, 1953.

[2] P. Morse and H. Feshbach, *Methods of Theoretical Physics, Part 2.* New York: McGraw-Hill, 1953.

[3] P. Moon and D. Spencer, *Field Theory Handbook.* Berlin: Springer, 1991.

[4] M. Abramowitz and I. Stegun, *Handbook of Mathematical Functions With Formulas, Graphs and Mathematical Tables.* New York: John Wiley & Sons, 1993.

[5] W. Magnus, F. Oberhettinger, and R. Soni, *Formulas and Theorems for the Special Functions of Mathematical Physics.* Berlin: Springer, 1966.

[6] I. Gradshteyn and I. Ryzhik, *Tables of Integrals, Series, and Products.* New York: Academic Press, 1965.

REFERENCES

Appendix C

Linear Algebra

This section recalls the basic definitions and the basic formulae of linear algebra. For a more detailed treatment see for example [1–3].

A *matrix* is a rectangular array of *scalars*, e.g., integer, real or complex numbers or integer-, real-, or complex-valued functions. The rectangular array

$$\boldsymbol{A} = \begin{bmatrix} A_{11} & A_{12} & \cdots & A_{1n} \\ A_{21} & A_{22} & \cdots & A_{2n} \\ \vdots & \vdots & \ddots & \vdots \\ A_{m1} & A_{m2} & \cdots & A_{mn} \end{bmatrix}, \tag{C.1}$$

is called a *matrix* of *m rows* and *n columns*. We denote matrices with boldface letters. We may use the index $<m \times n>$ to indicate the numbers of rows and columns. A matrix of type $<m \times n>$ is said to be an m by n matrix. A matrix of type $<n \times n>$ is called a square matrix of order n. A matrix of type $<m \times 1>$ is a *column vector*, and a matrix of type $<1 \times n>$ is a *row vector*.

For two matrices $\boldsymbol{A}$ and $\boldsymbol{B}$ of the same size $<m \times n>$ a *matrix sum*

$$\boldsymbol{C}_{<m \times n>} = \boldsymbol{A}_{<m \times n>} + \boldsymbol{B}_{<m \times n>} \tag{C.2}$$

is defined such that

$$C_{ij} = A_{ij} + B_{ij} \tag{C.3}$$

for each i and j. The product of a matrix $\boldsymbol{A}$ and $\boldsymbol{B}$ with a scalar b

$$\boldsymbol{C}_{<m \times n>} = b\, \boldsymbol{A}_{<m \times n>} \tag{C.4}$$

is defined by

$$C_{ij} = b \sum_{k=1}^{l} A_{ij} \,. \tag{C.5}$$

For two matrices $\boldsymbol{A}$ and $\boldsymbol{B}$ of the type $<m \times l>$ and $<l \times n>$ a *product*

$$\boldsymbol{C}_{<m \times n>} = \boldsymbol{A}_{<m \times l>} \boldsymbol{B}_{<l \times n>} \tag{C.6}$$

is defined. The elements of $\boldsymbol{C}$ are given by

$$C_{ij} = \sum_{k=1}^{l} A_{ik} B_{kj} \tag{C.7}$$

for each i and j. The matrix product has the following properties:

$$(\boldsymbol{A}\,\boldsymbol{B})\,\boldsymbol{C} = \boldsymbol{A}\,(\boldsymbol{B}\,\boldsymbol{C}) = \boldsymbol{A}\,\boldsymbol{B}\,\boldsymbol{C} \qquad \text{associative,} \tag{C.8}$$

$$(\boldsymbol{A} + \boldsymbol{B})\,\boldsymbol{C} = \boldsymbol{A}\,\boldsymbol{C} + \boldsymbol{B}\,\boldsymbol{C} \qquad \text{distributive,} \tag{C.9}$$

$$\boldsymbol{C}\,(\boldsymbol{A} + \boldsymbol{B}) = \boldsymbol{C}\,\boldsymbol{A} + \boldsymbol{C}\,\boldsymbol{B} \qquad \text{distributive,} \tag{C.10}$$

$$\boldsymbol{A}\,\boldsymbol{B} \neq \boldsymbol{B}\,\boldsymbol{A} \qquad \text{in general non-commutative.} \tag{C.11}$$

If all elements of a matrix are 0 the matrix is called the *null matrix* $\mathbf{0}$:

$$\mathbf{0}_{<m \times n>} = \begin{bmatrix} 0 & 0 & \cdots & 0 & 0 \\ 0 & 0 & \cdots & 0 & 0 \\ \vdots & \vdots & \ddots & \vdots & 0 \\ 0 & 0 & \cdots & 0 & 0 \\ 0 & 0 & \cdots & 0 & 0 \end{bmatrix} . \tag{C.12}$$

For any matrix $\boldsymbol{A}_{<m \times n>}$ the sum with the null matrix is given by

$$\boldsymbol{A}_{<m \times n>} + \mathbf{0}_{<m \times n>} = \boldsymbol{A}_{<m \times n>} . \tag{C.13}$$

and the product with the null matrix is given by

$$\mathbf{0}_{<k \times m>} \boldsymbol{A}_{<m \times n>} = \mathbf{0}_{<k \times n>} , \tag{C.14}$$

$$\boldsymbol{A}_{<m \times n>} \mathbf{0}_{<n \times k>} = \mathbf{0}_{<m \times k>} . \tag{C.15}$$

The *unit matrix* $\mathbf{1}$ is a quadratic matrix with the Konecker delta δ_{ij} as the matrix elements

$$\mathbf{1}_{<m \times m>} = [\delta_{ij}] = \begin{bmatrix} 1 & 0 & \cdots & 0 & 0 \\ 0 & 1 & \cdots & 0 & 0 \\ \vdots & \vdots & \ddots & \vdots & 0 \\ 0 & 0 & \cdots & 1 & 0 \\ 0 & 0 & \cdots & 0 & 1 \end{bmatrix} . \tag{C.16}$$

For any matrix $\boldsymbol{A}_{<n \times n>}$ the product with the unit matrix has the property

$$\mathbf{1}_{<m \times m>} \boldsymbol{A}_{<m \times n>} = \boldsymbol{A}_{<m \times n>} \mathbf{1}_{<n \times n>} = \boldsymbol{A}_{<m \times n>} \,. \tag{C.17}$$

A square matrix in which all nondiagonal elements are zero is a *diagonal matrix*. It can be written in the following abbreviated form:

$$\begin{bmatrix} D_1 & 0 & \cdots & 0 & 0 \\ 0 & D_2 & \cdots & 0 & 0 \\ \vdots & \vdots & \ddots & \vdots & 0 \\ 0 & 0 & \cdots & D_{n-1} & 0 \\ 0 & 0 & \cdots & 0 & D_n \end{bmatrix} = \mathrm{diag}[D_1, D_2, \ldots, D_{n-1}, D_n] \,. \tag{C.18}$$

A matrix $\boldsymbol{A}$ may be composed by submatrices $\boldsymbol{A}_{11}, \boldsymbol{A}_{12}, \boldsymbol{A}_{21}, \boldsymbol{A}_{22}$ in the following way:

$$\boldsymbol{A}_{<m \times n>} = \begin{bmatrix} \boldsymbol{A}_{11<p \times q>} & \boldsymbol{A}_{12<p \times (n-q)>} \\ \boldsymbol{A}_{21<(m-p) \times q>} & \boldsymbol{A}_{22<(m-p) \times (n-q)>} \end{bmatrix} . \tag{C.19}$$

For a quadratic matrix $\boldsymbol{A}_{<n \times n>}$ a *determinant* of nth order

$$\det \boldsymbol{A}_{<n \times n>} = \begin{vmatrix} A_{11} & \cdots & A_{1n} \\ \vdots & \ddots & \vdots \\ A_{n1} & \cdots & A_{nn} \end{vmatrix} \tag{C.20}$$

is defined by

$$\det \boldsymbol{A}_{<n \times n>} = \sum (-1)^{\alpha} A_{1,p_1} \cdots A_{n,p_n} \,, \tag{C.21}$$

where the summation has to be performed over all $n!$ permutations of the second indices $p_1 \ldots p_n$. The exponent α is the number of exchanges of indices for transforming the sequence $1 \ldots n$ into $p_1 \ldots p_n$. The determinant of a second order matrix is

$$\det \boldsymbol{A}_{<2 \times 2>} = A_{11} A_{22} - A_{12} A_{21} \,, \tag{C.22}$$

and the determinant of a third order matrix is

$$\begin{aligned} \det \boldsymbol{A}_{<3 \times 3>} &= A_{11} A_{22} A_{33} + A_{12} A_{23} A_{31} + A_{13} A_{21} A_{32} \\ &\quad - A_{13} A_{22} A_{31} - A_{11} A_{23} A_{32} - A_{12} A_{21} A_{33} \,. \end{aligned} \tag{C.23}$$

The determinant of the submatrix of a quadratic matrix obtained by deleting the ith row and the jth column is called the *minor* of the element A_{ij} and is given by

$$A_{ij}^{m} = \begin{vmatrix} A_{11} & \cdots & A_{1,j-1} & A_{1,j+1} & \cdots & A_{1n} \\ \vdots & \ddots & \vdots & \vdots & \ddots & \vdots \\ A_{l-1,1} & \cdots & A_{l-1,j-1} & A_{l-1,j+1} & \cdots & A_{l-1,n} \\ A_{l+1,1} & \cdots & A_{l+1,j-1} & A_{l+1,j+1} & \cdots & A_{l+1,n} \\ \vdots & \ddots & \vdots & \vdots & \ddots & \vdots \\ A_{n1} & \cdots & A_{n,j-1} & A_{n,j+1} & \cdots & A_{nn} \end{vmatrix} . \tag{C.24}$$

The *cofactor* A_{ij}^{a} of the element A_{ij} of a quadratic matrix is given by

$$A_{ij}^{a} = (-1)^{i+j} A_{ij}^{m} . \tag{C.25}$$

Due to (C.21) an nth order determinant may be expanded as sum of the products of the elements of any row or column of the appropriate matrix with their cofactors. This is the *Laplace expansion*

$$\det \boldsymbol{A}_{<n \times n>} = \sum_{i=1}^{n} A_{ij} A_{ij}^{a} = \sum_{j=1}^{n} A_{ij} A_{ij}^{a} . \tag{C.26}$$

A quadratic matrix is called *singular* if $\det \boldsymbol{A} = 0$. If a quadratic matrix is *non singular*, i.e., $\det \boldsymbol{A} \neq 0$, the inverse $\boldsymbol{A}^{-1}$ is defined by its property

$$\boldsymbol{A}^{-1} \boldsymbol{A} = \boldsymbol{A} \boldsymbol{A}^{-1} = \boldsymbol{1} , \tag{C.27}$$

and may be computed using *Cramer's rule*:

$$\boldsymbol{A}^{-1}_{<n \times n>} = \frac{1}{\det \boldsymbol{A}} \begin{bmatrix} A_{11}^{a} & A_{21}^{a} & \cdots & A_{n1}^{a} \\ A_{12}^{a} & A_{22}^{a} & \cdots & A_{n1}^{a} \\ \vdots & \vdots & \ddots & \vdots \\ A_{1n}^{a} & A_{1n}^{a} & \cdots & A_{nn}^{a} \end{bmatrix} . \tag{C.28}$$

The inverse of the product of two matrices is

$$(\boldsymbol{A}\, \boldsymbol{B})^{-1} = \boldsymbol{B}^{-1} \boldsymbol{A}^{-1} . \tag{C.29}$$

The inverse of a diagonal matrix is the diagonal matrix of the inverse elements:

$$(\mathrm{diag}[D_1, D_2, \ldots, D_{n-1}, D_n])^{-1} = \mathrm{diag}[D_1^{-1}, D_2^{-1}, \ldots, D_{n-1}^{-1}, D_n^{-1}] . \tag{C.30}$$

The *transpose* $\boldsymbol{A}^T$ of a matrix $\boldsymbol{A}$ is obtained by interchanging columns and rows of the matrix $\boldsymbol{A}$:

$$\boldsymbol{A}^T = \begin{bmatrix} A_{11} & A_{21} & \cdots & A_{m1} \\ A_{12} & A_{22} & \cdots & A_{m2} \\ \vdots & \vdots & \ddots & \vdots \\ A_{1n} & A_{2n} & \cdots & A_{mn} \end{bmatrix} . \tag{C.31}$$

If $\boldsymbol{A} = \boldsymbol{A}^T$ the matrix $\boldsymbol{A}$ is *symmetric*. The transpose of the product of two matrices is

$$(\boldsymbol{A}\,\boldsymbol{B})^T = \boldsymbol{B}^T\,\boldsymbol{A}^T . \tag{C.32}$$

If a matrix has the property $\boldsymbol{N}^T = \boldsymbol{N}^{-1}$ the matrix is called *orthogonal*.

The *complex conjugate* $\boldsymbol{A}^*$ of a matrix $\boldsymbol{A}$ is obtained by taking the complex conjugates of all elements of the matrix $\boldsymbol{A}$:

$$\boldsymbol{A}^* = \begin{bmatrix} A_{11}^* & A_{12}^* & \cdots & A_{1n}^* \\ A_{21}^* & A_{22}^* & \cdots & A_{2n}^* \\ \vdots & \vdots & \ddots & \vdots \\ A_{m1}^* & A_{m2}^* & \cdots & A_{mn}^* \end{bmatrix} . \tag{C.33}$$

If $\boldsymbol{A} = \boldsymbol{A}^*$ the matrix $\boldsymbol{A}$ is called a *real* matrix. The complex conjugate of the product of two matrices is

$$(\boldsymbol{A}\,\boldsymbol{B})^* = \boldsymbol{A}^*\,\boldsymbol{B}^* . \tag{C.34}$$

The *Hermitian conjugate* $\boldsymbol{A}^\dagger$ of a matrix $\boldsymbol{A}$ is the transpose of the complex conjugate of $\boldsymbol{A}$. It is obtained by taking the complex conjugates of all elements of the matrix $\boldsymbol{A}$ interchanging columns and rows of the matrix:

$$\boldsymbol{A}^\dagger = (\boldsymbol{A}^*)^T = (\boldsymbol{A}^T)^* = \begin{bmatrix} A_{11}^* & A_{21}^* & \cdots & A_{m1}^* \\ A_{12}^* & A_{22}^* & \cdots & A_{m2}^* \\ \vdots & \vdots & \ddots & \vdots \\ A_{1n}^* & A_{2n}^* & \cdots & A_{mn}^* \end{bmatrix} . \tag{C.35}$$

If $\boldsymbol{A} = \boldsymbol{A}^\dagger$ the matrix $\boldsymbol{A}$ is called a *Hermitian* matrix. The Hermitian conjugate of the product of two matrices is

$$(\boldsymbol{A}\,\boldsymbol{B})^\dagger = \boldsymbol{B}^\dagger\,\boldsymbol{A}^\dagger . \tag{C.36}$$

If a matrix has the property $\boldsymbol{U}^\dagger = \boldsymbol{U}^{-1}$ the matrix is called *unitary*. The inverses of the transpose, the complex conjugate and the Hermitian conjugate matrix are

$$(\boldsymbol{A}^T)^{-1} = (\boldsymbol{A}^{-1})^T , \tag{C.37}$$

$$(\boldsymbol{A}^*)^{-1} = (\boldsymbol{A}^{-1})^* , \tag{C.38}$$

$$(\boldsymbol{A}^\dagger)^{-1} = (\boldsymbol{A}^{-1})^\dagger . \tag{C.39}$$

Determinants have the following properties:

$$\det A^T = \det A \;, \tag{C.40}$$

$$\det A^\dagger = \det A^* = (\det A)^* \;, \tag{C.41}$$

$$\det(kA_{<n \times n>}) = k^n \det A \;, \tag{C.42}$$

$$\det(A\,B) = (\det A)(\det B) \;, \tag{C.43}$$

$$\det(A^{-1}) = 1/\det A \;, \tag{C.44}$$

$$\det(A + B) \neq \det A + \det B \;. \tag{C.45}$$

The determinant of a Hermitian matrix is real.

Let $A_{<n \times n>} = A^\dagger$ be Hermitian matrix, $x_{<1 \times n>}$ an n-dimensional column vector and $x^\dagger_{<n \times 1>}$ its Hermitian conjugate row vector. The expression

$$x^\dagger A\, x = \sum_{i,k} A_{ik} x_i^* x_k \tag{C.46}$$

is called a *Hermitian form*. For arbitrary vectors a the Hermitian form is real. If the Hermitian form is positive for arbitrary a it is called *positive definite*. If the Hermitian form is positive or zero for arbitrary a it is called *positive semidefinite*.

The determinant of the submatrix consisting of the first k rows and the first k columns of a matrix A is called its *leading principal minor* $\det_k A$ of order k.

$$\begin{aligned} \det{}_1 A_{<n \times n>} &= A_{11} \;, \\ \det{}_k A_{<n \times n>} &= \det \begin{bmatrix} A_{11} & \cdots & A_{1k} \\ A_{k1} & \cdots & A_{kk} \end{bmatrix} \text{ for } \quad 1 < k < n \;, \\ \det{}_n A_{<n \times n>} &= \det A \;. \end{aligned} \tag{C.47}$$

A Hermitian form is positive definite if all its leading principal minors are positive and it is positive semidefinite if all the leading principal minors are non negative.

REFERENCES

[1] G. Arfken, *Mathematical Methods for Physicists.* New York: Academic Press, 1985.

[2] C. Wong, *Introduction to Mathematical Physics.* Oxford: Oxford University Press, 1991.

[3] D. S. Watkins, *Fundamentals of Matrix Computations.* New York: John Wiley & Sons, 1991.

Appendix D

Fourier Series and Fourier Transform

D.1 THE FOURIER SERIES

Periodic signals may be expanded into a *Fourier series* [1–3]. A signal for which

$$s(t + nT_0) = s(t) \quad \text{for}\, n = \cdots -2, -1, 0, 1, 2, \ldots \tag{D.1}$$

is valid is called a *periodic signal* with the period T_0. The fundamental frequency f_0 is given by $f_0 = 1/T_0$ and the corresponding angular frequency is $\omega_0 = 2\pi f_0$. A signal

$$s(t) = \sum_{n=-\infty}^{\infty} a_n \mathrm{e}^{\mathrm{j}n\omega_0 t} \tag{D.2}$$

which is a superposition of harmonic signals at the fundamental frequency f and the harmonics nf is periodic with T_0. The series according to (D.2) is called a *Fourier series*. The expansion coefficients a_n are the complex amplitudes at the frequencies $n\omega_0$. The signal $s(t)$ is real if the condition $a_{-n} = a_n^*$ is fulfilled. For real signals $s(t)$ it follows from (D.2) and $a_{-n} = a_n^*$ that

$$s(t) = 2a_0 \,\Re \left\{ \sum_{n=1}^{\infty} a_n \,\mathrm{e}^{\mathrm{j}n\omega_0 t} \right\} . \tag{D.3}$$

Decomposing a_n in magnitude $|a_n|$ and phase φ_n by

$$a_n = |a_n| \,\mathrm{e}^{\mathrm{j}\varphi_n} \tag{D.4}$$

we obtain from (D.3)

$$s(t) = 2a_0 \sum_{n=1}^{\infty} |a_n| \cos(n\omega_0 t + \varphi_n) . \tag{D.5}$$

We introduce the real amplitudes

$$b_n = a_n + a_n^* \,, \tag{D.6}$$

$$c_n = \mathrm{j}\,(a_n - a_n^*) \,, \tag{D.7}$$

and obtain from (D.3)

$$s(t) = \frac{b_0}{2} \sum_{n=1}^{\infty} b_n \cos n\omega_0 t + c_n \sin n\omega_0 t \,. \tag{D.8}$$

Magnitude a_n and phase φ_n are obtained from the real amplitudes by b_n and c_n by

$$|a_n| = \frac{1}{2}\sqrt{b_n^2 + c_n^2} \,, \tag{D.9}$$

$$\varphi_n = -\arctan \frac{c_n}{b_n} \,, \tag{D.10}$$

$$\tag{D.11}$$

The functions $\mathrm{e}^{\mathrm{j}n\omega_0 t}$ form a complete orthogonal system of basis functions in the interval $[-T_0/2, T_0/2]$. Applying

$$\frac{1}{T_0} \int_{-T_0/2}^{T_0/2} \mathrm{e}^{\mathrm{j}(n-m)\omega_0 t}\, \mathrm{d}t = \delta_{mn} \,. \tag{D.12}$$

to (D.2) we obtain

$$a_n = \frac{1}{T_0} \int_{-T_0/2}^{T_0/2} s(t)\, \mathrm{e}^{-\mathrm{j}n\omega_0 t}\, \mathrm{d}t \,. \tag{D.13}$$

It can be shown that arbitrary periodic functions $s(t)$ with period T_0, which are smooth with the exception of a finite number of discontinuities and bounded within an interval of length T_0, may be represented by a Fourier series according to (D.2). At the points of discontinuity t_d of the function $s(t)$ the Fourier series converges to the mean value of the neighboring values on both sides of the discontinuity. Applying this equation to (D.2) we obtain

$$\lim_{\delta \to 0} \frac{1}{2} \left(s(t_d + \delta) + s(t_d - \delta) \right) = \sum_{n=-\infty}^{\infty} a_n\, \mathrm{e}^{\mathrm{j}n\omega_0 t_d} \,. \tag{D.14}$$

From (D.6), (D.7) and (D.8) it follows that

$$b_n = \frac{2}{T_0} \int_{-T_0/2}^{T_0/2} s(t) \cos n\omega_0 t\, \mathrm{d}t \,, \tag{D.15}$$

$$c_n = \frac{2}{T_0} \int_{-T_0/2}^{T_0/2} s(t) \sin n\omega_0 t\, \mathrm{d}t \,. \tag{D.16}$$

We can decompose any function $s(t)$ in an even part $s_e(t)$ and an odd part $s_o(t)$,

$$s(t) = s_e(t) + s_o(t)\,. \tag{D.17}$$

such that $s_e(t)$ and $s_o(t)$ are given by

$$s_e(t) = \frac{1}{2}(s(t) + s(-t))\,, \tag{D.18}$$

$$s_o(t) = \frac{1}{2}(s(t) - s(-t))\,\mathrm{d}t\,. \tag{D.19}$$

From (D.8), (D.17), (D.18) and (D.19) it follows that

$$s_e(t) = \frac{b_0}{2}\sum_{n=1}^{\infty} b_n \cos n\omega_0 t\,, \tag{D.20}$$

$$s_o(t) = \sum_{n=1}^{\infty} c_n \sin n\omega_0 t\,. \tag{D.21}$$

D.2 THE FOURIER INTEGRAL

In the case of impulsive and transient phenomena in many other cases we have to deal with non periodic signals. We start with the consideration of periodic signals and make the transition ΔT to infinity. In this way the distance of spectral lines in the frequency domain Δf given by

$$\Delta f = \frac{\omega}{2\pi} = \frac{1}{T_0} \tag{D.22}$$

is going to zero. We introduce a spectral amplitude density

$$\underline{S}(n\omega_0) = \frac{\underline{a}_n}{\Delta f} \tag{D.23}$$

and obtain from (D.2)

$$s(t) = \sum_{n=-\infty}^{+\infty} \underline{S}(n\omega_0)\mathrm{e}^{\mathrm{j}n\omega_0 t}\,\Delta f \tag{D.24}$$

and from (D.13)

$$\underline{S}(n\omega_0) = \int_{-T_0/2}^{T_0/2} s(t)\mathrm{e}^{-\mathrm{j}n\omega_0 t}\,\mathrm{d}t\,. \tag{D.25}$$

Making the transitions $T_0 \to \infty$ and with this $\Delta f \to \mathrm{d}f = \mathrm{d}\omega/2\pi$ and $n\omega_0 \to \omega$, we obtain from (D.24) and (D.25)

$$s(t) = \frac{1}{2\pi}\int_{-\infty}^{\infty} \underline{S}(\omega)\mathrm{e}^{\mathrm{j}\omega t}\,\mathrm{d}\omega\,, \tag{D.26}$$

$$\underline{S}(\omega) = \int_{-\infty}^{\infty} s(t)\mathrm{e}^{-\mathrm{j}\omega t}\,\mathrm{d}t\,, \tag{D.27}$$

These integrals represent linear functional transformations. The Fourier transformation is given by (D.27) and the inverse Fourier transformation is given by (D.26). The Fourier transformation and its inverse are denoted symbolically by

$$\underline{S}(\omega) = \mathfrak{F}\{s(t)\}\,, \tag{D.28}$$

$$s(t) = \mathfrak{F}^{-1}\{\underline{S}(\omega)\}\,. \tag{D.29}$$

Furthermore for the symbolic notation of a pair of Fourier transforms the correspondence symbol

$$s(t) \circ\!\!-\!\!\bullet\ \underline{S}(\omega) \tag{D.30}$$

is used. Replacing in (D.26) and (D.27) ω and $\mathrm{d}\omega$, respectively, with $2\pi f$ and $2\pi\,\mathrm{d}f$ respectively we obtain

$$s(t) = \int_{-\infty}^{\infty} \underline{S}(2\pi f)\mathrm{e}^{2\pi\mathrm{j}ft}\,\mathrm{d}f\,, \tag{D.31}$$

$$\underline{S}(2\pi f) = \int_{-\infty}^{\infty} s(t)\mathrm{e}^{-2\pi\mathrm{j}ft}\,\mathrm{d}t\,. \tag{D.32}$$

A sufficient condition for the existence of a Fourier integral is that $s(t)$ is smooth with the exception of a finite number of discontinuities and is absolutely integrable. The latter condition means that the integral

$$\int_{-\infty}^{\infty} |s(t)|\,\mathrm{d}t \tag{D.33}$$

exists.

The differentiation in the time domain yields a multiplication with $\mathrm{j}\omega$ in the frequency domain,

$$\frac{\mathrm{d}s(t)}{\mathrm{d}t} \circ\!\!-\!\!\bullet\ \mathrm{j}\omega\underline{S}(\omega)\,, \tag{D.34}$$

$$\frac{\mathrm{d}^n}{\mathrm{d}t^n}s(t) \circ\!\!-\!\!\bullet\ (\mathrm{j}\omega)^n\underline{S}(\omega)\,, \tag{D.35}$$

whereas multiplication with t in time yields a multiplication with j and differentiation in the frequency domain,

$$t\,s(t) \circ\!\!-\!\!\bullet\ \mathrm{j}\frac{\mathrm{d}\underline{S}(\omega)}{\mathrm{d}\omega}, \tag{D.36}$$

$$t^n\,s(t) \circ\!\!-\!\!\bullet\ \mathrm{j}\frac{\mathrm{d}^n}{\mathrm{d}\omega^n}\underline{S}(\omega). \tag{D.37}$$

A shift in the time domain by t_0 yields the multiplication with a phase factor in the frequency domain,

$$s(t-t_0) \circ\!\!-\!\!\bullet\ \mathrm{e}^{-\mathrm{j}\omega t_0}\underline{S}(\omega), \tag{D.38}$$

and a shift in the frequency domain by ω_0 yields

$$\mathrm{e}^{\mathrm{j}\omega_0 t}s(t) \circ\!\!-\!\!\bullet\ \underline{S}(\omega-\omega_0). \tag{D.39}$$

The complex conjugate functions transform as follows

$$s^*(t) \circ\!\!-\!\!\bullet\ S^*(-\omega), \tag{D.40}$$

$$s^*(-t) \circ\!\!-\!\!\bullet\ S^*(\omega). \tag{D.41}$$

Scaling a function in the time domain by a factor α yields a reciprocal scaling in the frequency domain

$$s(\alpha t) \circ\!\!-\!\!\bullet\ \frac{1}{\alpha}\underline{S}(\omega/\alpha), \tag{D.42}$$

Interchanging of the time and frequency variables yields

$$\underline{S}^*(\pi t/T^2) \quad \circ\!\!-\!\!\bullet \quad T^2 s^*(\omega T^2/2\pi). \tag{D.43}$$

D.3 THE DELTA DISTRIBUTION

The *delta distribution* has the properties

$$\delta(t) = \begin{cases} 0 & \text{for } t \neq 0 \\ \infty & \text{for } t = 0 \end{cases}, \tag{D.44}$$

$$\int_{-\infty}^{\infty} \delta(t)\,\mathrm{d}t = 1. \tag{D.45}$$

Since the delta distribution vanishes everywhere with an exception at $x = 0$, for an arbitrary function integrable within a neighborhood of $x = 0$, the relation

$$\int_{-\infty}^{\infty} f(x)\delta(x)\,\mathrm{d}x = f(0) \tag{D.46}$$

must hold. For arbitrary smooth functions $f(x)$ we obtain

$$\int_{-\infty}^{\infty} f(x)\delta(x - x_0)\,\mathrm{d}x = f(x_0)\,. \tag{D.47}$$

The proof follows by inserting $f(x + x_0)$ instead of $f(x)$ in (D.46). The delta distribution is not a function but can be considered as a symbol indicating the value of the integral (D.46) or (D.47), respectively. From (D.46) we obtain

$$\int_{-\infty}^{\infty} \delta(t)\mathrm{e}^{-\mathrm{j}\omega t}\,\mathrm{d}t = 1\,. \tag{D.48}$$

From (D.27), (D.28), (D.29) and (D.46) it follows that

$$1 = \mathfrak{F}\{\delta(t)\}\,, \tag{D.49}$$

$$\delta(t) = \mathfrak{F}^{-1}\{1\}\,. \tag{D.50}$$

We have to consider that the integral for the inverse transformation

$$\delta(t) = \frac{1}{2\pi}\int_{-\infty}^{\infty} \mathrm{e}^{\mathrm{j}\omega t}\,\mathrm{d}\omega \tag{D.51}$$

does not exist. However we can consider this integral as a symbol for the inverse Fourier transform according to (D.50). In symbolic notation we can write

$$\delta(t) \circ\!\!-\!\!\bullet\ 1\,, \tag{D.52}$$

$$\delta(t - t_0) \circ\!\!-\!\!\bullet\ \mathrm{e}^{-\mathrm{j}\omega t_0} \tag{D.53}$$

where (D.53) follows from (D.52) and (D.38) or from (D.27) and (D.47). The argument of the delta distribution may be scaled by

$$\delta(ax) = \frac{1}{a}\delta(x) \tag{D.54}$$

The proof follows from substitution $y = ax$ from

$$\int \delta(ax)\,\mathrm{d}x = \frac{1}{a}\int \delta(y)\,\mathrm{d}y = \frac{1}{a}\,. \tag{D.55}$$

From this we obtain

$$\delta(\omega) = \delta(2\pi f) = \frac{1}{2\pi}\delta(f)\,. \tag{D.56}$$

For the representation of line spectra the delta function may be used in the frequency domain. From (D.26), (D.48) and (D.56) it follows that

$$1 \circ\!\!-\!\!\bullet\ 2\pi\delta(\omega) = \delta(f)\,, \tag{D.57}$$

$$e^{j\omega_0 t} \circ\!\!-\!\!\bullet\ 2\pi\delta(\omega-\omega_0) = \delta(f-f_0)\,, \tag{D.58}$$

$$\cos\omega_0 t \circ\!\!-\!\!\bullet\ \pi\left[\delta(\omega-\omega_0)+\delta(\omega+\omega_0)\right]\,, \tag{D.59}$$

$$\sin\omega_0 t \circ\!\!-\!\!\bullet\ j\pi\left[\delta(\omega+\omega_0)-\delta(\omega-\omega_0)\right]\,. \tag{D.60}$$

Using the delta distribution we may also represent the Fourier transforms of harmonic signals. Also for such signals (D.33) is not fulfilled.

The sign(x) function is defined by

$$\mathrm{sign}(x) = \begin{cases} -1 & \text{for } x < 0 \\ 0 & \text{for } x = 0 \\ 1 & \text{for } x > 0 \end{cases}\,. \tag{D.61}$$

The uncertainty of the value of sign(0) has been removed by choosing the value 0. The Fourier transformation of the sign function is

$$\mathrm{sign}(t) \circ\!\!-\!\!\bullet\ \frac{2}{j\omega}\,. \tag{D.62}$$

The *step function* $\sigma(x)$ of the delta distribution is given by

$$\sigma(x) = \int_{-\infty}^{x} \delta(x_1)\,dx_1 = \begin{cases} 0 & \text{for } x < 0 \\ 1/2 & \text{for } x = 0 \\ 1 & \text{for } x > 0 \end{cases}\,. \tag{D.63}$$

The proof follows from (D.45) if we consider that the delta function gives only a contribution if 0 is within the interval of integration. Since

$$\sigma(x) = \frac{1}{2} + \frac{1}{2}\,\mathrm{sign}(x) \tag{D.64}$$

the Fourier transform of the σ function is given by

$$\sigma(t) \circ\!\!-\!\!\bullet\ \frac{1}{j\omega}\pi\,\delta(\omega)\,. \tag{D.65}$$

We also may introduce distributions representing the derivation of the delta function. Applying partial integration yields for a differentiable smooth function

$$\int_{-\infty}^{\infty} f'(x)\delta(x)\,dx = -\int_{-\infty}^{\infty} f(x)\delta'(x)\,dx\,. \tag{D.66}$$

From this we obtain

$$\int_{-\infty}^{\infty} f(x)\delta'(x)\,\mathrm{d}x = -f'(0)\,, \tag{D.67}$$

$$\int_{-\infty}^{\infty} f(x)\delta^{(n)}(x)\,\mathrm{d}x = (-1)^n f^{(n)}(0)\,. \tag{D.68}$$

The superscript $^{(n)}$ denotes the n-fold derivative with respect to the argument. The Fourier transformations of the derivatives of the delta function are given by

$$\delta^{(n)}(t) \quad \circ\!\!-\!\!\bullet \quad (\mathrm{j}\omega)^n\,. \tag{D.69}$$

REFERENCES

[1] G. Arfken, *Mathematical Methods for Physicists.* New York: Academic Press, 1985.

[2] R. Bracewell, *Fourier Transform and its Applications.* New York: McGraw-Hill, 1999.

[3] B. Friedman, *Lectures on Applications-Oriented Mathematics.* New York: John Wiley & Sons, 1969.

List of Symbols

Symbol	Description	Reference
A	Surface	(2.15)
A_e	Effective aperture	(11.148)
∂A	Boundary of a surface A	(2.13)
$\underline{a}$	Wave amplitude phasor	(8.40a)
$\underline{\boldsymbol{a}}$	Incident wave amplitude vector	(10.38)
$\boldsymbol{A}(\boldsymbol{x}, t)$	Magnetic vector potential	(3.1)
$\mathbf{A}$	Chain matrix	(10.11)
$\mathcal{A}(\boldsymbol{x}, t)$	Magnetic vector potential form	(3.1)
α	Attenuation coefficient	(3.62)
$\underline{b}$	Wave amplitude phasor	(8.40b)
$\underline{\boldsymbol{a}}$	Scattered wave amplitude vector	(10.38)
$\boldsymbol{B}(\boldsymbol{x}, t)$	Magnetic flux density vector	(2.24)
$\mathbf{B}$	Inverse chain matrix	(10.30)
$\mathcal{B}(\boldsymbol{x}, t)$	Magnetic flux density differential form	(2.30)
$\underline{\mathcal{B}}(\boldsymbol{x}, \omega)$	Magnetic flux density differential form phasor	(2.135)
β	Phase coefficient	(3.62)
C	Capacitance	(5.35)
C_{lk}	Partial capacitance	(5.25)
C'	Capacitance per unit of length	(5.50)
C''	Capacitance per unit of area	(5.54)
c	Speed of light	(2.74)
c_0	Free space speed of light	(2.56)
d	Differential operator	(2.112)
d	Exterior derivative	(2.112)
$\tilde{\mathrm{d}}$	Covariant derivative	(3.13)
δ_{mn}	Kronecker delta symbol	(A.8)
δ_{xyz}	Kronecker tensor	(A.12)
$\delta(x)$	Delta distribution	(D.47)
$\delta(\boldsymbol{x} - \boldsymbol{x}')$	Three-dimensional delta distribution	(5.9)

Symbol	Description	Reference
$\delta_e(\omega)$	Dielectric loss angle	(2.140)
$\delta_m(\omega)$	Magnetic loss angle	(2.141)
$\boldsymbol{D}(\boldsymbol{x},t)$	Electric flux density vector	(2.28)
$\mathcal{D}(\boldsymbol{x},\omega)$	Electric flux density differential form phasor	(2.133)
$\underline{\mathcal{D}}(\boldsymbol{x},t)$	Electric flux density differential form	(2.11)
$\boldsymbol{E}(\boldsymbol{x},t)$	Electric field vector	(2.2)
$\underline{\boldsymbol{E}}(\boldsymbol{x},\omega)$	Electric field vector phasor	(2.126)
$\mathcal{E}(\boldsymbol{x},t)$	Electric field differential form	(2.11)
$\underline{\mathcal{E}}(\boldsymbol{x},\omega)$	Electric field differential form phasor	(2.130)
ε	Permittivity	(2.23)
ε_r	Relative permittivity	(2.27a)
$\underline{\varepsilon}(\omega)$	Complex permittivity	(2.139)
$\underline{\varepsilon}_r(\omega)$	Complex relative permittivity	(2.142)
$\varepsilon'(\omega)$	Real part of complex permittivity	(2.139)
$\varepsilon''(\omega)$	Negative imaginary part of complex permittivity	(2.139)
f	Frequency	(2.85)
$F(\vartheta,\varphi)$	Element characteristics	(11.167)
$\boldsymbol{F}(\boldsymbol{x},t)$	Force vector	(2.7)
$\boldsymbol{F}^{el}(\boldsymbol{x},t)$	Electric force vector	(2.3)
$\boldsymbol{F}^{mag}(\boldsymbol{x},t)$	Lorentz force vector	(2.5)
$\mathfrak{F}\{s(t)\}$	Fourier transform	(D.28)
G	Conductance	(9.23a)
G_r	Radiation conductance	(11.226)
G'	Conductance per unit of length	(7.66)
$G_0(\boldsymbol{x},\boldsymbol{x}')$	Green's function	(5.13)
$\mathcal{G}(\boldsymbol{x},\boldsymbol{x}')$	Green's double form	(5.58)
γ	Propagation coefficient	(3.60)
$\boldsymbol{H}(\boldsymbol{x},t)$	Magnetic field vector	(2.5)
$\underline{\boldsymbol{H}}(\boldsymbol{x},\omega)$	Magnetic field vector phasor	(2.126)
$\mathcal{H}(\boldsymbol{x},t)$	Magnetic field differential form	(2.14)
$\underline{\mathcal{H}}(\boldsymbol{x},\omega)$	Magnetic field differential form phasor	(2.133)
$i(t)$	Electric current	(2.13)
$\underline{I}$	Electric current phasor	(10.1)
$\underline{\boldsymbol{V}}$	Current vector	(10.3b)
$\Im\{z\}$	Imaginary part of z	(3.61)
$\mathcal{I}(\boldsymbol{x},\boldsymbol{x}')$	Identity kernel	(5.60)
$\boldsymbol{J}(\boldsymbol{x},t)$	Current density vector	(2.21)
$\mathcal{J}(\boldsymbol{x},t)$	Current density differential form	(2.21)
$\underline{\mathcal{J}}(\boldsymbol{x},\omega)$	Current density differential form phasor	(2.133)

Symbol	Description	Reference
$\mathcal{J}_0(\boldsymbol{x}, t)$	Impressed current density differential form	(2.58)
$\underline{\mathcal{J}}_0(\boldsymbol{x}, \omega)$	Impressed current density differential form phasor	(2.137)
$\boldsymbol{J}_A(\boldsymbol{x}, t)$	Surface current density vector	(2.168)
$\mathcal{J}_A(\boldsymbol{x}, t)$	Surface current density differential form	(2.168)
$\underline{\mathcal{J}}_A(\boldsymbol{x}, \omega)$	Surface current density differential form phasor	(2.200)
g	Characteristic impedance square roots	(10.34)
k	Wave number	(3.55)
$\boldsymbol{k}$	Wave vector	(3.53)
L	Inductance	(5.76)
L'	Inductance per unit of length	(5.96)
$L_\square$	Inductance of a square parallel plate element	(5.103)
$\mathcal{J}(\boldsymbol{x}, t)$	Current density differential form	(2.21)
$\mathcal{J}_0(\boldsymbol{x}, t)$	Impressed current density differential form	(2.58)
$\mathcal{J}_A(u, v)$	Surface current density differential form	(2.168)
k	Wave number	(2.86)
λ	Wavelength	(2.86)
$M(\vartheta, \varphi)$	Array factor	(11.175)
M_{ij}	Mutual inductance	(5.75)
$\mathcal{M}_{e0}(\boldsymbol{x}, t)$	Impressed electric polarization form	(3.27)
$\underline{\mathcal{M}}_{e0}(\boldsymbol{x}, \omega)$	Impressed electric polarization form phasor	(3.32)
$\underline{\mathcal{M}}_{m0}(\boldsymbol{x}, \omega)$	Impressed magnetic polarization form phasor	(3.38)
$\mathcal{M}_{eA}(u, v, t)$	Electric surface polarization form	(2.176)
$\underline{\mathcal{M}}_{eA}(u, v, \omega)$	Electric surface polarization form phasor	(2.201)
$\mathcal{M}_{mA}(u, v, t)$	Magnetic surface polarization form	(2.178)
$\underline{\mathcal{M}}_{mA}(u, v, \omega)$	Magnetic surface polarization form phasor	(2.202)
μ	Permeability	(2.24)
μ_r	Relative permeability	(2.27b)
$\underline{\mu}(\omega)$	Complex permeability	(2.141)
$\underline{\mu}_r(\omega)$	Complex relative permeability	(2.141)
$\mu'(\omega)$	Real part of complex permeability	(2.143)
$\mu''(\omega)$	Negative imaginary part of complex permeability	(2.141)
n	Normal unit one-form	(A.107)
n	Refractive index	(2.75)
ω	Angular frequency	(2.85)
$p_L(\boldsymbol{x}, t)$	Power loss density	(4.9)
$\mathcal{P}_L(\boldsymbol{x}, t)$	Power loss density form	(4.9)
$p_{Le}(\boldsymbol{x}, \omega)$	Electric power loss density	(4.30)
$\mathcal{P}_{Le}(\boldsymbol{x}, \omega)$	Electric power loss density form	(4.30)
$p_{Lm}(\boldsymbol{x}, \omega)$	Magnetic power loss density	(4.32)

Symbol	Description	Reference
$\mathcal{P}_{Lm}(\boldsymbol{x}, \omega)$	Magnetic power loss density form	(4.32)
$\mathcal{P}_0(\boldsymbol{x}, t)$	Added power density form	(4.11)
$\overline{P}_0(\boldsymbol{x})$	Added time-average power density	(4.37)
P_c	Complex power	(9.1)
$P_{c0}(\boldsymbol{x}, \omega)$	Added complex power density	(4.35)
$\mathcal{P}_{c0}(\boldsymbol{x}, \omega)$	Added complex power density form	(4.35)
$\boldsymbol{\Pi}_e(\boldsymbol{x}, t)$	Electric Hertz vector	(3.19)
$\varPi_e(\boldsymbol{x}, t)$	Electric Hertz form	(3.19)
$\underline{\varPi}_e(\boldsymbol{x}, \omega)$	Electric Hertz form phasor	(3.34)
$\underline{\boldsymbol{\Pi}}_m(\boldsymbol{x}, t)$	Magnetic Hertz vector phasor	(3.43)
$\underline{\varPi}_m(\boldsymbol{x}, \omega)$	Magnetic Hertz form phasor	(3.44)
ϕ	Normalized potential	(5.21)
φ_0	Phase angle	(2.87)
$\Phi(t)$	Magnetic flux	(2.31)
$\Psi(t)$	Electric flux	(2.29)
$q(t)$	Electric charge	(2.3)
$\mathcal{Q}(\boldsymbol{x}, t)$	Electric charge density differential form	(2.36)
$\underline{\mathcal{Q}}(\boldsymbol{x}, \omega)$	Electric charge density differential form phasor	(2.136)
$\mathcal{Q}_A(\boldsymbol{x}, t)$	Electric area charge density differential form	(2.157)
$\underline{\mathcal{Q}}_A(\boldsymbol{x}, \omega)$	Electric area charge density differential form phasor	(2.199)
R	Resistance	(9.23b)
R_r	Radiation resistance	(11.53)
R_{ij}	Reaction of field $\mathcal{E}_i$, $\mathcal{H}_i$ on sources $\underline{\mathcal{M}}_{e0j}$, $\underline{\mathcal{M}}_{m0j}$	(4.62)
$\Re\{z\}$	Real part of z	(2.126)
$\rho(\boldsymbol{x}, t)$	Electric charge density	(2.36)
$\rho_A(\boldsymbol{x}, t)$	Electric area charge density	(2.157)
s_i	Unit one-form	(A.107)
$s(t)$	Signal	(D.26)
$\boldsymbol{S}(\boldsymbol{x}, t)$	Poynting vector	(4.13)
$\boldsymbol{S}$	Scattering matrix	(10.38)
$\mathcal{S}(\boldsymbol{x}, t)$	Poynting differential form	(4.13)
$\underline{S}(\omega)$	Signal spectrum	(D.27)
σ	Conductivity	(2.58)
$\boldsymbol{T}(\boldsymbol{x})$	Complex Poynting vector	(4.21)
$\boldsymbol{T}$	Transmission matrix	(10.52)
$\mathcal{T}(\boldsymbol{x})$	Complex Poynting differential form	(4.21)
$\boldsymbol{v}(\boldsymbol{x}, t)$	Velocity vector	(2.7)
$v(t)$	Voltage	(2.9)
$\underline{V}$	Voltage phasor	(10.1)

Symbol	Description	Reference
$\underline{V}$	Voltage vector	(10.3a)
$v_m(t)$	Magnetic voltage	(2.60)
V	Volume	(2.15)
∂V	Boundary of a volume V	(2.38)
$W(t)$	Energy	(2.8)
$w_e(\boldsymbol{x}, t)$	Electric energy density	(4.1)
$\overline{w}_e(\boldsymbol{x})$	Average electric energy density	(4.26)
$\mathcal{W}_e(\boldsymbol{x}, t)$	Electric energy density form	(4.3)
$\overline{\mathcal{W}}_e(\boldsymbol{x})$	Time-average electric energy density form	(4.26)
$w_m(\boldsymbol{x}, t)$	Magnetic energy density	(4.2)
$\overline{w}_m(\boldsymbol{x})$	Average magnetic energy density	(4.27)
$\mathcal{W}_m(\boldsymbol{x}, t)$	Magnetic energy density form	(4.3)
$\overline{\mathcal{W}}_m(\boldsymbol{x})$	Time-average magnetic energy density form	(4.27)
$\boldsymbol{Y}$	Impedance matrix	(10.6)
$\underline{y}$	Normalized admittance	(8.60b)
Z_0	Characteristic impedance	(7.37)
Z_F	Wave impedance	(2.78)
Z_{F0}	Free space wave impedance	(2.78)
Z_{mn}	Impedance matrix elements	(10.4)
$\boldsymbol{Z}$	Impedance matrix	(10.1)
$\underline{z}$	Normalized impedance	(8.60a)
$\boldsymbol{0}$	Null matrix	(C.12)
$\boldsymbol{1}$	Unit matrix	(C.16)
$\lrcorner$	Contraction	(2.181)
$\wedge$	Exterior product	(2.6)
$\circ\!\!-\!\!\bullet$	Fourier transform	(D.30)
$\in$	in	(2.159)
$\notin$	not in	(2.159)
$\star$	Hodge operator	(2.6)
Δ	Laplace operator	(3.14)
$\|\boldsymbol{a}\|$	Magnitude of the vector $\boldsymbol{a}$	(A.4)
∂	Partial differential operator	(2.94)
$\dot{\mathcal{D}}$	Partial derivative of $\mathcal{D}$ with respect to t	(4.5)
$P^{\triangleleft}\mathcal{A}$	Pullback of $\mathcal{A}$ to the path P	(A.173)
$\overline{\mathcal{S}(\boldsymbol{x}, t)}$	Time average of $\mathcal{S}(\boldsymbol{x}, t)$	(4.25)
$\perp_n$	Twist operator	(2.170)
$\times$	Vector product	(2.17)

About the Author

Peter Russer received his Dipl.-Ing. degree in 1967 and his Dr. techn. degree in 1971, both in electrical engineering, and from the Technische Universität Wien, Austria, where he was assistant professor from 1968 to 1971. In 1971 he joined the Research Institute of AEG-Telefunken in Ulm, Germany, where he worked on fiber optic communication, broadband solid-state electronic circuits, statistical noise analysis of microwave circuits, laser modulation and fiber optic gyroscopes. In 1979 he received the NTG Award for the publication "Electronic circuits for high bit rate digital fiber optic communication systems." Since 1981 he has been full professor and head of the Institute of High Frequency Engineering at the Technische Universität München (TUM), Germany. In 1990 he was visiting professor at the University of Ottawa and at the University of Victoria in 1993. From October 1992 through to March 1995 he was director of the Ferdinand-Braun-Institut für Höchstfrequenztechnik, Berlin, Germany. In 1994 he was elected Fellow of the IEEE for fundamental contributions to noise analysis and low-noise optimization of linear electronic circuits with general topology. His current research interests are electromagnetic fields, antennas, integrated microwave and millimeter-wave circuits, statistical noise analysis of microwave circuits and methods for the computer-aided design of microwave circuits. He has published more than 400 scientific papers in refereed journals and conference proceedings.

Peter Russer has developed a variety of courses in RF techniques, microwaves, quantum electronics and optical communications. He is the program director of the international graduate program "Master of Science in Microwave Engineering" at TUM. Over the years he has graduated more than 400 students of which more than 50 received their PhD degree.

Peter Russer has served as a member of technical program committees and steering committees of various international conferences (IEEE MTT-S, European Microwave Conference and as the member of the editorial board of several international journals *(Electromagnetics, International Journal of Numerical Modeling).* In 2002 he was elected as Chairman of U.R.S.I. Commission D for a three-year term. He is Fellow of the IEEE, member of the German Informationstechnische Gesellschaft (ITG) and the German Physical Society as well as the Austrian Physical Society.

Index

Admittance
matrix, 239
normalized, 210
per unit of length, 196
representation, 239
Ampère's law, 23–25, 143
Ampère, André, 2
Angle product, 52, 357
Antenna
logarithmic-periodic, 322
biconical, 321
broadband, 321
frequency-independent, 322
Antenna array, 301
multiplicative law, 303
array factor, 303
element characteristics, 302
linear, 302
microstrip, 319
Antenna gain, 296
Aperture antenna, 315
Area charge density, 47
Area charge differential form, 47
Array factor, 303
Attenuation coefficient, 64

Babinet's principle, 322
Basis
three-form, 45
two-forms, 45
Basis functions, 325, 332
complete set of, 331, 332
entire domain basis functions, 332
subsectional basis functions, 332
Bessel differential equation, 381
modified, 384
Bessel function, 381
modified, 384
modified of the first kind, 384
modified of the second kind, 384
of the first kind, 381
of the second kind, 381
spherical, 67, 386
Boundary condition, 46, 53
electric field intensity, 51
electric flux density, 48
magnetic field intensity, 51
magnetic flux density, 47
normal boundary condition, 46, 48
practical boundary condition, 120
tangential boundary condition, 49, 52

Capacitance, 93, 96
per unit of length, 140
partial capacitance, 94
per unit of area, 99
per unit of length, 98
Capacitor, 4, 22, 89
spherical capacitor, 96
Cartan, Elie, 5
Cartesian coordinate system, 10
right handed, 10
Cassegrain antenna, 314
Chain matrix, 241
Chain representation, 240
Characteristic equation

planar waveguide, 177
Characteristic impedance, 141, 149, 196
coplanar stripline, 193
coplanar waveguide, 193
microstrip line, 189
waveguide, 196
Charge
per unit of length, 140
Circulation, 25
Coaxial line, 131
Commutation relation, 355
Complete system of functions
spherical, 67
spherical harmonics, 393
Complex amplitude, 160
TEM wave, 198
Complex power, 213
Computer aided design, 325
Conductance
per unit of length, 146
Conductivity, 26
Conformal antenna, 317, 319
Conservative field, 58
Constitutive equations, 19
Continuity equation, 25
differential form, 41
integral form, 26
Contraction, 52, 357
Coordinate curve, 44
Coordinate surface, 44
Coordinate system
Cartesian coordinates, 351
circular cylindrical, 45, 370
curvilinear coordinates, 364
orthogonal, 366
spherical, 45, 373
Coulomb force, 9
Coulomb, Charles, 2
Coupling
critical coupling, 232
overcoupled, 232
undercoupled, 232
Covariant derivative, 59, 359
Cramer's rule, 398
Crank-Nicolson scheme, 344
Curl operator, 368, 372, 375
Current
conduction current, 24
displacement current, 24
generalized, 195
Current density
impressed surface current density, 51
surface current density, 51
surface current density, 49
Curvilinear coordinate system
orthogonal, 44
Cut-off frequency, 131, 153
material cut-off frequency, 44
Cut-off wavelength, 153, 169, 170, 186

Determinant, 397
Dielectric loss
dielectric loss angle, 43
dielectric loss factor, 43
Dielectric waveguide, 132
circular, 132
Differential form, 14
closed form, 358
degree, 40
double form, 100, 377
electric field differential form, 13
exact form, 358
exterior differential form, 17
magnetic field differential form, 15
one-form, 14
Differential operator
curl, 380
divergence, 380
gradient, 378
Dipole, 266, 271
electric dipole, 96
Dipole antenna, 265, 280
linear, 283
symmetric, 282
Dirac delta distribution, 47
three-dimensional, 91
Directional coupler, 259
Directivity, 266
Divergence operator, 369, 372, 375
Double form
double one-form, 100, 377
Duality, 62
Dyadic, 377

Effective antenna length, 290, 292
Effective aperture, 296
Effective area, 296
Electric charge, 21, 27

charge density, 25
charge density form, 21
electric charge density, 21
Electric current, 15, 27
current density, 16–18, 25
current density form, 18
Electric displacement, 19
Electric field, 10
force picture, 14
intensity, 10, 12, 25
Electric flux, 12, 20, 23, 27
density, 23, 47
electric flux density, 19, 20, 23, 25
electric flux form, 20
Electric force, 11
Electric polarization
impressed electric polarization, 60, 61
impressed electric polarization form, 60
impressed electric polarization phasor, 61
Electric potential, 13
electric potential difference, 13
Electric power, 77
Electromagnetic field, 9
Electromagnetic potentials, 57
Electromagnetic wave, 28
homogeneous wave equation, 60
plane wave, 28, 30, 31, 33
time-harmonic, 32
wave equation, 35, 57
wave front, 28, 29
Electroquasistatic approximation, 90, 91
Electrostatic field, 89
Element characteristics, 302
Energy, 13
Energy density
electric, 71
magnetic, 71
Energy density form
time-average electric, 75
time-average magnetic, 75
Energy picture, 5
Equivalence principle, 83
Equivalent circuit, 223
Equivalent sources, 83
Exterior derivative, 358
in curvilinear coordinates, 367
Exterior differential form, 355
p-form, 355
one-form, 20
three-form, 21
two-form, 17, 20
zero-form, 21
Exterior product, 5, 17

Factorial function, 382
Far-field, 273–275, 278–280, 283, 284, 287, 289, 299, 307
antenna array, 302
approximation, 68
horn antenna, 309
linear dipole antenna, 283, 292
loop antenna, 289
slot antenna, 316
surface emitter, 307
Faraday's law, 4, 24, 25, 41, 42, 143, 162
Faraday, Michael, 2, 24
Field concept, 9, 12, 78
Field intensity, 18
Field line, 14
Finite difference approach, 342
Finite difference method, 326
Finite element method, 326
Finite integration approach, 342
Flux density, 18
electric, 19
Flux linkage, 104
Force picture, 5
Foster equivalent circuit, 223
first kind, 223
second kind, 223
Foster representation, 223
first kind, 223
second kind, 223
Foster's reactance theorem, 215, 217
Fourier series, 401
Fractional expansion representation, 223
Frequency, 32
angular frequency, 32
band, 1
Fresnel lens, 315
Friis transmission formula, 301
Function
complete set of, 329, 330
inner product, 328
norm, 328
square integrable, 328

Gamma function, 382

Gauge transformation, 58
Gauss' law, 25
Generating function, 383, 385
Gradient, 367
Grassmann, Hermann, 5
Green's form
  double form, 267
  double one-form, 100
Green's function, 267, 331
  dyadic, 100
  microstrip line, 192
  scalar, 93
Group velocity, 135

Hankel function, 381, 382
  of the first kind, 381
  of the second kind, 381
  spherical, 386
Helmholtz equation, 61
  inhomogeneous Helmholtz equation, 62
  scalar, 376
  spherical coordinates, scalar, 67
  vector, 376
Hermitian form, 400
  positive definite, 400
  positive semidefinite, 252, 400
Hertz differential form
  electric, 59
  magnetic, 62
  retarded electric, 270
  retarded magnetic, 270
Hertz vector
  differential form phasor, 61
  electric, 59, 61, 62
  magnetic, 62
Hertzian dipole, 271, 273–275, 278, 279
  receiving antenna, 290
  time domain, 276
Hilbert space, 328
Hodge operator, 20, 356
  curvilinear coordinates, 45, 367
Horn paraboloid antenna, 314
Huygens' Principle, 2
Huygens, Christian, 2
Hybrid junction, 261

Identity kernel, 100, 268
Image line, 132
Impedance
  matrix, 239
  normalized, 210
  per unit of length, 196
  representation, 239
Impressed current density, 26
Inductance, 102, 103, 105
  mutual, 102, 103
  per unit of length, 105, 106, 140
  self-inductance, 103
Inductor, 4, 22, 90
Integral
  equation, 339
  Integral equation method, 325
  surface integral, 17
Irrotational field, 58

Jacobian determinant, 365

Kernel, 378
  identity kernel, 378
  Richmond form, 341
Kernel function, 328
Kirchhoff laws, 11, 37
  Kirchhoff current law, 11, 38
  Kirchhoff voltage law, 11, 39
Kronecker delta symbol, 352
Kronecker tensor, 353

Laplace equation
  two-dimensional, 137, 190
Laplace expansion, 398
Laplace operator, 59, 359
  for one-form, 63, 360
  for one-form, in circular cylindric coordinates, 373
  for one-form, in spherical coordinates, 376
  for one-forms in curvilinear coordinates, 281
  for one-forms in spherical coordinates, 281
  scalar, in circular cylindric coordinates, 372
  scalar, in curvilinear coordinates, 369
  scalar, in spherical coordinates, 375
Legendre differential equation, 388
  ordinary Legendre differential equation, 388
Legendre function

associated, 388
associated, first kind, 388
associated, second kind, 388
ordinary, first kind, 388
ordinary, second kind, 388
Legendre polynomials, 388
Lens, 313
antenna, 314
Line integral, 14
Loop antenna, 286
far-field, 289
radiated power, 289
receiving antenna, 291
Lorentz condition, 58, 59
Lorentz force, 10

Macroscopic field, 19
Magic T, 261
Magnetic field, 10
intensity, 15, 25
Magnetic flux, 12, 20, 27
continuity law, 25
magnetic flux density, 11, 19, 20, 25, 47, 57
magnetic flux form, 20
per unit of length, 141
Magnetic force, 11
Magnetic loss
magnetic loss factor, 43
magnetic loss angle, 43
Magnetic polarization
impressed magnetic polarization, 62
Magnetoquasistatic approximation, 90, 91
Magnetostatic field, 89
Material equations, 19
Material parameter
macroscopic, 19
Matrix, 395
cofactor, 398
complex conjugate, 399
diagonal, 397
Hermitian, 399
Hermitian conjugate, 252, 399
null matrix, 396
orthogonal, 399
product, 396
product with a scalar, 395
real, 399
singular, 398
sum, 395
symmetric, 399
transpose, 399
unit matrix, 396
unitary, 252, 399
Maxwell's equations, 11, 26, 35
differential form, 41
integral form, 25
local form, 40
phasor representation, 42
Maxwell, James C., 2, 24
Method of lines, 326
Method of moments, 325, 332
Metric coefficient, 45, 366
Microstrip antenna, 317
conformal, 319
Microstrip line, 132, 187
quasistatic approximation, 190
Minor, 398
leading principal minor, 252, 400
Mirror, 313
antenna, 313
principle, 128, 265
Mittag-Leffler expansion, 222
Mode degeneration, 174
Mode matching method, 326
Monopole antenna, 265
Moving frame, 367
Multiport, 237
linear multiport, 238
lossless multiport, 252
passive multiport, 252
port-number symmetric multiport, 240
reciprocal multiport, 254
source-free linear multiport, 238

Near field, 273
Network concept, 11, 12, 78
Network representation, 251
Neumann function, 381, 382
Normalized frequency, 177

Observation space, 92, 93, 100, 378
Ohm's law, 26
One-form
fundamental, 14
twisted one-form, 50
unit oneform, 367
One-port, 213

linear, 213
lossless, 252
passive one-port, 214
reactive one-port, 214
source-free one-port, 214
Operator equation, 332
Optical fiber, 133
Orthogonality, 160
Orthogonality relation
Bessel function, 383
Legendre function, 391, 392

Parabolic reflector antenna, 313
Parallel wire line, 131
Parasitic network elements, 39
parasitic capacitors, 39
parasitic inductors, 39
Partial derivation, 34
Partial wave synthesis method, 326
Patch antenna, 318
Periodic signal, 401
Permeability, 19
complex permeability, 43
relative permeability, 20, 44
Permittivity, 19
complex permittivity, 43
complex permittivity, 43
relative permittivity, 20, 43
Permutation, 353
even, 353
odd, 353
Phase
center, 311
coefficient, 64
velocity, 134
Phasor, 41
Planar antenna, 319
Planar dielectric waveguide, 174
Pocklington's integral equation, 341
Poincaré's lemma, 57, 358
Poisson equation, 91
vectorial, 100
Poisson, Siméon, 2
Polarization
circular, 36
circular polarization, 36
direction of polarization, 33
electric, 19
electric polarization, 270
electric surface polarization, 51
elliptic polarization, 36
impressed electric polarization, 85, 267, 271
impressed equivalent polarization, 82
impressed magnetic polarization, 267
linear polarization, 33, 35
magnetic, 19
magnetic area polarization, 82
magnetic polarization, 270
magnetic surface polarization, 51
Polarization losses, 43
Polarization potential
electric polarization potential, 61
magnetic polarization potential, 62
Port, 237
Potential
electromagnetic potential, 57
magnetic vector potential, 57
magnetic vector potential form, 57
scalar potential, 57
vector potential, 57
Power
active power, 214
complex power, 214
reactive power, 214
Power dissipation density
time-average electric , 75
Power loss density, 72
Poynting
differential form, 73
vector, 73
Poynting form
complex, 75
Poynting vector, 74
complex, 74
complex form, 74, 75
time-dependent, 74
Poynting's theorem
complex, 76
complex, integral form, 76
complex, local form, 76
integral form, 73
local form, 73
Product
cross-product, 353, 357
inner product, 352
scalar product, 352, 357
triple scalar product, 354, 357

vector product, 353
Propagation coefficient, 64, 196
Pseudoscalar, 354, 355
Pseudovector, 354, 355
Pullback
method of, 377

Quality factor
external quality factor, 232
loaded quality factor, 232
resonator, 227
unloaded quality factor, 231
Quasi TEM mode, 187
Quasi-conductor, 44
Quasi-dielectric, 44
Quasistatic field, 90

Radiation characteristics, 265
Radiation conductance, 317
Radiation diagram, 274
Radiation resistance, 276, 285, 286, 289, 297, 298, 300, 317
linear dipole antenna, 285
Reactance multiport, 252
Reactance one-port, 252
Reaction, 86
Reciprocity, 85, 86, 253
Lorentz Reciprocity Theorem, 85
network form, 88
reciprocity theorem, 86, 87
Reflection coefficient, 207
Reflector antennas, 313
Refractive index, 30
Resistance
per unit of length, 148
Resonant circuit, 213
damping, 220
relative bandwidth, 220
series, 217
Resonator, 213

Scalar, 354, 355
true scalar, 21
Scattering matrix, 245
Scattering representation, 245
Skin effect, 114
skin penetration depth, 114
skin penetration depth in copper, 114, 120
surface impedance, 121, 122
surface inductance, 122
Slot antenna, 315
Smith chart, 208
Solenoid, 89, 103
Solenoidal field, 57
Sommerfeld radiation condition, 69
Source space, 92, 93, 100, 378
Spectral domain method, 325
Speed of light, 9, 26, 31
Spherical harmonic, 67, 391, 392
Spherical wave
aperiodic, 276
Standing wave ratio, 209
Star operator, 20, 21, 356
curvilinear coordinates, 367
Static field, 89
Step function, 407
Step wave, 31
Stokes' theorem, 40, 364
Strip line, 77
Structure form
electric, 139
magnetic, 139
Structure function, 160
TEM mode, 139
TEM wave, 198
TE wave, 198
TM wave, 198
electric, 139
magnetic, 139
orthonormal, 250
Sturm-Liouville differential equation, 327
homogeneous, 327
inhomogeneous, 330
Surface current density, 120
Surface emitter, 307, 311, 312
Surface impedance, 122
Surface resistance, 122
Surface wave, 109, 116, 118, 122, 123
Symmetry, 255
mirroring, 255
rotation, 255
symmetry operations, 255

Tellegen's theorem
field form, 80
field representation, 79
general network form, 251

network form, 249
wave amplitude representation, 251
Three-form
fundamental, 21
TLM node
symmetric condensed, 346
Transformer, 102
Transmission factor, 301
Transmission line, 131
Transmission line equation, 195
Transmission line matrix method, 326, 342
TLM cell, 342, 344
TLM node, 344
Transmission matrix, 247
Transmission representation, 247
Twist operator, 50, 376
Twisted differential form, 50
Two-form
fundamental, 18
tube representation, 18

Uniqueness theorem, 82
Unit one-form, 45

Vector, 351, 355
axial vector, 354
column vector, 395
Hermitian conjugate, 252
magnitude, 352
orthonormal basis vector, 353, 366
polar vector, 352
row vector, 395
tangent vector, 366
Vector field, 10
Vector product, 11
Vector space
basis, 351
Voltage, 13, 14, 27
generalized, 195
electric voltage, 27
magnetic voltage, 27
Voltage standing wave ratio, 209

Wave
spherical, 67
transverse electromagnetic, 136
Wave equation, 33, 35, 59
scalar wave equation, 59
vector wave equation, 59
Wave impedance, 31, 62
TE mode, 155
TM mode, 158
free-space, 31, 115
quasi-conductor, 115
Wave number, 32
Wave packet, 134
Wave vector, 63
Waveguide
circular, 131
inhomogeneous, 187
junction, 256
rectangular, 131
wavelength, 169
Wavelength, 32
Wedge product, 17

Recent Titles in the Artech House Antennas and Propagation Library

Thomas Milligan, Series Editor

Adaptive Array Measurements in Communications, M. A. Halim

Advances in Computational Electrodynamics: The Finite-Difference Time-Domain Method, Allen Taflove, editor

Analysis Methods for Electromagnetic Wave Problems, Volume 2, Eikichi Yamashita, editor

Antenna Design with Fiber Optics, A. Kumar

Antenna Engineering Using Physical Optics: Practical CAD Techniques and Software, Leo Diaz and Thomas Milligan

Applications of Neural Networks in Electromagnetics, Christos Christodoulou and Michael Georgiopoulos

AWAS for Windows Version 2.0: Analysis of Wire Antennas and Scatterers, Antonije R. Djordjević, et al.

Broadband Microstrip Antennas, Girsh Kumar and K. P. Ray

Broadband Patch Antennas, Jean-François Zürcher and Fred E. Gardiol

CAD of Microstrip Antennas for Wireless Applications, Robert A. Sainati

The CG-FFT Method: Application of Signal Processing Techniques to Electromagnetics, Manuel F. Cátedra, et al.

Computational Electrodynamics: The Finite-Difference Time-Domain Method, Second Edition, Allen Taflove and Susan C. Hagness

Electromagnetic Modeling of Composite Metallic and Dielectric Structures, Branko M. Kolundzija and Antonije R. Djordjević

Electromagnetic Waves in Chiral and Bi-Isotropic Media, I. V. Lindell, et al.

Electromagnetics, Microwave Circuit and Antenna Design for Communications Engineering, Peter Russer

Engineering Applications of the Modulated Scatterer Technique, Jean-Charles Bolomey and Fred E. Gardiol

Fast and Efficient Algorithms in Computational Electromagnetics, Weng Cho Chew, et al., editors

Fresnel Zones in Wireless Links, Zone Plate Lenses and Antennas, Hristo D. Hristov

Handbook of Antennas for EMC, Thereza MacNamara

Iterative and Self-Adaptive Finite-Elements in Electromagnetic Modeling, Magdalena Salazar-Palma, et al.

Measurement of Mobile Antenna Systems, Hiroyuki Arai

Microstrip Antenna Design Handbook, Ramesh Garg, et al.

Mobile Antenna Systems Handbook, Second Edition, K. Fujimoto and J. R. James, editors

Quick Finite Elements for Electromagnetic Waves, Giuseppe Pelosi, Roberto Coccioli, and Stefano Selleri

Radiowave Propagation and Antennas for Personal Communications, Second Edition, Kazimierz Siwiak

Solid Dielectric Horn Antennas, Carlos Salema, Carlos Fernandes, and Rama Kant Jha

Switched Parasitic Antennas for Cellular Communications, David V. Thiel and Stephanie Smith

Understanding Electromagnetic Scattering Using the Moment Method: A Practical Approach, Randy Bancroft

Wavelet Applications in Engineering Electromagnetics, Tapan Sarkar, Magdalena Salazar Palma, and Michael C. Wicks

For further information on these and other Artech House titles, including previously considered out-of-print books now available through our In-Print-Forever® (IPF®) program, contact:

Artech House
685 Canton Street
Norwood, MA 02062
Phone: 781-769-9750
Fax: 781-769-6334
e-mail: artech@artechhouse.com

Artech House
46 Gillingham Street
London SW1V 1AH UK
Phone: +44 (0)20 7596-8750
Fax: +44 (0)20 7630 0166
e-mail: artech-uk@artechhouse.com

Find us on the World Wide Web at:
www.artechhouse.com